A Study Book for the NEBOSH National Diploma

in occupational Health and Safety Practice

Managing Health and Safety

Professional Membership (Grad IOSH)

Holders of the National Diploma may apply for Graduate membership (Grad IOSH) of the Institution of Occupational Safety and Health (IOSH), and on completion of a programme of Continual Professional Development (CPD) may apply for Chartered Safety and Health Practitioner status as a Chartered Member of IOSH (CMIOSH).
Chartered Member status reflects the competence demanded of professionals in health and safety management or leadership positions.

RMS Publishing Limited
Suite 3, Victoria House, Lower High Street, Stourbridge, DY8 1TA

First Published April 2005.
Second Edition August 2006.
Third Edition August 2008.
Third Edition April 2009 (reprint).
Third Edition August 2010 (reprint).
Fourth Edition March 2011.
Fourth Edition June 2012 (reprint).
Fourth Edition February 2013 (reprint).
Fifth Edition January 2014.
Fifth Edition September 2014 (reprint).
Fifth Edition November 2015 (reprint).
Sixth Edition August 2016.
Sixth Edition October 2016 (reprint).

Cover by Smudge Creative Design.
Printed and bound in Great Britain by CPI Antony Rowe.

ISBN-13: 978-1-906674-55-7

Contents

Preface

Publication users

This updated 6th Edition study book has been thoroughly updated in line with the current NEBOSH syllabus for the NEBOSH National Diploma 'Unit A Managing health and safety' providing an excellent reference for those looking to undertake a career as a health and safety practitioner.

This study book Unit A provides a thorough grounding in all major aspects of managing health and safety and topics covered include:

- Principles of health and safety management
- Principles of health and safety law
- Criminal law
- Civil law
- Loss causation and incident investigation
- Measuring and reviewing health and safety performance
- The assessment and evaluation of risk
- Risk control
- Organisational factors
- Human factors
- The role of the health and safety practitioner

It aims to prepare students for a career in health and safety by providing them with the ability to apply their knowledge and understanding of health and safety management issues in the workplace.

The NEBOSH National Diploma is the qualification for aspiring health and safety professionals building directly upon the foundation of knowledge provided by the NEBOSH National General Certificate. It also provides a sound basis for progression to postgraduate study.

Syllabus

The study book has been structured to reflect the order and content of the NEBOSH National Diploma in Occupational Health and Safety syllabus 'Managing health and safety'. In this way, the student studying for this qualification can be confident that this guide reflects the themes of the syllabus and forms an excellent study book for that purpose.

Each element of the study book has an element overview that sets out the learning outcomes, the contents and any connected sources of reference. In addition, the syllabus, and therefore this study book, is structured in a very useful way; focusing on hazards, their control and core management of health and safety principles which would be useful as reference for any health and safety practitioner.

Photographs and schematics

We have taken particular care to support the text with a significant number of photographs and schematics. The photographs have been selected to be illustrative of both good and bad working practices and should always be considered in context with supporting text. I am sure that students will find this a useful aid when trying to relate their background and experience to the broad based NEBOSH National Diploma syllabus. They will give an important insight into some of the technical areas of the syllabus for those who may not have a strong technical background.

Where diagrams/text extracts are known to be drawn from other publications, a clear source reference is shown and RMS wishes to emphasise that reproduction of such diagrams/text extracts within the Study Book is for educational purposes only and the original copyright has not been infringed.

Legal requirements

Legislation is referred to in context in the various elements that comprise the study book. This reflects the interest of the NEBOSH National Diploma syllabus and requirements to study new/amended legislation under the rule from NEBOSH that it has to have been in force for six months before it becomes examinable. In addition, the essential points of legislation relevant to this Unit of the Diploma syllabus are contained in the section of the study book under Relevant Statutory Provisions.

Decided cases have been included in Units A 'Managing health and safety' and B 'Hazardous substances / agents' and form part of the syllabus content; these cases are examinable.

Case studies / past incidents, as specified by the NEBOSH National Diploma syllabus, may be referred to in the Managing Health and Safety Study Book, it is important to note however that these cases will not be examinable. Additional case studies/past incidents may also be referred to, but although not specifically referred to in the syllabus, it is useful to be aware of them as they have an influence on the workplace and showing knowledge of them at time of examination may emphasise a greater depth of understanding of the topic. Further information on other case studies / past incidents can be found in the Unit C 'Workplace and work equipment safety' study book in the RMS series for the NEBOSH National Diploma.

Element A11 (the role of the health and safety practitioner) has been included within the content for Unit A but will not be examinable by the Unit A written examination. The content of this element will be assessed only through Unit DNI 'Application of health and safety management in the workplace'.

All statistics shown throughout this publication are the latest available at time of going to press.

National Vocational Qualification

We are confident that those working to national vocational qualifications in occupational health and safety will find this Study Book a useful companion. For students working towards the S/NVQ Level 5 in Occupational Health and Safety Practice they will find a good correlation between the scope of the Study Book series for NEBOSH National Diploma and the domain knowledge needs at that level.

Higher Level Qualifications

The structure, level and content of this study book is appropriate for those involved in study of health and safety at university level. The NEBOSH National Diploma is recognised by IOSH as fulfilling the academic requirements for application for Graduate Membership (Grad IOSH) of the Institution of Occupational Safety and Health (IOSH – www.iosh.co.uk). This is the first step to becoming a Chartered Health and Safety Practitioner as a Chartered Member of IOSH (CMIOSH). Chartered Membership (CMIOSH) conferring the official title of Chartered Safety and Health Practitioner. It is also accepted by the International Institute of Risk and Safety Management (IIRSM) as meeting their requirements for full membership (MIIRSM).

Relationship to other RMS Study Books

Students with limited experience may find the foundation knowledge in the RMS Publishing Study Books for Certificate level particularly useful, in particular the study book for the NEBOSH National General Certificate in Occupational Health and Safety.

Production of the publication

Managing Editor: ***Ian Coombes, Managing Director ACT, CMIOSH;*** member of NEBOSH Council, member of NEBOSH Board of Trustees, member of NEBOSH Qualifications and Technical Council and former NEBOSH examiner. Former member of IOSH Professional Affairs Committee and chairman of the Initial Professional Development sub-committee. Member of the Safety Groups UK (SGUK) management committee.

Acknowledgements

RMS Publishing wishes to acknowledge the following contributors and thank them for their assistance in the preparation of the publication:

Geoff Littley, Principal Consultant ACT, CMIOSH, NEBOSH Diploma, CSPA; experienced health and safety advisor, including manufacturing, NHS Trusts, Local Authorities and transport industries. Lead tutor for NEBOSH Diploma and Certificate level courses, in particular for the NEBOSH National General Certificate and Construction Health and Safety qualifications. Provides training and mentor support for Principal Designers.

Robert Weeks, Consultant ACT, Grad IOSH MIIRSM AMCIEH; an experienced NEBOSH examiner for National Diploma Unit A and various Certificate qualifications. A knowledgeable and skilled teaching/training practitioner with high skill levels in communication, management and business development.

Barrie Newell, Director ACT, FCMI; Lead Auditor OHSAS 18001, former senior manager in the chemical industry with over 20 years' experience in the management of high risk facilities processing highly flammable and toxic chemicals, including HAZOP implementation. Implemented waste management systems including, waste reduction, recycling, reuse, incineration, including energy recovery and disposal to land fill.

Kevin Coley, CMIOSH; a NEBOSH examiner with many years' health and safety experience in the private and public sector. Kevin has been a production manager in a large foundry, so has a heavy engineering background as well as a senior safety manager for large government bodies.

Paul Harvey, Grad IOSH LRIC; highly experienced trainer and consultant, delivering the NEBOSH Diploma qualification since 1999 in the UK, Middle East, South East Asia and the Caribbean.

Louise Crossman, BEng(Hons) Dip2.OSH CMIOSH OHSCR FRSPH; a NEBOSH examiner involved in the marking of both assignments and examination papers at Certificate and Diploma level. Started life as an Engineer and then a trainer in the use of various computer packages, moving into the field of Health and Safety in 1998.

Colin Willis, BA (Hons) Dip NEBOSH Grad IOSH; member of NEBOSH Board of Trustees and voted to become Vice-chair of the NEBOSH Qualifications and Technical Council (QTC).

Kerry Edwards BSc (Hons), CMIOSH, MIIRSM, OSHCR; a Chartered Safety and Health Practitioner, a member of the International Institute of Risk and Safety Management and other business and professional organisations. 22 years' experience within the health and safety field and a NEBOSH instructor and examiner, as well as providing vocational assessment to individual professionals.

Julie Skett, Design and Development Coordinator. Nick Attwood, Kris James, Alison Serdetschniy, Jack Aaron and Andy Taylor layout and formatting.

RMS Publishing also wishes to acknowledge the following contributors: Janice McTiernan.

Publications available from RMS:

Publication	Edition	ISBN
A Study Book for the NEBOSH National General Certificate in Occupational Health and Safety	Eighth	978-1-906674-44-1
A Study Book for the NEBOSH Certificate in Fire Safety and Risk Management	Fifth	978-1-906674-32-8
The Management of Construction Health and Safety Risk	Fourth	978-1-906674-37-3
The Management of Environmental Risks in the Workplace	Third	978-1-906674-24-3
The Management of Health and Well-being in the Workplace	First	978-1-906674-14-4
A Guide to International Oil and Gas Operational Safety	First	978-1-906674-19-9
A Guide to International Health and Safety at Work	Fifth	978-1-906674-47-2
Study Books for the NEBOSH National Diploma in Occupational Health and Safety:		
■ (Unit A) Managing health and safety	Sixth	978-1-906674-55-7
■ (Unit B) Hazardous substances/agents	Sixth	978-1-906674-56-4
■ (Unit C) Workplace and work equipment safety	Sixth	978-1-906674-57-1
Study Books for the NEBOSH International Diploma in Occupational Health and Safety:		
■ (Unit IA) Managing health and safety	Fourth	978-1-906674-52-6
■ (Unit IB) Hazardous substances/agents	Fourth	978-1-906674-53-3
■ (Unit IC) Workplace and work equipment safety	Fourth	978-1-906674-54-0
Controlling Skin Exposure (BOHS)	First	978-1-906674-00-7

Figure List (including tables and quotes)

Element A5

Element A6

Element A7

Element A8

Element A9

List of abbreviations

LEGISLATION

CAR	Control of Asbestos Regulations 2012
CDDA	Company Directors Disqualification Act 1986
CDM	Construction (Design and Management) Regulations 2015
CIMAH	Control of Industrial Major Accident Hazards Regulations 1984
CJA	Criminal Justice Act 2003
CLAW	Control of Lead at Work Regulations 2002
CLCA	Civil Liability (Contribution) Act 1978
CLER	Classification and Labelling of Explosives Regulations 1983
CMCHA	Corporate Manslaughter and Corporate Homicide Act 2007
CNWR	Control of Noise at Work Regulations 2005
COER	Control of Explosives Regulations 1991
COMAH	Control of Major Accident Hazards Regulations 2015
CoPA	Control of Pollution (Amendment) Act 1989
COSHH	Control of Substances Hazardous to Health Regulations 2002 (and as amended 2004)
CPA	Consumer Protection Act 1987
CSR	Confined Spaces Regulations 1997
CVWR	Control of Vibration at Work Regulations 2005
DA	Deregulation Act 2015
DAHSW	Deregulation Act 2015 (Health and Safety at Work) (General Duties of Self-Employed Persons) (Consequential Amendments) Order 2015
DDA	Disability Discrimination Act 1995
DSEAR	Dangerous Substances and Explosive Atmospheres Regulations 2002
ECA	European Communities Act 1972
EESR	Electrical Equipment (Safety) Regulations 1994
EIA	Environmental Impact Assessment Regulations 1999
EHSPS	Employers' Health and Safety Policy Statements (Exception) Regulations 1975
ELCIA	Employers' Liability (Compulsory Insurance) Act 1969
ELDEA	Employers' Liability (Defective Equipment) Act 1969
ERA	Employment Rights Act 1996
ERRA	Enterprise and Regulatory Reform Act 2013
ENVA	Environment Act 1995
EPA	Environmental Protection Act 1990
ESQCR	Electricity Safety, Quality and Continuity Regulations 2002
EWR	Electricity at Work Regulations 1989
EA	Equality Act 2010
FAA	Fatal Accident Act 1976
FIA	Freedom of Information Act 2000
GSIUR	Gas Safety (Installation and Use) Regulations 1998
HASAWA	Health and Safety at Work etc Act 1974
HSWAPU	Health and Safety at Work etc. Act 1974 (General Duties of Self-Employed Persons) (Prescribed Undertakings) Regulations 2015
HSWCER	Health and Safety at Work etc Act 1974 (Civil Liability) (Exceptions) Regulations 2013
HRA	Human Rights Act 1998
HSCER	Health and Safety (Consultation with Employees) Regulations 1996
HSEAR	Health and Safety (Enforcing Authority) Regulations 1998
HSFAR	Health and Safety (First-Aid) Regulations 1981
HSFR	Health and Safety (Fees) Regulations (HSFR) 2012
HSMAR	Health and Safety (Miscellaneous Amendments) Regulations 2002
HSOA	Health and Safety (Offences) Act 2008
HWR	Hazardous Waste (England and Wales) Regulations 2005
IER	Health and Safety (Information for Employees) Regulations 1989
IRR	Ionising Radiations Regulations 1999
LA	Limitations Act 1980
LASPO	Legal Aid, Sentencing and Punishment of Offenders Act 2012
LASPOFSC	Legal Aid, Sentencing and Punishment of Offenders Act 2012 (Fines on Summary Conviction) Regulations 2015
LOLER	Lifting Operations and Lifting Equipment Regulations 1998
LoWR	List of Wastes (England) Regulations 2005
LRCN	Law Reform (Contributory Negligence) Act 1945
MHSWR	Management of Health and Safety at Work Regulations 1999
MMHSPR	Mines Miscellaneous Health and Safety Provisions Regulations 1995
NIHHS	Notification of Installations Handling Hazardous Substances (Amendment) Regulations 2002
NOMAS	Dangerous Substances (Notification and Marking of Sites) Regulations 1990
OFAR	Offshore Installations and Pipeline Works (First-Aid) Regulations 1989
OLA	Occupiers' Liability Acts 1957 and 1984
OLSA	Occupiers' Liability (Scotland) Act 1960
PER	Pressure Equipment Regulations 1999

PPER	Personal Protective Equipment at Work Regulations 1992
PSSR	Pressure Systems Safety Regulations 2000
PIDA	Public Interest Disclosure Act 1998
PUWER	Provision and Use of Work Equipment Regulations 1998
QR	Quarries Regulations 1999
RA	Railways Act 2005
RIDDOR	Reporting of Injuries, Diseases and Dangerous Occurrences Regulations 2013
RRFSO	Regulatory Reform (Fire Safety) Order 2005
RTA	Road Traffic Act 1999
SARAH	Social Action, Responsibility and Heroism Act
SEA	Single European Act 1986
SMSR	Supply of Machinery (Safety) Regulations 2008
SPVSR	Simple Pressure Vessels (Safety) Regulations 1991 (as amended)
SRSC	Safety Representatives and Safety Committees Regulations 1977
SSA	Social Security Act 1975
SSAA	Social Security Administration Act 1992
SSACPR	Social Security Act (Claims and Payments) Regulations 1979
SSGA	Sale and Supply of Goods Act 1994
SSIIPDR	Social Security (Industrial Injuries) (Prescribed Diseases) Regulations 1985
SSSR	Health and Safety (Safety Signs and Signals) Regulations 1996
TULR	Trade Union and Labour Relations (Consolidation) Act 1992 (as amended)
UCTA	Unfair Contract Terms Act 1977
WAH	Work at Height Regulations 2005
WHSWR	Workplace (Health, Safety and Welfare) Regulations 1992

GENERAL

AALA	Adventure Activities Licensing Authority
ACOP	Approved Code of Practice
ADR	Alternate Dispute Resolution
AFCI	Arc-Fault Circuit Interrupter
AITT	Association of Industrial Truck Trainers
ALU	Asbestos Licensing Unit
ATP	Automatic Train Protection
AWS	Automated Warehousing Systems
AWS	Automatic Warning System
AZDN	Azodiisobutyronitrate
BASEEFA	British Approvals Service for Electrical Equipment for Flammable Atmospheres
BATNEEC	Best Available Techniques Not Entailing Excessive Cost
BAT	Best Available Techniques
BCEC	British Crane and Excavator Corporation Ltd
BEQ	Break-even Quantity
BLEVE	Boiling Liquid, Expanding Vapour Explosion
BPEO	Best Practicable Environmental Option
BP	British Petroleum
BS	British Standards
BSC	British Safety Council
BSI	British Standards Institution
CA	Competent Authority
CAT	Cable Avoidance Tool
CBA	Cost Benefit Analysis
CBI	Confederation of British Industry
CCF	Common Cause Failures
CCTV	Closed Circuit Television
CE	Conformité Européene
CENELEC	European Committee for Electro-technical Standardisation
CEN	European Standardisation Committee
CEO	Chief Executive Officer
CFC	Chlorofluorocarbons
CFDR	Centre for Dispute Resolution
CHaSPI	Corporate Health and Safety Performance Index
CITB	Construction Industry Training Board
CNC	Computer Numeric Control
COPD	Chronic Obstructive Pulmonary Disease
CPD	Continuous Professional Development
CPR	Cardio Pulmonary Resuscitation
CPS	Crown Prosecution Service
CSR	Corporate Social Responsibility
CST	Climate Survey Tool
DAs	Designating authorities
DTI	Department of Trade and Industry
EA	Environmental Agency
EC	European Community

ECCS	Emergency Core Cooling System
EEA	European Economic Area
EEBAD	Earthed Equipotential Bonding and Automatic Disconnection
EEC	European Economic Community
EFTA	European Free Trade Association
EHSR	Essential Health and Safety Requirements
EMA	Employment Medical Adviser
EMS	Environmental Management Systems
EN	European Standards
ESD	Electrostatic Discharge
ETA	Event Tree Analysis
ET	Employment Tribunal
EU	European Union
EVCs	Educational Visit Co-ordinators
EWC	European Waste Catalogue
FCCU	Fluidised Catalytic Cracking Unit
FEAB	Fartygsentreprenader AB
FID	Flame Ionisation Detector
FKAB	Fartygskonstructioner AB
FLT	Fork Lift Truck
FRC	Financial Reporting Council
FTA	Fault Tree Analysis
GB	Great Britain
GWT	Great Western Trains
HAZOP	Hazard and Operability Studies
HBC	High Breaking Capacity
HCFC	Hydrochlorofluorocarbons
HSC	Health and Safety Commission
HSE	Health and Safety Executive
HSMS	Health and Safety Management System
HST	High Speed Train
IBC	Intermediate Bulk Container
ICAEW	Institute of Chartered Accountants in England and Wales
ICC	Incident Contact Centre
ICBM	Intercontinental Ballistic Missile
IDLH	Immediately Dangerous To Life or Health
IEE	Institute of Electrical Engineers
ILO	International Labour Office
INSAG	International Nuclear Safety Advisory Group
IOD	Institute of Directors
IOSH	Institution of Occupational Safety and Health
IPC	Integrated Pollution Control
IPPC	Integrated Pollution Prevention and Control
IP	Index of Protection
ISO	International Organisation for Standardisation
LAAPC	Local Authority Air Pollution Control
LAPPC	Local Air Pollution Prevention and Control
LAR	Local Automatic Control System
LCA	Life Cycle Assessment
LEA	Local Education Authority
LEFM	Linear Elastic Fracture Mechanics
LEL	Lower Explosive Limit
LEV	Local Exhaust Ventilation
lm	Lumen
LPG	Liquefied Petroleum Gas
LPI	Liquid penetrant inspection
MAPP	Major Accident Prevention Policy
MATTE	Major Accidents To The Environment
MEGC	Multiple Element Gas Containers
MEL	Maximum Exposure Limit
MEP	Member of the European Parliament
MEWP	Mobile Elevated Work Platform
MIC	Methyl Isocyanate
MSDS	Material Safety Data Sheet
MSW	Municipal Solid Wastes
mV	Milli Volts
NCB	National Coal Board
NDT	Non-destructive Testing
NO_x	Nitrogen Oxide
OCPD	Overcurrent Protective Device
OES	Occupational Exposure Standard
OJ	Official Journal

PAS	Publicly Available Specification
PE	Photo-electric
PES	Programmable Electronic Systems
PPE	Personal Protective Equipment
PSLA	Pain, Suffering and Loss of Amenity
PTW	Permit to Work
QALY	Quality Adjusted Life Year
QMV	Qualified Majority Vote
RCS	Risk Control System
RES	Representatives of Employee Safety
RIAs	Regulatory Impact Assessments
RID	International Carriage of Dangerous Goods by Rail
RPE	Respiratory Protective Equipment
RTITB	Road Transport Industry Training Board
SADT	Self Accelerating Decomposition Temperature
SCBA	Self-Contained Breathing Apparatus
SELV	Separated Extra Low Voltage
SEPA	Scottish Environmental Protection Agency
SFAIRP	So Far As Is Reasonably Practicable
SO_2	Sulphur Dioxide
SSW	Safe Systems of Work
SWL	Safe Working Load
SWR	Steel Wire Rope
TEC	Treaty establishing the European Community
TFEU	Treaty on the Functioning of the European Union
TOR	Tolerability of Risk
TQM	Total Quality Management
TUC	Trades Union Congress
UKATA	UK Asbestos Training Association Ltd
UV	Ultra Violet
VDU	Visual Display Unit
VOC	Volatile Organic Compounds
VOSL	Value of a Statistical Life
WAC	Waste Acceptance Criteria
WHO	World Health Organisation
WRA	Waste Regulatory Authorities

This page is intentionally blank

Principles of health and safety management

Learning outcomes

On completion of this element, candidates should be able to demonstrate understanding of the content through the application of knowledge to familiar and unfamiliar situations. In particular, they should be able to:

A1.1 Explain the moral, legal and economic reasons for the effective management of health and safety.

A1.2 Outline the societal factors which influence an organisation's health and safety standards and priorities.

A1.3 Outline the uses of, and the reasons for, introducing a health and safety management system.

A1.4 Explain the principles and content of an effective health and safety management system including the reasons for integration with other management systems.

Contents

Relevant statutory provisions

Employers' Health and Safety Policy Statements (Exceptions) Regulations (EHSPS) 1975
Management of Health and Safety at Work Regulations (MHSWR) 1999 (as amended)
Management of Health and Safety at Work Regulations (Northern Ireland) 2000 (as amended)

Sources of reference

Reference information provided, in particular web links, was correct at time of publication, but may have changed.

IIRSM Code of Ethics, http://www.iirsm.org/code-ethics

IOSH Code of Conduct, http://www.iosh.co.uk/About-us/Regulating-the-profession.aspx

Managing for health and safety, HSG65 (third edition), HSE Books ISBN: 978-0-717664-56-6, http://www.hse.gov.uk/pubns/priced/hsg65.pdf

Occupational health and safety management systems – requirements (OHSAS 18000 series), OHSAS 18001 ISBN: 978-0-5805-9404-5, OHSAS 18002 ISBN: 978-0-5806-2686-9

Plan, Do, Check, Act, An introduction to managing for health and safety, INDG275, HSE Books, http://www.hse.gov.uk/pubns/indg275.htm

The above web links along with additional sources of reference, which are additional to the NEBOSH syllabus, are provided on the RMS Publishing website for ease of use - www.rmspublishing.co.uk.

A1.1 - Reasons for the effective management of health and safety

The reasons for risk management may be based on moral, legal, and economic considerations.

Moral

DUTY OF REASONABLE CARE

The common law duty of care is the duty placed by society on an individual to take reasonable care of those affected by what they do or do not do. This reflects the moral duty to take reasonable care to not hurt each other by one's actions.

UNACCEPTABILITY OF PUTTING HEALTH AND SAFETY OF PEOPLE AT RISK

It is unacceptable in our society to put the health and safety of people at risk, especially for one's own profit, if there are reasonable ways of reducing that risk to acceptable or tolerable levels. As an employer or manager, we would feel morally responsible if someone we had set to carry out a task suffered an injury as a consequence.

Considerations should include: the general well-being of employees and other workers; the interaction with the general public who either live near the organisation's premises or come into contact with the organisation's operations - for example, transportation, noise, effluent discharges, etc., and the consumers of the organisation's products or services, who ultimately keep the organisation in business.

SOCIETY'S ATTITUDE TO MORAL OBLIGATIONS

The attitude of society changes through time and can put pressure on the way employers conduct their business. The public outrage at rail, sea and air disasters shows that if the risk impinges on the general public, they are less likely to accept it as being part of life. Society sees employers as owing a moral obligation to the society in which they conduct their business. This is a long way from the attitude in the Industrial Revolution, where it seemed as though society was grateful for the chance to work and the prosperity an employer could bring, and did not express expectations to meet a moral duty to conduct a business safely. Generally speaking, ongoing legal decisions reflect the increasing awareness of society's attitude to moral obligations.

NATIONAL ACCIDENT/INCIDENT AND ILL-HEALTH STATISTICS

The effect of work on the health and safety of the working population is monitored via a number of sources. The HSE report annual accident and ill-health statistics using information from reports made under the Reporting of Injuries Diseases and Dangerous Occurrences Regulations (RIDDOR) 2013 and data supplied by the Labour Force Survey (run by the Office for National Statistics). Information reporting mechanisms for medical practitioners supplement these statistics providing details of new diagnoses of cases including those diagnosed in patients who are no longer in work.

The Health and Occupation Reporting network (THOR) holds statistics relating to patients seen by specialists since the early-1990s while the reporting was extended in 2005 via THOR-GP for GPs to notify new cases. Specific consideration is given, via SWORD (Surveillance of Work-Related and Occupational Respiratory Disease) and EPIDERM to conditions associated with lung and skin disorders respectively. The number of applications to the Industrial Injuries Disablement Benefit Scheme (IIDB) provide further useful statistics while Death Certificates can be used to gather figures for some work-related lung disease such as mesothelioma.

The Health and Safety Executive (HSE) reported the following statistics for 2014/15, the latest available at the time of going to press.

Accident/incident

- 142 workers were killed at work, a rate of 0.46 fatalities per 100,000 workers.
- 76,000 other injuries to employees were reported under RIDDOR, a rate of 293 per 100,000 employees.
- 611,000 injuries occurred at work; according to the Labour Force Survey, a rate of 2,030 per 100,000 workers. 152,000 led to over-7-days absence, with a rate of 500 per 100,000 workers.

Ill-health

- 1.2 million people who worked during the year were suffering from an illness (long-standing as well as new cases) they believed was caused or made worse by their current or past work, 500,000 of these were new conditions which started during the year.
- A further 0.8 million former workers (who last worked over 12 months prior to production of the statistics) were suffering from an illness which was caused or made worse by their past work.
- 2,538 people died from mesothelioma in 2013, and thousands more from other occupational cancers and diseases such as Chronic Obstructive Pulmonary Disease (COPD), including bronchitis and emphysema (around 15% may be work related).

EFFECT OF SIZE OF ORGANISATION ON ACCIDENT/INCIDENT RATES

The fatal injuries incidence rate for self-employed workers (small businesses) reported by the HSE for 2014/15 was over double that of employed workers in larger organisations. However, on further investigation, it can be seen that if the agricultural industry is considered separately, for other industry sectors the rates are higher overall for employed staff.

Year	Employees		Self -employed		Workers	
	Number	Rate (a)	Number	Rate (b)	Number	Rate (c)
2014/15	99	0.38	43	0.89	142	0.46
(a) per 100,000 employees (b) per 100,000 self-employed (c) per 100,000 workers						

Figure A1-1: Fatal injuries by class of worker (provisional). *Source: HSE.*

Main Industry	Employee	Self employed	Workers
Agriculture	8.04	10.12	9.12
Mining and quarrying	. .	. .	. .
Manufacturing	0.56	0.47	0.55
Gas, electricity and water supply: sewerage, waste and recycling	3.64	. .	4.31
Construction	1.86	1.28	1.62
Services	0.19	0.31	0.21
All industries	0.38	0.89	0.46

Figure A1-2: Rate of fatal injuries (per 100,000 employees or self-employed) by main industry - 2014/15. *Source: HSE.*

The data provided by the HSE do not, year on year, compare the injury and ill-health statistics for Small and Medium sized Enterprises (SMEs) with those of larger organisations. However, some analysis carried out in 2011 concluded that, with the exception of stress-related illnesses for which the reverse was true, the incidence of ill-health was lower in SMEs. The review also identified that injuries sustained within SMEs are substantially under-reported making it difficult to accurate comparisons; though not specifically stated, this may be a contributory factor to lower ill-health figures too.

Legal

The moral obligation for taking reasonable care of persons at work and those affected by work is not just reflected in common law, but also in statute law and is criminally enforceable. Health and safety legislation, in most cases, goes beyond taking reasonable care and may place duties of absolute, practicable or reasonably practicable care. The moral obligation for taking reasonable care of persons at work and others who may be affected is reflected within health and safety legislation and changes in legislation tracked over its development can be seen to essentially mirror the public opinion of the time.

Considerations should include possible consequences of failing to comply with health, safety and environmental legislation, approved codes of practice, guidance of regulators and accepted standards, plus other relevant legislation concerning fire prevention, pollution, and product liability. Loss may result from the preventive (by enforcement notices), punitive (through criminal sanctions), and compensatory effects of law. Failure to manage health and safety does not necessarily have to result in an accident to attract legal preventive or punitive measures. Notices and prosecutions can result from breaches of the law which create a risk, but have not caused a loss.

Where a loss has occurred, claims for compensation may be made for negligence and/or a breach of statutory duty; the introduction of the Enterprise and Regulatory Reform Act (ERRA) 2013 restricts the opportunities to claim for a breach of statutory duty.

PREVENTIVE

Where possible, the enforcing authorities will choose to work with organisations proactively to help them to achieve compliance before the need for reactive measures to be taken. As well as the raft of Approved Codes of Practice and Guidance documents published by the HSE, advice can also be sought from individual enforcing officers. Where this is ineffective or where the scenario requires a more formal approach, the regulators have the power to issue a range of enforcement notices: improvement notices for breaches of the law and prohibition notices where they feel that there is a risk of serious personal injury. Enforcement notices are issued to prevent situations from occurring that may result in loss. Failure to comply with the notices may result in criminal prosecution, in addition to prosecution for the original breach of law for which the notice was issued.

PUNITIVE

Prosecutions can be brought in the event of a legislative breach, whether or not an accident or case of ill-health has arisen. The purpose of such action, which takes place in the criminal courts, is meant to punish for not complying with the law of the land. For health and safety breaches, this punishment may take the form of fines, imprisonment or sometimes both. Health and safety legislation places duties on both the employer and employees themselves; as such the prosecution may be brought against an individual employer or contractor, the employer as an entity and/or a director or similar senior officer who acts on behalf of the body corporate.

COMPENSATORY EFFECTS OF LAW

Where an individual suffers a quantifiable loss due to a workplace accident or ill-health, that person may sue in the civil courts for compensation. The individual will usually pursue the claim on the basis of negligence (i.e. a failure to fulfil a common law duty). Generally, the awarded compensation will be settled under the organisation's Employer's Liability or Public Liability Insurance though the claims may cause an increase in the insurance premiums.

PRINCIPLE OF SELF-REGULATION

Self-regulation is concerned with establishing rules of conduct and performance for the operation of the organisation which go beyond meeting minimum requirements of the law. In line with Robens' initial philosophy of "let those who create the risk control it" it involves the identification of risks arising from the organisation's activities and establishing controls, even if the risk or controls are not specifically identified and prescribed in law. For example, many organisations have complex arrangements concerning worker welfare that are not necessarily covered by specific requirements of current legislation. As well as the risk assessment process, periodic monitoring, review and reflection play an important part in self-regulation.

Although the principle of self-regulation provides a sound basis on which health and safety should be managed within the workplace, evidence has shown that it is rarely successful in isolation. A study carried out by the HSE in 2005 showed that while many small and medium-sized businesses believed that they were compliant, further investigation showed that in fact they fell below expected standards. Input from partners such as clients and contractors can help with self-regulation but in many cases support and where necessary enforcement running in parallel is necessary to achieve ongoing compliance.

Organisations that self-regulate look holistically at issues that affect their workers and take into account not only the workplace stressors, but societal stressors (including domestic) which may affect them from time to time. These may manifest in poor attendance, sickness absence and poor performance. Such organisations will have established procedures to support worker well-being when problems with alcohol or drug dependency come to their attention. This may include the provision of professional counsellors to assist with their recovery. They will also proactively encourage worker health through the provision of advice on healthy eating and exercise.

Economic

IMPACT ON SOCIETY AND ORGANISATIONS

Moral, legal and economic reasons for managing health and safety are closely linked. An organisation that loses money through poor management of health and safety will have to answer to the board and stakeholders as to where this loss is occurring. A small company may lose much of its profits through poor health and safety management and not be able to break even; this could possibly lead to it going out of business. Economic reasons are considered to be powerful motivators.

Considerations should include the financial impact on the organisation of the costs of accidents, the effect on insurance premiums, possible loss of production/service and the overall effect on the 'profitability' of the organisation. Costs may be direct and indirect.

INSURED AND UNINSURED COSTS

The old adage, *"If you think health and safety is expensive, try having an accident"*, summarises the economic reason for ensuring effective health and safety management. Following an accident (or case of ill-health) there are inevitable costs which will result. Some of these will be clearly evident and as a direct consequence of the accident (such as fines from resultant prosecutions or replacement of damaged equipment) but there will be extensive associated costs which aren't immediately obvious (examples being consequences as far removed as drop in production due to reduced morale of staff or lost business due to late delivery of orders which had been due around the time of the accident.

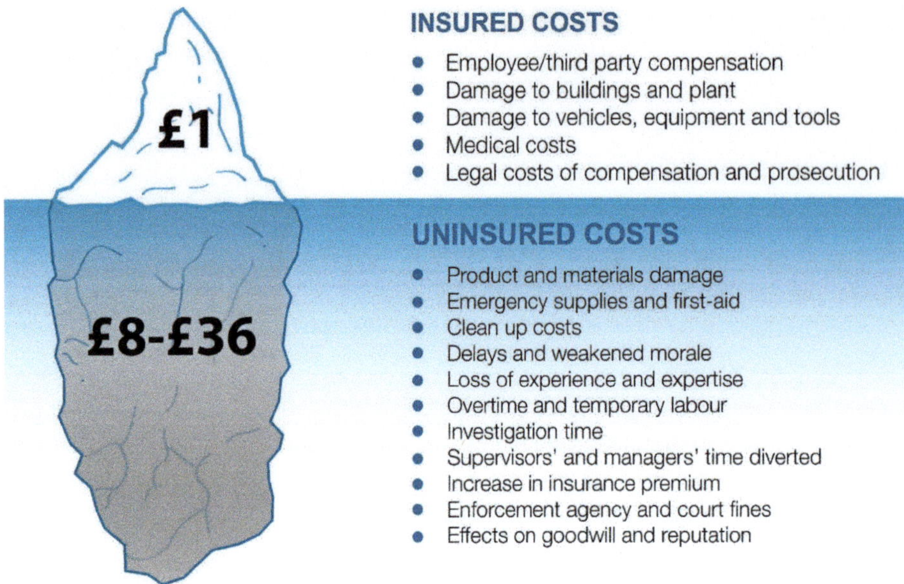

Figure A1-3 Costs incurred by the main contractor (1:11) during the building of a supermarket. *Source: RMS/HSG96 (no longer available).*

Some of these costs can be insured against so won't present an immediate loss to the business, though it should be remembered that following a claim it is normal for insurance premiums to be increased significantly on renewal. However, due to the wording of most policies and the unquantifiable nature of many of the costs, the insurance will only cover a small proportion of the real losses incurred. Many case studies over recent years have illustrated the difference between insured costs and uninsured costs. It has been shown that uninsured costs were often between 8 and 36 times greater than the costs of insurance premiums.

Costs relating to injury and ill-health may also be classified on the basis of whether they are direct costs or indirect costs.

Direct costs

- Fines.
- Fees for Intervention.
- Compensation payments.
- Legal fees associated with the defence of any prosecution or claims.
- Medical costs whether to fund private medical treatment or costs recovered by the NHS.
- Damage to the equipment, materials, property and environment.
- Insurance premiums increased.
- Lost time of injured person and continued employment payments for employee.

Indirect costs

- Time due to interference with service/production potentially leading to failure to fulfil orders on time, loss of bonuses, penalty schemes etc.
- Lost time by other employees who stop work or suffer reduced performance:
    - Out of curiosity.
    - Out of sympathy.
    - Involvement in investigation.
    - Weakened morale.
- Resultant overtime payments.
- Recruitment and training costs of replacement staff whether temporary or permanent.
- Lost time by supervisor or other managers:
    - Assisting injured employee.
    - Investigating the cause(s) of the accident.
    - Arranging for the injured employee's work to be continued by some other employee.
    - Selecting, training, and/or induction of a new employee to replace injured employee.
    - Preparing accident reports, attending hearings, inquests courts etc.
- Time and skill of specialists for:
    - First aid/firefighting/spill control.
    - Clean up/salvage/recovery.
    - Detailed investigation.
    - Legal advice/defence.
    - Customer/supplier/public confidence.
- The completion of any corrective action identified as necessary following the accident investigation and/or risk assessment reviews.

FINANCIAL BENEFITS OF EFFECTIVE HEALTH AND SAFETY MANAGEMENT

The financial benefits of effective health and safety management are derived from the advantages of not having to pay out unexpected costs, more consistent provision of service and better utilisation of workers and management, who can remain involved in productive activities. In addition, there is an argument that an organisation with effective health and safety management should expect to pay reduced insurance premiums, compared to other organisations with less effective health and safety management. Some organisations have identified the financial benefits from effective health and safety management to be in the order of 10% of their turnover, meaning that if their health and safety management was not as effective they would need to have 10% more turnover to pay for the costs. Where effective health and safety management is in place this means more financial resource is available to distribute to stakeholders or to achieve services.

The obvious financial benefits to society include reduced costs of emergency services, National Health Service (as these can only be recovered in the event of a personal injury claim being awarded) and other social support services and benefit payments. The HSE publish figures which give a more realistic figure of the costs to society taking into account not only these figures but the hidden costs and a valuation of the human costs incurred in terms of the impact of the victim's quality of life and that of the family; in the case of fatalities the cost of pain, grief and suffering is evaluated. Figures published by the HSE for 2013/14 (the latest available at time of publication) adjudged the average ***cost*** to society for ***each*** individual case as follows:

- Fatality: £1,575,000.
- Non-fatal injury: £7,500.
- Case of ill-health £18,700.

The benefits to society also include reduced demand on regulators, as well as the prosecution and court service.

A1.2 - Societal factors which influence an organisation's health and safety standards and priorities

Economic climate

It has long been recognised that the economic cycle of good and bad times influences health and safety in the workplace; when times are good organisations expand and the need for more workers increases. The introduction of new workers places challenges on training, induction, orientation and skill development. As a consequence of their introduction, accident rates may increase and these trends are well recorded in accident statistics drawn from Reporting of Injuries Diseases and Dangerous Occurrences Regulations (RIDDOR) reports.

However, when the economic climate is good organisations are more willing to invest in long term improvements that affect health and safety.

Conversely, when the economic cycle worsens and workers lose their jobs, those left in the workplace may take on more work but become more focused and therefore accident rates may fall; this effect was recognised in research undertaken by the HSE. However, longer working hours often result from economic downturn and the HSE has identified that working more than 60 hours almost doubles the risk compared to a more normal 30-49 hour week. In addition, when resources are stretched there is the potential for workers to cut corners to achieve difficult deadlines or because morale drops.

There is sufficient evidence for us to be concerned about the potentially negative effects of working long hours on physical health. The strongest evidence probably concerns the links with cardiovascular disorder.

Figure A1-4: Working long hours. *Source: HSE.*

When the economic climate is poor organisations are less willing to invest in long-term improvements that affect health and safety and they restrict any expenditure to that which is absolutely essential and delivers immediate benefit. Some organisations and workers may feel driven by a poor economic climate to take risks, allowing precautions to lapse.

Government policy and initiatives

Government policy to date concerning occupational health has been largely restricted to helping those in employment rather than supporting the health of all the working age population. A review by Dame Carol Black under the title of 'Working for a Healthier Tomorrow' was a major initiative reviewing the myriad of factors which influence health and well-being. The report concluded that 'work is a key determinate of self-worth, family esteem, identity and standing within the community, besides, of course, material progress and a means of social participation and fulfilment'. The report established a number of government initiatives to reduce the effects of work on worker health, encouraged a healthier lifestyle and pathways to return to work for those that had been out of work due to health reasons.

"It remains critically important to improve health at work and to enable workers with health problems to stay at work, but occupational health must also become concerned with helping people who have not yet found work, or have become workless, to enter or return to work."

Figure A1-5: Health and well-being. *Source: Dame Carol Black, Working for a Healthier Tomorrow.*

Government initiatives have included focusing on specific industries, such as construction through the 'safe site' campaign, or specific risks such as stress or asbestos. The 'simplification initiative' has led to the HSE's provision of sample risk assessments to reduce the time businesses spend on carrying out and recording risk assessments. In a similar way, an electronic template has been provided for those who need to prepare a written health and safety policy.

In 2010 Lord Young was commissioned to review the current position on health and safety, the results of which were published in the Common Sense, Common Safety report and led to the publication of Good Hand Safety, Good for Everyone by the government in 2011 proposing far-reaching reforms focusing on the primary concerns expressed of 'too much red tape' and unhelpful advice from apparent safety professionals. Professor Löftstedt's review in 2011 determined that the overall framework for health and safety was essentially appropriate but that the structure could be simplified to help compliance by businesses. The consequence of this, it would appear, was more far-reaching than Professor Löftstedt had envisaged. The HSE reduced the number of regulations by approximately 50% though a large proportion of these were outdated and consequently unnecessary. The red tape associated with reporting of incidents was reduced by amending RIDDOR to require accidents resulting in 7-days absence to be reported instead of 3-days while the Deregulation Act 2015 exempts up to 1.8 million self-employed occupations which do not present a risk from the requirements of health and safety law. The Executive has also reduced the number of Approved Codes of Practice, most noticeably the one for the Management of Health Safety at Work Regulations 1999, favouring guidance documents and a revamp of their website instead.

As a direct outcome of the report by Lord Young the Occupational Health and Safety Consultants Register was created. Independent safety consultants are encouraged to register but to remain listed the professional is required to demonstrate their ongoing development. This allows businesses to ensure that the consultant that they engage has the appropriate competence.

The emphasis on enforcement has moved away from proactive inspections of lower-risk businesses with a good safety record to concentrate resources on inspecting and enforcing within higher-risk industries

UK CORPORATE GOVERNANCE CODE

The first version of the UK Corporate Governance Code (the Code) was produced in 1992 by the Cadbury Committee. Its paragraph 2.5 is still the classic definition of the context of the Code:

"Corporate governance is the system by which companies are directed and controlled. Boards of directors are responsible for the governance of their companies. The shareholders' role in governance is to appoint the directors and the auditors and to satisfy themselves that an appropriate governance structure is in place. The responsibilities of the board include setting the company's strategic aims, providing the leadership to put them into effect, supervising the management of the business and reporting to shareholders on their stewardship. The board's actions are subject to laws."

Figure A1-6: Definition of the context of the code. *Source: Paragraph 2.5 of the UK Corporate Governance Code, Cadbury Committee 1992.*

The new Code published by the Financial Reporting Council (FRC) applies to accounting periods beginning on or after 17 June 2016 and applies to all companies with a Premium listing of equity shares regardless of whether they are incorporated in the UK or elsewhere.

A less serious initiative designed to illustrate that the government's health and safety focus is on a proportionate approach to managing real risks and not to impose irrational and unnecessary restrictions is the HSE's Myth Busting Challenge Panel which by February 2016 had considered 388 cases presented to them highlighting decisions which had allegedly been made on the basis of 'elf and safety'.

HEALTH AND SAFETY STATISTICS 2014/15

Workplace risks

'Dealing with difficult customers, patients, pupils etc.' and 'lifting or moving people or heavy loads' were the two most common self-reported risk factors in the workplace, in a 2014 survey of workplaces.

Working days lost

- 27.3 million days were lost due to work-related ill-health or injury (15 days per case).
- 23.3 million days were lost due to work-related ill-health.
- 4.1 million due to workplace injury.

Economic costs to Britain

Injuries and new cases of ill-health resulting largely from current working conditions cost society an estimated £14.3 billion in 2013/14 (based on 2013 prices).

Enforcement

- 586 cases were prosecuted by HSE in England and Wales.
- 70 cases were prosecuted by Local authorities in England and Wales.
- 72 cases were prosecuted by the Procurator Fiscal in Scotland.
- 12,430 enforcement notices were issued by all enforcing authorities.

For injuries and ill-health statistics - see earlier in the Element.

Industry/business risk profile

The UK industry/business risk profile has changed significantly since 1969; the free market in particular has resulted in many acquisitions, disposals and relocations of business to third world countries. The first businesses to be relocated were those engaged in high volume low profit manufacturing with associated high labour costs, for example, textiles. These industries were then followed by heavy industries such as ship building and steel manufacture, pottery and glass driven by not only lower labour cost, but often lower raw material and energy costs. In more recent times, the chemical and pharmaceutical sectors have also moved overseas, perhaps driven by the improved environmental standards required to operate within Europe compared with other territories. With each loss of business sector new work opportunities were created and progressively the UK has become more and more service based than manufacturing.

In more recent years injury rates have been overtaken by ill-health, for example, musculoskeletal diseases and psychosocial harm relating to stress in the workplace. Some current worker ill-health has resulted from adverse exposure in the past, for example, increased incidence of worker ill-health from exposure to asbestos some 30 years earlier resulting in the disease mesothelioma. This source of ill-health will be ongoing for many years, but the HSE is seeking to influence the system through the 'Help Great Britain Work Well' strategy:

> *"We have an unprecedented opportunity to keep building a 21st-century, world-class occupational health and safety system that will help Great Britain work well. If we can all come together to help achieve these things, maintain the gains made in safety, and seize the opportunity to give health the same priority, it will help improve productivity, keep business costs down, help keep workers safe and well, and protect members of the public."*

Figure A1-7: 'Help Great Britain Work Well' strategy. *Source: Dame Judith Hackitt DBE, HSE Chair.*

The strategy has six strategic themes:

1) There needs to be broader ownership of health and safety.
2) It is important to highlight and tackle the costs of work-related ill-health.
3) Wider recognition is needed of the business benefits that come with proportionate approaches to risk.
4) Too many SMEs are still unaware that straightforward advice and simple guidance is available that can help them manage their health and safety responsibilities.
5) We all need to horizon scan and 'design-in' effective risk management of new or emerging technologies and business models.
6) There are benefits for Great Britain and British industries by promoting our health and safety approach around the world.

Figure A1-8: 'Help Great Britain Work Well'. *Source: HSE.*

Globalisation of business

The ongoing acquisition of businesses has created organisations of an international scale and this has resulted in globalisation of businesses. Some organisations have grown to such a size and value they can now influence governments, by either providing work (together with the associated wealth it brings to a country) or by threatening to take it away. Not all global corporations operate in this way, but some have been known to influence governments, for example, through discouraging at government level improvements in worker rights and health, safety and welfare conditions. Globalisation can mean that standards of health and safety are 'imported' into the UK or 'exported' to other countries. As the UK and its organisations generally have good standards of health and safety it is hoped that this may result in standards in other countries being improved as standards are 'exported' to them. If organisations in the UK are taken over by new owners that are not familiar with the health and safety standards in the UK, the new owners may have difficulty understanding why so much effort is applied to health and safety in the UK. However, in most cases, the scale and standing of global organisations means that they have high standards of health and safety and they may improve the standards of organisations that they take over.

With globalisation of businesses comes management issues which must be overcome in relation to the health and safety of UK workers many of whom may be required to travel overseas to an organisation's sister site. With this travel comes the difficulties of managing health and safety remotely as well as travel health and security issues.

Migrant workers

Migrant workers in the UK often consist of workers from the former Soviet bloc countries, such as Poland, Romania, and Slovenia. As member countries of the European Union (EU) the citizens of these countries can access work in EU member states. They are drawn to work in countries such as Germany, France and the UK in search of improved work conditions and lifestyle; even if the country decides to leave the EU, the migrant workers who have moved to the UK in recent years will remain. Other migrant workers enter the EU illegally from countries such as Turkey, Africa and the Far East.

This mixture of ethnic backgrounds and languages of migrant workers has resulted in the organisations that recruit such workers having to deal with a wide range of communication and health and safety issues. Methods of communication to overcome any language barriers must be considered carefully not only for day-to-day instructions but also for safety information, training, management meetings, reporting mechanisms etc. Because of their difference in perception of risk, attitudes to work and motivation they may be prepared to take more risk than other workers. Migrant workers are more likely to accept high risk work or to work long hours without question or complaint, and this may result in their exploitation.

Twenty-one Chinese cockle picking workers were drowned by the incoming tide off the coast of Morecambe Bay, North West England on the evening of the 5 February 2004, when they were cut off by the incoming tide in the bay at around 9:30 in the evening. It would appear that the majority had attempted to swim to shore, but had been overcome by hypothermia.

At the subsequent hearing it was reported that British cockle pickers returning to shore on the same evening had attempted to warn the Chinese group by tapping their watches and trying to speak with them, but they did not heed their warnings possibly due to language barriers and the general animosity between the locals and the migrants. It is believed that the migrants were paid £5 per 25 kg of cockles. The disaster led to the Gangmaster Licensing Act 2004 and the formation of the Gangmasters Licensing Authority.

National level of sickness absence

In 2013,131 million working days (equating to 4.4 days/worker) were lost to ill-health which represents a drop of 27% since 1993. Many common diseases are directly linked to lifestyle factors, and while 30% of absence periods were attributed to minor illnesses (colds, sickness, etc.) these are generally not the conditions that keep people out of work for significant periods. Instead, common mental health problems and musculoskeletal disorders are the major causes of work-related sickness absence. Statistics published by the HSE suggest that approximately 23 million working days are lost as a result of work-related ill-health and about 4 million due to workplace accidents. The annual economic costs of sickness absence and worklessness associated with the working age population's ill-health are estimated to be over £15 billion. Studies undertaken by NICE (National Institute for Health and Care Excellence) suggest that the cost to the economy from sickness absence is £15 billion per year

Incapacity

Statistics published by the government office for disability indicate that in 2011/12 11.6 million people living within the UK have a disability with approximately half of these being of working age (5.7 million); these disabilities can be congenital or due to injury or ill-health and split between physical and mental health disabilities. Figures calculated by NICE suggest that 1.8 million of these (more than 30%) reported that the problem was caused or exacerbated by their work.

While approximately 5 million of these will be unable to work and as such are reliant on the available benefits more than 10% will have the ability to work in some capacity, even if not in their preferred or previous job role. In line with the societal expectations discussed below, businesses have a responsibility to help all employees and potential employees to work by making reasonable adjustments to ensure that they are not disadvantaged by their disability. The government initiative Access to Work is a benefit system designed to help individuals to start or stay in work. As well as providing financial support the benefits extend to the provision of other support mechanisms and advice which may be necessary to help individuals stay in work. Although designed to help the individual employee making the claim, there are benefits for the employer. The financial payments are intended to fund things such as adaptations to equipment, a job coach, disability training for colleagues as well as fares to travel to work for individuals who are unable to use public transport.

Societal expectations of equality

The importance of treating everyone equally has become increasingly recognised within our society. As early as the 1960s race relation legislation was introduced to reflect this and to advance equality with legislation for equal treatment of men and women following in the next decade. It wasn't until 1995 that the need to make adjustments to help people with disabilities was introduced. The Equality Act (EA) 2010 consolidated all of the discrimination legislation and largely standardised the approach to be adopted for all people with protected characteristics.

Although the Act gives an individual a cause for action if they feel that someone has discriminated against them, there is leeway for the health and safety professional which allows discrimination ***if*** it can be shown that health or safety would be otherwise compromised: it is lawful discrimination to prevent a female of child-bearing age from driving a tanker which contains a substance known to be harmful to an unborn child ***(ref Page v Freight Hire (Tank Haulage) Ltd 1981)***.

The 2010 Act defines someone as disabled if they have a physical or mental impairment that has a substantial and long-term negative effect on their ability to do normal day-to-day activities. In the same way as the Disability Discrimination Act (DDA) 1995 before it, the Equality Act (EA) 2010 requires organisations to make reasonable adjustments to accommodate the needs of people with disabilities to allow them to work and/or enjoy the services provided in the same way that someone without disabilities could. In a health and safety context this will include modifying or reorganising an employee's (or potential employee's) workplace and/or work activities to enable them to work in safety. The Act uses the term reasonable adjustments to prevent organisations from being held liable for failing to implement changes which would be disproportionately expensive and/or would have an unreasonable impact on the business and/or safety of others. However, there is an expectation that an employer will consider suitable adjustments from the recruitment stage and may include necessary modifications to the interview process or to an individual's work such as relocating their work area to a ground-floor office if use of stairs is difficult, provision of alternative equipment, offer additional training opportunities etc. As with the other characteristics protected by the Act, health and safety legislation takes precedent and where reasonable adjustments cannot be made without compromising the individual's safety it is not unlawful to discriminate ***(ref Farmiloe v Lane Group plc and North Somerset Council (2003)***.

A1.3 - The uses of, and the reasons for, introducing a health and safety management system

The legal requirements to manage health and safety

WRITTEN HEALTH AND SAFETY POLICY

Section 2(3) of the Health and Safety at Work etc. Act 1974 requires all employers to prepare a written statement of his general policy with respect to health and safety along with the organisation and arrangements in place to meet this policy. The policy must then be revised as often as necessary and the statement and all its revisions must be brought to the attention of all employees. It is the Employers' Health and Safety Policy Statement (Exception) Regulation 1975 which exempts employers of less than five employees of this duty.

THE ARRANGEMENTS FOR EFFECTIVE PLANNING, ORGANISATION, CONTROL, MONITORING AND REVIEW OF PREVENTIVE AND PROTECTIVE MEASURES

Although the 1974 Act makes reference to the need for defined arrangements to meet the requirements of the policy, it is Regulation 5 of the Management of Health and Safety at Work Regulations (MHSWR) 1999 which specify that the purpose these arrangements is for the effective planning, organisation, control, monitoring and review of the preventive and protective measures; for employers of five or more employees these arrangements must be recorded in writing.

The Regulation itself makes it clear that the arrangements must be suitable for the nature of the undertaking and the size of the organisation implying that less detailed arrangements may acceptable for smaller organisations with simple risks but that an 'off the shelf' set of arrangements is unlikely to meet the needs of any specific organisation. Although now obsolete in terms of its status, the information contained within the former Approved Code of Practice to the MHSWR remains valid and provides a useful explanation to the meaning of these five terms.

The health and safety management system will ensure effective:

- ***Planning*** - a systematic approach to the completion of risk assessments, identification of priorities, objectives and performance standards as well as the identification and implementation of appropriate risk control measures.

- ***Organisation*** - effective channels of communication between employees and their representatives. This may involve formal consultation, the use of health and safety committees and/or involving employees with the risk assessment process. Competence of those undertaking the risk assessments and identifying control measures must be ensured, with the support of competent health and safety assistance or advice if necessary.
- ***Control*** - the health and safety responsibilities of staff at all levels must be defined to ensure a co-ordinated approach. Expected performance standards should be defined and everyone given the adequate resources to allow them to achieve these standards. Suitable levels of supervision should be in place at all times with particular emphasis given to those who are new to a role.
- ***Monitoring*** - active monitoring techniques such as inspections must be employed to measure the adequacy and ongoing effectiveness of implemented controls prompting improvements to be made as required, thorough investigation of incidents will also help to identify weaknesses and the necessary steps to rectify them.
- ***Reviewing*** - as well as considering the results of monitoring exercises to priorities and instigate remedial action, a periodic review of the overall system will help to ensure that it remains effective.

ACCESS TO COMPETENT HEALTH AND SAFETY ADVICE

Unless one of the senior partners has the necessary competence themselves Regulation 7 of the Management of Health and Safety at Work Regulations (MHSWR) 1999 requires an employer to appoint an appropriate number of competent persons to assist in compliance with health and safety requirements; competence is defined as 'sufficient training and experience or knowledge and other qualities' to allow him to undertake the role. Employers may refer to the Occupational Safety and Health Consultants Register (OHSCR) when engaging external consultants to confirm their competence; those listed on the register must have demonstrated their competence and prove their participation in ongoing developmental activities to ensure that this competence is maintained.

Membership of a professional body such as the Institution of Occupational Safety and Health (IOSH) also requires a defined level of competence and ongoing development and may be used as a starting point to ascertain the suitability of a potential employed advisor. Wherever possible the appointment should be made from within the organisation in preference to the use of an external consultant and the number of advisors will be dependent upon the size of the undertaking, the associated risks and their distribution. Where more than one individual is appointed, the employer must ensure co-operation between them all and ensure that all individuals are given sufficient resources and have adequate means at their disposal to fulfil their role and responsibilities.

The responsibility lies with the employer to ensure that external consultants are made aware of all factors which may have an implication on health and safety as well as providing access to the health and safety information provided to employees as required by Regulation 10 of Management of Health and Safety at Work Regulations 1999. Internal advisors will need to be made aware of individuals who are employed on a fixed-term contract or consultancy basis.

What is a management system?

A health and safety management system provides a formalised framework to help organisations take a structured and logical approach to ensure and maintain effective health and safety provision and legal compliance. Although some of the elements which appear in a typical management system are required to ensure legal compliance, there is no legislative requirement to develop, implement or use a formal health and safety management system. As such there is no specific system which should be adopted: some organisations work with systems which have been developed and evolved internally while others opt for the model developed by the HSE and explained within their guidance document 'HSG65: Managing Health and Safety'.

Other, often larger organisations with more complex organisational structures opt for the Occupational Health and Safety Assessment Series (OHSAS) 18001 model developed by the British Standard Institute which emulates the International Organisation for Standardisation (ISO) models 9001 (to manage quality) and 14001 (to manage environmental safety). Organisations can earn certification against each of these standards which can be used to demonstrate their commitment to the discipline and standards achieved. However, while the two ISO models for quality and environmental safety have international recognition, the 18001 model for health and safety does not enjoy the same status as yet. In the forthcoming months ISO are to introduce the ISO 45001 standard for health and safety. It is expected to take the OHSAS 18001 model as its basis and it is presumed will ultimately offer organisations the opportunity to earn internationally recognised certification.

Key elements to any successful management system include good leadership and management, a trained/skilled/competent workforce and an environment where people are trusted which is underpinned by an understanding of the organisation's risk profile. Health and safety should be regarded as an integral aspect of managing all business risks.

Figure A1-9: The core elements. *Source: HSG65, HSE.*

Application of the Plan, Do, Check, Act cycle with reference to HSG65

The HSE has introduced a web based replacement for HSG65 - 'Successful health and safety management' which moves away from the POPMAR (Policy, Organising, Planning, Measuring Performance, Auditing and Review) model of managing health and safety to 'Plan, Do, Check, Act (PDCA)' which is closer to the framework used in management systems standards, such as OHSAS 18001 etc.

The new document is part of a set of guidance which replaced the Management of Health and Safety at Work Regulations Approved Code of Practice which was withdrawn in June 2013. The move towards Plan, Do, Check, Act is designed to achieve a better balance between the systems and behavioural aspects of management. It also introduces health and safety management which is an integral part of good management generally, rather than as a stand-alone system. Plan, Do, Check, Act is an on-going management system and should not be seen as a once only exercise.

While the comprehensive guidance document of HSG65 provides detailed explanation of the model and how it can be effectively implemented, the HSE's Industry Guide INDG275: Plan, Do, Check, Act - An Introduction to Managing for Health and Safety provides a useful summary of the model.

Figure A1-10: Plan, Do, Check, Act system. *Source: HSE.*

Part 3 of HSG65 describes the approach to successfully applying the Plan-Do-Check-Act model into an organisation. For each of the four elements, the document provides a list of key actions to be taken by senior personnel (referred to as 'Leaders' in HSG65) and managers as well as how workers can be involved in the process and the required level of competence.

PLAN

This early stage in the process requires a business to ascertain where they are and where they want to be in terms of health and safety management. This will allow the identification of the business aims and will lead to the steps to be taken, the identification of how this will be achieved and by who and how its success will be monitored; this will then form the basis of the policy.

Policy

Wherever possible the policy should be written by someone within the organisation. It should reflect the organisation's direction, values and beliefs and demonstrate an honest commitment to providing a health and safety environment. Commitment of the workforce can be secured, or at least enhanced, through consultation during its development, though the final version should be signed and endorsed by the person at the top of the organisation (director or owner) to confirm its status.

Planning

A policy is only as good as its implementation and it is important that staff at all levels act to reflect the statements made. Planning will be required to achieve the aims and objectives identified and consideration will need to be given to the processes which will be used to control risks, react to changing demands and to sustain positive health and safety attitudes and behaviours. Part of this planning process will involve an evaluation of the current position taking into account legal requirements and benchmarking of standards with equivalent organisations.

The competence required for the implementation and use of the controls will need to be taken into account. As well as designing, developing and implementing methods for achieving the required standard, consideration must also be given to how these standards will be maintained or enhanced and how ongoing management initiatives may be incorporated with other aspects of business management.

As well as developing control measures to manage risks, it is important to also have well-defined and well-developed emergency procedures which may be adopted if required. Like the development of the overall policy, involvement of the workforce can lead to increased commitment and useful input into the process.

DO

Risk profiling

An assessment of all risks should be undertaken including immediate risks to safety, potentially longer-term risks to health, low-severity: high-frequency as well as those with low-frequency but potentially catastrophic outcomes. The workers involved in the activity subject to the assessment can provide an invaluable resource and their input should be regarded as essential particularly when the assessor leading the process is not fully familiar with the scenario. Not only will the workers be able to ensure that all potential hazards are identified and existing controls understood but current short cuts and possible solutions may also be forthcoming. The level of detail and sophistication of the assessments will depend on the size of the business and nature of the risk: for a small, perhaps office-based business with simple risks, the process will be limited and straight-forward whereas a larger business with more complex risks may need to consult specialist advice or use more sophisticated analytical techniques. For large, high-hazard sites additional statutory requirements such the Control of Major Accident Hazard Regulations 2015 (COMAH) may apply.

Control measures should control the risk so far as is reasonably practicable which does not extend to disproportionate measures. Consideration of existing controls, legal standards and associated guidance and advice from industry bodies, trade-unions or safety advisors will help to determine the most appropriate approach. While for many risks specific controls may be identified and implemented, for some scenarios where the risk is more dynamic, a more generic approach may be appropriate supported by additional training of workers to ensure that they can recognise the need for additional controls when necessary.

Part of the control measures for some risks will require the use of health surveillance. This should be identified within the risk assessment and may be required as a result of specific legislation (for example, Control of Substances Hazardous to Health Regulations 2002) or if there is a reasonable likelihood of an onset of an identifiable disease or adverse health condition associated with the hazard in the particular condition of work for which valid monitoring techniques exist that could result in further protecting the worker. Organisations with five or more employees must ensure that the significant findings of the assessments are recorded.

A more detailed discussion of the risk assessment process is considered in 'Element A7 - The Assessment and Evaluation of Risk'.

Organising

The effective organisation of health and safety requires consideration to be given to four key elements, often referred to as the 4-Cs: Control, Cooperation, Communication and Competence.

Control

Adequate ongoing control of the mechanisms in place is reliant on an appropriate level of supervision. Those with a supervisory role, who may be team leaders, line managers or more senior personnel within smaller organisations, have a responsibility to lead by example and adhere to the rules that are expected of the workers to. As such it is important that they ensure they are familiar with the procedures in place and allocate adequate time and resources to allow them to be followed. Particular emphasis must be given to the safety-critical aspects of the job and supervisors must be familiar with any emergency procedures so that they can be instigated if required.

Supervisors should ensure that all staff understand and follow the rules in place. Staff should be encouraged to have a positive attitude and share their concerns or suggestions for improvement. Where several people have a supervisory role within a specific area a coordinated, consistent approach should be ensured.

The level of supervision required will be dependent upon the level of associated risk. It should be recognised that level may vary at certain time (for example, at shift-changeover) and that young or inexperienced staff will have a lesser understanding and as such require closer attention.

Of particular importance is the need to supervise contractors. Those who do not habitually work on site may not be familiar with the workplace risks and associated controls or their work may introduce new risks to the resident workforce. A coordinated approach with clear communication between the employer and contractor's organisation is essential. A pre-works meeting ensuring that the risks which the contractor will encounter and introduce are discussed and the control measures agreed, implemented and understood before work starts. Before each individual contractor commences work, their level of competence must be assured and they should be provided with the necessary information, instruction and training via induction training and/or toolbox talks and an appropriate level of supervision maintained.

The measures employed to ensure the effective control of contractors are considered in more detail in 'Element A9 - Organisational Factors'.

Cooperation

Legislation specifically requires consultation with employees about changes being made which may affect their health and safety provision with additional requirements required in organisations with union-appointed safety representatives. However, the level of co-operation advocated by HSG65 goes beyond these minimum standards. The benefit of involving staff in the whole management process has been considered previously but the benefits of a management/workforce partnership based on trust, respect and co-operation cannot be under-estimated. Where such values exist the culture of the organisation will become increasingly positive resulting in an environment where workers have a desire to work with managers to resolve problems.

In addition to the co-operation between managers and workers, it is important that co-operation be achieved between two organisations working closely together whether this be an organisation utilising the services of contractors, two organisations working within the same building or location or the interaction between customers and suppliers. Wherever the activities of one has a health and safety implication on the other, both parties have a responsibility to adopt a coordinated approach.

Such a coordinated approach is particularly important in the development and execution of emergency procedures. A consistent approach for all organisations on site is usually desirable with the possibility of sharing resources sometimes an option. Coordination with emergency services can also be important for organisations with significant risks on site for which advanced information may be invaluable in the event of an incident.

Communication

Effective communication throughout an organisation must be ensured. Organisations must inform workers of the policy for health and safety, the results of risk assessments and resultant controls and forthcoming changes which may impact on workers' health and safely. Workers must be given the opportunity to raise concerns and ideally offer suggestions for improvement while it is important that managers communicate effectively to ensure they adopt a co-ordinated approach.

The methods employed to achieve this level of communication will be dependent upon the size and nature of the organisation: large organisations on multiple sites may need to co-ordinate communications through safety representatives, formal meetings and written communications. Smaller organisations will often liaise directly with the workers. Regardless of the size of the organisation, when communication channels are being set up it is important to consider what information should be conveyed to who: information overload can be as obstructive as inadequate information. Particular emphasis must be given to the safety-critical information. The recipients of the information should be considered and appropriate measures put in place for communicating with those with limited literacy or any language barriers; it should be remembered also that inexperienced staff may need additional explanation. It is also important that the person giving the information has the necessary skills to communicate it clearly and be able to respond to queries as and when they arise. Where written communication is used, or where written information is available for reference following verbal communication, it is important to ensure that this is clear and remains accurate and updated.

The means of communication with the emergency services must also be planned; for higher risk facilities this may require a formal, planned, coordinated approach.

Competence

As discussed previously, employers have a statutory duty to ensure that they have access to competent advice. While for smaller organisations with lower risks this person may be someone from within the organisation who has an understanding of the business and the associated risks, for larger businesses and/or businesses with more significant risks, a specialist health and safety advisor would usually be required. Whoever is given the responsibility of the task, it is important that they be given the necessary resources not only to perform their everyday function but also to keep up to date with changes to legislation and industry good practice.

The risk profile undertaken within the 'Plan' element of the management model will have identified the risks present. In some circumstances, additional services of more specialist advisors may be required such as ergonomists, microbiologists, occupational hygienists, noise and vibration specialists, radiation protection advisers or specialist engineers. It may also be appropriate to engage an occupational health service including occupational health nurses, occupational health physicians and/or physiotherapists. When making such appointments it is essential that their competence in the form of knowledge, skill and experience within their specialist field be confirmed.

The importance of the competence of the whole workforce itself should not be underestimated. While attention is normally given to ensuring that individuals have the necessary skills to undertake their job role effectively, equivalent attention should be given to ensuring that they are able to undertake it safely. An evaluation of the workers competence in this area must be considered at recruitment, if they are moved to an alternative role within the business, as new processes or equipment are introduced and periodically to ensure that previous competence has not been eroded. It should be remembered that the competence of contractors to work safely should be confirmed as well as in-house employees.

Provision of suitable safety training forms an important element of risk control, though it should not be used as an alternative for more effective controls. In addition to the traditional 'classroom' training, effective training can also be achieved using techniques such as 'on the job training', written instructions, on-line information or for simpler tasks simply telling someone what to do. Care must be taken to ensure that the level of training is pitched appropriately for the learner taking into account any language or other barriers to learning and existing experience. The training may be provided by managers, in-house or external trainers but whichever option is used, the competence of the trainer must be ensured to ensure that up to date, accurate information is taught in a manner which results in effective learning. Following the training, it is important to confirm that it has been successful: this may be achieved by testing knowledge and by supervision back at work to ensure that the individual works safely.

Records of all training undertaken will provide a useful basis to monitoring when refresher training may be required. Skills used infrequently will become rusty whilst those undertaking routine tasks may become complacent and need reminding of the importance of particular working practices. Investigations into an incident may also indicate that refresher or top-up training is needed. The use of drills can be useful to reinforce and refresh the emergency procedures while for other skills more traditional refresher training techniques may be needed.

Implementing

Having established the health and safety risks, effective control measures must be effectively implemented. Control systems should be well designed, easy to understand and take into account existing practice and human capabilities and limitations. Care should be taken to ensure that the controls and associated documentation are proportionate to the level of risk: some lower risks can be managed using simple rules saving the more complex approach for higher or difficult to manage risks. In some cases detailed precautions such as a permit-to-work may be necessary. The requirements for routine and more specialist maintenance should also be determined at this stage so that this can be scheduled appropriately.

Having established the necessary controls, it is important that the necessary equipment and training is provided as part of the implementation process. Involving workers and undertaking consultation exercises will facilitate the implementation as will routine reviews of progress, perhaps as part of the Safety Committee's remit, to ensure that the process is not deviating. Once implemented a suitable level of management supervision should ensure that the controls are suitable, adequate and followed. Positive behaviour of workers should be praised and reinforced though it is essential to correct those who are not adhering to the required working practices.

CHECK

It is important to ensure that the controls implemented are and remain effective. It is therefore important to undertake monitoring processes to measure ongoing performance to allow improvements to be made if necessary. In the event of an incident, a thorough investigation can help to identify actions which will help to prevent recurrence

Measuring performance

Performance measurement techniques can be categorised as active and reactive. Active monitoring tend to be preventive in nature by identifying potential problems before they are realised; such techniques include routine inspections and safety tours, health surveillance, functional checks of equipment and formal audits. Reactive monitoring responds to adverse events after they have occurred to identify the cause and methods of preventing recurrence; examples would include investigations into accidents, near-misses or cases of work-related ill-health.

Using a variety of the above techniques will provide the best picture of an organisation's health and safety performance, though care must be taken to ensure that measurement is not undertaken simply for the sake of it. Some monitoring will be required by statute but otherwise should be selected based on the information required and the level of risk. Like that undertaken for other aspects of a business, all health and safety monitoring should add value. As such, the information gathered should be timely and give an indication not only of problems but also their cause. While determining the monitoring to be undertaken, it is normal to define performance indicators against which the results will be compared and it is also sensible to determine at this stage the action to be taken when results fall below the agreed standard. The monitoring and performance indicators may need to be changed periodically as the business evolves, the culture improves or legislative standards change. The management system should require results to be reported to the decision makers within the organisation whose commitment should prompt them to seek answers as well as to give support for the appropriate response to be instigated.

Accident investigation

Although an employer has a legal duty under the Reporting of Injuries, Diseases and Dangerous Occurrence Regulations (RIDDOR) 2013 to report accidents, near-misses and cases of work-related ill-health, except under specialist legislation, there is no specific duty to investigate such events. However, such an investigation can provide helpful information to improve health and safety and as such forms an important part of any management system. While a near-miss may be regarded as a 'lucky escape' the lessons which can be learnt from a thorough investigation can be as invaluable as those learnt from an accident investigation (without the need for anyone being injured first!)

The procedures in place for reporting and investigating incidents, including near-misses and cases of ill-health, should be defined within the development of the management system. It is important that those involved within the investigation itself are competent to do so while all employees must be aware of the reporting procedure. Where a trade union is recognised by an employer, the union representatives have the right to investigate potential hazards and examine the cause of workplace accidents. Whilst this may be undertaken independently, where the relationship between union and management is positive, the results will potentially be far better if a collaborative approach can be taken. For some incidents where more complex risks are involved it may be necessary to also utilise the services of a specialist within the investigation team.

High priority should be given to the investigation process so that staff recognise the importance allocated to it by the management. The purpose of the investigation is not only to identify the immediate cause, which may be quite obvious, but also the under-lying reasons behind the situation which may require more detailed consideration. The investigations may identify a weakness in control measures or training, deviations from the rules, including the development of short cuts or possibly identification of individuals who need to be held to account. It may also be appropriate to review the effectiveness of the emergency procedures as part of the investigation. Proactive action must follow based on the findings of the investigation.

A detailed record of the findings should be made. Not only will this provide an accurate summary for future reference but in the case of legal proceedings may be used as mitigation to demonstrate the employer's commitment to resolving the problems and also provide valuable information for the insurance company or legal team.

ACT

The first three elements of the model: Plan, Do and Check, form the basis of the management system. The Act element completes the loop. This encourages ongoing review of the overall system to ensure that it is as effective as possible with the outcomes providing the basis for the cycle to restart.

Reviewing performance

Like the provisions in place for monitoring, the scope, method and frequency for regular review should be defined at the Policy stage. The frequency will be dependent upon the speed of business evolution, maturity of the management system and risk profile but should be scheduled at least annually. The purpose of this formal review is to confirm legal compliance, that all risks are reduced so far as is reasonably practicable, that all systems and controls are working as originally envisaged and that the system remains current; it may be possible to stop doing things which are no longer necessary or respond to new risk.

The review process will need to consider a wide source of information including results of monitoring and investigation exercises, risk assessments, new and updated guidance and issues raised by the workforce. The inclusion in the review team of an individual independent to the company or business area may also provide impartiality and a fresh view.

To achieve maximum benefit from the review, the results must be formally published. This not only provides the information often sought by third parties, but also provides a reassurance to the workforce by publicising and crediting achievements while making commitments to improve in areas where shortfalls are identified.

Lessons learnt

It is important that maximum benefit is achieved by responding to the lessons learnt from incident investigations and the monitoring and review processes undertaken. It should be remembered that as well as ineffective risk control, the underlying barriers to effective health and safety management include poor leadership, failure to consider the implications of human factors and management implicitly, or in some cases explicitly, condoning poor behaviour.

To ensure that lessons can be learnt to continually improve health and safety provision within an orgaisation, it is important that senior management are receptive to all information and seek not to blame but to understand how problems can be rectified. A blame culture will result in individuals failing to report leading to the loss of valuable information while inapprorapite response from the board can result in filtered reports being submitted.

The effect of human behaviour should be carefully considered. Where problems arise which encourage a 'workaround' to be developed, the problem should be addressed to overcome the need for that deviation. Communication between all departments and sites of the organisation is also important: where an incident provides valuable lessons and opportunities for improvmenet, these should be shared to prevent similar incidents company-wide.

Trade associations often take the lead in collating data on a non-attributable basis and sharing the lessons learned with participating member organisations. The collective experience of what is learned enables common problem activities and hazards to be identified and underlying causes, human, organisational and technical, to be examined. In some cases the lessons are shared as an immediate alert, in other cases quarterly reports provide analysis of recent experiences and trends.

Reasons for the introduction of formal health and safety management systems

Organisations will have developed systems for managing the design, production, quality and financial aspects of their business and health and safety should be managed with the same emphasis and degree of control. A systematic approach such as that described within HSG65 or other models makes its management both easier and more effective. The policy statement provides the business with a clear aim and specific objection to work towards thus preventing it from progressing haphazardly towards a perceived but unclear goal.

The development of the system allows the role of all individuals, from senior managers to workers, as well as those with specific responsibilities to be defined along with the provision for interaction between them, thus encouraging all members of the organisation to deliver. The system for identifying, evaluating and controlling risks will provide a consistent approach as well as identifying and securing the resources necessary. As the system will also identify monitoring and review processes, the ongoing effectiveness of the system will be ensured which will result in ongoing legal compliance and reduction of risks thereby significantly reducing the incidence of injury and ill-health.

The benefits of an effective health and safety management system however go beyond this legal compliance and a safe workforce. The commitment demonstrated by such a robust system increases the morale of the workforce thus increasing their own commitment to the business which can have a positive impact on productivity and loyalty. For many businesses, public perception is important as is the view of third party suppliers and customers who will sometimes only do business with those organisations who can demonstrate a tight control over their health and safety.

As well as the financial benefits of well managed health and safety by the avoidance of adverse events, the existence and implementation of an effective health and safety management system can also sit favourably with insurance companies who may impose lower premiums as a result.

A1.4 - Principles and content of effective health and safety management systems

Health and safety policy

THE LEGAL REQUIREMENTS RELATING TO HEALTH AND SAFETY POLICIES AND ARRANGEMENTS

HASAWA 1974 Act - Section 2(3)

Section 2 (3) of the Health and Safety at Work etc. Act (HASAWA) 1974 requires employers to:

- Create a written health and safety policy statement.
- State the organisation and arrangements for bringing the statement into effect.
- Bring the policy to the attention of employees.
- Revise the policy as necessary.

The legal requirement to 'bring the policy to the attention of employees' is to ensure that employees are provided with information on all areas of the policy that affect them. For example, at induction of new employees they can be told the intention of the company regarding health and safety and the names of the persons with responsibility. Once the new employee is working, they can be given the information on the emergency procedures, arrangements for their particular tasks, including safe systems of work, permits to work and the relevant training. In this way, the health and safety policy is the overall control document and the vehicle for communication.

Management of Health and Safety at Work Regulations (MHSWR) 1999 - Regulation 5 (Health and Safety Arrangements)

Regulation 5 of MHSWR requires that appropriate arrangements must be made for the effective planning, organisation, control, monitoring and review of preventative and protective measures. Employers with five or more employees must set out their arrangements in writing. Many organisations have adapted their policy to ensure it covers the necessary requirements of this regulation.

Components of a health and safety policy document

Statement of intent

Overview

A particularly effective way of demonstrating management philosophy and commitment to health and safety is by setting out and communicating a policy statement. In order to accentuate this commitment the statement should be signed and dated by the most senior member of the management team, which will give it authority from the most senior level. Lack of firm management commitment of this kind leads to the belief that health and safety is not equal to other organisational goals.

It, and its revisions, must also be brought to the attention of employees, particularly as their involvement is essential in making the policy work.

Safety goals and objectives

Goals are seen to be long-term outcomes of an organisation's actions. Objectives are the steps necessary to reach that goal and tend to be shorter, time based, specific and measurable. Though they may be described as short, medium or long term objectives they remain steps towards the goal. The policy statement should establish a high standard of health and safety as a significant goal of the organisation. The extent of the goal should be well communicated and include health, not just safety.

The policy statement should establish a summary of the objectives that must be fulfilled to meet the goals. Often objectives are summated as being to prevent all accidents. This may be confusing, as this is better seen as a goal, rather than an objective. Objectives should include important management activities such as - plan, organise, control, monitor, and review.

Organisation

It is usual to express how the organisation is organised by showing an organisational chart, allowing reporting/communication structures of line and functional staff to be depicted. This is supported by the clear depiction of responsibilities of management and supervisors at all levels, duties of workers and the role and functions of health and safety practitioners. Setting out the organisation for health and safety will include:

- The individual ultimately responsible for health and safety within the organisation on whom the legislation places undeletable duties.
- The unbroken and logical delegation of responsibilities to assist the duty holder to fulfil his duties, through line management to first line managers (supervisors) who operate where the hazards arise.

- The identification of the managers, who are accountable to top management for ensuring that detailed arrangements for safe and healthy working are drawn up, implemented and maintained.
- The definition of the roles/function of line and functional management.
- Defining adequate support by all relevant functional management, not only by the health and safety practitioner, but also, according to need, by the medical staff (doctor and nurse), engineers, designers and chemists.
- Defining specific employees to assist with health and safety arrangements, for example, first aiders and fire wardens.
- Defining the responsibilities of employees and others.
- Stating employee involvement through representatives, briefing systems and committees.
- The nomination of persons with the authority and competence to monitor health and safety performance of the whole organisation, different locations and departments.

Job descriptions (or similar) should be drawn up to identify specific roles/functions and to avoid expensive and potentially dangerous overlapping. An involvement in health and safety should be seen as an essential part of good management. The responsibility for taking direct action lies with line management, rather than with the health and safety practitioner. The practitioner's role includes monitoring the effectiveness of the health and safety policy and the provision of accurate information, advice and support for senior management.

Duties and responsibilities in relation to health and safety

Duties placed on people, in order for the organisation to meet its objectives, must be clear and understood. Duties placed on subordinates do not remove the responsibility of the senior person to ensure that duties and the related actions are carried out. Indeed, it is a primary duty of the senior person to ensure that subordinates carry out their duties. Each person assigned duties is accountable for them being carried out. It follows therefore that a senior person is accountable for the duties they assign as well as their own.

It is important for managers to ensure the proper allocation of resources to health and safety and they may be held accountable in law for deficiencies within their area of control. Failures cannot be readily ascribed to 'circumstances beyond control'. The director or manager's job is to be in control. Duties and responsibilities should be defined for the following people:

- The senior manager of the organisation.
- The senior managers that report to the senior manager.
- The senior manager nominated to deal with health and safety matters.
- Middle managers.
- First line managers.
- Functional staff, in particular the health and safety practitioner.
- Employees.
- Others, such as volunteers.

Organisational structure in relation to health and safety

This should clearly show who is responsible to whom and the nature of the responsibility, line or functional responsibility. This can be visually presented by the use of a management structure diagram(s). This then summarises and supports the written definition of duties and responsibilities.

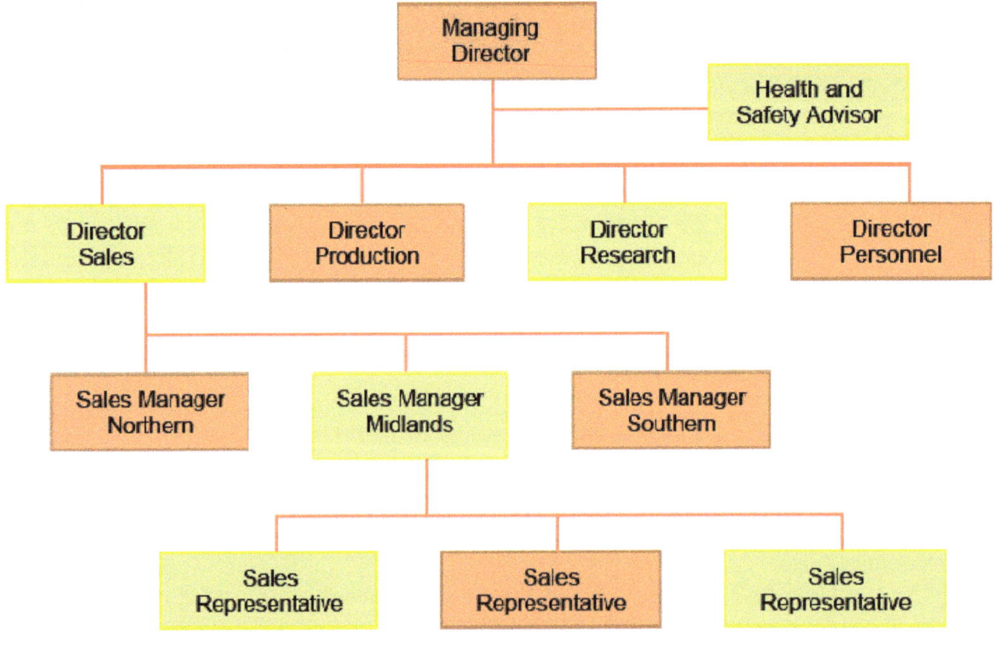

Figure A1-11: Organisational structure example. *Source: RMS.*

Specific employees

The organisation should define the essential people who will assist with the carrying out of specific, and safety arrangements, for example:

- First-aiders.
- Fire wardens.
- Risk assessors - general risks and specific risks.
- Employee representatives.
- Competent advisors.

Where the policy identifies the need for the support of ancillary services such as occupational health, occupational hygienist, radiation protection advisor, etc., reference should be made to these here also.

Arrangements

Scope

The scope of arrangements made by an organisation, and depicted in the policy, is dictated by the nature of the organisation (i.e. its hazards and controls). In summary, these tend to fall into two categories, general and specific arrangements. General arrangements will have an influence on a variety of hazards (for example, accident investigation, induction training, and first aid arrangements). General arrangements should cover the requirement to plan, organise, control, monitoring and review, as required by MHSWR Regulation 5. Specific arrangements will be established for specific hazards (for example, electricity, noise, manual handling, and display screen equipment). Arrangements to control a particular hazard will comprise a blend of technical, procedural and behavioural arrangements.

Detail

Inevitably the arrangements of an organisation are not necessarily set out in a single document. It is quite acceptable for the primary policy document to refer to other documents that depict these arrangements in detail. These connective documents are often known by a variety of names, for example, standing instructions, safe working practices, and safe systems of work, rules, and procedures.

Systems

Systems of work may be very general or specific and can be written in such a way as to make them advisory to work situations or very prescriptive. The formality of a system tends to be dependent on the level of risk arising from the work, perhaps the most formal being a permit to work system.

Rules

Rules are created to remind people of the correct (safe and healthy), mandatory behaviour expected of them. Rules should be written to avoid ambiguity and should not be confused with guidance. The clearest rules are those that can begin with the words 'always or never'.

Procedures

Procedures formalise the actions necessary within a work activity that will ensure health and safety. They tend to be written in the order that the actions are likely to be carried out. Procedures formalise and bind arrangements in place. If something needs to be done in a particular way the procedure would help to ensure it is done. They may be used during training or monitoring exercises. Procedures should set clear standards of performance and apply to:

- People selection, training and supervision.
- Equipment and substances - purchase, supply, transport, storage and use.
- Products and services - design, delivery, transport and storage.
- Environmental and place of work control.
- Management actions and the way work is done.

Standards

Standards help to build a positive culture and control risks. They should identify who does what, where, when, how and with what results. Above all the standards must be measurable, achievable and realistic.

Statements such as 'staff must be trained' are difficult to measure if it is not known exactly what 'trained' means and who is to do the work. Many industry-based standards already exist and can be adopted, where applicable. In other cases, advice will have to be taken in order to set standards applicable to the organisation. Standards should be specific enough to enable them to be measured, referring to numbers, quantities, levels and timing.

For example:

- Specifying levels of waste, effluent or emissions that are acceptable.
- Methods and frequency for checking guards on machines or access equipment.
- Specific levels, frequency and content of training.

Cross-reference to key documents

The health and safety policy of an organisation (statement of intent, organisation, arrangements) may be substantial. The documentation to depict this may also be extensive. It is usual to establish a system of manuals that depict and control the policy. The actual system used will depend on the organisation, but a primary document (manual) should contain all three of the above policy items, including any references to other documents that give additional information.

The purpose of the manual is to provide control of management actions and would therefore be made available or provided to them. Employees would usually gain access to the manual through a request system or library point and be provided with a summary of the main points of the manual. This includes a full copy of the statement, summary of the organisation and arrangements and reference to manuals. A leaflet or a booklet might achieve this.

Communication

The effective communication of the policy is important; it must be understood by all those affected by it. In order to achieve this, a good deal has to be done. Merely posting or distributing a copy to employees is not enough. Training and briefings will be necessary, as a minimum, to ensure effective communication. For new employees this is often done as part of the induction process.

The format, complexity and language(s) used should be considered. Experience shows that no one form of document is adequate to meet everyone's needs, therefore the document is usually produced in at least a summary and a detailed format. If it is likely to be revised frequently, a loose-leaf scheme will be advisable for the detailed version. This is especially true when the names and contact telephone numbers of staff are included.

The process of maintaining the document up to date is made easier where they are provided in a read only electronic version, particularly if a single version is provided on a central system such as an intranet. Arrangements must be put in place to enable revisions to be communicated effectively.

The role of the health and safety policy

Without active management involvement in health and safety, any attempt at organised accident prevention will be restricted and predominantly reactive. The overall role of a policy is to provide direction for an organisation, establishing a remit that will both guide the organisation to satisfy its aims and to bind it so that it does not stray from the standards that it sets to meet the aims. This will influence the decisions made by an organisation, in that they will need to fall within the intent of the policy.

Required by statute, the health and safety policy is an important element of any occupational health and safety management system (OHSMS). Its role therefore includes showing the intention of the employer (organisation) to managing health and safety and demonstrates a clear commitment to health and safety and the requirements of the OHSMS. It illustrates an acceptance of responsibility by the most senior manager and assignment of responsibilities to others. It defines the organisation's aims and reflects its values.

Where, as is preferable, the policy is established in consultation with workers, its role is also to establish shared values and aims. Its role is also to provide performance standards to measure achievement against. As such, its role is to provide direction and structure to the effort of the organisation in achieving successful health and safety management.

The health and safety policy provides a vehicle for the communication of important information on the position the organisation takes towards health and safety and its aims. Communication of the policy is also a legal requirement and it is used to give workers and other interested parties information on how the organisation's efforts regarding health and safety will affect them. For example, at the induction of new workers they can be told the intention of the employer regarding health and safety and the names of the persons with responsibility.

For managers, the policy sets aims and commitments that they are expected to fulfil on behalf of the organisation. Managers can use this to set objectives and management routines that help them to meet their responsibility. In turn the organisation can measure their performance against the expectations set out in the policy.

Required by statute, the health and safety policy provides an essential basis to any effective health and safety management systems. Its role is to define the employer's intention to effectively manage health and safety and to illustrate his commitment to provide the necessary resources for this to be achieved.

The policy should define the organisation's overall aims and values. Where, as is preferable, the policy is established in consultation with workers, the policy can be further enhanced by identifying shared values and aims.

Senior personnel must publicly accept their health and safety responsibilities as well as assign responsibilities to others at all levels of the organisation The policy will define the organisation's direction and ultimate health and safety goals.

The health and safety policy provides a vehicle for the communication of important information on the position the organisation takes towards health and safety and its aims. Communication of the policy is also a legal requirement and it is used to give workers and other interested parties information on how the organisation's efforts regarding health and safety will affect them. For example, at the induction of new workers they can be told the intention of the employer regarding health and safety and the names of the persons with responsibility.

For managers, the policy sets aims and commitments that they are expected to fulfil on behalf of the organisation. Managers can use this to set objectives and management routines that help them to meet their responsibility.

Requirements for a written health and safety policy and for recording arrangements

Section 2(3) of the Health and Safety at Work etc. Act 1974 stipulates that the policy statement required by the section must be in written form. By virtue of the Employers' Health and Safety Policy Statement (Exceptions) Regulations 1975, employers with less than five employees are not required to have a written statement.

Similarly, Regulation 5(2) of the Management of Health and Safety at Work Regulations 1999 requires employers of five or more employees to record the arrangements in place to ensure the effective planning, organisation, control, monitoring and review of the preventive and protective measures as required by regulation 5(1)

The general principles and objectives of a health and safety policy document

The principles of a health and safety policy document include:

- To ensure conformity to legal requirements and guidance.
- To be specific to the organisation.
- To be appropriate to the size and nature of the organisation's activities and risks.
- To be concise, clearly written in appropriate languages and communicated to those that work to it.
- To be reviewed to ensure it remains relevant and effective.

To comply with HASAWA Section 2(3) the structure for a health and safety policy should incorporate the following three sections:

1) Statement of Intent.
2) Organisation.
3) Arrangements.

For a smaller organisation with less complex risks it is likely that the policy will be quite straight forward while for larger organisations to address the arrangements in place a potentially detailed policy is often required. To encourage understanding and use of the policy, such larger policies are often better consolidated with cross-reference to separate documents which include the greater detail needed.

Merely distributing a copy of the health and safety policy document to managers or workers is not enough to ensure understanding. Training and briefings will be necessary, as a minimum, to ensure effective communication. For new managers and workers this is often done as part of the induction process.

Because it is useful to plan for change, it is worth producing the health and safety policy document in such a way that it is easy to identify changes. This can involve highlighting the different changes on the document, explaining the reasons for change, having a controlled number of copies of the document issued and recording that the changes have been issued to those that hold copies.

The process of maintaining the document up to date is made easier where they are provided in a read only electronic version, particularly if a single version is provided on a central computer system, such as an intranet. Arrangements must be put in place to enable revisions to be communicated effectively.

The objectives of the health and safety policy document include setting out the structure and content of the health and safety management system. By setting it out in writing it provides direction to the organisation's effort to ensure that all members are guided towards the ultimate goals thus preventing initiatives straying from their planned route.

The policy provides managers and workers with standards to work to and provides a framework to hold the organisation to account for its performance. By setting realistic and well defined performance indicators, ongoing performance can be measured by internal auditors and can be evaluated by regulators and third party certification organisations.

The key elements/components of BS OHSAS 18001:2007

Health and safety management systems are a formal and logical way of planning and controlling management of risk in an organisation. The models can be based on the HSE's model in the guidance 'HSG65 Managing for health and safety', BS OHSAS 18001 or the environmental model BS EN ISO 14001.

> *"Part of an organization's management system used to develop and implement its OH&S policy and manage its OH&S risks."*

Figure A1-12: Definition of an occupational health and safety management system. *Source: OHSAS 18001.*

All management systems have similar elements; typically focusing on policy, setting out intentions, carrying out these intentions, monitoring the success and effects of carrying them out, evaluating what has been learned and using it to refine policy and intentions. In this process there is a 'feedback' loop providing opportunity to make improvements to the way the system operates.

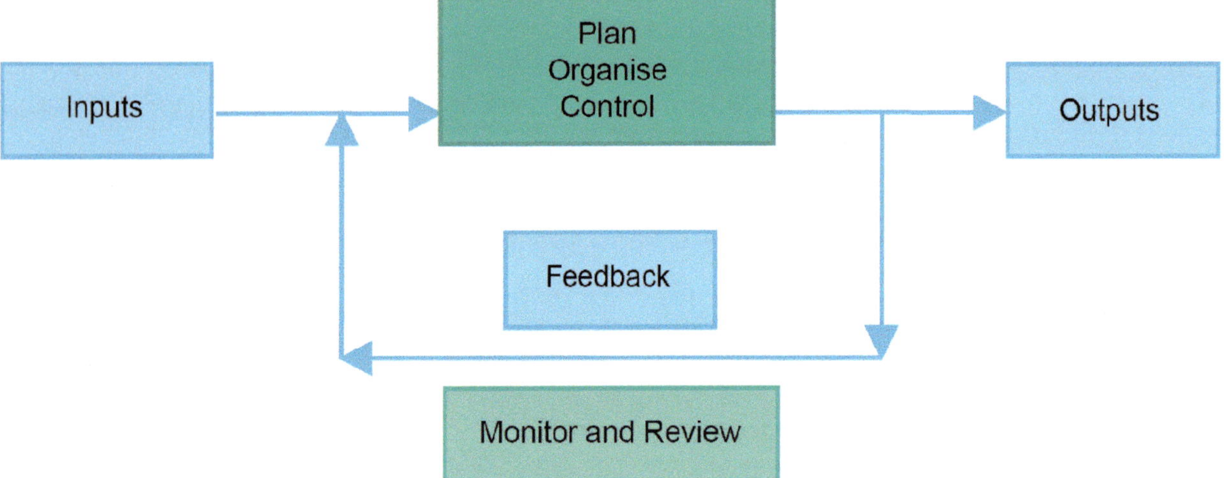

Figure A1-13: Main elements of a health and safety management system. *Source: RMS.*

BS OHSAS 18001

In response to demand for an Occupational Health and Management System (OHSMS) that provides a route to external certification the British Standards Institution developed the Occupational Health and Safety Assessment Series (OHSAS). This includes the standard, OHSAS 18001 and the accompanying guidelines for the implementation of OHSAS 18001, OHSAS 18002.

When seeking to comply with the standard for certification purposes OHSAS 18002 has a similar relationship to OHSAS 18001 as an approved code of practice has to regulations.

BS OHSAS 18001 was developed to be compatible with ISO 9001 (Quality) and ISO 14001 (Environmental) management systems standards in order to facilitate the integration of quality, environmental and occupational health and safety management systems by organisations, should they wish to do so.

Figure A1-14: Main clauses of BS OHSAS 18001. *Source: BSI/RMS.*

While the BS OHSAS 18001 standard gave organisations the opportunity to seek certification for their health and safety management in the same way as for their Quality and Environmental Management, it is not recognised internationally. This anomaly will be rectified with the advent of the ISO 45001 due in the near future after a period of consultation. The BS OHSAS 18001 model reflects the elements of the plan, do, check, act (PDCA) approach, now used in HSG65 and many other management models. ***Policy*** and ***planning*** reflects the '***Plan'*** element of PDCA; ***implementation*** and ***operation*** the '***Do'*** element and ***checking*** *and* ***corrective action*** the '***Check'*** element. ***Management review***, where the system is reviewed with a view to identifying improvements, is the '***Act'*** element which feeds into the ***continual improvement cycle***. Within most of the main clauses of the BS OHSAS 18001 model there are a number of sub-clauses that have to be satisfied, the requirements of the main clauses are outlined in the following section.

OCCUPATIONAL HEALTH AND SAFETY (OH&S) POLICY

The most senior management (called top management in OHSAS 18001) are responsible for writing and authorising the OH&S policy. The OH&S policy is a written statement demonstrating the commitment of an organisation to OH&S. It demonstrates the commitment of an organisation by establishing an overall sense of direction and setting the principles that underpin the actions to be taken. It is a formal commitment by the organisation's top management to deliver the necessary financial and human resources for effective OH&S management. It must also make a commitment to implementing a system for preventing harm to those who may be affected by their activities; this includes contractors, visitors and members of the public. The involvement of the top management in producing, authorising, communicating and promoting the OH&S policy is essential. This alone, however, is insufficient to develop a successful OH&S management system.

Ideally, the OH&S policy should be less than a page in length, written in simple language, dated and signed by the most senior manager of the organisation, for example, chief executive officer (CEO) or director with responsibility at board level for health and safety. The OH&S policy should be displayed as well as made available to all the relevant parties.

Planning

This may be the most challenging part of an OH&S management system. Unless all the OH&S risks to the organisation are identified and effective controls are implemented, the system will not deliver a workplace that minimises the risk of harm. Legal and other requirements need to be identified and the organisation's OH&S objectives established on the basis of the risks, legal issues and business requirements. A realistic programme needs to be put in place to deliver the objectives.

The aim is to plan for an OH&S management system capable of:

- Identifying and effectively controlling all occupational health and safety risks.
- Reacting smoothly to changing demands - legal requirements, new technology and innovation, organisational changes and reaction to regulatory body requirements.
- Sustaining a positive health and safety culture.

Planning for OH&S uses similar methodology to that used throughout other activities of the organisation and includes:

- Identifying specific requirements for the system - what needs to be done, where and with what priority.
- Setting clear performance criteria - what needs to be done.
- Identifying who is responsible - who gets it done and how.
- Setting time scales - when it is to be done by.
- Identifying the desired outcome - what the result should be.

The most important stage of the planning process is hazard identification, risk assessment and selection of appropriate risk control measures. If this stage has not been completed in a satisfactory manner the management system will almost certainly be incomplete. It is essential to effective planning that all risks and their controls are effectively determined.

Implementing and operating

The planning phase identifies what needs to be in place and the 'implementation and operation' phase is where all the necessary arrangements and actions are put in place. Having identified what risk controls need to be applied, the legal requirements, the objectives and programme, it is necessary for the organisation to ensure the arrangements are implemented and operational. For the successful implementation and operation of an OH&S management system BS OHSAS 18001 identifies seven important areas that need to be addressed:

- Resources, roles, responsibility, accountability and authority.
- Competence, training and awareness.
- Communication, participation and consultation.
- Documentation.
- Control of documents.
- Operational control.
- Emergency preparedness and response.

Checking and corrective action

It is often quoted that 'what you can't measure, you can't manage' and this statement is equally true with OH&S. Procedures need to be in place to measure and monitor the OH&S system, enabling the organisation to put right any deficiencies and to identify opportunities for improvement.

It is essential that systems are put in place to show that the OH&S management system is working and delivering the performance that the organisation is committed to achieve. The organisation needs to plan for monitoring, measuring, checking and correcting any deficiencies.

To ensure an effective checking and correcting stage in an OH&S management system, BS OHSAS 18001 identifies five areas that need to be addressed:

1) Performance measurement and monitoring.
2) Evaluation of compliance.
3) Incident investigation, nonconformity, corrective action and preventive action.
4) Control of records.
5) Internal audit.

The organisation must ensure that internal audits of the OH&S system are planned and conducted at intervals. An internal audit programme needs to be based on the results of the risk assessments of the organisation's activities, the results of previous audits and the organisation's performance objectives.

The internal audits should determine whether the OH&S management system conforms to planned arrangements which reflect the requirements of BS OHSAS 18001, that these arrangements have been fully implemented/maintained and are effective in meeting the organisation's policy and objectives.

The audit programme must be supported by an internal audit procedure which sets out who does internal audits, when they are conducted and how. Selection of internal auditors must ensure objectivity, impartiality and competence.

Reviewing

The management review is the opportunity for top management to carry out a strategic review of the OH&S management system over a previously determined period of time and to decide if any improvements need to be made. This should take place at least once a year. It should consider any feedback from interim reviews, audits, incidents, inspections and employee consultation, as well as information from external sources (such as regulators, trade associations and insurers). Where improvements will benefit the system and can sensibly be made, they need to be committed to and any required actions communicated.

Reviewing OH&S management systems is a fundamental requirement in any organisation. It ensures that processes and procedures are being applied effectively, as intended, and continue to meet the needs of the organisation.

Most importantly, it provides the mechanism to drive the continual improvement required of any management system. It should be a live process within the business in order to ensure a robust OH&S management system exists and is operational.

FORMAT OF BS OHSAS 18001 STANDARD

The BS OHSAS 18001 standard is extensive with consideration required for each detail. The word 'shall' is used throughout which means there is no choice in the matter and therefore when the organisation's OH&S management system is assessed it can clearly be measured against the requirements.

Figure ref A1-15 is representative of the style and content of the management system requirements depicted by the standard.

Clause 4.3.1 - Hazard identification, risk assessment and determining risk controls

The organization shall establish, implement and maintain a procedure(s) for the ongoing hazard identification, risk assessment, and determination of necessary controls.

The procedure(s) for hazard identification and risk assessment shall take into account:

a) Routine and non-routine activities.

b) Activities of all persons having access to the workplace (including contractors and visitors).

c) Human behaviour, capabilities and other human factors.

d) Identified hazards originating outside the workplace capable of adversely affecting the health and safety of persons under the control of the organization within the workplace.

e) Hazards created in the vicinity of the workplace by work-related activities under the control of the organization.

NOTE 1: It may be more appropriate for such hazards to be assessed as an environmental aspect.

f) Infrastructure, equipment and materials at the workplace, whether provided by the organization or others.

g) Changes or proposed changes in the organization, its activities, or materials.

h) Modifications to the OH&S management system, including temporary changes, and their impacts on operations, processes, and activities.

i) Any applicable legal obligations relating to risk assessment and implementation of necessary controls (see also the NOTE to 3.12).

j) The design of work areas, processes, installations, machinery/equipment, operating procedures and work organization, including their adaptation to human capabilities.

The organization's methodology for hazard identification and risk assessment shall:

a) Be defined with respect to its scope, nature and timing to ensure it is proactive rather than reactive.

b) Provide for the identification, prioritization and documentation of risks, and the application of controls, as appropriate.

For the management of change, the organization shall identify the OH&S hazards and OH&S risks associated with changes in the organization, the OH&S management system, or its activities, prior to the introduction of such changes.

The organization shall ensure that the results of these assessments are considered when determining controls.

Figure A1-15: Extract from BS OHSAS 18001. *Source: BSI, BS OHSAS 18001.*

Integration of quality, environmental, and health and safety management systems

BENEFITS

The quality and environmental management systems are very similar to the health and safety system described above. The models have the same general requirements, and in some cases, they only need a word replacement to be the same, for example, change 'Environmental Policy' in BS EN ISO 14000 to 'OH&S Policy' in BS OHSAS 18001. They all require strong leadership and commitment, worker participation, reduction of waste/errors, etc.

Recognition of the similarity in the basic requirements has led many people to believe that an integration of quality, environment and health and safety as one management system is the way ahead. A growing number of organisations have achieved integration of their management systems.

The British Standards Institute (BSI) has recognised the value of integration and produced a publicly available specification (PAS) 99; it is based on the six common requirements set out in the ISO guide 72, a standard for writing management systems.

The BSI states the following benefits to integration:

Reduced costs

- By avoiding duplication in internal audits, document control, training and administration, adopting future management systems will be much more effective.

Time savings

- By having only one management review.

A holistic approach to managing business risks

- By ensuring that all consequences of any action are taken into account, including how they affect each other and their associated risks.

Reduced duplication and bureaucracy

- By having one set of processes ensures the requirements of the specific standards are co-ordinated, workloads streamlined and disparate systems avoided.

Less conflict between systems

- By avoiding separate 'empires' for the likes of quality and environment, responsibilities are made clear from the outset.

Improved communication, both internal and external

- By having one set of objectives, a team approach culture can thrive and improve communication.

Enhanced business focus

- By having one system linked to the strategic objectives of the business contributes to the overall continual improvement of the organisation.

Improved staff morale and motivation

- By involving and linking roles and responsibilities to objectives, it makes change and new initiatives easier to implement and makes for a more dynamic and successful company.

Optimised internal and external audits

- By minimising the number of audits required and maximising the number of people involved.

The Japanese tend to have an integrated approach such that they put all aspects under the quality heading and say that any deviation, such as a minor injury accident, is a quality failure. The total quality management (TQM) approach they favour requires that they strive for no deviations, including those that could cause an injury.

Complete segregation of quality, environment and health and safety could result in one problem being tackled in three separate ways, for example, controlling noise, and some problems not being accepted as belonging to anybody, for example, poor leadership. A health and safety manager could recommend changing to plastic for milk bottles instead of glass to solve a noise problem, while the environmental manager recommends biodegradable cartons and the quality manager says the customer prefers glass.

LIMITATIONS

An area of concern for health and safety practitioners is that health and safety is often assumed to follow from other controls without being specifically addressed. For example, employees will be highly trained in all aspects of the operation and therefore guarding dangerous parts of machinery may not be seen as necessary as 'a highly trained operative would not put their hand in the dangerous parts of a machine'. The health and safety practitioner would consider what a machine operator may do when distracted, such as put their hand where they normally would not. The practitioner would require such danger to be guarded.

For some, the problem with an integrated system is that whoever manages it may have a vested interest in one area over the others. Many people believe that quality puts the product before the employees and others believe that health and safety issues slow the job down and get in the way of production. Environmental issues may take a secondary position to health and safety, for example, where a hazardous airborne substance is vented to atmosphere. Conversely, an environmentally hazardous substance may be replaced by one that is more harmful to health, for example, trichloroethylene instead of 1.1.1.trichloroethane, now banned because it depletes the ozone layer.

For each of the benefits set out by BSI a limitation may be identified. This does not mean that all limitations exist in all cases but that they can arise in some organisations over time.

Reduced costs..........but increased complexity?

- A single document to cover all aspects of the integrated system could be longer and more complex than each separate one. Having a document that addresses more aspects means that it is more likely to be influenced by changes in one of the aspects, thus resulting in revisions and re-issue more often.

Time savings................though will take longer than for a single discipline

- By having only one management review it will have to take longer in order to fully cover each of the aspects of the integrated system properly.

A holistic approach to managing business risks............but less attention to detail

- The wider the thinking related to the consequences of any action the more likely one aspect of the integrated system will get less attention. It still requires people to focus and champion a specific aspect of the integrated management system.

Reduced duplication and bureaucracy............but possibly not much less

- Each of the systems do have differences, in practice there will not be one set of processes for all circumstances. In practice therefore, the reduction could be minimal. Alternatively, the process will be developed to try to meet the requirements of all three disciplines, potentially not fully meeting the needs of any of them.

Less conflict between systems...............but what about the specialist 'fighting their corner'?

- Vested interests of each aspect may remain, even if hidden by an integrated system.

Improved communication, both internal and external............or pathway to confusion?

- One set of objectives will inevitably have to cover all the range of aspects of the integrated system; this may prove difficult or confusing. The objectives can be communicated in a cohesive way without system integration.

Enhanced business focus.................or out of focus?

- Integration diminishes business focus as the needs and objectives of each aspect become blurred.

Improved staff morale and motivation..................but do they need to be integrated to achieve that?

- It is not necessary to integrate the systems to link roles and responsibilities to objectives.

Optimised internal and external audits.............but probably more complex

- By minimising the number of audits required each audit will have to be of a longer duration in order to fully ensure that each aspect of the integrated system has been dealt with properly.

Considering both the benefits and the limitations, it could be argued that in many cases, the lack of belief in integration can be put down to anecdotal evidence and a poor understanding of the management systems. As in many cases, a compromise could be the answer.

A standard way of managing the systems with standardised monitoring, review and audit methods could be utilised. Many of the general requirements are the same and would only become specific at the detailed level.

For example, employee commitment and participation would be an over-riding requirement, and training is one way to help achieve this. The training itself would be specific to the needs of each management system.

This page is intentionally blank

Element

A2

Principles of health and safety law

Learning outcomes

On completion of this element, candidates should be able to demonstrate understanding of the content through the application of knowledge to familiar and unfamiliar situations. In particular, they should be able to:

A2.1 Explain the sources and types of law in force in the UK relevant to health and safety.

A2.2 Explain the concept of absolute and qualified duties in relation to health and safety legislation.

A2.3 Outline the influence and role of the European Union on UK health and safety legislation.

A2.4 Outline the status and procedure for the creation of UK Acts, Regulations and Orders.

A2.5 Outline the structure and functions of courts and related institutions in the UK.

A2.6 Outline the principles of the law of contract and its application to health and safety issues.

A2.7 Explain the principles of employment and discrimination law as it affects health and safety issues.

Contents

Relevant statutory provisions

Treaty of the Functioning of the European Union (2012/C326/01)

Equality Act (EA) 2010

Legal Aid, Sentencing and Punishment of Offenders Act (LASPO) 2012

Legal Aid, Sentencing and Punishment of Offenders Act 2012 (Fines on Summary Conviction) Regulations (LASPOFSC) 2015

Public Interest Disclosure Act (PIDA) 1998

Trade Union and Labour Relations (Consolidation) Act (TULR) 1992 (as amended)

Unfair Contract Terms Act (UCTA) 1977

Public Interest Disclosure (Northern Ireland) Order 1988

Courts Reform (Scotland) Act 2014

Sources of reference

Reference information provided, in particular web links, was correct at time of publication, but may have changed.

Health and Safety Offences, Corporate Manslaughter and Food Safety and Hygiene Offences, Definitive Guideline (effective from 1 February 2016), http://www.sentencingcouncil.org.uk/publications/item/health-and-safety-offences-corporate-manslaughter-and-food-safety-and-hygiene-offences-definitive-guideline/

The above web links along with additional sources of reference, which are additional to the NEBOSH syllabus, are provided on the RMS Publishing website for ease of use - www.rmspublishing.co.uk.

A2.1 - Sources and types of law

Sources of law

COMMON LAW

Nature and development

Common law refers to the body of law created by decisions made in cases heard. Common law follows the principle that similar facts should be treated the same on different occasions that they are considered, and not to do so would be unfair. Common law is often referred to as 'Judge made law' because of its dependency on the decisions of judges to make up the body of law. Common law impacts on both the criminal and civil branches of law. Although this source is relied upon mostly within civil law, some important offences within criminal law, including murder, which is still defined by the principles of common law. The common law duty of care where everyone has a duty to take reasonable of others plays an important in health and safety-related civil litigation.

Common law, rather than being laid down by Parliament, has traditionally grown up over the centuries. The origins of common law are vested in Anglo Saxon Customary Law and the administrative efficiencies of the Normans, after their conquest of England in 1066. The work of the administrators of William I developed into a judicial function. As they travelled around checking on local administration they began hearing cases in local courts. The administrators were called the Justices in Eyre 'a journey', because they had experience of various decisions they made in local courts, and were able to discuss their work on their return to Westminster, this began to establish a more unified system of law.

Henry II took steps to formalise the system further and introduced a new civil remedy for those who complained that their land had been seized, and in time a wider range of civil matters were heard. During the time of Richard II (1377-99) a system of circuit judges had been established by the use of more formally trained lawyers. By selecting the best customary rulings that could be applied outside the county where the case was heard, existing local customary laws were moulded into one unified law 'common' for the whole kingdom.

Judicial precedent

Judicial precedent, like common law, follows the principle that facts should be treated the same on subsequent occasions that they are considered. On considering a case the judge will determine if previous cases have been heard that related to the same facts. If no cases considering the same facts have been heard, the judge has the authority and responsibility to make a new decision on the new facts presented. In doing this the judge will set a precedent for other judges hearing cases involving the same facts.

When judges hear a case and make a judgement they will set out their ***ratio decidendi,*** the reasons for their decision. This is the legal reasoning within their judgement, on which the outcome of the case depends. The ***ratio decidendi*** is the binding principle set out for other judges to follow. The judges will often also set out their ***obiter dicta***, which are those things that do not set a binding precedent, but relate to the judgement of that particular case; they are 'by the way' comments.

The judgements made in this manner create binding precedents on subsequent similar cases on all lower courts. Decisions made by courts of equal status create persuasive precedent only. A decision made by the Supreme Court (previously the House of Lords) would be binding on all lower courts. The Civil Court of Appeal binds all courts lower than itself (i.e. High Court, County Court, Employment Appeals Tribunals and Tribunals).

There is scope for the courts to interpret such precedents in the light of changing circumstances. This does allow some deviation from an established precedent, but the decision making has to be justified. An example of this arose in the late 70's before the Road Traffic Act made it compulsory to wear front seat belts. A number of cases went to the High Court involving claimants who had not been wearing a seat belt. The High Court judges reduced their compensation by differing amounts. Finally the Court of Appeal decided the matter. In Froom v Butcher (CA) (1975) ***(not listed in the syllabus)***, it was held that compensation should be reduced by up to 25% if a person who was involved in a car accident was not wearing a seat belt.

Since 1966 the House of Lords (now Supreme Court) was no longer bound to follow its own decisions but it would be rare for the Supreme Court to overrule a previous precedent. Whereas common law is reliant on precedent for its existence and application, this does not apply to statute law. This does not mean that precedent and case law do not exist for statute law.

Statute law attempts to establish clear obligations to be followed. However, the application of the law can reveal some variance of perspective. It is sometimes only when a case is brought to court that the correct perspective can be confirmed.

The decision in a particular case sets a precedent. A precedent set in a civil claim may have some bearing on a case being heard for a criminal prosecution, and vice versa. A decision in a civil case, for example, involving personal injury, may be referred to in respect of the standard applicable to health and safety related statutes.

STATUTE LAW

Statute law is that law which is established by Parliamentary process. Parliament has progressively become involved in economic and social affairs, such that in more recent years the need to express complicated laws in statute form has become increasingly necessary. It would be inappropriate to rely on cases heard in court to determine the form and structure of all law.

Statute law is the 'ultimate' source of law; even if it conflicts with common law or equity it will prevail. It can abolish common law that is no longer useful or amend it to deal with changing circumstances and establish clarification. No court or other body can challenge the validity of an Act of Parliament and it is binding on those affected by it, however it cannot bind Parliament or its successors.

Statute law consists of primary legislation, Acts of Parliament, and delegated legislation, such as Regulations (for example, the Management of Health and Safety at Work Regulations (MHSWR) 1999) and Orders. Statute law is a source of both criminal and civil law. Some statutes, such as the Health and Safety at Work etc. Act (HASAWA) 1974 are entirely criminal law. Similarly some statutes are actionable only under civil law (for example, the Occupiers' Liability Act (OLA) 1957). It must be understood, however, that a small number of health and safety related statutes may be used not only as a basis for prosecution but also as a platform for civil actions relating to personal injury suffered in the workplace.

European Directives and Regulations

European Directives and Regulations both have a legal standing applicable to each of the EU Member states but each in a different way. EU Regulations apply to all member states in their entirety from the date of creation (or later date if defined). Effectively, incorporation into domestic law becomes automatic with the only consideration by the member state being how it will be enforced. An example of an EU Regulation is REACH (Registration, Evaluation, Authorisation and Restriction of Chemicals).

Directives however direct the member states to the standard which must be achieved. Although applicable within all states, the directive will define what needs to be incorporated into domestic legislation and by when; the method by which this is to be achieved is down to the individual member states to determine. The directives define minimum standards but it is permissible for an individual member state to impose a higher standard as long as there is no contradiction with the directive or other EU legislation.

In the case of the UK, EU Directives are implemented by the creation of appropriate statutes. For example, in the case of health and safety, EU requirements are implemented using the delegated facility of the HASAWA 1974. An example of this process was the introduction of six sets of Regulations which became law in 1993 (the so called 'six pack'). The EU Directives through this process are now an integral part of UK law. It should be noted that while for many directives, an equivalent UK regulation is created this need not be the case: the Physical Agents Directive was translated into the Control of Noise at Work Regulations 2005 and the Control of Vibration at Work Regulations 2005.

UK Acts of Parliament and Regulations

Parliament may make law as it sees fit, subject to the rules of continued membership of the European Union, and may, for example, repeal both earlier statutes or overrule common law (case law developed by the courts). Statute law often starts life as a consultative document (i.e. a green paper) and then proceeds as a 'Bill'. The Bill is debated in stages, namely:

- First Reading (permission for printing).
- Second Reading (policy debate).
- Committee Stage (clauses considered in detail).
- Report (report of amendments made at Committee Stage).
- Third Reading (finalised bill).
- Vote within the First House.

This procedure has to be followed in both the Commons and House of Lords before the Bill can receive Royal Assent; it usually happens within the House of Commons first though this is not mandatory. This is a lengthy procedure which explains why some Bills never become law (particularly Private Members' Bills). Once both houses have voted in favour of the Act it receives Royal Assent and either becomes law with immediate effect of on a deferred date which the Act provides (for example, although the Health and Safety at Work etc Act 1974 received Royal Assent in 1974, the statute did not become law until 1st April 1975).

The majority of bills eventually receive agreement in both houses, if some amendment or negotiation is required. For some propose legislation where the House of Commons and the House of Lords disagree, where the House of Commons regards it of particular importance they may overrule the House of Lords. The Lords may delay the Bill by up to one year, unless a 'guillotine action' is ruled by the House of Commons and a time limit for a House Lords agreement established. .Some Acts, the Health and Safety at Work etc Act 1974 being a suitable example allow delegated legislation such as Regulations to be made by Ministers of the Crown. In some cases, these statutory instruments can be laid before Parliament for 40 days before they can come into force rather than being positively voted upon. During this period a member may propose a negative affirmation to veto the statutory instrument. Delegated legislation is, therefore, particularly important especially in the changing field of health and safety in the workplace.

Prescriptive and goal-setting legislation

The approach to passing early health and safety legislation statutes and regulations tended to be mainly 'reactive', each designed to meet the need or accommodate the pressure of the moment without regard to any general pattern of development. Leading to 'prescriptive' legislation, this dealt with a particular health and safety issue in a particular industry sector. The nine different industry sectors included Petroleum, Alkali, Nuclear Installations, Mines, Quarries, Factories, Offices, Shops and Railway Premises, each of which was regulated by its own inspectors. This resulted in an infinite quantity of industry specific, ***prescriptive*** legislation leading to inconsistency and overlapping requirements.

Over time, whilst the vast array of health and safety legislation passed in the UK demonstrated the continued commitment to the health and safety of workers, the continually high accident rate in industry indicated that legislation was not effective. This led to the appointment of a Royal Commission on Safety and Health at Work, by the then Labour Government, in 1970, chaired by Lord Robens. Their Report, known as The Robens' Report was published in 1972 as the product of the first comprehensive review of 'the safety and health of persons in the course of their employment'. The Report recommended that in place of what was described as a 'haphazard and mass of ill-assorted and intricate detail' of existing legislation, there should be 'a comprehensive and orderly set of revised provisions under a new enabling Act. The new Act should contain a clear statement of the basic principles of safety responsibility. It should be supported by regulations and by non-statutory codes of practice'. The Report advocated a ***'goal-setting'*** approach to health and safety legislation rather than the more prescriptive and detailed approach previously taken.

The Health and Safety at Work etc. Act (HASAWA) 1974 introduced Robens concept of non-prescriptive, goal-setting legislation that laid down duties that must be achieved, but not stating in prescriptive detail how this was to be done. Unlike previous industry-specific legislation the HASAWA 1974 was not aimed at one particular sector of employment, but at protecting virtually all those at work and others who are affected by work activities. Established as an enabling Act, the HASAWA 1974 only contains general principles of responsibilities for health and safety, any detail is laid down in Regulations made through the provision established in Section 15 of the HASAWA 1974. Further advice on meeting the general requirements of the HASAWA 1974 is obtained from ACOPs, guidance notes and British Standards. The other health and safety legislation in place at the time, consisting of piecemeal prescription that was often complex and incomprehensible, was repealed by the HASAWA 1974.

The primary duties under the HASAWA 1974 are qualified by the phrase 'so far as is reasonably practicable', which allows the duty holder to consider the extent of the risk along with alternative methods available to control the risk. This non-prescriptive, goal-setting approach emphasises the need for duty holders to understand the risks and prevention methods and not just wait for legislation and enforcement to direct them to take specific prescribed action. This approach encourages involvement of a pro-active nature from employers due to the risk-based structure of the legislation, and promotes a culture where health and safety performance continues to improve, as there is no specific threshold at which to stop. Well-motivated, self-disciplined organisations adapt to working within goal-setting legislation, but organisations less focused on health and safety issues may never progress, finding it difficult to move health and safety initiatives forward without the direction of prescriptive legislation.

Immediately following the introduction of the HASAWA 1974 few regulations were made to elaborate on the general duties. As time has passed more specific regulations have been introduced under the HASAWA 1974 to introduce control measures for specific and general risks. These Regulations have, in the main, avoided specific prescription in order to remain applicable to all workplaces. Where prescription has been made it is often in setting standards of health and safety to be achieved, for example, noise exposure limits. The solutions as to how employers achieve these limits remain in the employers' hands. Though this legislation is more specific, it may be said that it remains largely goal-setting.

If legislation had remained only in the form of the HASAWA 1974 employers might not take the specific action expected by enforcing authorities to meet the goal of minimising risk (so far as is reasonably practicable). Although such things as risk assessments were implied in the HASAWA 1974, many organisations had no formal processes to evaluate them until the specific requirement for risk assessments were introduced in 1985 by the initial version of the Control of Substances Hazardous to Health Regulations (COSHH) and the original Management of Health and Safety at Work Regulations (MHSWR) in 1992 (the COSHH And MHSWR Regulations have since been reissued in 2002 and 1999 respectively). As more specific, Regulations are introduced as a response to EU Directives it may appear to many people that the legislative landscape is becoming more prescriptive. Goal setting legislation, due to its wide scope and adaptability to changing circumstances and conditions, allows social and technology advances to occur at a rapid rate without legislation hindering the process. In order for it to remain relevant, it is necessary to update prescriptive legislation in the light of technological advances and knowledge, for example, the change of the noise action levels in the Control of Noise at Work Regulations (CNWR) 2005.

Types of law

CRIMINAL LAW

Purpose

Society (the state) decides the standards which should be adhered to maintain a safe and secure existence for all in the community. Offences against the state are actionable through the criminal court system and guilt or innocence is decided in relation to the perceived offence.

Cases are decided 'beyond reasonable doubt', because the sanctions available to the courts may be severe and include financial punishment and or imprisonment. Those accused are presumed innocent until proven guilty for all offences. It should be noted that where offences are brought under the HASAWA 1974, though the principle of innocence still applies, after it is decided there is a case to answer the burden of proof is on the accused to prove they complied with the requirements.

Sanctions

A number of sanctions may be imposed on a person or organisation found guilty of a criminal offence. These were originally established in common law to ensure fairness (for example, manslaughter by an individual) but where offence relates to a breach of statute the possible sanctions are defined by the relevant legislation (for example, corporate manslaughter). They may include restriction on activities or punishment.

Prosecutions are usually brought about by the State, though there is some scope to bring private prosecutions. In a case involving contravention of health and safety legislation, such as HASAWA 1974, the prosecution is brought by the enforcing officer while Corporate Manslaughter cases will be the responsibility of the Crown Prosecution Service (CPS). In Scotland prosecutions are brought by the Procurators Fiscal and in Northern Ireland by the Director of Public Prosecutions.

In a criminal case the person charged is known as the accused. The outcome of a trial is a verdict of 'guilty' or 'not guilty' or, in Scotland, the outcome could also be 'not proven'. Penalties are intended as a punishment, although compensation may be ordered separately by the court.

Restrictions imposed such as a prohibition notice or requirements for improvement are often imposed by an enforcing authority without the need for a trial. Most legislation imposes sanctions such as fines and/or imprisonment. In some cases the courts will have the option to impose remedial orders and/or publicity orders.

There are three categories of offence which can lead to conviction:

Summary	Less serious/complex cases are usually heard in a Magistrates' Court (No Jury), or District Court or Sheriff Court ***summary*** in Scotland.
Indictment	More serious cases must be tried by a judge and jury in the Crown Court (Sheriff Court ***solemn*** in Scotland).
Triable either way	Defendant and prosecution can make representation for trial by jury and higher sentencing powers of the Crown Court (Sheriff Court ***solemn*** in Scotland) or in Magistrates' Court (District Court or Sheriff Court ***summary*** in Scotland). The defendant can choose the Court, but the Magistrates' (District Court or Sheriff Court ***summary*** in Scotland) can refer the case to a higher court for sentencing.

For more detailed information on this subject see 'Element A3.3 - The enforcement of health and safety law'.

CIVIL LAW

Purpose

Civil law exists in order to regulate disputes between individuals over the rights and obligations people have when dealing with each other. There are many branches of civil law including contract law and the law of tort.

Contract law deals with the legally binding relationships that individuals enter into, whereas tort law is aimed at redressing the wrongs committed by one person against another. Examples of torts include nuisance, defamation and negligence.

The branches of civil law applicable to health and safety at work are 'contract law', 'employment law' and the law of 'tort'. A tort is a civil wrong committed by one party against another. The tort most relevant to health and safety is that of 'negligence'. Negligence together with action for breach of statutory duty is the most common cause of action for injury caused at work.

Types of remedy

Once liability has been established, civil law provides compensation for an injured party known as the claimant. This compensation may be in the form of damages, other financial damages or an injunction (a court order prohibiting or demanding certain actions, for example to stop a noise nuisance).

Damages can take into consideration:

- Recovery of lost earnings.
- Recovery of resulting expenses.
- Compensation for pain, suffering, and other quantifiable loss.
- In the case of nuisance, an abatement of the nuisance (i.e. bringing it to an end).

Proof of negligence includes establishing some form of loss.

Burden of proof

CIVIL LAW

The burden of proof in most civil cases in which the case must be proved is on the 'balance of probability'. Here the claimant sues the defendant and, whilst the burden normally rests with the claimant to establish the facts of the matter, a presumption of liability is not made before the case is heard. The burden may therefore shift between the two parties as the case proceeds.

In order to prove a case a claimant is expected to set out the basis of their case and the defendant is similarly expected to express their defence by setting out the circumstances as they see them. Each party may bring witnesses, including expert witnesses, to illustrate the points of their case. After hearing both sides the judge will consider, on the balance of probability, which argument is the most likely and therefore who they find in favour of.

It is generally accepted that the 'balance of probability' is less onerous on either party to prove than that of 'beyond reasonable doubt', which is used in criminal proceedings. Therefore, the evidence presented by either party may not be complete, or fully robust, but the judge will make a decision based on the evidence that is presented. Colloquially, 'balance of probability' means that the judge must be at least 51% sure.

CRIMINAL LAW

The prosecutor in most criminal proceedings has to prove that the conduct of the accused led to an event forbidden by criminal law and that the conduct of the accused was accompanied by the prescribed state of mind i.e. guilty mind, their attitude being one of the following:

- ***Intention*** - the person intended the act or knew that there was a very high degree of probability of the consequences.
- ***Recklessness*** - whilst not desiring the consequences of his act, either the person can foresee the probability of it occurring, or a reasonable man would have foreseen the consequences.

It must be shown that the event not only occurred, but that the accused either intended it or should have perceived it occurring.

In the case of health and safety matters, such as a breach of Section 2 of the HASAWA 1974 by an employer, it is not necessary to show a guilty mind. The HASAWA 1974 carries an absolute duty to comply and as such the state of mind of the person found to be in non-compliance does not enter into the matter.

However, Section 37 of the HASAWA 1974 related to offences committed by senior officers of the body corporate, which refers to a breach occurring with the consent, connivance or neglect of the accused, and in such cases their intention or recklessness would have a bearing on the case.

In criminal proceedings there is a presumption of innocence on the part of the accused, and the prosecution (Crown) must prove guilt to the standard of 'beyond reasonable doubt' (i.e. the judge or jury must be 99% convinced). This will involve the Crown establishing enough suitable and sufficient evidence to persuade those hearing the case that the accused was guilty of the offence, beyond reasonable doubt.

The defence will try to establish doubt in order to weaken the prosecution's case; this may include the use of witnesses or other case law. The proof does not have to be beyond all doubt, but is subject to the decision of those hearing the case considering the facts as a reasonable person.

In the case of the HASAWA 1974 this requirement is modified by a duty placed on the accused to show that they were in compliance with the requirement that they were accused of breaching (Section 40). This differs from the traditional criminal doctrine where the accused has no need to prove compliance and the emphasis is on the prosecution bringing sufficient evidence to convince the court of the non-compliance.

This does not take away the over-all burden of proof for the prosecution to establish the case beyond reasonable doubt, in relation to the facts of the case, but this will clearly depend on whether the accused has the ability to illustrate that compliance existed.

In summary: the prosecution must prove that there is a valid case to be answered; the defendant must then convince the court, on the balance of probabilities, that they had complied with the duty imposed. In cases of gross negligence manslaughter the standard criminal doctrine of proving a guilty act and guilty mind applies.

A2.2 - Absolute and qualified duties

Concept of absolute and qualified duties

Civil and criminal law recognise that a person accused of wrong doing should be expected to have acted in a particular way and be expected to put effort into doing so. The level of expectation varies depending on the circumstances, in particular how defined or specific the law is in directing the person to act. This will often depend on how general or specific the legal requirement is or how critical it is to society that the person acts in a defined way. The most demanding duties are absolute and create the strictest expectation; they should only be applied where the duty created is so important and applicable that it is appropriate that in all circumstances it be met. Where these circumstances are not required a qualified duty may be more suitable.

Meaning of the terms

'ABSOLUTE'

Absolute duties are expressed by using words such as 'will', 'must' and 'shall'. By virtue of the Provision and Use of Work Equipment Regulations (PUWER) 1998, it is an absolute duty to maintain work equipment.

Regulation 5 of PUWER '98 requires: every employer shall ensure that work equipment is maintained in an efficient state, in efficient working order and in good repair. Absolute duties require compliance regardless of technical considerations or cost. This point is illustrated by the case of ***John Summers and Sons v Frost [1955]*** whereby the employer were found liable for the injury caused as a result of their failure to fully enclose the grinding wheel, despite their argument that this would have rendered the machine unusable.

Inadequate maintenance may be seen to be failure of an absolute duty. ***(See Stark v Post Office [2000] at the end of this Element)***.

'PRACTICABLE'

Where a measure to meet a practicable duty is technically possible, in the light of current knowledge and invention, then it must be taken ***(see Adsett v K&L Steelfounders and Engineers Ltd [1953] - discussed in detail at the end of this Element)***. Practicable requires the duty-holder to keep up with developing technology and current knowledge and invention and not be too slow to employ it. Practicable is a stricter standard than that of 'reasonably practicable' and may require the duty-holder to eliminate cost as a consideration.

'REASONABLY PRACTICABLE'

Reasonably practicable is a term used extensively within health and safety legislation and is a more qualified term than practicable and means more than physically possible. The duty holder considers both what can technically be achieved and the reasonableness of taking the available action in the particular circumstances. When deciding on what measures to implement, the duty-holder is allowed to balance the quantum of risk against the cost of the action, in terms of time, trouble and money. If it can be shown that cost is grossly disproportionate to the risk the action need not be taken. This does not usually mean that no action at all is taken but an action involving less cost should be taken.

The principle of balancing risk with cost was explained by Lord Asquith in the case of ***Edwards v National Coal Board (1949) - discussed in more detail at the end of this Element***.

A2.3 - Role of the European Union

Influence of the European Union and its main institutions as they affect UK health and safety legislation

INFLUENCE AND ROLE

The main role of the European Union (EU) is to enable the progressive integration of Member States' economic and political systems and the establishment of a single market based on the free movement of goods, people, money and services.

This move to integration depends on the establishment of equity within a legal structure and the availability of European standards. The EU influences UK legislation by creating a number of legal instruments including Treaties and Directives which cause the UK to establish domestic legislation on the matter that the Treaties and Directives focus on; this includes legislation affecting health and safety.

The EU Single European Act (SEA) identified the need to eliminate technical barriers to trade, such as the differing legal health and safety standards throughout the EU. The approach taken to eliminate the barriers was to develop a philosophy of essential health and safety requirements and harmonisation Directives that require the essential health and safety requirements to be met. In addition to this, the Single European Act recognised the need to encourage improvements in the working environment. Directives were created that required all Member States, including the UK, to meet specific requirements on controlling health and safety risks.

European standards are usually prepared by the European standards bodies: the European Committee for Standardisation (CEN) and the European Committee for Electrotechnical Standardisation (CENELEC), as a result of mandates being agreed with the European Commission. CEN and CENELEC are based in Brussels and they bring together the national standards bodies of the European Union and European Free Trade Association (EFTA). The UK member is the British Standards Institute (BSI).

Standards may be set by agreement on a text that has been developed or by agreeing to adopt an existing international standard as a European standard. All Member States have agreed to adopt the resulting standards as their national standards and to withdraw any existing and conflicting standards. The UK adopts the standards as British Standards by denoting them as ***BS EN XXXX***.

THE MAIN INSTITUTIONS OF THE EUROPEAN UNION

European Parliament

Members of the European Parliament (MEPs) are directly elected by EU voters every five years. The work of the European Parliament is important because in many policy areas decisions on new European laws are made jointly by Parliament and the Council of the European Union (Council of Ministers).

Parliament plays an active role in drafting legislation, including health, safety and environmental legislation. Parliament also has joint power with the Council over the annual budget of the EU. It was originally set up as a consultative assembly; the European Parliament has three fundamental powers:

- Legislative power on most issues (shared with Council).
- Budgetary power over EU spending.
- Supervisory power over other EU institutions.

It has the right to veto or amend legislation agreed by EU governments. It can also:

- Reject the EU's budget.
- Set up committees of inquiry.
- Sack the EU's executive Commission.
- Block EU expansions.
- Veto international agreements.

The European Parliament meets in full session in Strasbourg for one week every month. At other times, the MEPs work in Brussels and meet in special committees, similar to Members of Parliament in the UK.

In areas of legislation its role varies depending on the legislative procedure involved:

- Ordinary procedure (formerly called co-decision), where a proposal requires the agreement (or compromise) of both the Council and the Parliament.
- Consultation procedure, where the Parliament can influence draft legislation in that Council have to take account of amendments, but the Council has the final decision.
- Consent procedures, where Parliament considers draft legislation in its final proposed form as forwarded by the Council. It decides whether to approve the draft (it cannot amend it) by an absolute majority of the votes cast.

The Parliament also approves the appointment of the European Commission, and approves international agreements.

Since the Single European Act (SEA), all treaties marking the accession of a new Member State and association treaties are subject to Parliament's assent. The Maastricht Treaty introduced a requirement for Parliament's assent for agreements establishing a specific institutional framework or entailing modifications to legislation adopted under the ordinary (formerly co-decision) procedure.

Council of Ministers

The Council of Ministers is where national ministers meet to discuss and, together with the European Parliament, adopt EU laws.

The Council of the European Union (or informally called the Council of Ministers) is made up of representatives of each of the Member States and represents the individual Member States. The relevant ministers attend specialised meetings as appropriate, for example, transport ministers meet to discuss proposed transport legislation. Presidency of the Council of the European Union rotates between Member States and changes every six months. This is the main decision and law-making body of the EU. The Council can take decisions either by simple majority, by qualified majority, or unanimously on the most important issues. In most cases, the Council uses qualified majority voting, a weighted system which means that each member state has a certain number of votes reflecting its size of population. A majority of the member states must be in favour, as well as a minimum number of overall votes.

The European Council consists of heads of Member States, foreign ministers and the President of the European Commission. The European Council is basically a 'super meeting' of the Council of the European Union which formulates and reviews overall EU policy at 'European summits', which are held four times a year. They may also negotiate new treaty obligations requiring unanimity and ratification.

European Commission

The European Commission is the executive body of the EU; it comprises appointed Commissioners and the EU's civil service. It implements the agenda set by the European Council. The Commission proposes EU legislation and checks it is properly applied across the EU. It works in the interests of the EU as a whole.

The Commission consists of 28 members appointed by the member governments for periods of 5 years; they are led by the Commission President. Once appointed the Commissioners must act independently of the member governments and of the Council, and they must act in the interests of the EU as a whole.

Individual members have special responsibilities (for example, transport and agriculture) and cannot be removed by the Council. The European Parliament can, by a two thirds majority, pass a motion of censure compelling the whole group to resign.

The Commission's role is:

- To initiate Community legislation and policy.
- To enact legislation where the power to do so has been delegated by the Council.
- To enforce the treaties.
- To be the executive arm of the Community.
- To initiate Community policy.
- To defend Community interests in the Council.

The Commission is responsible for ensuring that EU law is correctly applied. Consequently, where a Member State fails to comply with EU law, the Commission has powers of its own (action for non-compliance) to try to bring the infringement to an end and, where necessary, may refer the case to the European Court of Justice. The Commission takes whatever action it deems appropriate in response to either a complaint or indications of infringements which it detects itself.

Under the non-compliance procedure started by the Commission, the first phase is the pre-litigation administrative phase also called 'infringement proceedings'. The purpose of this pre-litigation stage is to enable the Member State to conform voluntarily to the requirements of the Treaty. There are several formal stages in the infringement procedure, including investigation and the letter of formal notice for the Member State to respond to the identified problem regarding application of EU law, within a given time limit. If the Member State does not satisfy the Commission, the Commission has the option of referral to the European Court of Justice, which opens the litigation procedure. In this respect, the Commission enjoys a discretionary power in deciding whether or not to commence infringement proceedings and to refer a case to the Court of Justice.

The status of instruments in EU law - Treaties, Regulations, Directives and Decisions

European Union primary law deals with the powers of the EU to pass laws on certain topics and establishes the institutions that will produce the actual legislation. Primary law defines and limits areas of power. Primary legislation in the EU consists of Treaty provisions that establish its legal framework and set out the fundamental policies and operational procedures.

EU secondary law is the actual legislation produced under the provisions detailed in the primary law. Secondary legislation relates to the legislation passed by the Commission, Council and Parliament working together and includes Regulations, Directives and Decisions.

TREATIES

A treaty is a binding agreed by each of the member states within the EU. They impose no direct implication on the individual citizens of the member states in the way that regulations and ultimately directives do but they place controls on the relevant governments. Their purpose are to set out the objectives and rules for EU institutions as well as defining how decisions are made and the relationship between the EU and its member countries. The requirement for member states to enforce EU regulations and incorporate directives into their law, as described in more detail later, are defined at treaty level.

Treaty of the Functioning of the European Union (originally the Treaty of Rome) - overview of main Articles affecting health and safety

The European Union, (then called the European Economic Community) was established by the Treaty of Rome in 1957. The Treaty contained a number of articles which defined how the community would operate. The Treaty of Amsterdam resulted in the renaming of the Treaty (to the 'Treaty establishing the European Community') and renumbering of the Articles contained within it, a process which was repeated by the Lisbon Treaty. The Treaty is now referred to as the Treaty of the Functioning of the European Union, but the articles, though numbered differently, carry the same importance

Article 114 (Article 100A of the Treaty of Rome, 'Treaty establishing the European Community', and subsequently Article 95) gives the legal authority for the imposition of EU Directives on Member States. Article

95A was added to the Treaty of Rome in order to accelerate the removal of trade barriers between Member States, many of which have health and safety implications.

Article 153 (originally Article 118A and subsequently 137) was added to the Treaty of Rome and deals with the harmonisation of conditions in the working environment, including the maintenance and improvement of measures to ensure health and safety.

Article 130 was added to the Treaty of Rome by the Single European Act 1987 and is primarily concerned with the environment, but does have health and safety implications.

The Single European Act

The Single European Act (SEA), which came into force on 1 July 1987, extended the EU's (then called European Economic Community) field of competence and brought about significant changes in relations between the institutions and in their operating rules. It also gave formal legal status to European political co-operation.

The Treaty on European Union (The Maastricht Treaty)

This Treaty came into force on 1 January 1993 following ratification of the provisions agreed at Maastricht in December 1991. It envisages monetary union and closer co-operation on foreign, security, justice and home affairs policies. The stated aim of the 135 page treaty is 'to organise relations between the member states' and create 'an ever closer union among the people of Europe'.

Provisions of the Maastricht Treaty include the following:

- The EEC is renamed the European Community.
- Provision is made for certain intergovernmental structures, notably the European Central Bank in Frankfurt.
- The term 'European Union' is introduced to encompass both the EC and the new intergovernmental structure.
- Common policies will be decided in Brussels.
- Qualified majority voting (QMV) is extended to more than 70 policy areas.
- The principle of 'subsidiarity', first enunciated by Jacques Delors in 1989, is affirmed (i.e. the Community should only act to achieve objectives which cannot be sufficiently achieved by the member states).

Effect of the Lisbon Treaty

The main effects of the Lisbon Treaty of interest here are:

- Amended qualified voting to take account of new Member States.
- Increased the equality of Parliament with regard to Council in the legislative process.
- Made the ordinary legislative procedure the primary procedure for the introduction of legislation.
- Renamed the 'Treaty establishing the European Community' (TEC) to the 'Treaty on the Functioning of the European Union' (TFEU) and renumbered associated Articles.

REGULATIONS

A Regulation is similar to a national law with the difference that it is applicable in all EU member states. Regulations are the most direct form of EU law. As soon as they are passed they have binding legal force throughout every Member State, equal with national laws.

Member States do not have to take action themselves to implement EU Regulations. Regulations are passed either jointly by the EU Council and European Parliament, or by the Commission alone.

Regulations are applicable to Member States and the individuals within the Member States. They enter force on the date given within them or twenty days after publication in the EU Official Journal. An example of an EU Regulation is Council Regulation (EC) No 2062/94/EEC, which established the European Agency for Health and Safety at Work. The only aspect where the member state has discretion is the method by which it will be enforced.

DIRECTIVES

A Directive is a legal requirement provided for in an EU Treaty; directives are made within the bounds of the Treaty. EU law has supremacy over national law; therefore a Directive is binding on Member States, who must give effect to the Directive by transposing it into national law within a set time. The UK is accountable as a Member State for any failure/fault in transposition or meeting requirements in practice, for example, in ensuring the enforcement of requirements. Non-compliance in this way could lead the UK to be subject to infringement proceedings brought by the European Commission, which can result in the UK being fined.

Under the ordinary legislative procedure (formerly co-decision) the European Commission prepares the text of a draft Directive after consultation with its own and Member States' experts. The draft is presented to the Council and Parliament using the ordinary legislative procedure, set out later in this element. When it is agreed it is published in the EU Official Journal (OJ) and becomes law.

Directives are used to harmonise the laws of the Member States, including those covering occupational health and safety. Directives set out general rules, in the form of results to be achieved, to be transposed into national laws by each Member State, in a manner that they deem appropriate. Because the formats of Directives are results orientated they allow the UK the facility to implement the Directive in a way that suits UK legislation.

This is very important in situations where the UK already has legislation in place relating to the intent of the Directive. As an active Member State, the UK contributes to the forming of Directives and can make use of its experience in drafting its own legislation. In addition, where the UK does not have legislation in place relating to a Directive that is being drafted it enables the UK to influence the outcome of the Directive and therefore the transposition into UK legislation.

Framework Directives set out the overall objectives and specific matters are dealt with individually in the form of Daughter Directives. In the UK the Directives are translated into Acts and Regulations. For example, the Working Time Directive was translated into the Working Time Regulations in the UK.

DECISIONS

Decisions are EU laws relating to specific cases. They can come from the Council of the European Union (sometimes jointly with the European Parliament) or the EU Commission. They can require authorities and individuals in Member States either to do something or stop doing something, and can also confer rights on them. EU decisions are addressed to specific parties (unlike EU Regulations) and are fully binding.

Establishing Directives

ROLE OF PARLIAMENT, COUNCIL AND COMMISSION

The procedures for the introduction of EU legislation have been amended and its extent of use varied over time, in particular by the Treaty of the Functioning of the European Union (TFEU). This Treaty has reflected the increasing power of Parliament to influence legislation developed by the Commission. For many occasions Parliament is placed on an equal level with Council in considering, amending and confirming legislation.

The procedures in use to establish legislation are:

1) Ordinary procedure (Article 289 and 294 TFEU) (formerly called the co-decision procedure).
2) Consultation procedure.
3) Consent procedure.

The ***ordinary*** legislative procedure (formerly co-decision procedure) relies on the ability of the Council and Parliament to agree by placing both on an equal level. This is now the procedure used for the establishment of the majority of directives.

The ***consultation*** legislative procedure requires that before taking a decision the Council must take note of the opinion of Parliament. Failure to consult Parliament makes the legislation illegal and capable of annulment by the European Court of Justice. When the Council intends to substantially amend the proposed legislation it is required to consult Parliament again.

In the ***consent*** legislative procedure Parliament considers draft legislation in its final proposed form as forwarded by the Council. It decides whether to approve the draft (it cannot amend it) by an absolute majority of the votes cast. The Lisbon Treaty does not give Parliament any formal role in the preceding stages of the procedure to consider the Commission's proposal, but as a result of inter-institutional arrangements it has become the practice to involve Parliament informally. This procedure is used for legislation related to discrimination and membership of the European Union.

The co-operation legislative procedure, not listed above, obliged the Council to take into account at second reading those of Parliament's amendments that were adopted by an absolute majority, which were accepted by the Commission. Its importance diminished by the extended use of the co-decision procedure under the Amsterdam Treaty and its use abolished by the Treaty of Lisbon.

ORDINARY LEGISLATIVE PROCEDURE (FORMERLY CO-DECISION PROCEDURE)

Since the Treaty of Nice the ordinary legislative procedure (formerly the co-decision procedure) has been applied to 45 legal bases, the Lisbon Treaty added a further 40 legal bases, in particular justice, freedom, security and agriculture.

It is now used in most areas where a qualified majority in Council is required, including those relating to the free movement of workers, environment, freedom of establishment and the freedom to provide services, the harmonisation of legislation for the operation of a single market and consumer protection including much of the legislation designed to protect workers. It does not apply to several important areas that require unanimity in Council, taxation and trans-national aspects of family law.

Since the Treaty of Lisbon the ordinary legislative procedure is the usual default procedure used for the introduction of legislation. The procedure gives Parliament an effective veto and power-share. If Council and Parliament agree the legislation is adopted at first reading. However, if there is disagreement between Parliament and the Council at this stage, a Conciliation Committee is convened, and, if there is still no agreement, the proposal fails.

Outline of the ordinary legislative procedure (Article 289 and 294 TFEU)

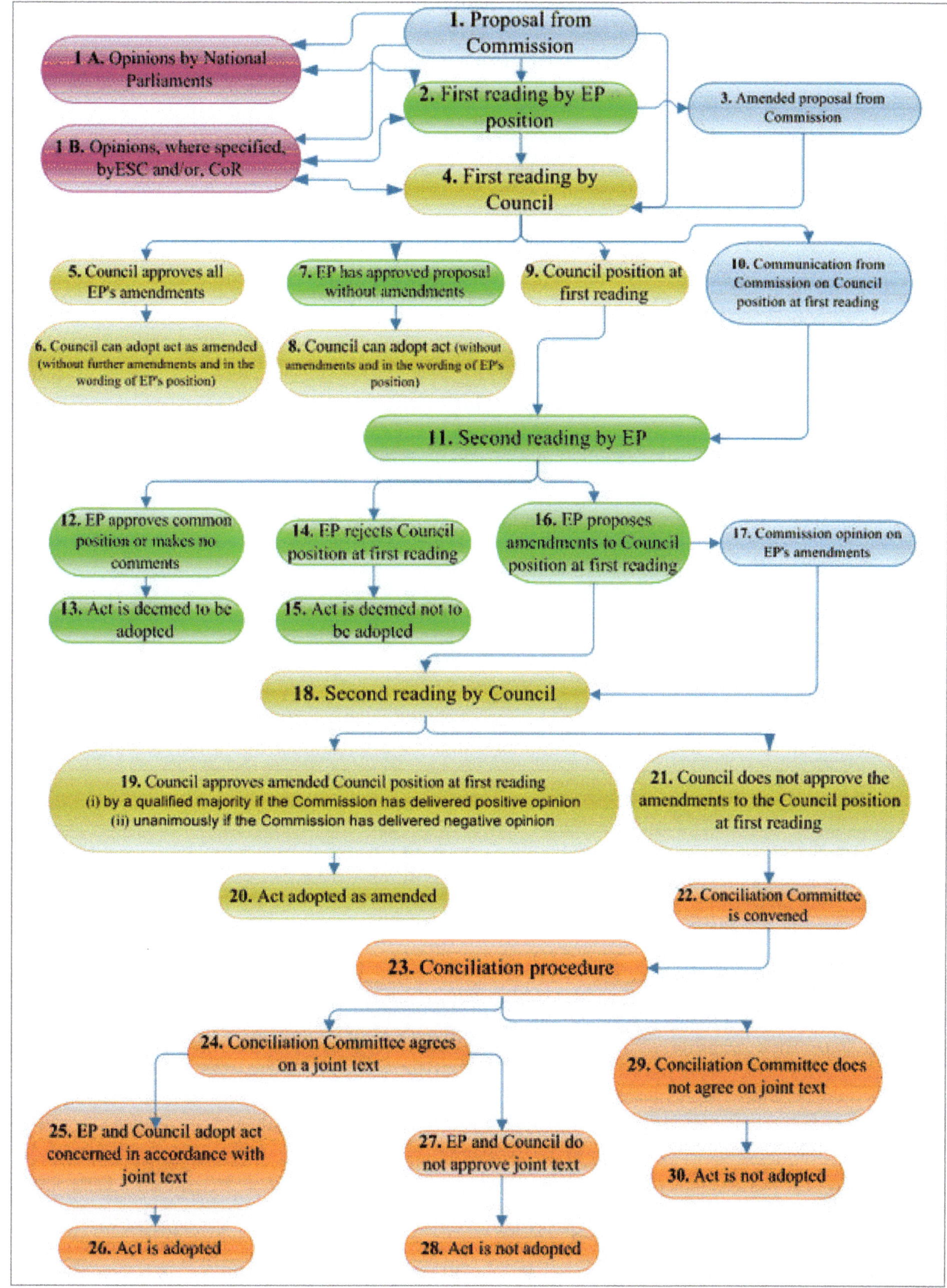

Figure A2-1: Ordinary (co-decision) procedure. *Source: EU Commission.*

The ordinary legislative procedure follows the same steps as the former co-decision procedure. However, the wording of the procedure has changed considerably, notably to underline the equal role of Council and Parliament in this procedure.

1) Commission proposal.
2) First reading in Parliament. Parliament adopts its position by a simple majority.
3) First reading in the Council.

The Council adopts its position by a qualified majority vote (QMV). If the Council approves Parliament's position the legislation is adopted.

4) Second reading in Parliament.

Parliament receives the Council's position and has three months to take a decision. Parliament may:

- Approve the proposal as amended by Council or take no decision, in both cases, the legislation as amended by the Council is adopted.
- Reject the Council's position by an absolute majority of its Members, the legislation is not adopted and the procedure ends.
- Adopt, by an absolute majority of its Members, amendments to the Council's position, which are then put to the Commission and the Council for their opinion.

5) Second reading in the Council.

The Council considers the Parliament's amendments and votes on them. To accept the amendments the vote must be a qualified majority on the Parliament's amendments that are supported by the Commission and unanimously on those that were not supported by the Commission. If the Council approves all of Parliament's amendments no later than three months after receiving them, the legislation is adopted; otherwise the Conciliation Committee is convened within six weeks.

6) Conciliation.

The Conciliation Committee consists of an equal number of Council and Parliament representatives, assisted by the Commission. It considers the common position of Parliament and Council and has six weeks to draft a joint text. If the Committee does not approve, by a qualified majority of the representatives of Council and by a majority of the representatives of Parliament, a joint text by the deadline, the procedure stops. If the Committee approves the joint text it goes to the Council and Parliament for approval.

7) Third reading and conclusion of the procedure.

The Council and Parliament have six weeks to approve the joint text. The Council decides by a qualified majority and Parliament by a majority of the votes. The legislation is adopted if Council and Parliament approve the joint text, if either of the institutions has not approved it by the deadline, the procedure stops and the legislation is not adopted. It is important to bear in mind that outside the formal procedure there is an option for informal negotiations between the Council and Parliament to enable first reading adoption of legislation.

QUALIFIED MAJORITY VOTING

The qualified majority voting process entails more votes than a simple majority. In the case of QMV there is no longer equality of voting rights. Each country has a certain number of votes in line with its size of population. Since 1st January 2007, a new weighting of votes was introduced and qualified majority is obtained if (with 27 Member States):

- The decision receives at least 255 votes of a total of 345 (73.91%).
- The decision is approved by a majority of Member States.
- The decision is approved by at least 62% of the EU's population (the check on this latter criterion must be requested by a Member State).

If a proposal does not come from the Commission, adoption of legislation by the Council requires at least 255 votes in favour, cast by at least two-thirds of the Council's members. The Treaty of Nice (EU Enlargement) increasing the number of Member States to 27 and votes were distributed as follows.

Member State	***Votes (per Member State)***
Germany, France, Italy, United Kingdom	29 each
Spain, Poland	27 each
Romania	14
Netherlands	13
Belgium, Czech Republic, Greece, Portugal, Hungary	12 each
Austria, Bulgaria, Sweden	10 each
Denmark, Finland, Ireland, Lithuania, Slovakia	7 each
Cyprus, Estonia, Latvia, Luxembourg, Slovenia	4 each
Malta	3
Total	***345***

Figure A2-2: Qualified majority voting. *Source: EU official website.*

Note: The above chart will change as the EU grows in membership.

Distinction between directives made under Articles 114 and 153 of the TFEU

As discussed previously, the original numbers allocated to the Articles within the Treaty of Rome signed in 1957 have since changed. For these discussions, the ***current*** numbers are used, as referenced within the NEBOSH Diploma Guide.

Article 114 is concerned with the harmonisation of the internal market. These are important in terms of health and safety because they relate to the supply of goods used at work, such as personal protective equipment and machinery. The intention of the Articles was to reduce barriers to trade by establishing agreed standards throughout the EU, this included setting out essential health and safety requirements to be complied with when goods were manufactured and supplied.

Directives issued under these Articles include:

- Machinery Directive 2006/42/EC, which led to the Supply of Machinery (Safety) Regulations 2008 (SI 2008/1597) in the UK, replacing an earlier version of the same regulations made to meet an earlier Directive. Machinery supplied must comply with these regulations, which set out essential health and safety requirements.
- Personal Protective Equipment Supply Directive (89/686/EEC) which led to the Personal Protective Equipment (EC Directive) Regulations in the UK, subsequently replaced by the Personal Protective Equipment Regulations 2002 (SI 2002/1144) - not to be confused with the Personal Protective Equipment Regulations 1992 implementing one of the daughter directives and focusing on the provision, use and storage of the equipment rather than its design and manufacture. Personal protective equipment supplied must comply with this Regulation, which sets out essential health and safety requirements.

Article 153 provides as follows:

The Member States shall pay particular attention to encouraging improvements, especially in the working environment, as regards the health and safety of workers, and shall set as their objective the harmonisation of conditions in this area, while maintaining the improvements.

In order to help achieve the objective laid down in the first paragraph, the Council, acting by a qualified majority on a proposal from the Commission and after consulting with the European Parliament and the Economic and Social Committee, shall adopt, by means of directives, minimum requirements for gradual implementation, having regard to the conditions and technical rules obtaining in each of the Member States. Such Directives shall avoid imposing administrative, financial and legal constraints in a way which would hold back the creation and development of small and medium-sized undertakings.

The provisions adopted pursuant to this Article shall not prevent each Member State from introducing more stringent measures for the protection of working conditions compatible with the Treaty.

Important health and safety Directives were made under Article 153, including the 'Framework Directive' 89/391/EEC and a series of 'Daughter Directives'. UK Regulations were derived from directives made under Article 137, examples include:

- Framework Directive (89/391/EEC) led to the Management of Health and Safety at Work Regulations in the UK, originally in 1992 but amended and reissued in 1999.
- Work with Display Screen Equipment Directive (90/270/EEC) led to the Display Screen Equipment Regulations 1992 in the UK.
- Physical Agents (Noise) Directive (2003/10/EC) led to the Control of Noise at Work Regulations (2005) in the UK, which revised an earlier UK Regulation made to meet an earlier Directive.
- Artificial Optical Radiation Directive (2006/25/EC) led to the Control of Artificial Optical Radiation at Work Regulations 2010 in the UK.

UK LAW IN RESPONSE TO EU DIRECTIVES

Most directives associated with health and safety in the workplace are implemented in the UK where possible by regulations made under Section 15 of the HASAWA 1974. Section 15 lays down the framework that must be adhered to when making health and safety regulations. In most cases, these statutory instruments (Regulations) are laid (placed for members to read) before Parliament for 40 days before they can come into force, rather than being positively voted upon.

Regulations start life in the form of a consultative document, which is distributed to any interested parties such as the Confederation of British Industry (CBI) and Trades Union Congress (TUC) for comment. The consultative document outlines the background and aims of the proposed legislation together with a cost/benefit analysis (effect on society of the proposal, i.e. the reduction of loss compared to the cost of implementation) as well as the draft regulations themselves. Comment is sought on the proposed regulations as well as the assumptions and costs in the cost benefit assessment. The Secretary of State must consult the HSE when making or modifying regulations made under the HASAWA 1974. The UK is obliged to implement EU directives in full; therefore the consultative role can only have a limited effect in this instance.

The role of the European Court of Justice (ECJ) and procedure for referring cases

THE ROLE OF THE EUROPEAN COURT OF JUSTICE (ECJ)

The role of the European Court of Justice is to ensure that interpretation of the treaties is in accordance with the rule of law.

Judges are drawn from a pool of judges provided by each Member State; each Member State sends one judge to Luxembourg who sits for a term of six years. The President of the Court of Justice is elected from and by the judges and serves for a period of three years. The President presides over hearings and controls both the judicial and administrative business of the Court. The advocates-general, of which there are eight, provide impartial, legal opinion to the judges in order to assist them in their decision. A registrar to the Court is appointed for a period of six years.

The European Court of Justice meets in a number of different ways, it may sit in chambers (consisting of three or five judges), in a Grand Chamber (thirteen judges) or as a full Court.

The case is considered first by the Advocates General who assist the Court by giving an independent view of the case prior to the Court's decision. However, the Court is not bound by these opinions. The European Court of Justice issues a single judgement based on a majority decision. No dissenting views are given. Enforcement of judgements is through the national courts of all Member States, and fines can be imposed on Member States failing to apply the ruling made in the judgement.

The Single European Act empowered the European Court of Justice to set up a Court of First Instance (formerly called General Court), which it did in 1988. This Court of First Instance is designed to relieve the pressure on the European Court of Justice. The Court of First Instance comprises one judge per Member State, currently 27. The Court usually sits as 3 or 5 judges; it does not have permanent advocates, and in exceptional circumstances this role may be carried out by a judge. As with the European Court of Justice, it may sit as a Grand Chamber or full court if the importance or legal complexity warrants it. More than 80% of cases are heard by a Chamber of three judges. A registrar to the Court is appointed for a period of six years.

The Court of First Instance deals with all direct action by citizens and organisations against Community institutions (except in anti-dumping matters) and appeals against Commission Decisions in competition cases. Right of appeal from the Court of First Instance is to the European Court of Justice.

The Court of Justice should not be confused with the European Court of Human Rights. The Court of Human Rights' function is to interpret the European Convention for the Protection of Human Rights which operates completely outside of the European Union and recognised by all countries in the continent of Europe, not just those within the EU. The articles of the Convention provide for matters such as the right not to be subjected to inhuman or degrading treatment and freedom of peaceful assembly.

The European Court of Justice (ECJ) gives rulings on cases brought before it. The five most common types of case are:

1) Requests for a preliminary ruling.
2) Actions for failure to fulfil an obligation.
3) Actions for annulment.
4) Actions for failure to act.
5) Actions for damages.

1) The preliminary ruling procedure

The national courts in each EU Member State are responsible for ensuring that EU legislation is properly applied in that Member State. However, there is a risk that courts in different Member States might interpret EU legislation in different ways. To limit the effects of this there is a 'preliminary ruling procedure'. This means that if a national court is in any doubt about the interpretation or validity of an EU law it may, and sometimes has to, ask the European Court of Justice for advice. This advice is given in the form of a 'preliminary ruling'. Though this is referred to as a 'preliminary ruling' it is the only and final ruling of the ECJ and is binding on Member States.

2) Proceedings for failure to fulfil an obligation

The Commission can start these proceedings if it has reason to believe that a Member State is failing to fulfil its obligations under EU law. These proceedings may also be started by another EU Member State. In either case, the ECJ investigates the allegations and gives its judgment. The accused Member State, if it is found to be at fault, must put things right at once. If the Court finds that the Member State has not complied with its judgment, it may impose a fine.

3) Actions for annulment

If any of the Member States, the Council, the Commission or (under certain conditions) Parliament believes that a particular EU law is illegal they may ask the Court to annul it. These 'actions for annulment' can also be used by private individuals who want the Court to cancel a particular law because it directly and adversely affects them as individuals. If the Court finds that the law in question was not correctly adopted or is not correctly based on the Treaties, it may declare the law null and void.

4) Actions for failure to act

The Treaties require the European Parliament, the Council and the Commission to make certain decisions under certain circumstances. If they fail to do so, the Member States, the other EU institutions and (under certain conditions) individuals or companies can lodge a complaint with the ECJ in order to have this failure to act officially recorded.

5) Actions for damages

Any person or company who has suffered damage as a result of the action or inaction of the EU or its staff may bring an action seeking compensation before the Court of First Instance. Appeal from a judgement in the Court of First Instance is to the ECJ.

PROCEDURE FOR REFERRING CASES

The UK courts retain jurisdiction to review the administrative implementation of EU legislation within the UK. The courts ensure the effective and uniform application of EU legislation and prevent differing interpretations.

In the UK the High Court and the Court of Appeal (Court of Session and High Court of Judiciary in Scotland) have a jurisdiction to interpret EU legislation and can receive requests to refer a case to the European Court of Justice (ECJ) on a point of law, they are however not obliged to grant a right of referral. If however the case goes to the Supreme Court on appeal, the Supreme Court is obliged to refer it to the European Court of Justice if either of the parties involved wishes it.

In effect, the referral to the European Court of Justice will only usually be made once the final court of appeal within the UK has been reached, usually the Supreme Court, but High Court of Justiciary for criminal cases in Scotland.

Article 267 of the Treaty on the Functioning of the European Union (TFEU) states:

"The Court of Justice of the European Union shall have jurisdiction to give preliminary rulings concerning:

- *The interpretation of the Treaties.*
- *The validity and interpretation of acts of the institutions, bodies, offices or agencies of the Union.*
- *Where such a question is raised before any court or tribunal of a Member State, that court or tribunal may, if it considers that a decision on the question is necessary to enable it to give judgment, request the Court to give a ruling thereon.*
- *Where any such question is raised in a case pending before a court or tribunal of a Member State against whose decisions there is no judicial remedy under national law, that court or tribunal shall bring the matter before the Court.*
- *If such a question is raised in a case pending before a court or tribunal of a Member State with regard to a person in custody, the Court of Justice of the European Union shall act with the minimum of delay."*

Figure A2-3: Article 267 of TFEU. *Source: Treaty on the Functioning of the European Union (TFEU).*

Cases are submitted to the registry of the ECJ and a specific judge and advocate-general are assigned to each case. The procedure that follows is in two stages, a written and then an oral phase.

At the first stage, all the parties involved submit written statements and the judge assigned to the case draws up a report summarising these statements and the legal background to the case. At the second stage, the oral phase, a public hearing takes place. Depending on the importance and complexity of the case, this hearing can take place before a chamber of three, five or 13 judges, or before the full Court.

At the hearing, the parties' lawyers put their case before the judges and the advocate-general, who can question them. If the Court considers that this particular case raises a new point of law, the advocate-general then gives their opinion. The judges deliberate and deliver their judgment; the Court does not necessarily follow the advocate-general's opinion. Judgments of the Court are decided by a majority and pronounced at a public hearing. Dissenting opinions are not expressed. Decisions are published on the day of delivery.

The Court of First Instance deals with all direct action by citizens and organisations against Community institutions (except in anti-dumping matters) and appeals against Commission Decisions in competition cases. Cases of this type will be taken in the Court of First Instance using a similar procedure to that outlined above. Right of appeal from the Court of First Instance is to the European Court of Justice.

HOW DECISIONS OF THE EUROPEAN COURT OF JUSTICE ARE ENFORCED THROUGH COURTS OF MEMBER STATES

The European Court of Justice enforces its ruling through the principles of 'supremacy' and 'direct effect'. The ECJ has ruled on a number of occasions that EU law is ***'superior'*** to the national laws of a Member State. Where conflict arises the EU law takes 'precedence' and the law of the Member State has to be disapplied.

The UK accepts that EU law takes 'precedence', provided that it continues to respect fundamental constitutional principles of the UK. The ECJ is, in effect, the interpreter and enforcer of the various Treaties of the EU, and UK judges are obliged to take note of their decisions - even when they are in conflict with UK law. There is no right of appeal from an ECJ decision as the ECJ is the most senior court in the EU; a ruling of the ECJ related to EU law would be applicable to all Member States.

EU law, like UK law, covers a wide range of issues, some that affect individuals. The provisions of Treaties and Regulations can have a 'direct effect' enabling individuals to take action against each other regarding the rights granted in the provisions.

In the case of fundamental obligations in a Treaty, such as free movement of workers or discrimination, which is not subject to exemption or conditions, this may have an immediate and 'direct effect' without the need for the creation of legislation in the UK, under 'supremacy' principles. A ruling from the ECJ may confirm or clarify these rights, and enable the case to be heard in the light of this ruling in the UK Court that asked for clarification.

EFFECT OF DECISIONS OF THE EUROPEAN COURT OF JUSTICE ON UK LAW

If the decision of the ECJ relates to the application by the UK of an EU Directive through its own legislation, the legislation will need to be amended in order to establish conformity with the Directive, as clarified by the ECJ ruling. Not to take this action would leave the UK open to fines from the ECJ or the EU Commission through its non-compliance procedure. Where decisions relate to individuals' rights the ruling will be used to decide the case in the UK court requesting the ruling.

A2.4 - UK Acts, Regulations and Orders

The status and procedure for making UK Acts of Parliament, Regulations and orders

Acts of Parliaments are defined as Primary Legislation. Their creation follows a detailed and robust system requiring consent of both Houses of Parliament before they receive Royal Assent.

The duties imposed by Acts of Parliament are often general in nature, as can be seen by Health and Safety at Work etc Act 1974, with many of them being created as Enabling Acts which allows subordinate legislation to be made under them: Section 15 of the 1974 Act enables regulations to be created by the Secretary of State. Some Acts, including the Equality Act (EA) 2010, are Consolidating Acts bringing together two or more previous statutes into one.

Regulations are an example of subordinate legislation. Although they have the same standing within the legal system as an Act, their creation is limited by the scope of the associated Act and as such, with the basic principles already agreed by Parliament, the process is quicker and less cumbersome with the majority of regulations needing to avoid a negative veto rather than pass a full vote within the Houses. Penalties for offences involving a breach of the more specific duties defined within Regulations are usually less than those imposed for the more general duties of the Act.

Orders are also subordinate legislation which can be created under an Act. While Regulations can impose additional specific duties (within the overall scope of the general duties) Orders have an administrative function as seen in the Health and Safety at Work etc. Act 1974 (Application outside Great Britain) Order 1989.

Regulatory reforms within recent years have extended the role of the Order with the most obvious example being the Regulatory Reform (Fire Safety) Order 2005. Regulatory reform legislation (initially in the form of the Regulatory Reform Act 2001 and more recently the Legislative and Regulatory Reform Act 2006) allows orders to be created to effectively consolidate and reorganise existing statute.

Although requirements exist by virtue of the Fire Safety (Regulator Reform) Order 2005, the Order itself imposes no new duties. Instead it has reorganised existing legislation relating to fire safety, removed some duties and presented legal requirements in a more coherent manner.

ACTS OF PARLIAMENT

The functions of green and white papers

During passage through Parliament and before Royal Assent an intended Act is called a Bill. Green papers are consultation documents, which propose a strategy to be implemented in legislation or set out proposals that the Government wishes to obtain a public opinion on. They are tentative in nature, having a very open ended timescale and few firm intentions.

When considering legislation the Government may publish a white paper, which contains policy statements and explanations for proposed legislation. A white paper signifies a clear intention to pass legislation and sets out the structure intended. This provides early parliamentary consideration of what is intended.

Progression of a Bill through Parliament

A Bill goes through the following process:

First House - usually the House of Commons

First Reading: Bill introduced into the House of Commons for formal reading. The First Reading of a Bill is a formality, which involves a member reading the title of the Bill; the First Reading takes place without debate. The First Reading is essentially an announcement that the bill has been introduced. Soon after the first reading copies of the bill are published and made available for members to read; they are also made available on the Parliament website.

Second Reading: The Second Reading provides the first real occasion for debate on the general principles of a Bill though the detailed discussion is reserved for the committee stage. Unless the Bill is non-controversial, in the Commons, the Opposition will usually decide to vote against a Government Bill on its second reading, or to move an amendment to the motion that the Bill is read a second time..

Committee Stage: When a Bill has passed its second reading in the House of Commons, it is usually referred to a General Committee 'upstairs' for detailed examination. At the committee stage, MPs study the Bill in depth, dealing with each clause in turn, line by line. Whilst individual provisions may be altered, the main principle of the Bill is preserved. MPs are often 'briefed' by interest groups who wish to see amendments favouring their position.

Report Stage: Any amendments made during the committee stage must be approved (or rejected) by the whole house during the report state. The report stage is a detailed debate where further amendments may be moved.

Third Reading: The third reading of a Bill often follows on immediately after the report stage. It is generally quite short (unless it is of constitutional importance) and the bill is reviewed in its final form including amendments made at earlier stages. Substantive amendments cannot be made at this stage to a Bill in the Commons. Assuming that the final version of the bill is approved, it is then passed by hand, bound in green ribbon, to the Lords.

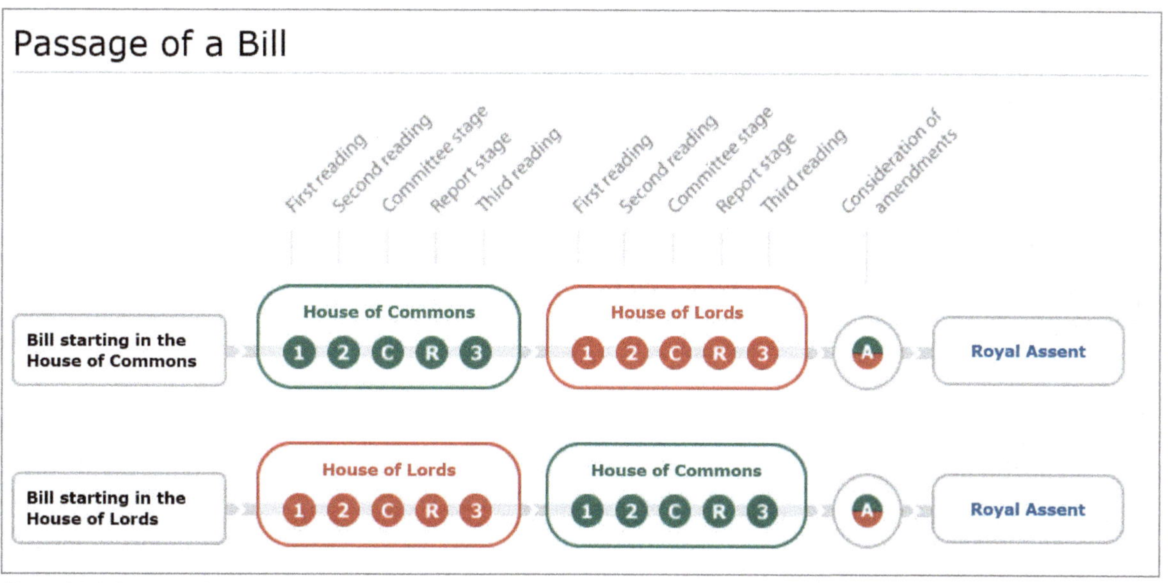

Figure A2-4: Passage of a Bill. *Source: Parliament UK.*

Second House - usually the House of Lords

The Bill follows similar procedures in the Lords. Any amendments by the Lords must be returned to the Commons for consideration. If the Lords reject the Bill for 2 sessions, it may receive Royal Assent without the Lords agreement as the House of Lords can only delay a Bill for a maximum period of 1 year. When the Lords return the bill it is bound in red ribbon.

Royal Assent

After being passed by both Houses, the Bill receives Royal Assent and becomes an Act, or Statute, and is listed in the 'Statute Book', where it remains until it is repealed.

REGULATIONS

Acts can require a great deal of detailed work to implement and operate them, so many are drafted in order to provide a 'broad framework' with a provision included that provides powers to Ministers to fill in the 'detail' by means of delegated (or subordinate) legislation, for example, Regulations. This was the case with regard to the

HASAWA 1974. Many of the EU Directives are incorporated into UK legislation using Regulations as a vehicle under the suitable Act.

The use of Regulations to establish law has a number of advantages:

- It saves Parliamentary time as Ministers and the civil service are left to produce the detailed legislation to meet the need of the 'Parent Act'.
- It is faster; the Parliamentary process for producing an Act is slow whereas Regulations can be made quickly.
- It enables unforeseen problems to be dealt with. Parliament cannot foresee all issues to be dealt with over the lifetime of an Act; Regulations can deal with them as they arise.
- It is less rigid in that Regulations can be withdrawn or amended quickly by another statutory instrument (Regulation) if it proves impracticable.

Procedure under Section 15 of the HASAWA and permissible subject matter of Regulations

Some Acts allow delegated legislation such as Regulations to be made by Secretaries of State. The Health and Safety at Work etc. Act (HASAWA) 1974 is an 'enabling' Act which allows the Secretary of State and other designated Ministers to make health and safety Regulations (Section 15).

After consultation the Secretary of State with responsibility for health and safety prepares draft legislation in the form of a Statutory Instrument, which will contain the proposed Regulations. In most cases, other than national emergency, these statutory instruments are laid before Parliament for 40 days before they can come into force rather than being positively voted upon. During this period a member may propose a negative affirmation to veto the statutory instrument.

Regulations should be drawn up with the aim of maintaining or improving standards of health, safety and welfare. Section 15 lays down the framework that must be adhered to when making health and safety regulations. Failure to follow this framework would allow the courts to treat the Regulations as void under the legal principle of ***'ultra vires'*** that is, that it exceeds the prescribed limits or that it has been made without due compliance with the correct procedure.

Role of the Secretary of State in making Regulations

As previously discussed, the Secretary of State is empowered by Parliament, via primary legislation such as the HASAWA 1974, to make Regulations. The nominated Secretary of State ensures the consultation and drafting of the Regulations. As they are actually drafted by the HSE the role of the Secretary of State is to facilitate the process and includes the acceptance of the draft and approval of its final version to go forward as a statutory instrument. The Secretary of State takes responsibility for ensuring that the proposed Regulation does not contravene or extend beyond the provisions of the parent Act and that it is laid before parliament in the correct manner. Section 15 of the HASAWA 1974 - Health and Safety Regulations says:

1) Subject to the provisions of Section 50, the Secretary of State shall have power to make Regulations under this section for any of the general purposes of this Part (and Regulations so made are in this Part referred to as 'Health and Safety Regulations').

Role of the Health and Safety Executive in making Regulations

The Secretary of State must consult with the Health and Safety Executive (HSE) when making or modifying Regulations made under the HASAWA 1974. The Health and Safety Executive (HSE) has a pivotal role to play in the drawing up, consulting and drafting of Regulations. The general functions of the HSE are laid down in Section 11 of the HASAWA 1974 and enable the HSE to conduct research, explore where Regulations may be required and propose this to the Secretary of State.

Various stages of consultation

Regulations start life in the form of a consultative document, which is distributed to any interested parties such as the CBI and TUC for comment. The consultative document outlines the background and aims of the proposed legislation together with a cost/benefit analysis as well as the draft regulations themselves. Comment is sought on the proposed regulations as well as the assumptions and costs in the cost/benefit assessment. The results of the consultation are used to guide the formulation of the final version of the Regulations, which is presented to the Secretary of State for approval.

Socio-technical cost-benefit analysis

When a proposal is made to introduce a change in legislation the HSE are required to conduct a cost benefit analysis. The results of this analysis accompany the proposed Regulations as part of the consultative document. The analysis seeks to establish the socio-technical costs and benefits of the introduction of the Regulations, and will consider the various options under discussion.

A good example of the application of this process was the cost benefit analysis included with the consultation on the then proposed Construction (Health Safety and Welfare) Regulations 1996 (since superseded by the

Construction (Design and Management) Regulations (CDM) 2007 and now 2015), which had the following costs and benefits:

Negative effects - costs to industry - arising from:

- The requirement to fit an intermediate guard rail to prevent falls from heights.
- The segregation of vehicles from pedestrians on sites.
- The provision of additional welfare facilities.
- Additional inspections of scaffolding and fire and emergency precautions.

Estimated first year cost:	£45-£54 million
Subsequent additional costs:	£6 million occasionally rising to £19 million per year
Total cost over 25 years (present value):	£113-£130 million (0.2% of the industry's annual turnover (est. £50 billion))

Positive effects - benefits:

About 60% of accidents on construction sites were as a result of falls from height or being struck by vehicles on site. The additional protection measures are expected to achieve a 10-15% reduction which, in turn, will mean a 6-9% overall reduction in construction site accidents.

Estimated annual benefits:	£18-£27 million
Total savings over 25 years (present value):	£127-£210 million

Thus, although the cost of implementing the Regulations will outweigh the benefits in the first year, the overall benefits to society over 25 years (£127-£210 million) are expected to outweigh the costs over the same period (£113-£130 million). As in the above case a socio-technical cost benefit analysis should consider advantages given to society by the implementation of a change and the technical and social costs of that change.

A2.5 - Structure and function of the courts and related institutions

The function, jurisdiction and powers of courts

EMPLOYMENT TRIBUNALS

Functions

An Employment Tribunal usually consists of a legally-qualified chairperson appointed by the Lord Chancellor and two lay members, one representing management interests and the other employees (for example, one from a trade union). These are selected from panels kept by the government department responsible, after nominations from employers' organisations and trades unions. All hearings are heard by the chairperson and either two other members or, with the consent of the parties, one other member. When all three members of an Employment Tribunal are sitting the majority view prevails.

Jurisdiction

Employment Tribunals deal with complaints relating to health and safety at work matters, as well as a wide spectrum of other employment matters. Employment tribunals deal with the following matters relevant to health and safety:

1) Victimisation (automatic grounds for unfair dismissal, Section 100 the Employment Rights Act (ERA) 1996).
2) Dismissal, actual or constructive, following a breach of health and safety law, regulation and/or term of an employment contract.
3) Appeals against improvement and prohibition notices (Section 24 HASAWA 1974).
4) Safety representatives and safety committees [time off, payment functions and training of representatives Safety Representatives and Safety Committees Regulations (SRSC) 1977].
5) Representatives of employee safety [pay and time off under the Health and Safety (Consultation with Employees) Regulations (HSCER) 1996].
6) Suspension from work for medical reasons [Employment Protection (Consolidation) Act (EPCA) 1978].
7) Suspension from work on maternity grounds.
8) Appeals against served Improvement or Prohibition Notices.

Powers

They may make financial awards to employees, unlimited for issues concerning discrimination and public interest disclosure, but limited to £50,000 for all other cases. They may direct that certain rights must be given to employees performing duties under specific legislation, for example, Disputes over Appointed Trade Union Safety Representatives rights to time off to perform their duties. They can uphold, modify or amend requirements under appeal for Improvement or Prohibition notices.

An arbitration scheme involving ACAS is proposed for cases relating to unfair dismissal. Both parties must agree to the scheme and the employee must be advised by a 'relevant independent adviser' (for example, solicitor, barrister, competent trade union officials or workers at advice centres). Any award made is binding and enforceable in the County Court. Appeals against decisions made by an Employment Tribunal are normally referred to an Employment Appeal Tribunal. Appeals against a decision in relation to an appeal against health and safety improvement or Prohibition Notices are referred directly to the Queens Bench of the High Court.

MAGISTRATES' COURTS

Functions

Magistrates' Courts are the first court to deal with all summary or indictable offences. Magistrates' Courts are mostly staffed by lay magistrates who are not legally qualified and sit part-time. District Judges (Magistrates' Court) are full time members of the judiciary. They normally hear the longer and complex cases that come before a Magistrates' Court. Lay Magistrates sit two or usually three to a court, a District Judge sits alone. The Magistrates' Court deals with minor (summary) health and safety offences. Appeal from this court is usually to the Crown Court, but, in some cases might be to the High Court (on a point of law).

The Magistrates Court also has the role of the 'Court of First Instance'. All criminal cases are initially referred to this court which is the defendant's first formal opportunity to make their position known and they will ask how they plead. The court will determine whether there is sufficient evidence to support a charge, and the severity of the case to determine whether it should return to the Magistrates Court at a later date or be referred to the Crown Court for a trial by jury.

Jurisdiction

Magistrates' Courts have criminal and some civil (for example, matrimonial maintenance orders and questions regarding adoption) jurisdiction and deal with all offences in the first instance; summary offences are normally heard by the Magistrates' Court. The magistrates may determine that more serious cases should be indicted to the Crown Court for trial or sentencing.

On hearing a criminal case, a Magistrates' Court can impose a compensation order as well as a penalty. This may be particularly useful where the loss is relatively small and easily quantifiable or the injured party is in need of immediate financial help arising from the offence. The victim may prefer a compensation award from the criminal proceedings instead of taking separate civil action. Where the loss is not easily quantifiable or disputed the Magistrates' Court can require additional evidence, like a medical report, or may decide not to make a compensatory order.

Powers

The general extent to the powers of a Magistrates' Court to sentence is limited to six months in prison and a limit on most fines to £5,000 or up to 12-months imprisonment. For health and safety offences, which are defined within Section 33 of the Health and Safety at Work etc Act 1974. Since 1974, the potential fines have been changed by virtue of the Health and Safety (Offences) Act 2008 and Section 85 of the Legal Aid, Sentencing and Punishment of Offenders Act 2012 which came into force on 12th March 2015. Now, the maximum penalty for some health and safety offences is an unlimited fine and/or up to 12-months imprisonment.

COUNTY COURTS

Functions

The County Court is usually presided over by a Circuit judge who will have had at least 10 years advocacy experience as a barrister in the Crown or County Court. He or she normally sits alone, although in a small number of civil cases (for example, libel) there may be a jury. Appellants to the County Court have right of appeal to the Civil Division of the Court of Appeal.

Jurisdiction

County Courts have civil jurisdiction only but deal with almost every kind of civil case arising within the local areas for which the courts are established. In practice, they deal with the majority of the civil litigation in England and Wales.

Powers

To award damages up to their limit. County Courts can hear cases in contract up to £25,000 and tort claims of up to £50,000 in the case of actions for damages relating to personal injury. They may exceed the limit with consent of both parties.

HIGH COURT

Functions

The High Court is staffed by judges (Justices of the High Court) who must be persons who have had right of audience in the High Court for at least ten years or a Circuit judge who has held office for at least two years. When hearing a case at first instance a High Court judge sits alone, for example, when hearing a case for personal injury damages. A Divisional Court of two or more High Court judges sits to hear appeals on a point of law from Magistrates' and Crown Courts.

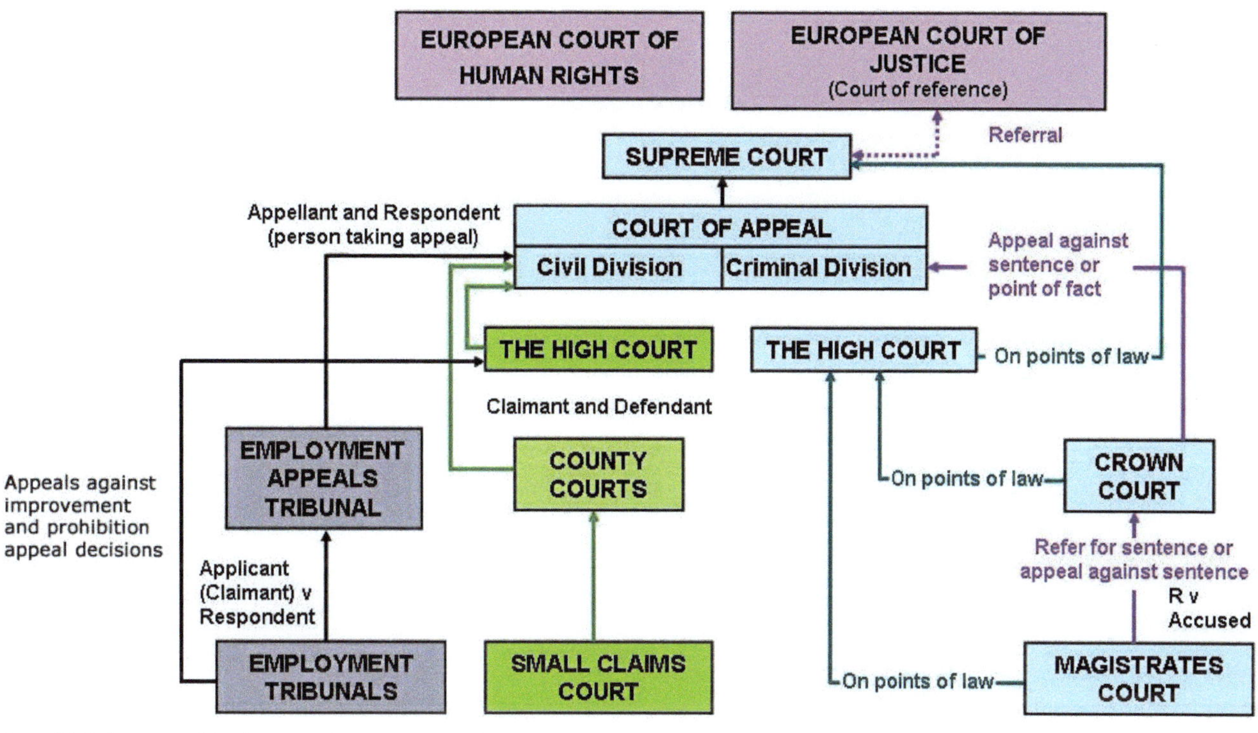

Figure A2-5: Structure of the Courts and tribunal system (England and Wales). *Source: RMS.*

Jurisdiction

The High Court will deal with the more high value, complex or important criminal and civil cases. The High Court is organised into three divisions. These are the Queen's Bench, the Chancery, and the Family Divisions. It also has supervisory jurisdiction over all lower courts and many tribunals, in the form of judicial reviews.

The Queen's Bench division is presided over by the Lord Chief Justice and is the part of the High Court that relates most to health and safety matters. The Queen's Bench Division also contains an Admiralty Court and a Commercial Court. Except where other special courts have exclusive jurisdiction, the High Court Queen's Bench Division can deal with any common law civil matter. In relation to criminal matters, the Divisional Court of the Queen's Bench Division hears appeals on points of law from the Magistrates' Court and Crown Court.

The Chancery division deals with traditional equity matters such as disputed wills, company matters, trusts and mortgages. The Family division deals with matters of family law, including divorce and probate.

Powers

There is no upper limit to the sum that may be claimed for damages relating to personal injury in a High Court. When dealing with a point of law referred by a criminal court the High Court gives a ruling, which the lower courts are bound to follow.

CROWN COURT

Functions

The Crown Court is a superior court of record created by the Courts Act (CA) 1971. In theory, the Crown Court is a single court; however, in practice it comprises a number of courts in large towns and cities and the Old Bailey in London. A circuit judge, a recorder, or a High Court judge may sit in the Crown Court. Sometimes lay magistrates also sit. Very serious offences, such as murder and treason, may only be heard by a High Court judge in the Crown Court. All prosecutions for indictable offences are heard by a judge with a jury of 12 persons.

Jurisdiction

All indictable offences are tried in the Crown Court. An indictment is a formal statement of the serious offences with which the accused is charged, prepared for trial by jury. The Court also hears appeals from the Magistrates' Court and committals for sentence from the Magistrates' Court. In addition, it deals with a few types of civil cases on matters arising from the Magistrates' Court.

Powers

The Crown Court may impose 'an unlimited fine', or up to two years imprisonment, or both, for each breach of health and safety law (HASAWA 1974 and subordinate Regulations), manslaughter and murder up to 25 years and life imprisonment respectively. From the Crown Court appeal on criminal matters is made to the Criminal Division of the Court of Appeal. As with the Magistrates' Court, an appeal on a point of law may also be made to a Divisional Court of the Queen's Bench Division (High Court).

COURT OF APPEAL

Functions

The Court of Appeal is divided into Criminal and Civil Divisions. The Court consists of 35 Lords Justices of Appeal. Normally, three judges will sit together to hear appeals from county courts, the High Court and the Employment Appeal Tribunal. It does not conduct a complete rehearing of the case but reviews the record of the evidence in the lower court and the legal arguments put before it. A majority decision is given and dissenting judgments are stated.

Jurisdiction

The Civil Division of the Court of Appeal hears appeals from the County Court, High Court (with regard to civil matters) and Employment Appeals Tribunal. The Criminal Division of the Court of Appeal hears appeals against conviction and sentence from the Crown Court. An indictable case heard in the Crown Court may be referred on appeal to increase the sentence, where the Crown Prosecution Service feels the Crown Court has been too lenient.

Powers

The Court of Appeal may uphold or reverse the earlier decision, in the County Court, High Court, and Employment Appeal Tribunal. In the case of criminal appeals the court can dismiss or allow the appeal (including increase a sentence) or order a new trial.

SUPREME COURT

Functions

Apart from the very limited jurisdiction of the European Court of Justice (ECJ), the Supreme Court is the highest court of appeal in the United Kingdom. The Court, an independent institution, hears appeals on arguable points of law of the greatest public importance, for the whole of the United Kingdom in civil cases, and for England, Wales and Northern Ireland in criminal cases.

Historically, from its days as being the House of Lords, the Supreme Court consists of between nine and twelve independently appointed Law Lords (Justices) who are life peers; as time progresses, as the Law Lords retire from their role they may be replaced by some of the UK's most senior judges. A minimum of three (but normally five) Law Lords constitutes a court. A majority decision is given and dissenting judgements are stated.

Under the Treaty on the Functioning of the European Union (Article 267 TFEU), the Court must refer to the European Court of Justice (ECJ) in Luxembourg any question of European Union law, where the answer is not clear and it is necessary for it to give judgment. There is no means to challenge a Supreme Court decision.

Jurisdiction

Civil

The Court hears appeals from the Court of Appeal (Civil Division), the Court of Session in Scotland and the Supreme Court of Northern Ireland. In all cases, either the lower court or the Appeal Committee of the Supreme Court must certify that a point of law of general public interest is involved.

There is a 'leapfrogging' method, whereby a direct appeal may be made from the High Court or Divisional Court to the Supreme Court bypassing the Court of Appeal. All parties must consent and a point of law of general public interest has to be involved. This process is most often used for revenue appeals, where the construction of statutes may be involved.

Civil cases coming to the Supreme Court for appeal from Scotland may only establish a persuasive precedent for the rest of the UK, as some parts of the law of Scotland are derived from Roman law rather than Common law. If the legal principles involved are the same in both systems of law it can become binding for the UK, as was the case of Donoghue v Stevenson (1932), which was a fundamental case on the law of negligence and was an appeal from the Scottish Court of Session.

Criminal

The Court hears appeals from the Court of Appeal (Criminal Division) and the Divisional Court of the Queen's Bench (High Court). As with Civil matters, the court will only hear a point of law of general importance. The final court to hear criminal case appeals in Scotland is the High Court of Justiciary, not the Supreme Court.

Powers

The Court may uphold or reverse the earlier decision of lower courts or order a new trial. A majority decision is given and dissenting judgments are stated. There is no means for appealing a Supreme Court decision.

COURT OF JUSTICE (EUROPEAN)

Functions

The European Court of Justice (ECJ) hears legal actions connected with European Union (EU) law. The ECJ has a President, twenty seven judges, eight Advocates-General and a Registrar, all of whom are appointed by

Member States, but act independently of the Member State that appointed them. A single majority judgement is given and the views of dissenting judges are not published.

Jurisdiction

The ECJ's jurisdiction is limited to that set out in the EU Treaties. The ECJ can hear disputes between Member States or between the Commission and a Member State relating to matters covered by an EU Treaty. It can also consider the legality of acts or omissions of the Council of Ministers, the European Commission or the European Parliament. It will hear complaints that legislation created has not been constituted correctly.

It will also hear cases referred by the national courts of Member States for 'preliminary ruling' on matters of interpretation of EU laws into Member State laws. The ECJ will hear appeals from the Court of First Instance.

Powers

The ECJ can:

- Decide on disputes between Member States or the Commission.
- Declare that the offending institutions have violated the EU treaties.
- Uphold or dismiss appeals by persons, for example, against fines imposed by the Commission for breaches of EU law.
- Provide a ruling on the interpretation of EU law and its application to Member States.
- Null an incorrectly constituted EU law.
- Give an advisory opinion on whether their proposals are compatible with existing EU law at the request of the Council of Ministers or of the European Commission.

COURT OF SESSION, HIGH COURT OF JUSTICIARY, SHERIFF COURTS, DISTRICT AND JUSTICE OF THE PEACE COURTS IN SCOTLAND

Introduction

As in England and Wales there are two legal systems in Scotland, one for criminal and one for civil cases. The court structure differs from those in England, Wales and Northern Ireland (Northern Ireland is similar to that of England and Wales).

Court of Session

The Court of Session is Scotland's supreme civil court; it sits in Parliament House in Edinburgh as a court of first instance and a court of appeal. Appeal from the Court of Session is to the Supreme Court. For the purposes of hearing cases, the Court is divided into the Outer House and the Inner House.

The Outer House consists of 22 Lords Ordinary sitting alone or, in certain cases, with a civil jury. They hear cases at first instance on a wide range of civil matters, including cases based on delict (tort) and contract. The Inner House is the appeal court; cases are usually heard by three judges. The Inner House hears cases on appeal from the Outer House, the Sheriff Court and Employment Appeals Tribunals.

On occasion, if a case is particularly important or difficult, or if it is necessary to overrule a previous binding authority, a larger court of five or more Judges may be convened.

High Court of Justiciary

The High Court of Justiciary is Scotland's supreme criminal court and will hear appeals from lower courts on criminal matters; there is no higher court of appeal for criminal cases in Scotland. When sitting as a court of first instance it deals with the most serious crimes, including murder and homicide. Cases are presided over by a single judge and a jury of 15 people.

When sitting as an appeal court, the High Court of Justiciary consists of at least three judges when hearing appeals against conviction and two when hearing appeals against sentence. More judges may sit when the Court is dealing with exceptionally difficult cases or those where important matters of law may be considered.

Appeals are heard from the High Court (for cases of first instance heard by the High Court of Justiciary Outer House), Sheriff Court and Justice of the Peace Courts. In addition, the Lord Advocate may refer a point of law which arises in the course of a case to the High Court (Inner House) for an opinion. This allows the High Court of Justiciary to give directions which set out the law for future similar cases.

Sheriff Courts

The Sheriff in a Sheriff Court is a Judge and is usually a qualified advocate or solicitor. Sheriffs deal with Solemn and Summary Criminal cases and civil actions under Ordinary (over £1,500), Summary cause (up to £1,500) and Small Claim Procedures. Small claims include up to £3,000.

In hearing solemn criminal proceedings (on indictment) the Judge will sit with a jury of 15 people; when sitting for summary proceedings the judge sits alone. The Sheriff Court deals with the majority of civil cases and the Sheriff Principal can hear appeals at a local level as an alternative to taking a case to the Court of Session.

Justice of the Peace Courts

A Justice of the Peace (JP) Court is a lay court where a JP, who is not a legally qualified person, sits with a legally qualified Clerk. The Clerk provides advice to the Justices on matters of law and procedure. The maximum sentence that a JP can impose is:

- 60 days imprisonment or a fine not exceeding £2,500.

In Glasgow only, some courts are presided over by a legally qualified Stipendiary Magistrate. The maximum sentence that a Stipendiary Magistrate may impose is:

- 12 months imprisonment or a fine not exceeding £10,000.

Justice of the Peace Courts were created by the Criminal Proceedings etc. (Reform) (Scotland) Act 2007 and they replaced former District Courts which were operated by local authorities. This change resulted from Summary Justice Reform proposals in 2005 which sought to unify the administration of Sheriff and District Courts. The changes were introduced on a Sheriffdom by Sheriffdom basis from 2008-2010.

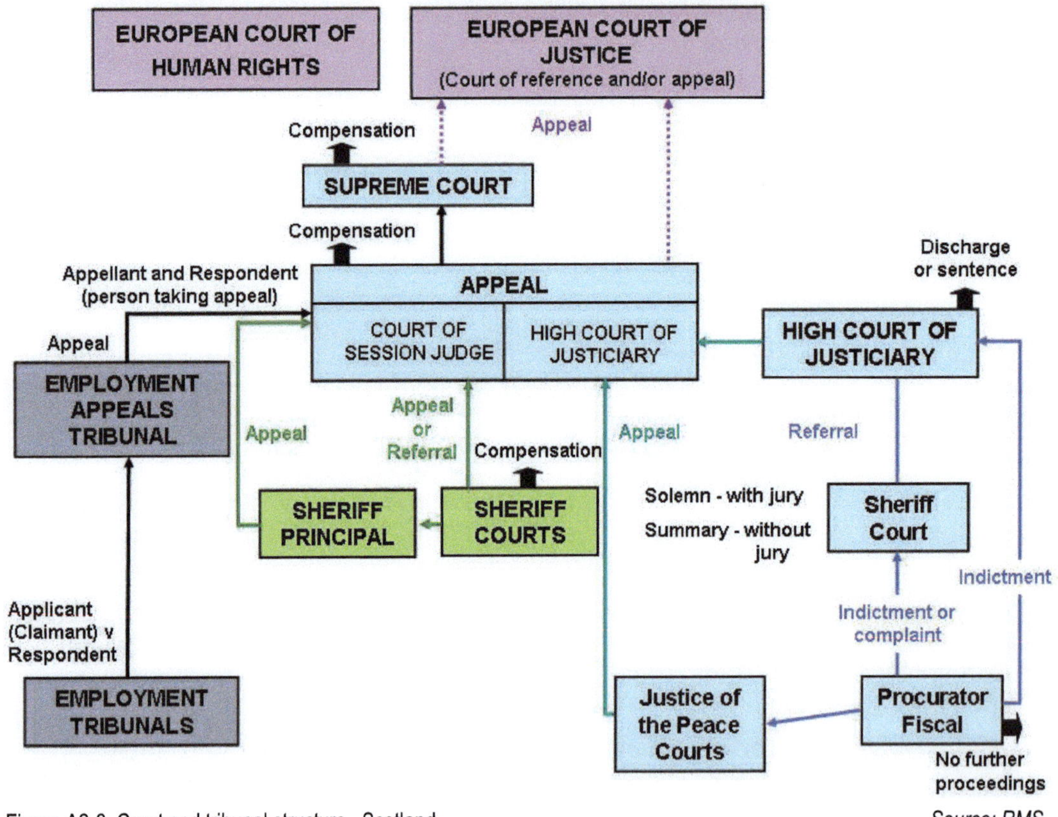

Figure A2-6: Court and tribunal structure - Scotland. *Source: RMS.*

Basic procedures for bringing prosecutions for breaches of health and safety legislation

PROSECUTIONS FOR BREACHES OF HEALTH AND SAFETY LEGISLATION

Plea in the Magistrates' Court

In England and Wales a HSE Inspector may take a prosecution in a Magistrates' Court personally. This power is authorised under section 39 of the HASAWA 1974. Otherwise, the case will be taken by the Crown Prosecution Service in England and Wales.

All criminal cases are initially presented to the Magistrates' Court in their function as the 'Court of First Instance' where the accused will be asked to enter a plea of either guilty or not guilty to the charges. If a plea of not guilty is entered, for summary offences, or for those offences that may be tried 'either way' where the accused has opted for the Magistrates' Court to deal with their case, a trial date convenient for both defence and prosecution witnesses will be fixed. The trial is usually listed for some weeks later. The magistrates, having adjourned the case, may decide to remand the accused on bail. Effectively this means that if they did not attend court on the trial date they would have committed an offence under the Bail Act, for which they may be sentenced.

Bail means the granting of freedom subject to the accused having a duty to surrender to custody or to attend court at a future date and time stated to them. This is known as unconditional bail (i.e. the only condition attached is that the accused reappears at the stated date), but it is also possible for the magistrates to grant conditional bail, which imposes further conditions.

Magistrates also have the power to grant bail to an offender who appears before them while in custody after arrest (this will therefore apply only to fairly serious offences, not summary cases where the offender is free until the date of his/her appearance in court).

In the case of indictable offences or offences that may be tried 'either way' and the accused has elected to be tried on indictment the Magistrates' Court performs committal proceedings, in order to enable the case to proceed to the Crown Court. This stage requires the Magistrates' Court to determine that there is a case to answer and that the matter should be dealt with in the Crown Court.

Trial procedure in the Magistrates' Court

- Charge read out - accused pleads. If pleads not guilty, then...
- Opening speeches by prosecution and defence.
- Prosecution witnesses - examined by prosecution, defence and possible re-examination by prosecution.
- Defence witness.
- Summary by prosecution and defence.
- Magistrates retire to reach a verdict.
- Verdict is given.
- Sentence is given, if the Magistrates feel that their powers are adequate, or they may commit the defendant to the Crown Court for sentencing if they feel their powers are not adequate.

A similar procedure applies in the Crown Court where trials are conducted before a Judge and jury:

- Judge advises and directs on points of law.
- Jury decides guilt or innocence.

Appeals

Appeals relating to conviction and sentencing by the Magistrates' Court may be made to the Crown Court. Appeals can be made to the High Court:

- On a point of law (for example, where a specific interpretation of a legal issue is involved, such as the meaning of a particular section of the HASAWA 1974).
- On the grounds that the magistrates have acted outside their jurisdiction.

Appeals from the Crown Court to the Court of Appeal can be made on questions of law and fact, and against conviction or sentence. Finally, the Court of Appeal may grant leave to appeal to the Supreme Court if the case involves a legal point which is of public interest.

PURSUING CIVIL ACTIONS

See also - 'Element A4 - Civil Law'.

Procedures for the bringing of civil actions (civil procedures rules)

The procedures for personal injury cases were radically overhauled as a result of a review by Lord Woolf (the Woolf Report). This was done in an attempt to speed up and simplify the process as well as keeping costs within reasonable limits. The resulting Civil Procedures Rules were introduced on 26 April 1999 and have the following aims:

- To ensure that all parties are dealt with on an equal footing so that wealthier parties do not have an unfair advantage.
- To save expense.
- To deal with cases in a way that is proportionate to the sums involved the complexity and importance of the case and the financial position of the parties.
- To ensure that cases are dealt with expeditiously and fairly in ways which promote an open and full exchange of information.
- To allot an appropriate share of the Court's resources, while taking into account the needs of other cases.

The rules apply to both the High Court and County Court. Many of the obscure Latin phrases have been replaced and the terms used simplified. For example, the person who brought an action was called a 'plaintiff' and is now known as the 'claimant' and the term 'interlocutory applications' is replaced with 'court applications'. Cases are managed by the Court and alternate dispute resolution (ADR) is encouraged. A new body, the Centre for Dispute Resolution (CFDR) has been formed to deal with this. This recognises that about 94% of cases are settled out of court and that it is far less expensive to engage in an, on average, two day mediation process than an eight year battle for compensation suffered by one widow relating to the death of her husband! CFDR estimates that the ADR process saves £250,000 per case in professional fees alone (averaged over 40 cases).

Expert witnesses

The role of the expert witness is to assist a party to establish the facts, assess the merits of the case and help with its preparation. The overriding duty of the expert is to the court rather than to the person who pays for his or her services. Where two parties wish to submit expert evidence, the court may direct that evidence on the particular issue be given by one expert only. The rules also allow the court to direct on matters relating to the fees and expenses paid to experts.

Pre-action protocol

The process begins with a pre-action protocol, which aims to reduce the need for litigation and encourages contact between the parties prior to the action starting. Pre-action protocols have been developed for a number of events including personal injury cases and clinical negligence.

The protocol for personal injury involves a 3 stage process:

1) The letter of claim.
2) The reply.
3) The investigation.

The letter of claim

The action is started on a claim form. This contains a summary of the facts of the cases together with the details of the loss (for example, financial; injuries suffered).

The reply

The defendant must reply within 21 days after service together with details of any insurer that he or she has.

The investigation

The defendant has a maximum of three months to investigate the claim. A reply must then be sent to the claimant (or the claimant's solicitor) which:

- Admits liability.
- Denies liability in full.
- Gives a partial admission of liability.

Reasons for full or partial denial of liability must be given together with any relevant documents. These are the same documents that the defendant would have to disclose anyway under court proceedings. Thus, the defendant will not be able to issue a denial of responsibility without giving reasons or to delay the decision as to whether to settle or fight.

- The claimant will be able to make a decision, based on the defendant's reply, about whether or not to proceed with the claim. If the defendant fails to make a proper reply, then the claimant will be able to start proceedings without being liable for costs.

Commencement of proceedings

Once the proceedings have been started the defence has up to 28 days to file a response. Extensions may be granted but this is rare particularly if the defendant has failed to comply with the pre-action protocol.

The response consists of:

- Those facts which are admitted to.
- Those facts which are denied with reasons.
- Those facts which cannot be admitted or denied and which the claimant must therefore prove.
- The defendant's version of the events.

The statement of truth

The defendant's case must be supported by a 'statement of truth' which must be signed by a senior officer of the organisation, a partner or a legal representative. The statement begins: *"The defendant believes the facts stated in this defence are true."*

Although not sworn, a signatory who signs the statement without honestly believing it to be true may be held in contempt of court. If a legal representative signs the document, then they are deemed to have briefed the organisation as to its meaning and the penalties will be the same as if the defendant had signed it originally.

Disclosure of documents

All documents that could support or adversely affect the case must be disclosed. Many of these may have been previously disclosed under the pre-action protocol procedure.

The disclosure statement

Documents sent to the opposing party must be accompanied by a formal statement of disclosure confirming that this has been done with full knowledge of the requirements and to the best of their ability.

Offers to settle

An offer to settle may be made by either party and a payment made into the court. If the other side decides not to accept, this has a bearing on the awarding of costs.

Case management

The case will be managed by the Court who will give 'directions' on how it is to be handled, based on an allocation questionnaire filed by the parties involved.

A summary judgement may be made after giving 14 days' notice to both parties against either the claimant or the defendant because:

- The claim has no prospect of success.
- The defence has no real prospect of success.

Claims will be dealt with according to their value:

Name	*Sum*	*Points*
Small claims	Up to £5,000	
Fast-track	£5,000 - £25,000	Standard time between giving of directions and trial no more than 30 weeks. Trial date fixed at time of direction. Court discretion to limit disclosure of documents. Oral expert evidence limited to two fields of expertise and to one expert per party.
Multi-track	Over £25,000	Court gives directions for management of case and timetable for trial. Alternatively a Case Management and/or a Pre-Trial Review may be fixed.

Figure A2-7: Case management. *Source: RMS.*

After allocation to the appropriate track the Court will set a timetable or fix a case management conference or a pre-trial review or a combination of these. Under the rules, trials must be held 'as soon as is practicable'. ***See figure ref A2-8***.

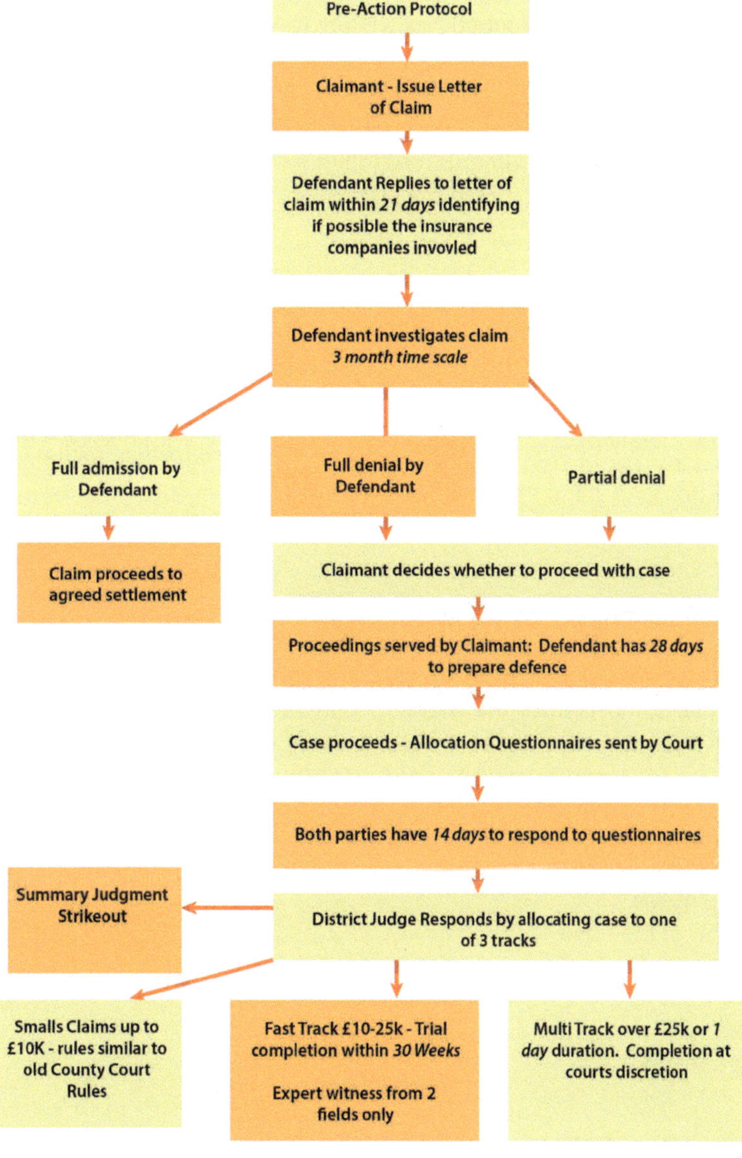

Figure A2-8: Progress of a case. *Source: RMS.*

The role of Employment Tribunals in matters of health and safety

ROLE OF EMPLOYMENT TRIBUNALS IN HEALTH AND SAFETY MATTERS

An employment tribunal usually consists of a legally qualified chairperson appointed by the Lord Chancellor and two lay members - one representing management interests and the other employees. These are selected from panels kept by the Department of Employment after nominations from employers' organisations and trades unions. Employment tribunal's deal with the following issues related to health and safety:

- Victimisation or other discrimination.
- Dismissal, actual or constructive, following a breach of health and safety law, regulation and/or term of an employment contract.
- Appeals against improvement and prohibition notices.
- Safety representatives and safety committees (time off, payment functions and training of representatives under the Safety Representatives and Safety Committees (SRSC) Regulations 1977).
- Representatives of employee safety (pay and time off under the Health and Safety (Consultation with Employees) Regulations (HSCER) 1996).
- Suspension from work for medical reasons (Employment Protection (Consolidation) Act (EPCA) 1978).
- Suspension from work on maternity grounds.

Appeals are made against improvement and prohibition notices on the grounds of:

1) Time limits for compliance.
2) Substantive law involved.

PRESENTATION OF HEARINGS

A case begins with an applicant or claimant (i.e. employee making a complaint) filing an originating application, setting out the nature of the claim and the redress sought. A copy of this document is then sent to the 'offending' employer (known as the 'respondent') who submits a defence in another document, known as a notice of appearance. If one party to the proceedings wishes to know more about allegations made against him he can request 'further particulars'; if this request is refused when such particulars are relevant to the case the tribunal can order 'discovery' of them (i.e. their production).

'Privileged' documents do not have to be made available to the other party, though their existence must be disclosed to the tribunal, for example, documents predominantly concerned with litigation, such as counsel's opinion, solicitors'/clients' letters.

PROCEDURE

Current law relating to procedure in Employment Tribunals is contained in regulations made under the Employment Tribunals Rules of Procedure 2004.

Importance of documents

If a party proposed to refer to a document at the hearing they should supply the other side with copies in advance (a week before the hearing), as well as the Chairperson and the two lay members. Failure to do so can mean that the other party might ask for an adjournment in order to study them, especially if they are complicated. It should be borne in mind that documentary evidence at a hearing is much more important and convincing than oral evidence.

WITNESSES

Persons present at all material times and occasions relating to the complaint should be available to give evidence as witnesses. Witnesses should commit to writing their recollections of events and conversations; this will facilitate answers to questions at the hearing. Witnesses cannot refer to notes when giving evidence, unless the notes were made at the time of the event or conversation. If a witness is reluctant to appear, the tribunal can order him to do so if it is felt that his evidence would be material.

TIMETABLE

Both parties to the proceedings must be informed at least 14 days before the date fixed for the hearing. The hearing itself consists of two parts:

1) The applicant (the employee):
    - Presents the outline of the case.
    - Calls their evidence.
2) The respondent (the employer):
    - Calls their evidence.
    - Makes a closing speech.

FULL HEARING

Proceedings are always heard in public unless there is a good reason for not doing so, for example, national security.

In his opening speech an applicant should preferably mention:

- His role in the organisation.
- The facts on which he bases his case.
- Any relevant statutes and case law.

Employers may be represented by a specialist member of management or may decide to instruct lawyers if the case is complicated. Any witnesses are then called and examined on oath. This is then repeated by the respondent. Then in a closing speech each side's representative should sum up the relevant facts and law, answering any questions raised by the other side or the chairperson or members of the tribunal.

EVIDENCE

The strict 'hearsay' evidence rules do not apply in employment tribunals, though of course first-hand evidence of what the applicant/witnesses actually saw or heard carries much more weight.

CONCLUSION OF HEARING

After completion of the final speeches, the tribunal will give its decision - normally on the day of the hearing, but if the case is difficult it can be postponed. The decision is oral, consisting of a summary of the facts and the result. At a later date the decision is set out formally.

COSTS

Costs are not normally awarded to a successful party; and in order for costs to be awarded it must be shown either that:

- One party acted 'frivolously or vexatiously' (i.e. there never was a valid claim).
- The proceedings were brought or conducted unreasonably (for example, failure by one party to warn witnesses to attend, thereby resulting in delay).

Moreover, an order for costs would be in keeping with the ability of a party to pay; if a party was unemployed this would be taken into account. Applications for costs must be made either at the hearing, or within a reasonable time afterwards.

REVIEW OF DECISION

A decision of an Employment Tribunal can be challenged, or 'reviewed' on the following grounds:

- The decision discloses an error on the part of the tribunal.
- A party did not receive notice of proceedings.
- The decision was made in the absence of a party entitled to be heard.
- New evidence has come to light.
- The interests of justice require a review.

Moreover, the Divisional Court of the Queen's Bench Division of the High Court can overturn a decision of an Employment Tribunal, disclosing an error of law. For instance, if there was a relevant conflicting judgement of the Employment Appeal Tribunal (EAT) which the tribunal had not taken into account, then the tribunal can be reconvened to rectify the oversight.

APPEALS

Appeals from Employment Tribunals lie to the Employment Appeal Tribunal (EAT), except in specific health and safety matters (including appeals against enforcement notices), where appeal lies to the Divisional Court of the Queen's Bench Division of the High Court. If either party is dissatisfied on a point of law with the decision of the Employment Tribunal, it may appeal the decision to the High Court within 42 days of the date of the decision. Like the EAT, the High Court can normally only interfere where the tribunal has erred in respect of the view of law (not facts) which it took. While employers are unlikely to succeed in most cases of appeal on substantive law, Employment Appeal Tribunals and the High Court are more flexible about time limits.

PRECEDENT

The decisions of an Employment Tribunal bind no one other than those parties involved in the dispute. The EAT is not bound to follow its own decisions and decisions of the High Court, though not binding, have a persuasive authority. Only the decisions of the Court of Appeal and the Supreme Court are binding on the High Court, EAT and Employment Tribunals.

The appeals system - routes and grounds of appeal

The appeals system for Courts and Tribunals has been covered, in context and, in the above explanation of the functions of the Courts and Tribunals. In Summary:

In the Criminal Courts: Appeals may be made against a guilty verdict by the defendant though there is no right for an appeal by the prosecution where the accused is acquitted unless it is believed that the law has been interpreted incorrectly by a magistrate. However, since the Double Jeopardy protection was removed, if significant and substantial evidence comes to light subsequent to the trial, the prosecution can seek a retrial.

An appeal against the sentence may be instigated by either the prosecution or convicted if considered to be too lenient or excessive respectively. Appeals to decisions made by the magistrates are presented to the Crown Court while Crown Court decisions are referred to the Criminal Division of the Court of Appeal. The basis of the appeal must be presented for consideration by the relevant court before the right to appeal is granted.

Within the Civil Court system either party has the right to request to the Civil Division of the Court of Appeal the right to appeal a decision.

Employment Tribunal decisions are normally referred to the Employment Appeal Tribunal (EAT) with the exception of decisions about Improvement or Prohibition Notice appeals which are referred directly to the Queens Bench of the High Court. Appeals to decisions made by the EAT are heard by the Civil Division of the Court of Appeal.

The ultimate court for all appeals to Court of Appeal decisions is the Supreme Court. Although the Supreme Court may be required to refer to the European Court of Justice for a Preliminary Ruling interpreting a relevant point of EU law, there is no right to appeal to any Supreme Court decision.

A2.6 - Principles of the law of contract

Definition and meaning of contract

The law of contracts is at the heart of most business dealings, it is one of the three or four most significant areas of legal concern and can involve variations on circumstances and complexities.

A contract may be defined as:

> *"An agreement with specific terms between two or more persons or entities in which there is a promise to do something in return for a valuable benefit known as consideration (benefit)."*

Figure A2-9: Definition of a contract. *Source: Law.com.*

A contract is more than the mere exchange of promises - there must be an intention to create a legal relationship. Both parties giving something, or have promised to give something, of value in consideration for any benefit accrued from the agreement.

The agreement requires a 'meeting of minds' whereby all parties concerned are in agreement about the subject of contract. They must therefore have common understanding of that subject. A contract can take a number of forms, it can be written or verbal and it can be express or implied.

WRITTEN

Many contracts are constructed in written form and some are very complicated. The written form of contract will often be used where what is being contracted is complicated or is required to be bound by a number of conditions.

However, written contracts do not have to be complicated, a simple written contract may be established where a person writes down the cost of doing work for someone and the person writes down that they accept it. Many contracts, in practice will be a combination of written and verbal. A 'statute of frauds' requires that certain contracts be in writing, and that they be signed by all parties to be bound by the contract.

The most common types of contracts to which a statute of fraud applies are:

- Contracts involving the sale or transfer of land.
- Contracts to answer for the debt or duty of another.
- Contracts that, by its terms, cannot be completed within one year, for the statute to apply, the actual terms of the contract must make it impossible for performance to be completed within one year.

VERBAL

It is not illegal to enter into a verbal contract, unless the nature of the contract itself is illegal. Simple arrangements may be capable of being agreed verbally, but they will rarely suit complicated situations. A verbal contract may be difficult for a court to uphold since the parties to the contract will disagree and contradict each other, with no way for either side to prove their version of the contract unless it is witnessed by someone.

Written contracts are often amended in practice by verbal contracts; though this can be problematic they can be upheld. A verbal contract may be an extension or amendment to a written contract to deal with situations that were not foreseen. Verbal contracts are still common on the floor of the London Stock Exchange between brokers, whose bids and acceptance for share transfer are made verbally, 'in good faith'.

EXPRESS

A contract may be express in that it establishes clear limits to the contract, for example, how long it is to last or who it relates to. Express contracts are those where terms are expressly stated, either verbal or written. Contracts of employment will contain express terms, including things like pay, hours and holidays.

IMPLIED

An implied contract is one that may be deduced from circumstances alone. Custom and practice may establish an implied contract over time. Implied contracts may derive from such things as services that one person gains from another, which may commence as a one off occasion and through habituation are established as a continuing implied contract.

An implied term, such as with a contract of employment, is not normally provided in writing, but is considered to be part of it. For example, a duty by the employer to provide a healthy and safe working environment is an implied term owed to an employee as part of their contract of employment.

Principles of the law of contract

Civil law exists in order to regulate disputes between individuals over the rights and obligations people have when dealing with each other. There are many branches of civil law including contract law. Contract law deals with the legally binding relationships that individuals enter into.

APPLICATION

Components of a contract

The following elements are essential items of a valid (and legally enforceable) contract:

- An offer and an acceptance.
- An intention to create a legal relationship.
- Appropriate capacities. All parties must have the appropriate (legal) capacity to undertake the contract.
- A consideration must be present (unless the agreement is by deed).
- The object of the contract must be legally acceptable.
- Consent must be genuine (implying full knowledge).
- It must be possible to fulfil the contract.
- The contract must not be contrary to public policy.

There are three classes of contract:

Contracts of record:	These are usually judgements of courts, which merge with the original contract which was the subject of court action. They may also be recognisance, where one party agrees to be bound by the terms offered by the Court.
Speciality contracts:	These are for specific transactions where the law requires deeds to be 'signed, sealed and delivered'. Examples are: conveyancing, property leases over 3 years, and legal articles of partnership.
Simple contracts:	The majority of contracts, and the ones to which occupational health and safety liabilities generally apply, are simple contracts. They may be made orally, or in writing, or may be implied by conduct. ***Express contracts*** are those whose terms are stated. They may be verbal or written down. ***Implied contracts*** are those whose terms are not specifically expressed, but are implied by the actions of the parties concerned. ***Executed contracts*** are those which have been wholly performed by the parties concerned. ***Executory contracts*** are those which are either wholly or partly unperformed.

THE CONTRACT PROCESS

The offer

A contract must begin with an offer, made by the offeror, to the offeree(s). The offer must be made clearly to the offeree, and may be made orally, in writing or by conduct. The offer must be communicated to the offeree, a person cannot be said to have accepted an offer of which they are unaware.

An offer of contract is not the same as an invitation to treat. An offer, or invitation to treat, means that an invitation to make an offer of contract has been made. If an auctioneer requests bids, he is making an invitation to treat, which is accepted by those who bid. The bids are offers of contract, and the auctioneer accepts one of the offers by the fall of his hammer.

Similarly, goods displayed in a shop are construed as invitations to treat, not as offers of contract being made by the shop. When a customer acts on that offer to treat, the customer's actions are taken as an offer to purchase (i.e. an offer of contract).

The acceptance

Acceptance of an offer may be made orally, in writing, or by conduct ***but only by the offeree.*** To be valid, an acceptance must be unqualified and must conform entirely to the terms of that offer.

In commercial terms, when a contract is 'put out to tender', that is an invitation to treat. The tenders are offers of contract, one of which will be accepted by the client, to form a contract. Whilst this simple form of contract applies to specific 'one off' services etc., tenders to supply on-going quantities of goods or services receive a separate acceptance each time an order is placed.

Intention

The agreement made between the offeror and the offeree, whether orally, in writing, or by action, must have the intention of forming a legally enforceable relationship. Whilst some contracts - notably those between private individuals for private purposes, may well be enforceable by the court without evidence, the courts will require written or oral evidence in most commercial or business agreements.

Consideration

The 'consideration' is a fundamental part of any simple contract. It has been defined as 'same right interest, profit or benefit accruing to one party, or some forbearance, detriment or loss suffered or undertaken by the other, as part of the contract'. Consideration may be regarded albeit simplistically, as 'the price' involved in the contract. It need not be monetary. It must be real (i.e. definite rather than vague) and genuine; must be legal; must move from the promise; need not be adequate; must be possible; and must not be past.

Appropriate capacities

In general, any person may enter into a binding contract. However, there are special rules governing some 'persons' (for example, corporations).

There are specific restrictions aimed particularly at the protection of minors (i.e. those under 18) and those considered to be particularly vulnerable as a result of mental capacity, insanity or drunkenness.

Consent

Genuine consent, based on a full and accurate knowledge of the matters of the contract is essential to the validity of the contract. That validity may be vitiated (harmed) by a number of factors, and where any one of them exists, there can be no true consent and so the contract may be voided, or may be void.

Vitiating factors include:

- Mistakes.
- Misrepresentation.
- Duress.
- Undue influence.
- Illegality.

Voiding of contracts

Grounds for a contract being declared void include:

Mistakes - in general, at common law, mistakes do not invalidate a contract. However, some classes of mistake do - 'operative' mistakes of fact undermine a contract to such an extent that they invalidate it. An example is where one party to a contract has one item in mind, as the subject of the contract, and the other has something entirely different.

Misrepresentation - misrepresentation is a false statement of fact which induces a party to enter a contract and thereby results in that party suffering a loss. Misrepresentation may be fraudulent, negligent, or innocent, and the term applies only to statements of fact, not to opinions.

Duress and undue influence - in general rule of law, a valid agreement can only be made where both parties exercise their own free will without constraint or pressure.

Duress at common law means violence (or the threat of violence) to a party to a contract, or to members of their family, or the threat of unlawful imprisonment.

Undue influence is a parallel but more subtle vitiating factor, where a party uses their influence to gain undue benefit for themselves either as a result of persuading someone to enter into a contract with them or by persuading the influenced party to enter into a contract with a third party, for their own benefit. Such contracts, induced by undue influence are voidable, and may be void.

Illegality - a contract is illegal if it contravenes a statute or common law.

Illegal contracts include those which involve the commission of a crime or tort, and those which contravene or offend against certain fundamental rules of common law or morality and collectively termed 'public policy'. Examples are those contracts which result in restraint of trade, extortionate repayment terms or sexual immorality (for example, a 'wife swapping' contract).

Contracts are also illegal if they prejudice:

- Foreign relations (for example, a contract to carry out acts which are illegal in a foreign and friendly country).
- The administration of justice (for example, a contract to circumvent the bankruptcy laws).
- Standards in public life (for example, a contract to purchase a knighthood).
- The revenue (a contract to defraud the Inland Tax Office or to avoid local government rates).

DISCHARGING OF CONTRACTS

Contracts may be discharged by: ***Mutual agreement***.

Performance - where both parties have completely fulfilled their obligations.

Breach - where one party repudiates their liability before performance is due, or where they fail to perform the obligations they have undertaken. (The injured party is therefore able to sue for damages).

Impossibility or frustration - where circumstances alter before a hitherto entirely valid contract can be performed (for example, outbreak of war, or a change in the law), the contract may be voided.

Operation of law - where events result in legal consequences that render the original contract discharged, for example, the passage of time, and the operation of the Limitation Act 1980, which sets time limits for various actions (particularly 6 years in respect of torts, or of 3 years following the death or injury of a relevant party).

THE RELATIONSHIP BETWEEN PRODUCER AND VENDOR

Since the introduction of the HASAWA 1974, legislation has more precisely focused on matters of safety in design. This has included the important contribution made by the Consumer Protection Act (CPA) 1987. The CPA 1987 amended some features of the HASAWA 1974 contained in Section 6. Both the HASAWA 1974 and CPA 1987 define articles and substances for use at work, in Sections 53(1) and 45 respectively. While the two Acts define similar duties, Section 6 of HASAWA 1974 relates to articles and substances for use at work while the CPA 1987 is designed to protect the consumer whilst not at work.

Duties

Articles for use at work

Section 6 HASAWA 1974, as amended, imposes on any person who designs, manufactures, imports and supplies any such article (or any article of fairground equipment) a general duty to:

- Ensure, so far as reasonably practicable, it is so designed and constructed as to be safe at all times when set, used, cleaned or maintained by a person at work.
- Carry out or arrange necessary testing or examination for that purpose.
- Ensure the provision of adequate information about design use and conditions of use (Section 6) and keep it updated (CPA 1987 Schedule 3).

Substances for use at work

Section 6 as amended by CPA 1987 Schedule 3 imposes on manufacturers, importers and suppliers of any such substance a similar general duty to that with respect to articles:

- Ensure, so far as reasonably practicable, it will be safe at all times when used, handled, processed, stored or transported by a person at work.
- Carry out or arrange necessary testing or examination for that purpose.
- Ensure the provision of adequate information about risks from its inherent properties, tests and necessary conditions of use, and keep it updated.

There are additional duties under HASAWA 1974 Section 6 and CPA 1987 Schedule 3 on:

- Designers and manufacturers of articles, and manufacturers of substances, to carry out research to discover and, so far as is reasonably practicable, eliminate risks.
- Installers of articles for use at work to ensure, so far as is reasonably practicable, safe installation.

It is important to note that importers and suppliers do not need to repeat testing, examination or research already completed by designers or manufacturers. It is accepted that a person who designs, manufactures, imports or supplies an article for or to another person on the basis of a written undertaking of safety, that other person may rely on that undertaking.

Liability of importers

The CPA 1987 Schedule 3 added Section 6 (8A) of the HASAWA 1974, this emphasised that nothing in the HASAWA 1974 relieves importers of articles or substances from any of their duties as regards anything done (or not done) within the control of a foreign designer or manufacturer. This made the organisation that imported goods into the UK responsible for ensuring articles and substances met the requirements set out for design and manufacture in Section 6. Importers cannot use as a defence that that the duty was held by a foreign designer or manufacturer rather than the importer.

Unsafe imported goods

The HASAWA 1974 gives no power to enforcing authorities to stop the supply of unsafe products at source or prevent the sale of unsafe products; so enforcing officers have been given powers to interrupt the supply/distribution chain by the use of a prohibition notice. Additionally, Her Majesty's Customs and Excise [HMCE] officers have power to:

- Detain articles and substances (cf. Health and Safety Executive (HSE) inspectors).
- Transmit information to the HSE about unsafe imported products; HASAWA 1974 Section 25(A), added by CPA 1987 Schedule 3.

The HSE enforces the law relating to products purely used at work under Section 6 of the HASAWA 1974, which was amended by the CPA 1987.

THE RELATIONSHIP BETWEEN VENDOR AND CONSUMER

The Consumer Rights Act 2015

The Consumer Rights Act 2015 replaced three major pieces of consumer legislation:

1) The Sale of Goods Act.
2) Unfair Terms in Consumer Contracts Regulations.
3) The Supply of Goods and Services Act.

The Consumer Rights Act 2015 sets out a framework that consolidates in one place key consumer rights covering contracts for goods, services, digital content and the law relating to unfair terms in consumer contracts.

In addition, the Act introduces easier routes for consumers and small and medium sized enterprises ('SMEs') to challenge anti-competitive behaviour through the Competition Appeal Tribunal ('CAT').

The Act clarifies the maximum penalties that the regulator of premium rate services can impose on non-compliant and rogue operators. It also consolidates enforcers' powers to investigate potential breaches of consumer law and clarifies that certain enforcers (Trading Standards) can operate across local authority boundaries. It will also give the civil courts and public enforcers greater flexibility to take the most appropriate action for consumers when dealing with breaches or potential breaches of consumer law.

HSE's role as a market surveillance authority

Public authorities, such as the HSE, are responsible for monitoring the safety and conformity of products against European product supply law. This is an activity known as market surveillance.

HSE is one of the UK's market surveillance authorities, which monitors and enforces legal requirements on the safety (including risks to health) of most products used at work. This work is undertaken in addition to HSE's other functions, including its traditional role as a workplace health and safety regulator. Other UK market surveillance authorities such as trading standards, which are locally based, deal mainly with the safety of consumer products - although they take the lead role (instead of HSE) on the safety of some products used at work (for example, personal protective equipment).

What does HSE do in the field of product safety?

HSE leads on the monitoring of the health and safety of most products used at work in the UK. HSE allocates resources to undertake both reactive and proactive market surveillance.

Investigation of product safety issues may be undertaken in response to concerns about products, which fall within HSE's enforcement remit when raised by users, regulators, or as a result of information received from the European Commission or other member states.

Proactive market surveillance is also undertaken on selected products based on the level of risk involved. This is often done in co-operation with other European market surveillance authorities, in order to apply the requirements of European product safety law in a consistent and co-ordinated way.

HSE's policy on market surveillance

HSE will undertake market surveillance work following the same principles and policy as it does for its other activities, acting independently, impartially and proportionately on the basis of risk. This may also include taking appropriate action in cases of lower risk, where inaction could undermine the integrity of the European system for product safety (for example, incorrect CE marking, missing documentation). However, these activities may not be treated with the same urgency as high-risk issues and, if co-operation is given by the parties involved, such administrative issues can usually be dealt with without formal enforcement action.

Enforcement by HSE

HSE's product safety enforcement powers are broadly similar to, and arise from, those given by the Health and Safety at Work etc. Act 1974. All warranted HSE staff may exercise these powers in accordance with the law and HSE's own procedures.

HSE can investigate product safety issues, requiring co-operation and information from those who may be involved. HSE inspectors can require that a product is left undisturbed or take possession of dangerous products, and can direct - usually through formal Enforcement Notices - that an activity such as the supply of a product presenting serious risk of injury is stopped immediately, or that improvements are made by a specified date. Those who are served these Notices may appeal to the Employment Tribunal. HSE can apply to the courts to request the forfeiture of certain products, subject to the court's ruling following judicial consideration of the facts.

HSE may instigate proceedings in the magistrates' court, although this is usually reserved for the most serious matters or where Notices have not been complied with. Penalties for proven breaches of the law are mostly financial and subject to the same principles as for other health and safety matters. Costs may be awarded by the court to HSE. As most offences under UK product safety law are triable either way - meaning they can be heard at either magistrates' or Crown court - unlimited fines may be imposed on those found guilty (by Crown courts).

How does HSE undertake market surveillance?

HSE has a number of small, geographically-dispersed teams of inspectors and administrative staff, supported by senior managers, who undertake product safety/market surveillance work on behalf of the organisation.

They are supported by policy staff who maintain contact with the other market surveillance authorities in the UK and throughout Europe, and by specialists in key areas - particularly mechanical, electrical and control systems safety. Other specialists can also be called upon and, where necessary, research and testing can be commissioned, usually through HSE's agency, the Health and Safety Laboratory (HSL). In some cases, HSE works with officials from other market surveillance authorities, and those responsible for border control.

All HSE visiting staff may act in an 'eyes and ears' role when visiting workplaces. However, unless part of a specific proactive market surveillance project, they will only normally raise matters of concern arising during inspections and investigations.

A limited amount of proactive market surveillance is also undertaken by operational policy and specialist staff, for example, at trade shows or to undertake specific market surveillance projects.

A2.7 - Principles of employment and discrimination law

Purpose of contract of employment

Section 53 of the HASAWA 1974 defines an 'employee' as an individual who works under a contract of employment. It further defines a 'contract of employment' as a contract of employment or apprenticeship (whether express or implied and, if express, whether oral or in writing). Thus, a formal written contract is not necessarily required in order for someone to be considered an employee for the purposes of health and safety.

In addition, the case of Ferguson v John Dawson and Partners [1976] showed that in civil law, if an individual is working 'like an employee' the same duty of care will apply; in this case Ferguson, for purposes of payment, was working as a sub-contractor though in reality he was working full-time for Dawson, under his direct instruction.

The purpose of a contract of employment is to establish a legally binding agreement between an employer and an employee. The contract sets out what both parties can expect from each other in the relationship created by the contract of employment. The contract highlights important employment terms and conditions, such as an employee's job title, start date, pay rate, benefits, retirement options, grounds for termination and dispute resolution policies. It may also set out critical conditions of employee conduct, for example, with regard to health and safety. The employer may similarly be bound, including by reference to its obligations set out in supplementary documents referred to in the contract of employment.

Employment contracts offer protection for both employers and employees. A contract of employment will seek to minimise disagreement between the parties involved by helping to establish a clear conscious commitment to the contract by both parties. To minimise confusion, employers may require employees to sign to confirm their acceptance of a legally binding contract prior to beginning or early in their employment. Even if not formally written, the implied term for both the employer and employer to act legally, with honesty and integrity and to fulfil their common law duty of care will always apply.

Employment Rights Act (ERA) 1996

The Employment Rights Act (ERA) 1996, Part 1, states that employers must give their employees written information about the terms of employment, ***'statement of particulars of employment'***, within two months of starting work (provided that the duration of the job is more than one month).

Details should include:

- The names of the employer and employee.
- The date when employment began.

- The duration of temporary contracts.
- Details of any previous employment which counts as continuous service.
- Job title or brief job description.
- Pay details (including holiday and sick pay).
- Details of working hours and conditions.
- Length of notice to terminate the contract.
- If work outside the UK for longer than one month is required: details of duration, currency of payments, pay and benefits.
- Work location, or locations, and the address of the employer.
- Collective agreements with trades unions affecting the job.
- Disciplinary rules and details of possible disciplinary actions, or reference to an available document.
- Who to apply to in respect of disciplinary and grievance procedures and how this is done.
- This requirement does not extend to rules, disciplinary decisions, decisions to dismiss, grievances or procedures relating to health and safety.

An employee who fails to undertake all the duties agreed in the contract will be in breach of contract.

Employment law as it relates to health and safety issues

DISCIPLINARY PROCEDURES

The Employment Relations Act (ERA) 1999 exempts employers with fewer than 20 employees from the requirement to provide details of the disciplinary and grievance procedures, but they must provide the name of the person to whom application can be made.

Precise grievance and disciplinary procedures are therefore usually part of the contract of employment and employers must adhere to them if they wish to avoid liability. The procedure usually comprises a series of oral and written warnings. The ERA 1999 gives workers a right to be accompanied in certain disciplinary and grievance hearings.

FAIR AND UNFAIR DISMISSAL

There are many acceptable grounds for fairly dismissing an employee including genuine redundancy, retirement, misconduct and lack of capability or qualifications. Dismissal must however be reasonable (fair) and the ERA 1999 requires the employer to give these reasons in writing. The ERA 1999 also includes a test of reasonableness when determining if a dismissal was fair in relation to the 'size and administrative resources of the employer's undertaking'. In order for an employee to claim unfair dismissal they must first have to be dismissed.

There are two types of dismissal:

1) ***Actual*** - a formal dismissal of the employee by the employer either orally or in writing.
2) ***Constructive*** - where the employee is compelled to leave due to the conduct of the employer. The employer's conduct is considered to have been a fundamental repudiation of the contract of employment.

Dismissal can occur even when a fixed term contract has not been renewed. The ERA 1999 prohibits the use of waivers for unfair dismissal, but not redundancy payments in fixed term contracts. Grounds for unfair dismissal include:

Health and safety dismissals and detriments - detriment in relation to safety representatives and representatives of employee safety (for example, loss of overtime). This category includes ordinary employees who leave or refuse to return to the workplace because of a health or safety hazard which they reasonably thought to exist.

Dismissal in connection with trade unions - dismissal is automatically unfair if it is principally due to membership of, or the taking part in at an appropriate time, the activities of a trades union which is independent of the employer. This applies equally to non-membership as well as membership. Under the terms of the Trade Union and Labour Relations (Consolidation) Act (TULRCA) 1992, dismissal relating to refusal to join a trades union is automatically unfair even if there is a closed shop.

Dismissal for asserting statutory rights - dismissal of an employee trying to enforce their employment rights under the ERA 1999 is also automatically unfair.

Unfair selection for redundancy - where selection for redundancy has been made unfairly (for example, trades union membership, complaints about safety standards) or where the employer has failed to comply with an agreed prior arrangement (for example, last in - first out).

Transfer of business - if part or the whole of a business undertaking is transferred and an employee is dismissed as a result then dismissal is automatically unfair.

Dismissals in connection with pregnancy or childbirth - where the principal reason for dismissal is that the worker is pregnant or any reason connected with childbirth.

Pressure on employer to dismiss unfairly - for example, pressure put on the employer by a worker because of a refusal to join a trade union in a closed shop. Dismissal is unfair even if there is industrial action or the threat of action. The trade union as well as the employer may be sued by the dismissed employee.

Dismissal for making a protected disclosure - this relates to disclosures brought in by the Public Interest Disclosure Act (PIDA) 1998 (discussed in more detail later).

PROTECTED CHARACTERISTICS

In general, it is unlawful for employers to discriminate on the grounds of age, colour, race, nationality or ethnic or national origins, gender or marital status, disability or trades union status. The exception to this is where there is ***a genuine occupational qualification*** or the job is one of two held by a married couple. Much of the discrimination legislation has been repealed in total and replaced by the Equality Act (EA) 2010. The Equality Act (EA) 2010 is designed to bring discrimination law into a common position and to extend some equality rights.

The following are now protected characteristics under the Equality Act (EA) 2010:

- Age.
- Disability.
- Gender reassignment.
- Marriage and civil partnership.
- Pregnancy and maternity.
- Race.
- Religion or belief.
- Sex.
- Sexual orientation.

The Equality Act (EA) 2010 covers the following areas:

- Employment.
- Access to goods, facilities and services.
- Selling, letting and managing of premises.
- Education.
- Public Transport.

People with protected characteristics are given general and specific rights to be treated equally and a right of action if they are not. The Equality Act (EA) 2010, Section 136, establishes the burden of proof such that: *"If there are facts from which the court could decide, in the absence of any other explanation, that a person contravened the provision concerned, the court must hold that the contravention occurred."*

The Act prohibits discrimination of which there are two forms:

- ***Direct discrimination*** is where people are treated less favourably because of their protected characteristic: as an example a female employee may claim discrimination if she believes that she was overlooked for promotion purely because she was a woman.
- ***Indirect discrimination*** arises where and organisation has a policy which consequently treats a protected group less favourably: an example in this case would be if a supermarket advertises for shop assistants over 2 metres tall thereby disadvantaging shorter persons, women in particular.

The Equality Act (EA) 2010 as prohibits other specific conduct:

- Harassment - participating in, allowing or encouraging behaviour that offends someone or creates a hostile atmosphere (for example, making sexist jokes at work).
- Victimisation - treating someone less favourably because they have complained or been involved in a complaint about sexual discrimination (for example, taking disciplinary action against someone for complaining about discrimination against themselves or another person).

The Equality Act (EA) 2010 established the following remedies; the tribunal may:

- Make a declaration as to the rights of the complainant and the respondent in relation to the matters to which the proceedings relate.
- Order the respondent to pay compensation to the complainant.
- Make a recommendation that within a specified period the respondent takes specified steps for the purpose of eliminating or reducing the adverse effect on the complainant or on any other person.

If a respondent fails, without reasonable excuse, to comply with an appropriate recommendation the tribunal may make an order for compensation or increase the amount of compensation to be paid. There are a small number of situations when it is acceptable to discriminate.

Many of the cases which can be used to understand when discrimination is lawful are based on the old legislation each protecting a separate characteristic. However, as the protection is largely unchanged, it is believed that this case law will remain relevant under the 2010 Act.

Age

Section 5 of the Equality Act (EA) 2010 specifies the characteristic 'age' as:

In relation to the protected characteristic of age:

- *A reference to a person who has a particular protected characteristic is a reference to a person of a particular age group.*
- *A reference to persons who share a protected characteristic is a reference to persons of the same age group.*
- *A reference to an age group is a reference to a group of persons defined by reference to age, whether by reference to a particular age or to a range of ages.*

Figure A2-10: Section 5 of EA 2010. *Source: Equality Act (EA) 2010.*

The Equality Act (EA) 2010 protects against age discrimination in the workplace. The law ensures that people are not denied jobs or harassed because of their age and, in most cases, will have an equal chance of training and promotion. Age discrimination can be explained as occurring when someone treats a person less favourably because of their age, and uses it as a basis for prejudice and unfair treatment against that person. However, discrimination has not occurred if it can be shown that the action was a proportionate means of achieving a legitimate aim.

Employers should ensure they have policies in place which are designed to prevent discrimination in:

- Recruitment and selection.
- Determining pay.
- Training and development.
- Selection for promotion.
- Discipline and grievances.
- Countering bullying and harassment.

There is no statutory upper age limit on the right to claim unfair dismissal or to receive redundancy payments. The default retirement age of 65 was removed by the Government on 6th April 2011, making compulsory retirement at any age unlawful unless objectively justified.

Sex

Section 11 of the Equality Act (EA) 2010 specifies the characteristic 'sex' as:

In relation to the protected characteristic of sex:

- *A reference to a person who has a particular protected characteristic is a reference to a man or to a woman.*
- *A reference to persons who share a protected characteristic is a reference to persons of the same sex.*

Figure A2-11: Section 11 of EA 2010. *Source: Equality Act (EA) 2010.*

The Equality Act (EA) 2010 makes it unlawful to treat anyone on grounds of sex less favourably than the opposite sex is, or would be, treated in the same circumstances. This provision extends to protect a women who is treated less favourably because she is breast feeding. However, a man may not allege discrimination if he is treated less favourably than a woman in connection with her pregnancy or childbirth.

The Equality Act (EA) 2010 also protects those who suffer discrimination by way of victimisation for bringing a complaint under its provisions. The Equality Act (EA) 2010 applies equally to men as well as women.

It is lawful to discriminate on the grounds of marital status or gender when there is a genuine occupational qualification (for example, a fashion model), for reasons of privacy or decency (for example, a single sex school) or where the job is one of two held by a married couple. There are some other non-occupational exceptions relating to the appointment of head teachers in schools or colleges that may be restricted to members of a religious order where this is contained in a trust deed or other founding instrument. In addition, women are allowed to receive special treatment when they are pregnant.

Sex discrimination may also be lawful in the interests of health and safety. In the case of ***Page v Freight Hire (Tank Haulage) Ltd [1981]*** a 23 year old female driver was taken off a job involving the transport of chemicals which involved a risk to women of child bearing age. She brought an action under the Sex Discrimination Act (SDA) 1974 for unlawful sex discrimination.

The SDA 1974 did not allow the employer to stop employing women at the first hint of danger; however the employer's duty under Section 2 of the HASAWA 1974 takes precedence. In this case, there was evidence that the chemical was dangerous to women of Mrs Page's age and the action of stopping her from carrying out the work was therefore reasonable. Thus, there was no unlawful discrimination.

Race

Section 9 of the Equality Act (EA) 2010 specifies the characteristic race as:

Race includes:

- Colour.
- Nationality.
- Ethnic or national origins.

In relation to the protected characteristic of race:

- A reference to a person who has a particular protected characteristic is a reference to a person of a particular racial group.
- A reference to persons who share a protected characteristic is a reference to persons of the same racial group.
- A racial group is a group of persons defined by reference to race; and a reference to a person's racial group is a reference to a racial group into which the person falls.

The fact that a racial group comprises two or more distinct racial groups does not prevent it from constituting a particular racial group.

It does not apply to religious beliefs unless that constitutes racial discrimination. Thus, racial discrimination would exist if the religion was Jewish, but not Protestant or Catholic since these last two are not a matter of race. Less favourable treatment would include the segregation for reasons of race, while an example of indirect discrimination would include the imposition of a dress code, without good reason, which contradicts the requirements of some ethnic groups.

In the case of ***Panesar v Nestlé and Co Ltd [1980]*** an orthodox Sikh who naturally wore a beard as part of his religious belief was refused employment on hygiene grounds. The strict rules required that no beards or excessively long hair was allowed. The claimant claimed indirect discrimination, which was rejected. The Court of Appeal held that as the defendants had supported their rule with scientific evidence there was in fact no discrimination.

Disability

The Equality Act (EA) 2010 aims to protect disabled people and prevent disability discrimination. It provides legal rights for disabled people in the areas of:

- Employment.
- Education.
- Access to goods, services and facilities including larger private clubs and land based transport services.
- Buying and renting land or property.
- Functions of public bodies, for example, the issuing of licences.

The definition of 'disability' is set out under the Equality Act (EA) 2010. A person has a disability if:

- They have a physical or mental impairment.
- The impairment has a substantial and long-term adverse effect on their ability to perform normal day-to-day activities.

For the purposes of the Equality Act (EA) 2010, these words have the following meanings:

- 'Substantial' means more than minor or trivial. A severe disfigurement is to be treated as having a substantial adverse effect on the ability of the person concerned to carry out normal day-to-day activities. This includes impairments controlled by measures that mean, but for the measure, the impairment would prevent them conducting day to day activities; this includes medical treatment or prosthetics.
- 'Long-term' means that the effect of the impairment has lasted or is likely to last for at least twelve months (there are special rules covering recurring or fluctuating conditions - treated as continuing if it is likely to recur), or for the rest of the life of the person affected.
- 'Normal day-to-day activities' include everyday things like eating, washing, walking and going shopping.

People who have had a disability in the past that meets this definition are also protected by the Equality Act (EA) 2010. There are additional provisions relating to people with progressive conditions. People with HIV, cancer or multiple sclerosis are protected by the Equality Act (EA) 2010 from the point of diagnosis. People with some visual impairment are automatically deemed to be disabled.

Some conditions are specifically excluded from being covered by the disability definition:

- Alcohol, nicotine or drug dependency.
- Hay fever.
- Pyromania, kleptomania.
- Physical/sexual abuse.
- Exhibitionism.
- Voyeurism.
- Disfigurement from tattooing or piercing.

However, the consequences of some of these conditions are covered by illnesses like cirrhosis of the liver caused by alcohol misuse or lung cancer from smoking, which could lead to a recognised disability.

Under the Equality Act (EA) 2010 an employer has a duty to make ***'reasonable adjustments'*** for disabled applicants and employees. Adjustments should be made to avoid them being put at a substantial disadvantage compared to non-disabled people. The need to make reasonable adjustments can apply to the working arrangements or any physical feature of the workplace.

Adjustments to working arrangements include adjusting working hours. Adjustments to physical features might include replacing steps with a ramp. In addition, if the person is at a substantial disadvantage and it is reasonable, the employer should provide an auxiliary aid, for example, a visual or personal vibration fire alarm for the deaf.

The Equality Act (EA) 2010 defines a physical feature as:

a) A feature arising from the design or construction of a building.
b) A feature of an approach to, exit from or access to a building.
c) A fixture or fitting, or furniture, furnishings, materials, equipment or other chattels, in or on premises.
d) Any other physical element or quality.

Action to deal with a physical feature causing a substantial disadvantage is prescribed as:

a) Removing the physical feature in question.
b) Altering it.
c) Providing a reasonable means of avoiding it.

The Equality Act (EA) 2010 prohibits the employer from requiring payment for the reasonable adjustments provided.

The Equality Act (EA) 2010 also provides rights for people not to be directly discriminated against or harassed because they have an association with a disabled person. This can apply to a carer or parent of a disabled person. In addition, people must not be directly discriminated against or harassed because they are wrongly perceived to be disabled. A person without disability cannot be discriminated against merely because an employer treats someone with disability more favourably with regard to their disability.

Under the Equality Act (EA) 2010, section 138, there is a process for people to obtain information from the person or organisation they think may have discriminated against them. The process involves working through the 'discrimination and other prohibited conduct' forms. They are offered as a guide; there is one form for the person's questions and another form for the organisation's answers.

Other forms or a letter may be used to set out the questions, if preferred. Answers are expected within eight weeks of the questions being asked, the answers are admissible in evidence in proceedings and a tribunal may draw an inference from failure to answer questions effectively.

ASSOCIATIVE DISCRIMINATION

The Equality Act (EA) 2010 covers exactly the same groups of individuals that were protected by the previous legislation. However, the headings of age, disability, gender reassignment, race, religion or belief, sex, sexual orientation, marriage and civil partnership, and pregnancy and maternity are now to be known as 'protected characteristics'.

This wording is designed to overcome the weakness seen in the wording of the Disability Discrimination Act 1995 in the 2008 case of Sharon Coleman against her employer: Attridge Law. Her son had some serious health conditions which brought him within the remit of the definition of Disability within the Act. As primary carer, at times his mother often needed to request time off or flexible working hours.

Although these were afforded to other employees who were parents of children without disabilities, but not to her, the response to her requests often being derogatory and offensive. Her claim for direct discrimination and harassment was unsuccessful as the wording of the 1995 Act was such that the prohibition applied only to protect those with the protected characteristic.

However, when referred to the European Court of Justice, it was held that the associated EU Directive should be interpreted as prohibiting associated direct discrimination and harassment, hence the careful wording of the Equality Act (EA) 2010.

UNDERTAKING HEALTH AND SAFETY ROLES AT WORK

The need to prevent dismissal for those conducting health and safety roles at work is both a moral and legal responsibility.

If people fear that they will be dismissed for being involved in health and safety matters or reporting a problem they will avoid this action. This will relate to persons formally appointed by the employer to assist the employer, employee health and safety representatives and employees.

Section 44 of the Employment Rights Act (ERA) 1996 specifies protection for matters concerning 'Health and Safety Cases', this prohibits employees suffering detriment.

1) An employee has the right not to be subjected to any detriment by any act, or any deliberate failure to act, by his employer done on the ground that:

a) Having been designated by the employer to carry out activities in connection with preventing or reducing risks to health and safety at work, the employee carried out (or proposed to carry out) any such activities.

b) Being a representative of workers on matters of health and safety at work or member of a safety committee:

(i) In accordance with arrangements established under or by virtue of any enactment.

(ii) By reason of being acknowledged as such by the employer.

The employee performed (or proposed to perform) any functions as such a representative or a member of such a committee, the employee took part (or proposed to take part) in consultation with the employer pursuant to the Health and Safety (Consultation with Employees) Regulations (HSCER) 1996 or in an election of representatives of employee safety within the meaning of those Regulations (whether as a candidate or otherwise).

c) Being an employee at a place where:

(i) There was no such representative or safety committee.

(ii) There was such a representative or safety committee but it was not reasonably practicable for the employee to raise the matter by those means.

He brought to his employer's attention, by reasonable means; circumstances connected with his work which he reasonably believed were harmful or potentially harmful to health or safety.

d) In circumstances of danger which the employee reasonably believed to be serious and imminent and which he could not reasonably have been expected to avert, he left (or proposed to leave) or (while the danger persisted) refused to return to his place of work or any dangerous part of his place of work.

e) In circumstances of danger which the employee reasonably believed to be serious and imminent, he took (or proposed to take) appropriate steps to protect himself or other persons from the danger.

2) For the purposes of subsection (1)(e) whether steps which an employee took (or proposed to take) were appropriate is to be judged by reference to all the circumstances including, in particular, his knowledge and the facilities and advice available to him at the time.

3) An employee is not to be regarded as having been subjected to any detriment on the ground specified in subsection (1) (e) if the employer shows that it was (or would have been) so negligent for the employee to take the steps which he took (or proposed to take) that a reasonable employer might have treated him as the employer did.

Figure A2-12: Section 44 of ERA 1996. *Source: Employment Rights Act (ERA) 1996.*

As can be seen by subsection 1a, employees appointed by the employer are afforded protection; this would include a health and safety practitioner, occupational health nurse, first aider, fire warden, or safety engineer appointed to inspect equipment for health and safety reasons.

Subsection 1b relates to employee representatives appointed for consultation reasons. Subsection 1c relates to employees reporting hazards related to their work; we should note the Management of Health and Safety at Work Regulations (MHSWR) 1999 places a duty on employees to report shortcomings in health and safety arrangements to their employer. The remaining subsections deal with circumstances of danger in which employees act to protect themselves and others. It is important to note that if the employee has created the danger by not following agreed practices this does not prevent the employer taking disciplinary action regarding the creation of the danger in the first place.

Section 48 of the Employment Rights Act (ERA) 1996 enables an employee to make a complaint to an Employment Tribunal if they have been subject to a detriment in contravention of Section 44. When an employee makes a complaint it becomes the employer's duty 'to show the ground on which any act, or deliberate failure to act, was done'. Complaints must normally be presented promptly, within 3 months of the detriment taking place.

Section 49 of the Employment Rights Act (ERA) 1996 enables, where the case is 'well-founded' the Employment Tribunal to make:

- A declaration to that effect.
- An award of compensation to be paid by the employer.

The amount of compensation awarded would be what the tribunal considers 'just and equitable in all the circumstances' having regard to:

- The infringement.
- Any loss which is attributable to the infringed, including any expenses reasonably incurred and loss of any benefit which might reasonably have been expected.

DISCLOSURE OF WRONG-DOING AT WORK

The Public Interest Disclosure Act (PIDA) 1998 (dubbed the 'whistleblowers charter') amended Section 43 of the Employment Rights Act (ERA) 1996 such that it allows employees who are penalised or dismissed for revealing workplace malpractice (including dangers to health, safety or the environment), to claim compensation or reinstatement for unfair dismissal.

With 'whistleblowing' there is a potential conflict of interest between the employer's legitimate interest in preserving confidentiality and the greater public interest. In the case of Initial Services Ltd v Potterill [1968], where there was disclosure of a price cartel to the press, Lord Denning said that the employee need not be confined to the disclosure of crimes but 'any misconduct of such a nature that it ought to, in the public interest be disclosed to others'.

The Public Interest Disclosure Act (PIDA) 1998 states that an employee must not be subject to detriment or dismissal for making a ***'protected disclosure'***. A protected disclosure means a 'qualifying disclosure', which is made (disclosed) in accordance with the manner specified in the Public Interest Disclosure Act (PIDA) 1998.

The Public Interest Disclosure Act (PIDA) 1998 defines a ***'qualifying disclosure'*** as 'any disclosure of information which, in the reasonable belief of the worker, tends to show one of the following':

- Criminal offences.
- Failure or likelihood to fail to comply with a legal obligation.
- Miscarriage of justice is likely to occur.
- The health and safety of an individual is endangered.
- Any environmental harm.
- Deliberately concealing information relating to other protected disclosures.

The relevant failure, which the disclosure relates to, is not restricted to those occurring in the UK and the legal obligation does not have to be a legal obligation in the UK. It is important to note that disclosures that would cause the person disclosing it to commit an offence are not protected disclosures, this may relate to information controlled by the Official Secrets Act. Similarly, if information is provided to someone in order to gain legal advice and a position of legal privilege is established the disclosure of this information is not a qualifying disclosure.

Disclosure is ***'made in accordance with the manner specified'*** in the Public Interest Disclosure Act (PIDA) 1998 where it is:

- Made, in the public interest, to the employer, other person responsible for the matter or other person authorised by the employer to receive disclosures.
- Made in the course of obtaining legal advice.
- Made to a Minister of the Crown, where the worker's employer is appointed by an enactment.
- Made to a prescribed person, this includes the HSE and Local Authority Inspectors (prescribed in the Public Interest Disclosure (Prescribed Persons) Order (PIDPPO) 1999).

Disclosure can be made ***'in other cases'***, outside the above methods, perhaps to the charity 'Public Concern at Work', the Police or 'the media', provided:

- It is in the public interest.
- The person making it reasonably believes it is substantially true.
- It is not made for personal gain.
- Specified conditions are met:
  - The worker reasonably believes they will suffer detriment if they make the disclosure to their employer.
  - Where no other person is prescribed, they reasonably believe the evidence will be concealed or destroyed by the employer.
  - That previous disclosure of substantially the same information has been made to the employer or prescribed person.
- It is reasonable in all circumstances to have made the disclosure.

In addition, a disclosure is made in accordance with the Public Interest Disclosure Act (PIDA) 1998 if it is ***'an exceptionally serious failure'*** and it is made:

- In the public interest.
- The person making it reasonably believes it is substantially true.
- It is not made for personal gain.
- It is reasonable in all circumstances to have made the disclosure, taking regard to who the person was that it was disclosed to.

Dismissal is automatically unfair for a qualifying disclosure made in accordance with the manner specified by the Public Interest Disclosure Act (PIDA) 1998. The Employment Tribunal's £50,000 limit for compensation awards does not apply in these cases.

Situations where it is lawful to discriminate and protection for those undertaking safety roles at work

The situations where it is lawful to discriminate and the protection for individuals undertaking safety roles or disclosing wrong-doings are discussed in detail within the previous section. In summary:

EMPLOYMENT RIGHTS ACT (ERA) 1996

The Employment Rights Act (ERA) 1996 protects the employee from being treated inappropriately or being unfairly dismissed.

As well as making it unlawful for an employer to dismiss without a justified reason, it also protects an employee against constructive whereby the only option left to the employee is to resign as a result of the unacceptable conditions placed on the employee whilst at work; one example of such circumstances may be requiring an individual to work in unsafe conditions which clearly brings the employer outside of the terms of any contract of employment.

The Act also prevents employers for taking disciplinary action against an employee who refuses to undertake activities they know to have an unacceptable associated health and safety risk. It is also unlawful for an employer to dismiss where the primary reason is associated with pregnancy or childbirth.

Section 44 of the Act protects those who have a specific health and safety role within the organisation which will include Health and Safety Advisors, Occupational Health Practitioners, Safety Representatives, First Aiders etc. The rules to protect employees in the event of redundancies of transfer of business are also defined within the 1996 Act.

TRADE UNION AND LABOUR RELATIONS (CONSOLIDATION) ACT (TULRCA) 1992

The Trade Union and Labour Relations (Consolidation) Act (TULRCA) 1992, Section 152, establishes as unfair dismissal under the Employment Rights Act (ERA) 1996 action taken against employees because they are member of, or take part in at an appropriate time, the activities of a trades union which is independent of the employer.

This applies equally to non-membership as well as membership. Under the terms of the Trade Union and Labour Relations (Consolidation) Act (TULRCA) 1992 dismissal for failing to join a trades union is automatically unfair even if there is a closed shop.

Section 240 of the Trade Union and Labour Relations (Consolidation) Act (TULRCA) 1992 also provides that any breach of contract which endangers human life or causes serious bodily injury or exposes valuable property to destruction is a criminal offence.

Section 241 establishes that it is also an offence to intimidate another person and prevent that person from going about their lawful business.

EQUALITY ACT (EA) 2010

Previous discussions have shown that while it is unlawful to directly or indirectly or indirectly discriminate against an individual group of people on the basis of their protected characteristic (or characteristics) there are some circumstances where action which may be deemed discriminatory can be justified and as such is lawful.

Examples include:

- Situations where genuine occupational qualification applies such as fashion models, or actors/actresses for a specific role.
- For reasons of privacy or decency.
- For reasons of hygiene (see the case of Nestle v Pastor discussed earlier).
- In the interests of health and safety ***(see the case of Page v Freight Hire (Tank Haulage) Ltd discussed earlier)***.

PUBLIC INTEREST DISCLOSURE ACT (PIDA) 1998

Often referred to as the 'Whistleblowers Act' this legislation affords protection to those who expose bad practice, failure to comply with health and safety legislation being one example, where such exposure is regarded as being in the public interest.

Where such a disclosure is made it is unlawful for the employer to then discipline the person who has made the disclosure for his action or to discriminate against him.

Decided cases

JOHN SUMMERS AND SONS V FROST 1955 ALL ER 870 AC.740, [F.A.1937 S14 (1), F.A.1961 S14 (1)]

Securely fenced (absolute duty)

Facts:

Mr Frost was injured whilst using a grinding wheel. His thumb was caught between the small (1cm) gap between the guard and the grinding wheel. Although the wheel was partially guarded, section 14 of the Factories Act imposed an absolute duty to guard the dangerous parts of machinery.

Decision:

The House of Lords ruled that there had been a breach of statutory duty.

Relevance:

The obligation to guard securely those parts of a machine which are dangerous must be fulfilled in such a manner, as to afford 'complete protection' to the 'careless or inattentive worker as well as the meticulous, careful one'. Moreover, the dangerous part is securely fenced if and only if the presence of the fence makes it no longer dangerous.

The duty was to provide fencing which absolutely prevented any part of any person working on the premises from coming into contact with any dangerous part of the machine. This illustrates the literal interpretation of statutes that the courts are required to make - if the effect of this is to make the machine unusable then so be it.

STARK V THE POST OFFICE [2000] ICR 1013

[Specific wording to define an absolute duty]

Facts:

Mr Stark suffered injuries when the brakes of his bike, provided by the Post Office, failed and he fell off. The basis of his claim was that the Post Office were in breach of Regulation 5 of Provision and Use of Work Equipment Regulations 1998 which stated that the employer shall ensure that work equipment is maintained in an ***efficient state, in efficient working order and in good repair***.

Decision:

The Post Office were found liable. The evidence that the bicycle had been maintained was not in question. However, it was held that if the brakes failed, the bike was not in an efficient state, efficient working order and good repair.

Relevance:

This case illustrates the importance of the specific wording used to define an absolute duty and that there is no scope for discretion when determining compliance.

ADSETT V K AND L STEELFOUNDERS AND ENGINEERS LTD (1953) 1 ALL ER 97 AND 2 ALL ER 320 [MEANING OF PRACTICABLE]

Statute Factories Act (FA) 1937, s.47; later Factories Act (FA) 1961, s.63

Facts:

Mr Adsett contracted pneumoconiosis in a foundry. The main cause was inhalation of silica dust, which became airborne as he shovelled various casting sands and compounds through a grate onto a conveyor below. To control this, the employers installed a dust extractor near the conveyor. This was done as soon as they thought of it, but it was well after the onset of Mr Adsett's disease and too late to save him from disablement.

Decision:

The question concerned the meaning of the word 'practicable' in section 47 of the Factories Act (FA) 1937, which required employers to take all practicable measures to protect employees from inhalation of dust, etc. Dictionary definitions said, 'possible to be accomplished with known means or resources', 'capable of being carried out in action' and 'feasible'. 'Practicable' is a stricter standard than 'reasonably practicable' so the question of cost should be eliminated. No measure could be 'practicable' if it was not within current knowledge and invention. The employers were held not to have been in breach of statutory duty.

Relevance:

There was an appeal, as it was argued, on behalf of Mr Adsett, that the technology to install an extractor did exist, but that the employers (and everyone else in the industry) had not thought of this particular application for an extractor. It was said that the employers could not escape liability just because they did not think of this particular use for an extractor. The appeal was rejected on the grounds that 'practicable' meant that a measure had to be known for its application by people in the industry and especially by the experts.

EDWARDS V NATIONAL COAL BOARD (1949) [ALL ER 743 KB 704]

Reasonably practicable - the 'quantum of risk' test

Facts:

Mr Edwards was killed when a section of roadway collapsed. The National Coal Board (NCB) argued that the cost of shoring up would have been prohibitive and was not justified.

Decision:

The NCB were found liable on the basis that the cost of making this particular road section safe was not great when compared to the risk.

Relevance:

The 'Quantum of Risk' test should be applied in determining whether or not the steps that were taken in the circumstances of the incident were sufficient to satisfy the requirements of the term 'reasonably practicable' where such an obligatory term is stipulated.

> *"This 'Quantum of Risk' test is the analogy of a scale balance whereupon the degree of risk is placed on one pan and the sacrifice involved in the measures necessary for averting the risk (whether in money, time or trouble) is placed on the other; and that, if it be shown that there is a gross disproportion between them - the risk being insignificant in relation to the sacrifice - the defendants discharge the onus upon them."*

Figure A2-13: 'Quantum of risk' test definition. *Source: Lord Asquith.*

This page is intentionally blank

Element

A3

Criminal law

Learning outcomes

On completion of this element, candidates should be able to demonstrate understanding of the content through the application of knowledge to familiar and unfamiliar situations. In particular, they should be able to:

A3.1 Explain the key requirements of the Health and Safety at Work etc. Act 1974 and the Management of Health and Safety at Work Regulations 1999.

A3.2 Explain the status of Approved Codes of Practice and guidance and the statutory procedures for making Approved Codes of Practice.

A3.3 Explain the responsibilities and powers of enforcing agencies and officers and the range of options related to enforcement action, their implications and appeal procedures.

Contents

Relevant statutory provisions

Biocidal Products and Chemicals (Appointment of Authorities and Enforcement) Regulations 2013

Corporate Manslaughter and Corporate Homicide Act (CMCHA) 2007

Deregulation Act (DA) 2015 (Section 1)

Deregulation Act (DAHSW) 2015 (Health and Safety at Work) (General Duties of Self-Employed Persons) (Consequential Amendments) Order 2015

Employment Rights Act (ERA) 1996

Health and Safety at Work etc. Act HASAWA) 1974

Health and Safety at Work etc. Act 1974 (General Duties of Self-Employed Persons) (Prescribed Undertakings) Regulations (HSWAPU) 2015

Health and Safety (Fees) Regulations (HSFR) 2012

Social Action, Responsibility and Heroism Act (SARAH) 2015

Biocidal Products and Chemicals (Appointment of Authorities and Enforcement) Regulations (Northern Ireland) 2013

Sources of reference

Reference information provided, in particular web links, was correct at time of publication, but may have changed.

Enforcement Management Model (EMM), HSE, http://www.hse.gov.uk/foi/internalops/ocs/100-199/130_5/

Enforcement Policy Statement, HSE41, http://www.hse.gov.uk/enforce/enforcepolicy.htm

Guidance on the application of Fees for Intervention (FFI), HSE47, HSE Books, ISBN: 978-0-7176-6456-6, http://www.hse.gov.uk/pubns/hse47.htm

Health and Safety Offences, Corporate Manslaughter and Food Safety and Hygiene Offences, Definitive Guideline (effective from 1 February 2016), http://www.sentencingcouncil.org.uk/publications/item/health-and-safety-offences-corporate-manslaughter-and-food-safety-and-hygiene-offences-definitive-guideline/

The above web links along with additional sources of reference, which are additional to the NEBOSH syllabus, are provided on the RMS Publishing website for ease of use - www.rmspublishing.co.uk.

A3.1 - The Health and Safety at Work etc. Act 1974 and the Management of Health and Safety at Work Regulations 1999

The Health and Safety at Work etc. Act (HASAWA) 1974

APPLICATION OF THE HEALTH AND SAFETY AT WORK ETC. ACT 1974

The Health and Safety at Work etc. Act (HASAWA) 1974 covers all employment activities, apart from private domestic workers. Consequently, when HASAWA 1974 came into force some years ago, it brought, for the first time, twelve million people under the protection of health and safety legislation. These included laboratories, academic institutions, research establishments and many other areas which did not comply with the definition of a 'factory' given in Section 175 of the Factories Act, which basically required manual labour or gain, and the manufacture of articles and substances.

HASAWA 1974 applies to employers, self-employed persons, subcontractors, and visitors to places of employment, members of the public affected by the employer's activities, designers, suppliers, importers, employees, directors and managers. It also provides the enforcers, including the Health and Safety Executive (HSE) inspectors and the Local Authorities' environmental health officers, with various enforcement powers. Sections 2-4, 6 and 9 are the duties on employers and others, Section 7 is the duty on employees while Section 8 places a duty on all people. The duties are expressed in general terms so that they apply to all workplaces and work activities. The absolute duties placed on the employer are qualified by 'so far as is ***reasonably practicable***' and the duty on the employee is to take ***'reasonable care'***.

Section 15 of HASAWA 1974 allows for Regulations to be made to stricter standards than the reasonably practicable duties provided for in the Sections of HASAWA 1974. As an example, the employers' duty to make an assessment under the Control of Substances Hazardous to Health Regulations (COSHH) 2002 is absolute and amplifies the general reasonably practicable duty of the employer to make arrangements to ensure that substances and articles can be used, handled, stored and transported safely under Section 2 (2)b of HASAWA 1974.

RELATIONSHIP BETWEEN GENERAL AND SPECIFIC DUTIES

HASAWA 1974 sets out a series of general duties under Sections 2-4 and 6-9:

General duties

- Section 2 - General duties of employers to the employees.
- Section 3 - General duties of employers and self-employed to persons other than their employees.
- Section 4 - General duties of persons concerned with premises to persons at work other than their employees.
- Section 6 - General duties of manufacturers etc. as regards articles and substances for use at work.
- Section 7 - General duties of employees at work.
- Section 8 - Duty not to interfere with or misuse things provided pursuant to certain provisions.
- Section 9 - Duty not to charge employees for things done or provided pursuant to certain specific requirements.

HASAWA 1974 Section 2, subsection (1) expresses the general duty on employers to their employees in order to meet the broad goal of ensuring the employees' health, safety and welfare. This is deliberately general in order to avoid a narrowing of the duty through over prescription. In this way, the duty will stand the test of time and societal expectations of what is reasonably practicable.

HASAWA 1974 Section 2, subsections (2), (3) (4), (6) and (7) are the more 'specific' duties. These duties are additional to the general duty expressed in Section 2 (1) but do not set a limit to what the employer has to achieve in meeting the duty under Section 2 (1). The duties in these subsections are still listed under the heading of general duties of employers, but require the employer to accept 'specific' obligations that will contribute to meeting their general duty under Section 2 (1) and reduce health, safety and welfare risks. Further 'specific' duties are placed on the employer under Section 2 subsections (3), (4), (6) and (7). These include requirements for a health and safety policy and consultation arrangements.

REQUIREMENTS OF THE HEALTH AND SAFETY AT WORK ETC ACT 1974

General duties of HASAWA 1974

Section 2 - general duties of employers to employees

Section 2 (1) general duties

An employer shall ensure, so far as is reasonably practicable, the health, safety and welfare of all employees at work. This general duty requires the employer to determine and ensure the carrying out of a wide range of actions to benefit the health, safety and welfare of employees at work.

Additional to this general duty, HASAWA 1974 Section 2 specifies duties with regard to particular matters to be considered to satisfy the general section 2(1) duty.

Section 2 (2) 'specific' duties

Subsection 2(2) of HASAWA 1974 identifies five specific areas that in particular the employer's general duty includes. The fact that these five areas are specified does not mean that they are the only issues with which the employer should be concerned. Without prejudice to the general duties expressed in Section 2(1) the duty is extended to the following particular matters:

The provision and maintenance of plant and systems of work that are, so far as is reasonably practicable, safe and without risks to health.

Arrangements for ensuring, so far as is reasonably practicable, safety and absence of risks to health in connection with ***the use, handling, storage and transport of articles and substances***.

The ***provision of such information, instruction, training and supervision*** as is necessary to ensure, so far as is reasonably practicable, the health and safety at work of his employees. Clearly there will be some circumstances where some information and simple instruction will suffice whereas in other circumstances more detail would be necessary. The wording of this subsection is important as it does not specifically refer to the provision of such information, instruction and training to employees but to ensure employees' safety. The accident leading to the prosecution of Swan Hunter in 1982 illustrated the importance of passing information to all relevant parties: failure to tell the contractors details of the well-established system of work of returning all oxygen cylinders to the top deck of the ship overnight led to a massive explosion as a welding arc was struck the following morning due to the small, foreseeable, oxygen leak within the confines of the lower decks. ***See R v Swan Hunter Shipbuilders Ltd and Another [1982] 1 All ER 264 in 'Relevant decided cases' at the end of this Element.***

So far as is reasonably practicable as regards any ***place of work*** under the employer's control, the ***maintenance of it in a condition*** that is safe and without risks to health and the ***provision and maintenance of means of access to and egress from it*** that are safe and without such risks.

The ***provision and maintenance of a working environment*** for his employees that is, so far as is reasonably practicable, safe, without risks to health, and adequate ***as regards facilities and arrangements for their welfare*** at work.

Section 2 (3) safety policies

An employer must prepare, and revise when necessary, a written statement of policy with respect to health and safety at work and, in particular, the organisation and arrangements which have been implemented to ensure that the policy is being carried out. Such a statement should be drawn to the attention of the employees. Under separate Regulations, employers with fewer than five employees are exempt from having a written policy statement.

Section 2 (4) appointment of safety representatives

This subsection allows the Secretary of State to make Regulations dealing with consultations between employers and recognised trade unions, and their method of representation. This led to the Safety Representatives and Safety Committee Regulations (SRSC) Regulations 1977 and in more recent years the Health and Safety (Consultation with Employees) Regulations 1996 to meet this requirement in non-unionised organisations.

Section 2 (5) repealed

Section 2 (6) consultation

This subsection established a duty on the employer to consult recognised trade union representatives (under SRSC Regulations 1977).

Section 2 (7) safety committee

This subsection requires employers to establish a 'safety committee', if requested to do so by recognised trade union representatives (in accordance with SRSC Regulations 1977).

Section 3 - general duties of employers and self-employed to persons other than employees

Section 3 (1) general duty

Employers have a duty to conduct their undertakings in such a way as to ensure, so far as is reasonably practicable, that persons not in their employment, who may be affected by the undertaking, are not exposed to risks to their health and safety.

See R v Associated Octel Co Ltd [1996] 4 All ER 846 (as well as R v Swan Hunter Shipbuilders Ltd and Another [1982] 1 All ER 264 and R v Nelson Group Services (Maintenance) Ltd [1998] All ER 420), in 'Relevant decided cases' at the end of this Element.

Section 3 (2) self-employed

A similar duty was placed on self-employed persons, though the Deregulation Act (DA) 2015 has restricted those to which the duty applies. The section now requires: self-employed persons who conduct a prescribed undertaking to conduct the undertaking so that neither themselves, nor others, are affected by their activities and exposed to health and safety risks. The above is qualified by 'reasonably practicable'.

The implications of the Deregulation Act 2015 and the defined prescribed undertakings are discussed in greater detail within section ***'A3.3 - The enforcement of health and safety law'***.

Section 4 - general duties of persons concerned with premises to persons other than their employees

Anyone in control of premises or plant used as a place of work by persons not in their employment must:

- Ensure safe access and egress to premises and plant.
- Ensure that plant or substances in the premises, or provided for their use, are safe and without risk to health.

Both are qualified by reasonable practicable.

Section 4 (3) tenancies

The obligations under Section 4 are transferred to the tenant, or person under contract, if the terms of the agreement are related to maintenance and repair of premises, access or egress there from, safety of plant or substances and any health risks arising from these.

Section 5 - repealed

Section 6 - general duties of those who manufacture etc.

The duties of those who design, make, import or supply anything for use at work to ensure, so far as is reasonably practicable, that the article is so designed and constructed, that it will be safe and without risks to health at all times when it is being set, used, cleaned or maintained by a person at work; additionally there is a duty:

- To carry out any necessary testing and examination to ensure the duty imposed by the preceding paragraph.
- To take any necessary steps to provide adequate information about the safe use and to ensure safe dismantling and disposal.
- To revise provided information if it becomes known that anything gives rise to a serious risk to health or safety.

Section 6 of HASAWA 1974 also establishes similar duties for those that manufacture, import or supply substances for use at work.

This Section was substantially altered by the Consumer Protection Act (CPA) 1987 in order to incorporate fairground equipment into the scope of HASAWA 1974. The CPA 1987 also introduced a defence of 'foreseeablity'. If the lack of health and safety cannot reasonably be foreseen by the duty holder it 'shall be disregarded'. Also, when determining if the duty has been met, regard must be given to the information and advice provided by duty holders to each other and the end user. In order to avoid confusion when articles and substances are supplied involving hire purchase agreements or similar finance arrangements, Section 6 of HASAWA 1974 clarifies that the supplier duty is on the 'effective' supplier (who carries on the business of actual supply) not the 'ostensible' supplier (who carries on the business of finance).

Section 7 - general duties of employees

This Section places general obligations upon all employees.

Briefly, the general duties require that employees should:

- Take reasonable care of their own health and safety and that of others who may be affected by their acts or omissions.
- Co-operate with the employer so as to ensure that the employer can comply with statutory obligations.

Note: Regulation 14 of the Management of Health and Safety at Work Regulations (MHSWR) 1999 places additional duties on the employee.

Section 8 - duties not to interfere with or misuse things pursuant to certain provisions

No person shall intentionally and recklessly misuse or interfere with anything provided under HASAWA 1974, and Regulations made under it, in the interests of health, safety or welfare.

Section 9 - duty not to charge employees for things done or provided pursuant to certain specific requirements

Employers must not levy, or permit to be levied, any charges for anything done or provided to meet any specific requirement in HASAWA 1974 or Regulations made under it. This includes, for example, personal protective equipment (PPE), first aid equipment, and safety-related training.

Management of Health and Safety at Work Regulations (MHSWR) 1999

APPLICATION OF THE MANAGEMENT OF HEALTH AND SAFETY AT WORK REGULATIONS 1999

The Management of Health and Safety at Work Regulations (MHSWR) were introduced under Section 15 of HASAWA 1974 and are intended to set out duties that apply to all workplaces, which are aimed at improving health and safety management. Many of the regulations of MHSWR 1999 have a broad, goal-setting format but others are quite specific. Their purpose is to add 'clarification' to the duties expressed in HASAWA 1974. As the title suggests, MHSWR 1999 create mandatory duties to ensure that health and safety is managed in a formal, systematic way. They create extended duties on employers, employees and others.

MHSWR 1999 are supported by guidance known as 'HSG65 - Managing for Health and Safety'. ***See also - 'A3.2 - Approved Codes of Practice and guidance' - later in this Element.***

Duties placed on employers

One of the most significant duties placed on the employer by MHSWR 1999 is the requirement to undertake suitable and sufficient risk assessments. Risk assessments should give particular consideration to the specific vulnerabilities of pregnant employees, and of young people and children due to their lack of perception and experience. Following the risk assessment, appropriate control measures must be put in place. This might include the development of emergency procedures, including procedures for summoning external assistance, and the provision of information to those who could be affected. Where identified as necessary by the risk assessment, the employer must provide employees with the necessary health surveillance to monitor for signs of the onset of adverse health effects.

In order to manage health and safety successfully, employers must plan, organise, control, monitor and review the management arrangements. To ensure that they have access to the necessary support and advice to assist in the management of health and safety, employers must appoint one or more competent person who can provide this assistance. Although such an appointment may be an external contractor, the regulations require appointments to be made internally if possible.

Duties placed on employees

The general duties placed on employees by Section 7 of HASAWA 1974 is extended by regulation 14 of MHSWR 1999. All employees are required to:

a) Use machinery, equipment, dangerous substances, transport equipment, means of production and safety devices in accordance with any training or instructions that have been given.
b) Report those things that may reasonably be considered to be a danger to health and safety and also, any shortcomings in the employer's arrangements for health and safety.

While the duty lies with the employer to take action to protect the health and safety of a new or expectant mother (and her baby) the responsibility lies with the employee to notify her employer in writing of her pregnancy (or if she has given birth in the preceding 6-months or is breast-feeding).

Duties placed on others

MHSWR 1999 require that employers who jointly occupy a worksite, such as a construction site or office block, co-operate and co-ordinate their activities. It is important that both parties are made aware of any risks which may arise as a result of the other's undertaking at the site to ensure that adequate controls can be implemented. Examples of issues that are subject to co-operation and co-ordination are the provision of health and safety training, emergency procedures, parking facilities, first aid provision and welfare facilities, etc.

MHSWR 1999 require liaison between a host employer and temporary employment businesses to ensure that the workers placed with the host employer receive relevant information.

Suitable and sufficient risk assessments

Regulation 3 of MHSW 1999 requires every employer to make a suitable and sufficient assessment of the risks to the health and safety of both his employees and of those not in his employment.

These assessments must be reviewed following any significant change or if there is another reason to believe that it is no longer valid; any additional controls identified as necessary must then be implemented. Employers of five or more people must record the significant findings of the assessments and identify those individuals particularly at risk.

Where the persons working within the business are women of child-bearing age, the risk assessment must identify and assess any hazards which present a specific or increased risk to new and expectant mothers.

Particular attention must also be given to any additional risks to young persons (anyone under the age of 18) or children (anyone not over the compulsory school age). Before employing a young person, risk assessments must be completed which takes into account, in particular, the young person's lack of awareness and inexperience and the extent of training which will be required for the young person, as well as other issues with specific consideration to their implication with young persons. The risk assessment must take into account that young persons may not be employed for work which involves a risk of accident which may not be obvious to a young person due to their inexperience, that is beyond their physical or psychological capacity, involves toxic, carcinogenic or mutagenic substances, radiation, extreme cold, heat, noise or radiation unless it is necessary for the purposes of their training.

Regulation 4 requires that the basis for any preventive and protective measures identified by the assessment must be in line with the 'General Principles of Prevention' detailed within Schedule 1 to the regulations. Offering nine principles for consideration, the schedule starts with consideration as to whether the risk could be avoided and if not the need to evaluate and the possibility of combating at source before resorting to other suitable but less definitive control options.

The regulations require that the risk assessment is suitable and sufficient and that all significant findings are recorded. Although the regulations themselves do not explain what would be regarded as suitable and sufficient, HSE document 'HSG65: Managing Health and Safety' indicates that a suitable and sufficient risk assessment will show that:

- A proper check was made.
- Those who might be affected were identified.
- All of the obvious risks were dealt with, taking into account who may be affected.
- The precautions are reasonable with the any remaining risk being low.
- The workers or their representatives were involved in the process.

In addition, the level of detail should be proportionate to the risk and appropriate for the nature of the work. No risk assessment is expected to identify risks which would not be reasonably foreseeable, though it should be remembered that once a previously unforeseeable risk has become evident, this will now need to be assessed.

Current good practice suggests that a suitable and sufficient risk assessment should involve reference to appropriate information sources and the record should indicate the time period for which it is likely to remain valid. Some risk assessments are required by virtue of other more specific regulations (for example, Control of Major Accident Hazards: COMAH) which impose additional requirements for the risk assessments.

The requirement to record only significant findings indicates that trivial risks need not be assessed in detail and noted. Under normal circumstances, insignificant risks encountered in normal day to day living can be ignored unless the activity compounds or significantly alters those risks. The case of ***R v Porter 2008*** illustrates this point: the headmaster of a private primary school was prosecuted for failing to ensure the safety of his pupils when a young child jumped down a small flight of steps connecting the two playgrounds and fell, bumping his head. His Crown Court conviction was overturned by the Court of Appeal who found that a boy of that age would encounter such steps regularly within his daily living and as such did not constitute a significant risk; the court emphasised that employers are required to manage real, not fanciful or hypothetical risks. It is surmised that had the steps been slippery due to ice, or if the child had suffered with limited sight or mobility, this would have constituted a situation which significantly altered the level of risk necessitating a risk assessment.

Providing employees with information on significant risks

Although it is Regulation 10 of MHSWR which focuses specifically on the need to provide employees with information, the requirement is embedded within several of the other regulations.

Every employer must provide his employees with comprehensible and relevant information on the results of any risk assessment, highlighting the risks identified and the preventive and protective measures in place. Where a child is to be employed, this information must also be conveyed to the parents of the child along with details of the risks arising from the undertaking as a whole.

Regulation 8 requires every employer to identify procedures to be followed in the event of serious and imminent dangers and for danger areas. Perhaps not surprisingly, this duty extends to informing anyone who may be exposed to these potential dangers about their nature and the protective measures in place.

Where an employer is to utilise the services of temporary workers (for example, agency staff), information must be conveyed to those individuals about any identified occupational skills or qualifications required to ensure that the work may be undertaken safely as well as any health surveillance identified as necessary associated with the work to be undertaken or the relevant workplace. As both the temporary workers and their normal employers must be provided this information, in some circumstances, it may be more appropriate for this information to be conveyed via the parent employer.

Although regulations 11 and 12 do not relate specifically to the provision of information to employees, it should be remembered that employers must also communicate any relevant risks and the associated control measures to employers who will have staff working on site for a period of time as well as to employers of other organisations with whom the worksite is shared as part of the co-operation and coordinated approach to effectively managing health and safety. This sharing of information should result in the employees of all organisation being provided with all of the relevant information.

Providing instruction and training for employees

Regulation 13 requires an employer to consider the employee's capabilities before allocating work to them. Health and safety training must be provided to all employees on recruitment or where the risks to which they are exposed change either as a result of a change of role or due to the introduction of new equipment or processes. The training should be adapted to meet the learning needs of the trainees (in particular specific requirements for young persons) and be refreshed periodically. In line with the general requirement defined in Section 9 of the Health and Safety at Work etc. Act 1974, this training must be provided within normal working hours. It is however recognised that in some circumstances it is not possible to deliver training in line with all shift patterns; in such cases overtime should be paid or time off given in lieu.

Schedule 1 of the regulations includes the provision of training to employees as one of the principles of prevention and, in addition to the need to provide information to employees, Regulation 8 also identifies the need to provide training to those who are identified as competent persons to assist with the execution of emergency procedures.

Providing adequate and appropriate supervision

Section 2(2)(c) of the Health and Safety at Work etc. Act 1974 requires employers to provide the necessary information, instruction, training and supervision. While the need for information, instruction and training is reiterated specifically within MHSWR 1999, no specific regulation is dedicated to supervision. That does not reduce its importance, however, Regulation 5 requires employers to have arrangements in place for the effective planning, organisation, control, monitoring and review of the preventive measures. To achieve adequate control it is necessary to ensure adequate supervision, particularly for those who are learning or new to the job. A more explicit, albeit very specific requirement, is defined within Regulation 19(3)9b). As discussed previously, there are defined risks to health to which young persons must not be exposed except strict controls; one of the controls defined for situations where exposure is necessary and justifiable is that the young person must be supervised by a competent person.

Consultation with employees

The need for employers to consult with their employees on matters of health and safety is defined within section 2(4) of the 1974 Act. This is reiterated and expanded within the Safety Representatives and Safety Committee Regulations 1977 and the Health and Safety (Consultation with Employees) Regulations 1996, though not specifically within MHSWR 1999. However, once again, to achieve the requirements of Regulation 5 (specifically the need to organise), it is necessary to establish effective means of communication and consultation to demonstrate a positive approach to health and safety.

REQUIREMENTS OF THE MANAGEMENT OF HEALTH AND SAFETY AT WORK REGULATIONS 1999

Regulation 3 - risk assessment

MHSWR 1999 require employers (and the self-employed) to assess the risk to the health and safety of their employees and to anyone else who may be affected by their work activity. This is necessary to ensure that the preventive and protective steps can be identified to control hazards in the workplace.

Consideration must be given to the population at risk, for example, the number and type of persons who might be exposed to harm, the likelihood of harm and the severity (consequence) of harm. The risk assessment must be suitable and sufficient.

Where an employer employs or is about to employ young persons (under 18 years of age) they must carry out a risk assessment that takes particular account of:

- The inexperience, lack of awareness of risks and immaturity of young persons.
- The layout of the workplace and workstations.
- Exposure to physical, biological and chemical agents.
- Work equipment and the way in which it is handled.
- The extent of health and safety training to be provided.
- Risks from agents, processes and work listed in the Annex to Council Directive 94/33/EC on the protection of young people at work, available in the HSE publication 'Young people at work: A guide for employers'.

Where five or more employees are employed, the significant findings of risk assessments must be recorded in writing (the same threshold that is used in respect of having a written health and safety policy). This record must include details of any employees being identified as being especially at risk.

Regulation 4 - principles of prevention to be applied

Regulation 4 of MHSWR 1999 requires an employer to implement preventive and protective measures on the basis of general principles of prevention specified in Schedule 1 to the Regulations. These are:

- Avoiding risks.
- Evaluating the risks that cannot be avoided.
- Combating the risks at source.

- Adapting the work to the individual, especially as regards the design of workplaces, the choice of work equipment and the choice of working and production methods, with a view, in particular, to alleviating monotonous work and work at a predetermined work-rate and to reducing their effect on health.
- Adapting to technical progress.
- Replacing the dangerous by the non-dangerous or the less dangerous.
- Developing a coherent overall prevention policy that covers technology, organisation of work, working conditions, social relationships and the influence of factors relating to the working environment.
- Giving collective protective measures priority over individual protective measures.
- Giving appropriate instructions to employees.

Regulation 5 - health and safety arrangements

Appropriate arrangements must be made for the effective planning, organisation, control, monitoring and review of preventative and protective measures (in other words, for the management of health and safety). Again, employers with five or more employees must have their arrangements in writing.

Regulation 6 - health surveillance

In addition to the requirements of specific regulations such as COSHH and the Control of Noise at Work Regulations, consideration must be given to carry out health surveillance of employees where there is a disease or adverse health condition identified in risk assessments.

Regulation 7 - health and safety assistance

The employer must appoint one or more competent persons to assist him in complying with the legal obligations imposed on the undertaking. The number of persons appointed depends on the size of the establishment and the range and severity of the risks. If more than one competent person is appointed, then arrangements must be made for ensuring adequate co-operation between them.

The competent person(s) must be given the necessary time and resources to fulfil their functions. This will depend on the size of the undertaking, the risks to which employees are exposed and the distribution of those risks throughout the undertaking.

The employer must ensure that competent person(s) who are not employees are informed of the factors known (or suspected) to affect the health and safety of anyone affected by business activities. Competent people are defined as those who have sufficient training and experience or knowledge and other qualities to enable them to perform their functions. Persons may be selected from among existing employees or from outside though, where there is a suitable person in the employer's employment, that person shall be appointed as the 'competent person' in preference to a non-employee.

Regulation 8 - procedures for serious and imminent danger and for danger areas

Employers are required to set up emergency procedures and appoint competent persons to ensure compliance with identified arrangements, to devise control strategies as appropriate and to limit access to areas of risk to ensure that only those persons with adequate health and safety knowledge and instruction are admitted.

Regulation 9 - contacts with external services

Employers must ensure that clear procedures exist to allow, where necessary, contact can be made with external services. This particularly applies with regard to first-aid, emergency medical care and rescue work.

Regulation 10 - information for employees

Employees must be provided with relevant information about hazards to their health and safety arising from risks identified by the assessments. Clear instruction must be provided concerning any preventative or protective control measures including those relating to serious and imminent danger and fire assessments.

Details of any competent persons nominated to discharge specific duties in accordance with MHSWR 1999 must also be communicated to employees, as should risks arising from contact with other employer's activities ***(see Regulation 11)***.

Before employing a child (a person who is not over compulsory school age) the employer must provide those with parental responsibility for the child with information on the risks that have been identified and preventative and protective measures to be taken.

Regulation 11 - co-operation and co-ordination

Employers who work together in a common workplace have a duty to co-operate to discharge their duties under relevant statutory provisions. They must also take all reasonable steps to co-ordinate measures and inform their respective employees of risks to their health or safety that may arise out of their work.

Regulation 12 - working in host employers' or self-employed persons' undertakings

This extends the requirements of Regulation 11 to include persons working as sole occupiers of a workplace under the control of another employer. Such persons would include those working under a service of contract and employees in temporary employment businesses under the control of the first employer. The controlling employer has a duty to give relevant information to the employer of the persons. The second employer has a duty to give that information to his employees.

Regulation 13 - capabilities and training

Employers need to take into account the capabilities of their employees before entrusting tasks. This is necessary to ensure that they have adequate health and safety training and are capable enough at their jobs to avoid risk. To this end consideration must be given to recruitment including job orientation when transferring between jobs and work departments. Training must also be provided when other factors such as the introduction of new technology and new systems of work or work equipment arise. Training must be repeated periodically where appropriate, be adapted to take account of any new or changed risks to the health and safety of the employees concerned and take place during working hours.

Regulation 14 - employees' duties

Employees are required to use machinery, equipment, dangerous substances, transport equipment, means of production and safety devices in accordance with any training or instructions that have been given. They must also inform their employer (and other employers) of any dangers or shortcomings in the health and safety arrangements, even if there is no risk of imminent danger.

Regulation 15 - temporary workers

Consideration is given to the special needs of temporary workers, in particular to the provision of particular health and safety information such as qualifications required to perform the task safely or any special arrangements such as the need to provide health screening.

Regulation 16 - risks assessment in respect of new or expectant mothers

Where the work is of a kind that would involve risk to a new or expectant mother or her baby, then the assessment required by MHSWR 1999 Regulation 3 should take this into account.

If the risk cannot be avoided, then the employer should take reasonable steps to:

- Adjust the hours worked.
- Offer alternative work.
- Give paid leave for as long as is necessary.

Regulation 17 - certificate from a registered medical practitioner in respect of new or expectant mothers

Where the woman is a night shift worker and has a medical certificate identifying night shift work as a risk then the employer must put her on day shift or give paid leave for as long as is necessary.

Regulation 18 - notification by new or expectant mothers

The employer need take no action until he is notified in writing by the woman that she is pregnant, has given birth in the last six months, or is breastfeeding.

Regulation 19 - protection of young persons

Employers of young persons shall ensure that they are not exposed to risk as a consequence of their lack of experience, lack of awareness or lack of maturity. No employer shall employ young people for work which:

- Is beyond his physical or psychological capacity.
- Involves exposure to agents that chronically affect human health.
- Involves harmful exposure to radiation.
- Involves a risk to health from extremes of temperature, noise or vibration.
- Involves risks which could not be reasonably foreseen by young persons.

This regulation does not prevent the employment of a young person who is no longer a child for work:

- Where it is necessary for his training.
- Where the young person will be supervised by a competent person.
- Where any risk will be reduced to the lowest level that is reasonably practicable.

Regulation 20 - exemption certificates

The Secretary of State for Defence may, in the interests of national security, by a certificate in writing exempt the armed forces, any visiting force or any headquarters from certain obligations imposed by the regulations.

Regulation 21 - provisions as to liability

Employers cannot submit a defence in criminal proceedings that contravention was caused by the act or default either of an employee or a person appointed under Regulation 7 of MHSWR 1999. The implications of this regulation are discussed in greater detail later in this element.

Regulation 22 - exclusion of civil liability

The changes to the legal position regarding breach of statutory duty made by Section 69 of the Enterprise and Regulatory Reform Act (ERRA) 2013 do not apply to claims by new and expectant mothers. The exemptions made by the Health and Safety at Work etc. Act 1974 (Civil Liability) (Exceptions) Regulations 2013 mean that new and expectant mothers will continue to be able to obtain compensation for breach of statutory duty where there has been a breach of relevant health and safety legislation, in particular Regulations 16-17A of MHSWR 1999.

Regulation 22 specifies:

"(1) Breach of a duty imposed by Regulation 16, 16A, 17 or 17A shall, so far as it causes damage, be actionable by the new or expectant mother.
(2) Any term of an agreement which purports to exclude or restrict any liability for such a breach is void."

Figure A3-1: Regulation 22 of MHSWR 1999. *Source: Management of Health and Safety at Work Regulations (MHSWR) 1999.*

Regulations 24-29 - revocations and amendments

Implications of Sections 36 and 37 of HASAWA and Regulation 21 of MHSWR

IMPLICATIONS OF SECTION 36 OF HASAWA 1974

Offences due to fault of other person (Section 36)

Section 36 of HASAWA 1974 provides that if any person commits an offence due to the act or default of some 'other person', that other person shall be guilty of the offence, and a person may be charged with and convicted of the offence, whether or not proceedings are taken against the first-mentioned person. The first person may be a corporate body and the 'other person' may be an employee or local manager within the company, a contractor from another company, insurance engineer or even a health and safety consultant. In other words: if an employer commits an offence as a result of another person, the employer, the other person or both may be prosecuted. The case brought by the HSE against Hooper in 2004 illustrates the point: Hooper, a health and safety consultant, was engaged by a woodworking company to undertake its risk assessments, a task that, it transpires, he completed woefully inadequately. As a result an accident happened resulting in an operator losing part of his finger. The woodworking company was in breach of the requirement to undertake a suitable and sufficient risk assessment, an offence which was due to the default of Hooper, the 'other person'. In this event, by virtue of Section 36, both the woodworking company and Hooper were successfully prosecuted for their part in the breach.

IMPLICATIONS OF SECTION 37 OF HASAWA 1974

Offences by the body corporate (Section 37)

Section 37 of HASAWA 1974 provides that where an offence has been committed by a body corporate and is proved to have been committed with the consent or connivance of, or to have been attributable to any neglect on the part of, any director, manager, secretary of the body corporate or a person who was purporting to act in any such capacity, they as well as the body corporate shall be guilty of that offence and shall be liable to be proceeded against and punished accordingly.

Thus, if a company, Local Authority or other corporate body is in breach of a health and safety requirement, and if it can be proved that the breach of the requirement was due to a Director, Manager or Company Secretary, etc. consenting (allowing something wrong to take place or continue), conniving (perhaps along with others 'turning a blind eye' or even instigating or encouraging something wrong) or neglecting (failing to take care of health and safety) then they, as well as the corporate body, can be prosecuted for the same offence. The Scottish case of Armour v Skeen demonstrated that a Local Authority employee with the role of Director of Roads would, for the purposes of this section, be regarded as representing the body corporate. Despite Armour's claims that it was the policy makers on whom the section placed a duty; this was not an opinion shared by the Judge.

The pre-trial hearing of R v Port of Tilbury Docks also demonstrates that to avoid being guilty of negligence in this context, a director has a responsibility to proactively learn about any health and safety breaches. A section 37 prosecution was brought against Mr Gladding, a senior manager, with the allegation that through his negligence, the practice of an unsafe activity with a fork lift truck continued, ultimately leading to the accident resulting in a young boy being crushed. Gladding argued that he was unaware of the practice so was not in a position to prevent it and as such had no case to answer. The HSE stood by their allegation and the point was referred for a pre-trial hearing. It was confirmed that it was Gladding's business to know what was going on and as such he had a responsibility to find out. The case for the prosecution was therefore confirmed and the full trial of Gladding was allowed to proceed. ***See 'R v P [2007] EWCA Crim 1156' in the 'Relevant decided cases' section.***

IMPLICATIONS OF REGULATION 21 OF MHSWR 1999

Provisions as to liability and defence

Regulation 21 of the Management of Health and Safety at Work Regulations 1999 prevents employers from submitting a defence in criminal proceedings that a contravention was caused by the act or default either of an employee or a person appointed under Regulation 7 of MHSWR 1999 (for Health and Safety Assistance). As such an employer cannot assert a fundamental defence that lays blame on an employee or appointed person.

However, it does not prohibit employers from referring to these parties as part of the employer's defence in showing that they acted outside their training and normal operating procedures, despite reasonably practicable efforts by the employer. One of the primary benefits of this regulation is to prevent employers trying to delegate their duties. Although the cases of R v British Steel and R v Nelson Group predate the introduction of the regulation, the cases help to illustrate its purpose. British Steel were prosecuted for a breach of Section 3 of HASAWA following an accident during the relocation of a platform which was being moved by subcontractors supervised by a British Steel Engineer. Part of their defence focused on the supervisory role of the engineer which they unsuccessfully argued absolved the company of their responsibility. Nelson Group however successfully defended their prosecution: a gas fitter (employee of Nelson Group) left a house in an unsafe state after working on the gas system in the property. Nelson argued that they had provided the necessary training, systems of work and supervision and as such had done everything reasonably practicable, a view with which the Appeal Judges agreed.

Initially, it was anticipated that this legal principle may have been superseded by the introduction of Regulation 21, the wording of which suggests that a reasonably practicable defence may not be permissible. However, despite the prosecutions assertion that Regulation 21 prevented such a defence, Hatton Transport Management (HTM) were able to successfully show that the extent of what is reasonably practicable could still be argued.

In this case, two road workers inexplicably moved a telescopic lighting tower without lowering it first and were electrocuted when they hit an overhead power cable. They had received all of the necessary training and systems of work and HTM argued that the deviation from this system of work was not foreseeable. The judge confirmed that as the duty which HTM were alleged to have breached was a duty qualified by reasonable practicable, a reasonably practicable defence was permissible. ***See 'R v British Steel Plc [1995] IRLR 310', 'R v HTM [2006] EWCA CRIM 1156' and 'R v Nelson Group Services (Maintenance) Ltd [1998] All ER 420' in 'Relevant decided cases' at the end of this Element***.

A3.2 - Approved Codes of Practice and guidance

Purpose, role, structure, application and status of approved codes of practice and HSE guidance notes

PURPOSE

The purpose of an Approved Code of Practice (ACOP) and guidance is to provide advice that has professional standing and weight in order to drive appropriate action by duty holders. The Workplace (Health, Safety and Welfare) Regulations (WHSWR) 1992 have both an ACOP and guidance. The publication containing the ACOP and guidance is called '***Workplace Health, Safety and Welfare. Workplace (Health, Safety and Welfare) Regulations 1992. Approved Code of Practice and Guidance (L24)'***. The ACOP is an interpretation of the legal requirements and states how the duty holder can comply with the law. The guidance gives more practical help in fulfilling the legal requirements.

ROLE

The role of the ACOP and guidance is to provide advice, derived by extensive consultation, in a format that will lead to better understanding and compliance with legal requirements set out in legislation. The role of the ACOP can also be said to be to provide interpretation of legal requirements and advice on how they ***should*** be complied with, though it should also be noted that while an ACOP may define a preferred or recommended means of compliance, an alternative approach which meets the requirements of the statute would be equally acceptable. In comparison the role of the guidance is to provide advice on how the legal requirements ***could*** be complied with by offering detailed and practical advice and options for consideration. This can be demonstrated by looking at Regulation 6 of the WHSWR 1992, relating to ventilation. Regulation 6 (1) states:

> *"Effective and suitable provision shall be made to ensure that every enclosed workplace is ventilated by a sufficient quantity of fresh or purified air."*

Figure A3-2: Regulation 6 of WHSWR 1992. *Source: Workplace (Health, Safety and Welfare) Regulations (WHSWR) 1992.*

The ACOP explains what effective and suitable provision for enclosed workplaces ***should*** involve.

> *"Enclosed workplaces should be sufficiently well ventilated so that stale air, and air which is hot or humid because of the processes or equipment in the workplace, is replaced at a reasonable rate."*

Figure A3-3: ACOP comment on ventilation of enclosed workplaces, WHSWR 1992 Regulation 6. *Source: WHSWR ACOP and Guidance L24.*

The ACOP introduced the phrase 'at a reasonable rate'. The guidance that accompanies the Regulation and ACOP provides advice on what this ***could*** mean.

> *"The fresh-air supply rate should not normally fall below 5 to 8 litres per second, per occupant. When establishing a fresh-air supply rate, consider the following factors:*
>
> - *The floor area per person.*
> - *The processes and equipment involved.*
> - *Whether the work is strenuous."*

Figure A3-4: Guidance advice on fresh-air supply rate, WHSWR 1999 Regulation 6. *Source: WHSWR ACOP and Guidance L21.*

The guidance is designed to be good, practical advice and reduces the barriers employers may perceive in meeting the legal requirements.

STRUCTURE AND APPLICATION

The '***Workplace Health, Safety and Welfare. Workplace (Health, Safety and Welfare) Regulations 1992. Approved Code of Practice and Guidance (L24)'*** is the publication that contains the Workplace (Health, Safety and Welfare) Regulations 1999, the ACOP and the guidance.

The Regulations are numbered sequentially, each followed by the appropriate ACOP and guidance, sometimes just the ACOP and sometimes just the guidance. The information is colour coded in columns and boxes down the left hand side of the page.

The Regulations have white columns and boxes, the ACOP has pale green and the guidance has darker green. Furthermore, the regulations are written in italics, the ACOP in standard font but in bold, and the guidance in standard font. This clear visual differentiation makes it easier to follow and to see at a glance whether a requirement is a legal duty, an interpretation of a legal duty or a practical way to comply with a legal duty.

In the ACOP and Guidance publication, at the end of the WHSWR 1992, there are 2 schedules, two appendices and references for further reading. ***Schedule 1*** relates to ***provisions applicable to factories which are not new workplaces, modifications, extensions or conversions***, and relates to Regulation 10 and 20. ***Schedule 2, Repeals and Revocations*** shows where the WHSWR 1992 have amended other legislation.

STATUS OF ACOPS

Approved Codes of Practice (ACOPs) are provided for in HASAWA 1974, in Section 16. They are instruments intended to give people more detailed information on compliance with their duties under the law. As such they have been given careful consideration in their development and only receive approval by the Health and Safety Executive (HSE) after consent is obtained from the Secretary of State.

Section 17 of HASAWA 1974 sets out how ACOPs may be used in criminal proceedings.

> *"(2) Any provision of the code of practice which appears to the court to be relevant to the requirement or prohibition alleged to have been contravened shall be admissible in evidence in the proceedings; and if it is proved that there was at any material time a failure to observe any provision of the code which appears to the court to be relevant to any matter which it is necessary for the prosecution to prove in order to establish a contravention of that requirement or prohibition, that matter shall be taken as proved unless the court is satisfied that the requirement or prohibition was in respect of that matter complied with otherwise than by way of observance of that provision of the code."*

Figure A3-5: Use of ACOP in criminal proceedings. *Source: HASAWA 1974, Section 17.*

Whilst failing to follow an ACOP is not in itself an offence, the accused would have to show that the measures they took were at least as effective.

It is, therefore, strongly advisable to follow the provisions of an ACOP where they apply. An ACOP is 'quasi-law', which means although it is not itself law, it may be used in evidence against someone or as a defence to show that the defendant did comply with the law.

Section 40 of HASAWA 1974 allows for the defendant to prove that they did everything that was practicable or reasonably practicable, and they can do this by showing that they did everything in accordance with the ACOP.

STATUS OF HSE GUIDANCE MATERIAL

HSE guidance material is produced in the form of guidance notes, other advisory literature, website guidance and guidance that accompanies regulations. This is produced in order to provide information on how legal requirements ***may*** be complied with and on best practices. Guidance provided in regulations is designed to amplify and clarify further the advice provided in the ACOP.

They are not mandatory, as such guidance is provided principally to be useful to the reader rather than to establish strict lines of action; the guidance may not be appropriate or easily applicable to every work situation in the form given. As such it has no legal standing.

However, if an offence were alleged, compliance with HSE guidance would tend to demonstrate the application of good practice to a court.

Statutory procedures for making approved codes of practice

Section 16 of HASAWA 1974 sets out the statutory procedure for making approved codes of practice. Authority is vested in the HSE to:

- Approve and issue such codes of practice (whether prepared by it or not) as in its opinion are suitable for that purpose.
- Approve such codes of practice issued or proposed to be issued otherwise than by the Executive as in its opinion are suitable for that purpose.

Approval of a code of practice by the HSE must be with the consent of the Secretary of State responsible for health and safety at work. Before presenting the proposed ACOP to the Secretary, the HSE must consult with any relevant government departments and other agencies, in particular the Public Health England for any codes relating to such agents as electromagnetic radiation or any other specific departments as instructed by the Secretary of State.

The executive may revise an ACOP, or, as has been seen recently with the now obsolete ACOPs associated with MHSW 1999 and CDM 2015, withdraw the approval of a code of practice; the same process of consultation and consent from the Secretary of State must be followed for this process as for its original approval. Whenever an ACOP is approved, revised or withdrawn, the Executive must issue a written notice which identifies the code and date when it will take effect.

As can be seen by the wording of Section 16 of HASAWA 1974 the HSE can establish an ACOP from whatever source it wishes - it does not have to be associated with specific regulations. In the past HSE has established an ACOP for rider-operated lift trucks, though no specific regulation relates. Section 16(1)(b) allows the executive to approve codes or sections of codes prepared by other organisations; as an example it has approved a British Standard associated with personal protective equipment.

A3.3 - Enforcement of health and safety law

Identification of authorities empowered to enforce health and safety legislation

Section 18(1) of HASAWA 1974 places the primary responsibility for the enforcement of health and safety legislation on the Health and Safety Executive, though sub-section 18(2) allows the Secretary of State to make regulations that set out the extent to which the HSE and local authorities are responsible for enforcement of relevant statutory provisions in the workplace.

The Secretary of State may prescribe and assign or transfer enforcement responsibilities between the HSE/Local Authority as necessary. The enforcement of health and safety of railway undertakings was transferred from the HSE to the Office of Rail Regulation and its enforcing officers under Section 2 of the Railways Act (RA) 2005.

Division of responsibilities between enforcing authorities

The division of responsibilities between enforcing authorities is principally made under the Health and Safety (Enforcing Authority) Regulations (HSEAR) 1998. In general the HSE is responsible for enforcement in premises such as factories, fairgrounds, quarries, mines, radio/television/film production and construction sites as well as the enforcement of specific legislation, such as Section 6 of HASAWA 1974. The Local Authority is generally responsible for premises such as offices, launderettes, most shops, premises where food is prepared and distribution premises such as warehouses.

The HSE is responsible for enforcement with regard to local authority activities and the local authority is responsible for enforcement in HSE offices. The Office of Rail Regulation is responsible for health and safety of railway undertakings. Following the report into the failings by Mid Staffordshire NHS Foundation Trust, since 1st April 2015 the Care Quality Commission (CQC) is responsible for the inspection and enforcing the requirements for safety (and quality of care) of all patients and service users in CQC-registered services.

However, the responsibility remains with the HSE (in hospitals and nursing homes) or Local Authority (for residential care) for inspection and enforcement in relation to the safety of employees, visitors and contractors.

Powers of enforcing authorities and their inspectors (HASAWA 1974 Sections 20 and 25)

Inspectors are those people appointed by enforcing authorities under Section 19 of HASAWA 1974; they are issued with an instrument of appointment a copy of which may be used, in a similar way to a police officer's warrant card, to demonstrate their position and authority. Section 20 of HASAWA 1974 grants health and safety inspectors a number of powers to allow them to undertake their duties.

The powers are summarised below:

- The right of entry (without having to gain permission from a Magistrate). An inspector may enter any premise which he considers to be a place of employment at any reasonable time of day or night.
- An inspector may take with him a police constable if he has reasonable cause to think that there would be any serious obstruction to prevent him from carrying out his duties.
- An inspector can also take with him any other person who has been authorised by the HSE/Local Authority.
- Equipment and materials required to carry out tests etc. may also be taken into places of employment.
- An inspector may require premises, parts of premises or anything in the premises to be left undisturbed until any examination or investigation is completed.
- An inspector has the right to take measurements, photographs and make sound recordings as he thinks fit.
- Section 20 allows an inspector to obtain samples of articles and substances, either in the premises or in the atmosphere, or in the vicinity of the premises.
- An inspector may order equipment, plant, substances, etc. to be dismantled or destroyed if they are a cause of immediate danger.
- An inspector may take into possession the articles or substances so that they may examine them, prevent them from being tampered with or retained for use as evidence in any potential prosecution.
- An inspector has the right to require the production of, and take copies of, books and documents which are relevant to his work.
- An inspector has the right to examine any person, either alone or with a second person, as he thinks fit, and to require that person to provide a written statement of fact. It should be remembered however that any information obtained in this way may only be used in future proceedings in the form of a witness statement. Where the information potentially incriminates the person giving it, for the information to be permissible for use at a later date, the inspector must suspend the interview and reconvene under the auspices of the Paces and Criminal Evidence Act 1984 (PACE) so that the individual is aware that the information being given is being given under caution, and as such the right to remain silent applies.
- An inspector may request the use of any facilities and for the assistance from the representative of the business to allow him to undertake his task.

To further the power defined within Section 20 to require an article or substance to be dismantled or destroyed, under Section 25 he may himself seize and render harmless an article or substance he believes to be of imminent danger of serious personal injury. In such circumstances a sample will be retained if possible to allow identification and a written report detailing the particulars of the seizure be written and given to a responsible person representing the business. In addition to the powers discussed above, an inspector has the authority to conduct legal proceedings before a Court of Summary Jurisdiction (Magistrates' Court) in England and Wales, and serve improvement and prohibition notices under Sections 39, 21 and 22 respectively.

Obligations of enforcing officers: duty to give information to employees or their representatives

Under Section 28 (8) of HASAWA 1974 inspectors are required to provide information to employees or their representatives where this is necessary to keep them adequately informed about health and safety matters affecting them. The information the inspector is required to supply relates to both factual information regarding their workplace and any actions the inspector proposes to take in respect of their work premises. Information provided to employees or their representatives must also be given to employers. Under section 28 (8) of HASAWA 1974 inspectors may furnish to a person who appears to them to be likely to be a party to any civil proceedings arising out of any accident, occurrence, situation or other matter, a written statement of relevant facts observed by them in the course of exercising their powers.

DUTY NOT TO DISCLOSE INFORMATION (HASAWA 1974 SECTION 28)

Section 27 of HASAWA 1974 empowers inspectors (with the Secretary of State's consent) to require the provision of specified information in order for them to carry out its functions. Under section 28 this information, which must only be used for the purposes of the inspector's functions, must remain confidential - particularly the disclosure of trade secrets. Information must not be disclosed without the consent of the person who provided it. There are, however, some exceptions, including the provision of information to public bodies such as the police to enable them to carry out their statutory functions and release of information held under the Freedom of Information Act (FIA) 2000.

Offences and maximum penalties under the law (HASAWA 1974 Section 33)

OFFENCES

Section 33 of HASAWA 1974 outlines the provisions as to offences under relevant health and safety statutory provisions. The main offences are:

- To fail to discharge a duty to which he is subject by virtue of Sections 2 to 7.
- To contravene Section 8 or 9.

- To contravene any health and safety regulations or any requirement or prohibition imposed under any such Regulations (including any requirement or prohibition to which he is subject by virtue of the terms of or any condition or restriction attached to any licence, approval, exemption or other authority issued, given or granted under the regulations).
- To contravene any requirement imposed by or under regulations under Section 14 or intentionally to obstruct any person in the exercise of his powers under that Section.
- To contravene any requirement imposed by an inspector under Section 20 or 25.
- To prevent or attempt to prevent any other person from appearing before an inspector or from answering any question to which an inspector may by virtue of Section 20(2) require an answer.
- To contravene any requirement or prohibition imposed by an improvement notice or a prohibition notice (including any such notice as modified on appeal).

A number of other offences are also identified by Section 33:

- The obstruction of an inspector.
- Making a false statement.
- Falsifying records.
- Impersonation of an inspector.

MODE OF TRIAL AND PENALTIES

The Health and Safety Offences Act (HSOA) 2008 amended Section 33 and Schedule 3A of the Health and Safety at Work etc. Act (HASAWA) 1974 relating to modes of trial and maximum penalties. HSOA 2008 raised the maximum penalties available to the courts in respect of certain health and safety offences. It introduced imprisonment and increased the maximum fine for certain offences. It raised the maximum penalties that could be imposed on summary trial for breaching health and safety regulations from £5,000 to £20,000.

Section 85 of the Legal Aid, Sentencing and Punishment of Offenders Act 2012 was then brought into force on 12th March 2015 by the Legal Aid, Sentencing and Punishment of Offenders Act 2012 (Fines on Summary Conviction) Regulations 2015.

This had the effect of increasing the level of most fines available for magistrates' courts to an unlimited fine and resulted in the maximum fine available on summary conviction under Section 33 of HASAWA 1974 becoming an unlimited fine.

Offence	***Mode of trial***	***Penalty on summary conviction***	***Penalty on conviction on indictment***
An offence under Section 33(1)(a) consisting of a failure to discharge a duty to which a person is subject by virtue of Sections 2 to 6.	Summary or indictable.	Imprisonment for a term not exceeding 6 months*, or an unlimited fine, or both.	Imprisonment for a term not exceeding two years, or an unlimited, or both.
An offence under Section 33(1)(a) consisting of a failure to discharge a duty to which a person is subject by virtue of Section 7.	Summary or indictable.	Imprisonment for a term not exceeding 6 months*, or an unlimited fine, or both.	Imprisonment for a term not exceeding two years, or an unlimited, or both.
An offence under Section 33(1)(b) consisting of a contravention of Section 8.	Summary or indictable.	Imprisonment for a term not exceeding 6 months*, or an unlimited fine, or both.	Imprisonment for a term not exceeding two years, or an unlimited, or both.
An offence under Section 33(1)(b) consisting of a contravention of Section 9.	Summary or indictable.	An unlimited fine.	An unlimited fine.
An offence under Section 33(1)(c). To contravene any health and safety Regulations.	Summary or indictable.	Imprisonment for a term not exceeding 6 months*, or an unlimited fine, or both.	Imprisonment for a term not exceeding two years, or an unlimited, or both.
An offence under Section 33(1)(d). To obstruct any person in the exercise of his powers under Section 14 of HASAWA 1974.	Summary only.	An unlimited fine.	
An offence under Section 33(1)(e) - to contravene any requirement imposed by an inspector under Section 20 or 25. (F) - to prevent	Summary or indictable.	Imprisonment for a term not exceeding 6 months*, or an unlimited fine, or both.	Imprisonment for a term not exceeding two years, or an unlimited, or both.

Offence	*Mode of trial*	*Penalty on summary conviction*	*Penalty on conviction on indictment*
any person from appearing before an inspector, or (G) - to contravene an improvement or prohibition notice.			
An offence under Section 33(1)(h). To intentionally obstruct an inspector.	Summary only.	Imprisonment for a term not exceeding 6 months** (in England and Wales) or 6 months* (in Scotland), or an unlimited fine, or both.	
An offence under Section 33(1)(i). Failure to provide information.	Summary or indictable.	An unlimited fine.	An unlimited fine.
An offence under Section 33(1)(j). Disclosing restricted information.	Summary or indictable.	Imprisonment for a term not exceeding 6 months*, or an unlimited fine, or both.	Imprisonment for a term not exceeding two years, or an unlimited, or both.
An offence under Section 33(1)(k) - to make a false statement. (I) - to make a false entry in a register, or (M) - to forge or use forged documents.	Summary or indictable.	Imprisonment for a term not exceeding 6 months*, or an unlimited fine, or both.	Imprisonment for a term not exceeding two years, or an unlimited, or both.
An offence under Section 33(1)(n). To pretend to be an inspector.	Summary only.	An unlimited fine.	
An offence under Section 33(1)(o). Failure to comply with a court order.	Summary or indictable.	Imprisonment for a term not exceeding 6 months*, or an unlimited fine, or both.	Imprisonment for a term not exceeding two years, or an unlimited, or both.
An offence under the existing statutory provisions for which no other penalty is specified.	Summary or indictable.	Imprisonment for a term not exceeding 6 months*, or an unlimited fine, or both.	Imprisonment for a term not exceeding two years, or an unlimited, or both.

Figure A3-6: Mode of trial and maximum offences under the Health and Safety at Work etc. Act 1974. *Source: HSE.*

**6 months imprisonment will increase to 12 months on commencement of Section 154(1) of the Criminal Justice Act (CJA) 2003.*

***Will increase to 51 weeks on commencement of Section 281 (5) of the Criminal Justice Act (CJA) 2003.*

Table current at time of print.

The amendment to Schedule 3A of HASAWA 1974 and the maximum prison sentences allowed on summary trial is influenced by the Criminal Justice Act (CJA) 2003 and commencement orders under it. For example, provision was made in the CJA 2003 for a maximum penalty for imprisonment on summary trial of 12 months; however, at the time of going to print the commencement order for Section 154 (1) of the Criminal Justice Act (CJA) 2003 had not been made, which means the maximum term of imprisonment for summary offences is six months.

Offences for which imprisonment is, and is not, a form of sanction

Since the introduction of the Health and Safety Offences Act (HSOA) 2008 imprisonment for health and safety offences is now possible for a wider number of offences on summary conviction.

Only the following do not have the potential for imprisonment conviction:

- A breach of section 9 of HASAWA 1974 (relating to charging employees for health and safety provisions).
- Obstructing an inspector exercising their power under Section 14 of HASAWA 1974 (with regard to formal enquiries).
- Failure to provide information.
- Pretending to be an inspector.

Implications of amendments to Section 3 of HASAWA 1974 by the Deregulation Act (DA) 2015

One of the Government's responses to the report by Professor Loftstëdt identifying that the regulation of self-employed persons was considerably more stringent than elsewhere within the European Union, was to exempt up to 1.8 million such persons from the duties defined by Section 3 of the Health and Safety at Work etc. Act 1974.

The Deregulation Act (DA) 2015 rewords subsection 3(2) which now reads: 'It shall be the duty of every self-employed person who conducts an undertaking of a prescribed description to conduct the undertaking in such a way as to ensure, so far as is reasonably practicable, that he and others (not being his employees) who may be affected thereby are not thereby exposed to risks to their health and safety'. The undertakings of a 'prescribed description' are defined with the Health and Safety at Work etc. Act 1974 (General Duties of Self-Employed Persons) (Prescribed Undertakings) Regulations 2015.

The schedule to the Regulations identify six specific industry sectors which are identified as of a prescribed description namely:

- Agriculture.
- Asbestos.
- Construction.
- Gas.
- Genetically modified organisms.
- Railways.

As this list is contained within a schedule, it may amended following the pre-planned review of the effectiveness of the regulations 5-years from its commencement.

In addition to the prescribed undertakings listed above, like many other health and safety regulations, it overcomes the problem associated with prescriptive legislation by also stating that the exemption will not apply if the undertaking presents a risk to those other than the self-employed person (or his employees) even if not listed within the schedule. By virtue of this changed self-employed persons within other sectors where their activities do not present a risk to themselves or others no longer have a duty under Section 3.

The principles of enforcement with reference to the HSE's 'Enforcement policy statement' (HSE41)

To help ensure a consistent approach to enforcement across all areas and sectors, the HSE produced Enforcement Guide provides health and safety inspectors details the range of enforcement actions available to them and how they should be best applied. As part of the Enforcement Guide, the HSE issued an Enforcement Policy Statement, which set out the general principles and approach expected to be followed by the enforcing authorities (mainly HSE and local authorities). All inspectors who take enforcement action are required to follow the principles of the Statement.

Quoting from the Statement:

> *"The enforcing authorities have a range of tools at their disposal in seeking to secure compliance with the law and to ensure a proportionate response to criminal offences."*

Figure A3-7: Enforcement action. *Source: HSE enforcement guide.*

Other than a proportionate approach, inspectors are required to appropriately target enforcement, adopt a consistent approach and ensure transparency and accountability of all actions taken. It goes on to say that inspectors may offer duty holders advice, both verbal and in writing, which may include a warning that they are, in the inspector's opinion, not complying with the law. They may issue improvement and prohibition notices, withdraw approvals, vary licence conditions or exemptions, issue formal cautions (only in England and Wales), and they may prosecute (or report to the Procurator Fiscal with a view to prosecution in Scotland).

The policy also recognises that it is impossible and inappropriate to investigate all reported accidents. Instead the policy requires consideration to be given to the scale of the potential or actual harm, severity of the possible breach, enforcement priorities, practicality of achieving results and the wider relevance of the investigation. Careful consideration must be given during the investigation of workplace deaths as to whether a manslaughter (or corporate manslaughter/homicide) prosecution may be relevant.

In such cases, though investigation of potential health and safety offences will be conducted by the health and safety inspector, the manslaughter considerations would be investigated by the police who would ultimately recommend to the Crown Prosecution Service the suitability for prosecution. As such a co-ordinated approach with the police, in line with the 'Work related deaths: practical protocol', will be necessary.

Quoting from the statement regarding the purpose and method of enforcement:

> *"The ultimate purpose of the enforcing authorities is to ensure that duty holders manage and control risks effectively, thus preventing harm. The term 'enforcement' has a wide meaning and applies to all dealings between enforcing authorities and those on whom the law places duties (employers, the self-employed, employees and others)."*

Figure A3-8: Purpose and method of enforcement. *Source: HSE enforcement guide.*

Proportionality

The importance of a proportionate approach to health and safety by all interested parties has been widely advocated by the HSE for several years. The policy statement identifies eight principles to be considered by inspectors to ensure that they take a proportionate approach during enforcement. It recognises the need for businesses to both comply but also grow and as such not only the seriousness of any breach but also the economic impact of any proposed enforcement will be considered. In return, duty holders are expected to place an emphasis on managing significant risks. Inspectors are required to take account of how far short of expectations a duty holder falls and the consequential risk. Consideration will need to be given to the qualification of any duty breached and unless there is a gross disproportion to the risk and cost of control duty holders must accept the need to incur costs to reduce risks and comply with duties or expect enforcement action to be brought. In some scenarios where the risk cannot be reduced but remains serious, it may be decided that the only option is to prohibit such a situation to continue. However, in contrast, a sensible approach should be made in the initial stages following an emergency to give duty holders the opportunity to respond appropriately.

Consistency

The policy emphasises that consistency does not mean uniformity: a similar approach in similar circumstances is adopted, also recognising the importance of fair treatment and sustaining economic growth. Although consistency is important multiple variations can be found in two seemingly similar situations; as such discretion by the inspector to take into account all factors must be allowed. The enforcement management model and peer review are therefore valuable tools to ensuring that approaches remain consistent.

On occasions there will be conflict between the approach and requirements of health and safety enforcement and that of other agencies. In such cases, all regulators involved must work together to determine a suitable outcome.

Transparency

Although it is important to ensure that duty holders are advised what action they need to take, it is equally appropriate to ensure that they realise what they need not do (or indeed should not). Enforcing officers may rely on the raft of guidance information produced by the HSE and available on the website. In the event that non-compliance is identified, enforcing officers must explain the action they are to take, the reasons behind that decision, the actions required to achieve compliance, potential consequences if they fail to comply and, ideally in discussion with the duty-holder, a suitable timescale for this to be achieved. They must also make it clear to duty holder what must be achieved to ensure compliance and what would be good practice to prevent unnecessary burdens on businesses. Within the legal constraints in relation to disclosure of information, wherever possible enforcing officers should keep employees, their representatives and, where appropriate families, involved in the process.

Choice of enforcement option linked to the HSE's 'Enforcement Management Model (EEM)'

While the policy statement defines the principles to which enforcing officers should follow, it makes reference to the enforcement management model to provide a more detailed explanation to help achieve the strived for consistent approach. The EMM is designed to be used by inspectors in all regulatory action, though the model uses a serious of linear flow charts which it readily acknowledges will require discretion by the enforcing officer when considering more complex scenarios. The model also forms the basis of routine monitoring of inspectors' performance. Although the primary purpose of the EMM is for reference by inspectors, there are benefits for duty holders to be familiar with its content as it provides an understanding of the approach they can expect inspectors to take. The process requires the inspector to consider a serious of factors relevant to the ultimate decision of the action to be taken. Consideration must be given to the priorities for action, the risk of serious injury and the risk gap (how far short of the expected standard).

An initial assumption can then be made as to the most probable action to be taken. While in some circumstances it may be decided to opt for prosecution at this early stage, more common approaches include provision of advice or the issue of improvement or prohibition notices. The option of a formal caution is available to inspectors in some circumstances though this is not addressed within the EMM. Having established the provisional plan of action, inspectors are required to consider 'duty-holder factors' including past performance in terms of incidents and compliance, whether the minimum legal standard is being missed for commercial gain and general standard of compliance and willingness to improve. Before a final decision is made the strategic factors must also be considered: whether the proposed action is in the public interest, whether the action will protect vulnerable groups and/or lead to sustained compliance as well as the broader socio-political impact.

Consequences of material breach

FEE FOR INTERVENTION

The Fee for Intervention was introduced in October 2012, by virtue of Regulations 23-25 of the Health and Safety (Fees) Regulations 2012. This stemmed from the government's approach to health and safety reforms defined within the 2011 document 'Good health and safety, good for everyone' which encouraged a change in the culture towards health and safety, one principle being to shift the cost of enforcement from the public purse to the non-compliant duty-holder. The regulations state that where an inspector identifies a 'Material Breach', the duty-holder will be required to cover the costs resulting from the associated investigation and enforcement. A Material Breach is defined as a contravention of health and safety law which necessitates a notice in writing to be issued. As such technical or trivial breaches for which verbal advice would suffice would not be subject to the fee.

However, where a written notification of contravention is sent, an improvement or prohibition notice is served or where a prosecution is to be brought, the fee will be payable for all work associated with the alleged breach. This would include the time charged at an hourly rate (of £129/hour at the time of writing) for the time taken for carrying out site visit(s), preparation of written notices, statements, reports and the time dedicated by support staff involved in the case. To ensure consistency and inappropriate charging of the fees, inspectors must ensure that their enforcement approach is consistent with that defined within the enforcement management model. Any time taken by the HSE for the preparation of a prosecution case, once the charge has been made or for defending an appealed notice would not be chargeable. Indeed, where a notice is overturned, any fees paid would be refundable. Similarly, self-employed persons whose work doesn't put other people at risk where there fees are already payable to the HSE (for example, those who deliberately work with certain biological agents) will not be required to pay a fee for intervention.

On issue of the written notice by the inspector, as well as identifying the legislation, which in the inspector's opinion has been breached and the justification of that opinion, attention must also be drawn to the fact that a fee is payable. The invoice is issued separately and must clearly detail a breakdown of the work undertaken and the time allocated to each activity. The invoice is payable by the duty-holder within the 30-days unless there is a dispute, in which case appeals must be presented to the HSE for consideration within 21-days of the invoice.

Disputes may arise if the duty-holder believes that they are not guilty of the alleged material breach and/or if they are of the opinion that the time charged for is excessive. Initially, businesses are encouraged to query the invoice with the HSE but where still in disagreement and dispute may be raised. HSE staff plus an individual representative considers the case. It should be noticed that for disputes which are not upheld, the fee payable will be increased to include the time involved in handling the dispute. Unresolved invoices are then payable within 10-days of notification (unless the initial 30-day payment period has not expired).

At present, the fee is only payable by duty-holders for whom health and safety is enforced by HSE (i.e. businesses where the responsibility for enforcement lies with the Local Authority etc. will not be issued with a fee for intervention invoice. The regulations specify a validity period of five years. However the initial review conducted one year after its introduction concluded that the initiative had been a success and it is anticipated that legislation will be passed for it to remain in place beyond 2017. In addition the review recommended it being extended to beyond just the HSE and as such it may be that in future years the local authorities too will be able to recover their costs in this way.

ENFORCEMENT NOTICES - TYPES, PURPOSE, STATUS, CONDITIONS FOR BEING SERVED, GROUNDS FOR APPEAL, APPEAL PROCEDURES, EFFECTS OF APPEAL (HEALTH AND SAFETY AT WORK ACT SECTIONS 21-24 AND 39)

Enforcement notices

Before issuing an enforcement notice an enforcing officer will have considered alternative enforcement action, like providing comment in oral and written form. There are two types of enforcement notice, an improvement notice and a prohibition notice. It should be remembered that both notices are a confirmation of non-compliance with legislation. When choosing to use a notice the enforcing officer has opted not to prosecute for the offence at that point in time, though failure to comply with the notice may lead to prosecution for the original breach at a later date. When enforcing officers serve a notice they have a duty to inform a relevant employee health and safety representative of the circumstances, in addition to the person the notice is served on.

Prohibition and improvement notices are in a standard format, which is on a form completed on site by the enforcing officer. The top copy of the notice is given to the person who is, in the enforcing officer's opinion, in breach and the carbon copy is retained. Both types of notice contain a schedule, where a written description of the steps to be taken to comply with the enforcing officer's requirements, and in particular outlines of practical guidance, relevant codes of practice or other accepted standards may be included. Failure to comply with an enforcement notice puts the duty holder in breach of the notice, and consequently in breach of Section 33 of HASAWA 1974.

The duty holder committing this breach is guilty of two offences:

- Failing to comply with the enforcement notice.
- The original breach of the particular legislation in question.

The responsibility to confirm compliance with a notice remains with the employer; the enforcing officer may or may not return to determine compliance with the notice. Notices are not 'lifted' by the enforcing officer, but the act of being in compliance satisfies the notice. The enforcing officer has the power to withdraw or amend a notice and to extend the time period for compliance, where relevant.

Improvement notices (HASAWA 1974 Section 21)

Section 21 of HASAWA 1974 allows inspectors to use the improvement notice procedure. It is aimed at situations where there is not a serious risk, but a breach of legislation relates.

Enforcing officers can serve an ***improvement notice*** if they are of the opinion that there:

- Is a ***contravention*** of one or more of the relevant statutory provisions.
- Has been a contravention of one or more of those provisions in circumstances which make it likely that the contravention will continue to be repeated.

Some examples of where an improvement notice may be used include:

- Incomplete or no health and safety policy.
- Incomplete or no general risk assessment.
- Inadequate or no general training of managers.
- Restricted walkways or trailing cables in offices.
- Storage of oils causing risk of slipping.

The effect of the notice is to require a specified improvement to take place, within a specified time, in order to bring the situation back into legal compliance.

In the improvement notice the enforcing officer must:

- State that the enforcing officer is of the opinion that there is or has been a contravention.
- Specify the provisions, in their opinion, which are contravened.
- Give particulars of the reasons for their opinion.
- Specify a period of time within which the person is required to remedy the contravention.

Section 23 of HASAWA states that an inspector may, though need not, provide the duty-holder on whom the notice is served guidance as to how compliance with the notice may be achieved. This may make reference to one or more Approved Code of Practice and/or offer more than one option that may be taken.

An improvement notice allows a period of time for improvements to be carried out which should be realistic for the scale and complexity of the corrective action required. To ensure sufficient time for the employer to appeal; the minimum timescale for any Improvement Notice is 21 days allowed to appeal. If the employer felt the time allowed was inadequate this may be grounds for appeal.

Prohibition notices (HASAWA 1974 Section 22)

If enforcing officers are of the opinion that a workplace activity involves, or will involve, the ***risk of serious personal injury***, they can serve on the person responsible for the activity a ***prohibition notice***. The notice will usually take immediate effect and require the activity to cease. In circumstances where the enforcing officer considers the immediate stopping of the activity to be inappropriate (it may present its own risk) the notice can come into effect at a fixed date. A notice can relate to a system of work, equipment, workplace or a person.

A prohibition notice must:

- State that the enforcing officer is of the opinion that there is a risk of serious personal injury.
- Specify the matters which create the risks.
- Direct that the activities must not be carried out, unless the matters are remedied.

Examples of where a prohibition notice may be used:

- Unguarded machinery.
- Incomplete scaffold.
- Untrained personnel using high risk equipment, for example, rough terrain, fork lift truck or a person hoist.
- Inadequate procedures for entry into a confined space, for example, sewers.

Appeal against enforcement notices

If identified as appropriate, Section 23 allows the enforcing officer to withdraw or extend a notice before the end of the specified period.

Formal appeals are provided for by section 24 of HASAWA 1974. A person on whom either type of notice is served can appeal to an Employment Tribunal (ET) within 21 days from the date of service of the notice. The tribunal can extend this period on written application that it was not reasonably practicable for the appeal to be brought within 21 days. A copy of the standard appeal form is always issued with the notice.

An appeal has the effect of suspending an improvement notice until a decision has been made by the ET or the appeal is withdrawn. In the case of a prohibition notice, the notice remains in force, unless requested by the person making the appeal and directed by the Employment Tribunal at the time of the application for appeal.

When the appeal is heard, the ET will consider the facts of the appeal and may make a number of responses in view of their findings. The options are to affirm the notice as originally defined, amend the notice (for example, the time to comply with an improvement notice) or cancel it. Appeal from an Employment Tribunal's decision regarding appeal against an enforcement notice is to the High Court.

Options for enforcement action with reference to HSE Enforcement Guide

SIMPLE CAUTIONS

A formal caution, now referred to as a 'simple' caution, can be used by non-health and safety enforcement bodies and has been adopted for health and safety by local authorities. The HSE in England and Wales uses it as an additional procedure for dealing with certain offenders in limited circumstances where an offence has been committed but a prosecution is not appropriate, perhaps as it would not be regarded as within the public interest. However, a caution may be only used where prosecution would be an option, i.e. where there would be a realistic prospect of conviction if the prosecution option had been chosen.

"A simple caution is a statement by an Inspector that is accepted in writing by the duty holder, that the duty holder has committed an offence for which there is a realistic prospect of conviction. A formal caution may only be used where a prosecution could be properly brought. 'Formal cautions' are entirely distinct from a caution given under the Police and Criminal Evidence Act by an inspector prior to asking questions of a suspect concerning an alleged offence. Enforcing authorities should take account of current Home Office guidelines when considering whether to offer a formal caution."

Figure A3-9: Formal cautions. *Source: HSE Enforcement Policy Statement HSE41.*

The simple caution is an option rarely used by the HSE and unlikely to be used for more serious offences. It should not be seen as letting the accused off; evidence is gathered in the same way and to the same extent as when a prosecution is to be taken. Once the evidence shows that a prosecution is appropriate, if specific circumstances exist, (perhaps illness of the accused or at the request of the victim) a caution may be chosen instead. The offender must admit the offence and, after seeking legal advice if required, agree to accept the written caution; failure to accept the caution is likely to result in a prosecution.

A simple caution should not normally be considered an option if one has been issued and accepted previously. If following the acceptance of the caution the breach continues or is repeated, prosecution will normally follow.

For completeness and clarity, it should be noted that HSE is not authorised to issue a conditional caution which was introduced by the Criminal Justice Act 2003.

Prosecution options

Section 39 of the Health and Safety at Work etc. Act 1974 allows England-based HSE inspectors to bring a prosecution of a health and safety offence. Within Scotland this responsibility lies with the Crown Office and Procurator Service (COPFS) who may take guidance from the HSE inspector or conduct its own investigation before deciding whether to proceed with the prosecution. The enforcement policy states that prosecution will normally be only recommended for cases where the evidence available offers a realistic prospect of conviction and the prosecution is in the public interest.

The process begins with the prosecuting officer presenting an Information to the Justice's Clerk detailing the facts of the case which is then issued to the accused in the form of a summons. Criminal offences can be categorised according to their seriousness with three types of offence. Reference to the table showing offences, their mode of trial and maximum penalties illustrates which offences fit into these.

Summary offence

Section 33 defines several of the more minor offences as being summarily only an example being the obstruction of an enforcing officer; such cases will be heard within the Magistrate's Court. There is a 6-month time limit (from the date that the alleged offence was committed) in which summary-only cases may be presented to the court to be allowed unless extenuating circumstances exist.

Indictable offence

More serious offences must be tried in higher courts with the authority to provide penalties in-keeping with an offence of such magnitude. Manslaughter cases, for example, will be heard within the Crown Court before a judge and jury. Although there is no time limit for bringing an indictable (or either-way) offence, it is expected that cases will be progressed and presented as quickly as possible. The court has the right to dismiss a summons on the grounds of 'abuse of process' if the time lapsed is considered excessive. These cases are first considered by the lower court in their capacity of a Court of First Instance, such as a Magistrates' Court, in order to establish a 'prima facie' case that is that there is a case worthy to proceed. Since the Health and Safety (Offences) Act 2008 amended Section 33 of the Health and Safety at Work etc. Act 1974, there are no indictable only offences under the jurisdiction of the Health and Safety Executive.

Hybrid or 'either way' offence

The majority of health and safety are triable 'either way'. As such they can be tried by a lower court or higher court, for example, a Magistrates Court or Crown Court, depending on circumstances. As with all cases, they begin their journey within the lower court for a plea to be entered and a decision made as to whether it is to be referred to the higher court. At present all defendants have a right to a trial by jury and as such if they request that the case be heard by the higher court, this request must be allowed. Although cases are sent to the higher court if the lower court considers it does not have sufficient powers of sentence, since the increased penalties introduced by the Health and Safety (Offences) Act 2008 and the Section 85 of the Legal Aid, Sentencing and Punishment of Offenders Act 2012, fewer cases need to be referred for this reason. However, where the case is considered to be unusually serious or complex, the higher court will be used.

On the conviction of directors and similar persons on indictment for offences in connection with the management of a company (Sections 36 and 37 of HASAWA 1974), the courts may also make a disqualification order by virtue of the Company Directors Disqualification Act (CDDA) 1986, Sections 1 and 2. A conviction in a lower court could lead to a maximum of five years of disqualification or in the higher courts a maximum of 15 years of disqualification.

Implications of the Social Action, Responsibility and Heroism Act 2015

The Social Action, Responsibility and Heroism Act 2015 was introduced to protect those acting for the general good even though outside of defined procedures. There has been concern for some time, often unfounded, that if an individual acts on their initiative in an emergency situation which either puts themselves at risk deviating from agreed protocols as they are unworkable in the specific scenario, they may be vulnerable to future criminal or civil liability. This Act defines when such legal action would not be appropriate with protection extending to those whether acting as part of their employment (for example, as a firefighter) or in a voluntary capacity (perhaps as a first-aider).

□he 2015 Act defines how the individual's action should be considered if they are alleged of negligence or breaching a statutory duty: the term ***social action*** relates to the individual acting ***for the benefit of society***'. ***Responsibility*** refers to the individual demonstrating a ***predominantly responsible approach towards protecting others***. ***Heroism*** occurs when the individual ***intervenes in an emergency to assist an individual in danger***. Where action is taken that meet these criteria legal action should not be pursued against the individual.

The HSE has explained its approach to the incorporation of the new Act in the context of the fire service, though principles would apply to all sectors where an unforeseeable emergency situation arises. The responsibility for the organisation to manage all foreseeable risks, and for the individual employee to take reasonable care of themselves and others is not compromised in any way: employers may not expect employees to put themselves at unreasonable risks, even if the public unreasonably expect it, nor should employees act recklessly.

However, where an individual has decided to act entirely of their own volition, putting themselves at risk for the benefit of others and the action is unlikely to put others at risk, the HSE would regard this as a true act of heroism and as such would not take any further action against the individual(s) involved to pursue action against them.

The effect on criminal proceedings of Section 40 of the Health and Safety at Work Act

ONUS OF PROOF (SECTION 40)

Section 40 of HASAWA 1974 establishes that in any proceedings for an offence under any of the Relevant Statutory Provisions of HASAWA 1974, involving a failure to comply with a duty or requirement to do something so far as is practicable, reasonably practicable or to use the best means to do something 'it shall be for the accused to prove that it was not practicable, or not reasonably practicable to do more than was in fact done to satisfy the duty or requirement, or that there was no better practicable means than was in fact used to satisfy the duty or requirement'.

This onus of proof is thought by many to mean that the accused is guilty until they prove themselves innocent and that this appears to go against the basic principle in law that the accused is innocent until proved guilty. However, it should be remembered that before this is required, the prosecution must first demonstrate a prima facie case to which the defence must answer; that is the prosecution must convince the court that the defence have apparently failed in their duty before the need arises for the defence to demonstrate that what they had done met the required standard. In addition, the onus of proof laid on the defendant is less onerous than that on the prosecutor as regards proving the offence 'beyond reasonable doubt' and can be done by satisfying the court 'on the balance of probabilities'.

The High Court of Judiciary stressed: *"Once there is a prima facie case against the accused in that he has not ensured the health, safety and welfare at work of his employees then the onus under Section 40 is on the accused. Thus the Crown does not have to prove that it was reasonably practicable to comply with the Act."*

This onus of proof was tested in the Court of Appeal (Criminal Division) in the case R. v Davies (David Janway) 2002. Davies, the owner and occupier of a plant hire yard and workshop appealed against his conviction for a breach of duty under HASAWA 1974 Section 3 (1) and Section 33, relating to the failure to ensure that people not in his employment were not exposed to danger. A self-employed subcontractor had been crushed to death in Davies' yard by a reversing JCB.

The trial judge ruled that Section 40 of HASAWA 1974 imposed a legal burden of proof on a defendant to show, on the balance of probabilities, that it would not have been reasonably practicable for him to have done more to satisfy the duty to ensure safety. Davies appealed on the grounds that a reverse burden of proof under Section 40 of HASAWA 1974 was incompatible with the presumption of innocence enshrined in the Human Rights Act (HRA) 1998.

The appeal was dismissed, stating that Section 40 is compatible with the Human Rights Act (HRA) 1998 as it is justified. It was said that HASAWA 1974 was motivated by the need to protect public safety and its regulatory nature meant that those operating within it had to conform to certain standards. It was acceptable in health and safety to impose absolute duties on employers and it did not follow that a reverse onus of proof, within reasonable limits, would infringe Article 6 (2) of the Human Rights Act (HRA) 1998.

This was further clarified in the prosecution of Chargot in 2008. In this case the prosecution presented to the court the facts that the death of a driver caused when his dumper truck over-turned was sufficient to show that Chargot had failed in his duty to ensure their employee's safety and health. Chargot argued that this was insufficient and to present a prima facie case the prosecution must explain what should have been done to discharge the duty. Law Lords hearing Chargot's final appeal disagreed with this notion, stating the incident itself was sufficient for the onus of proof to turn to the defence.

Application of common law manslaughter and the Corporate Manslaughter and Corporate Homicide Act 2007 to work related accident/incidents

CORPORATE MANSLAUGHTER: A BRIEF HISTORY

Prior to the introduction of the Corporate Manslaughter and Corporate Homicide Act 2007, corporate manslaughter was a common law offence. For an organisation to be successfully convicted, it was first necessary for a senior employee with a controlling mind for the company to be convicted of Manslaughter by Gross Negligence and then it would be possible for proceedings to be brought against the company the for corporate offence.

Case law shows that this offered the possibility for conviction of smaller employers with a simple organisational structure: the first employer to be convicted of common law corporate manslaughter was OLL Ltd after its managing director was convicted of Manslaughter by Gross Negligence for the deaths of four children on a canoeing trip organised by the centre.

The cases involving Great Western Trains following the Southall Rail Crash and of P&O Ferries following the ferry disaster at Zeebrugge, however, demonstrated that it was very difficult to secure a conviction for larger organisations. Despite the judge describing the management of P&O as having 'a disease of sloppiness' and if you considered the negligence of the team as a whole it constituted gross negligence, no one individual was sufficiently at fault to be convicted of the manslaughter offence. As such, although prosecuted for health and safety offences, the company could not be held accountable with a Corporate Manslaughter conviction.

It was this weakness in the common law offence which finally brought the introduction of the 2007 Act which supersedes it and focuses on the action of the organisation's management as a whole and not just individuals. It must be noted however, that the common law offence of Manslaughter by Gross Negligence still exists and may still be used against an individual.

CORPORATE MANSLAUGHTER AND CORPORATE HOMICIDE ACT (CMCHA) 2007

The Corporate Manslaughter and Corporate Homicide Act (CMCHA) 2007 came into force on 6 April 2008 and established an offence enabling the conviction of an organisation where the gross failure in the way activities were managed or organised resulted in a person's death. The offence is not dependent on finding the individual controlling mind or designed to prosecute individuals, but is intended to be used when management failures lie across an organisation and it is the organisation itself that will face prosecution while individuals will remain liable for the common law crime of gross negligence manslaughter as before but have ***no*** liability under the Act itself.

Legal Criteria for prosecution

Section 1 of the Act defines the offence. Prosecution may be brought against an organisation if:

- The way they manage and organise their activities results in a fatality.
- This constitutes a gross breach of any relevant duty of care owed to the deceased.
- The activities managed and organised by senior management is a substantial element of the breach.

The section goes on to define the term 'gross breach' as the conduct falling far below what can be expected of the organisation within the circumstances.

Enforcement and prosecution

All fatalities will normally be reported to the enforcing authority and investigation instigated by the Health and Safety Executive (or Local Authority). The responsibility for enforcement of potential manslaughter cases however lies with the police. As such, in line with the guidance for the investigation of a death at work, as defined within the Enforcement Policy Document, if there is an indication that Corporate Manslaughter, or indeed gross negligence manslaughter, the HSE will refer to the police to take over the investigation while continuing to work closely with them.

If all of the criteria for prosecution exist, the police will refer the case to the Crown Prosecution Service (or the Crown Office and Procurator Fiscal Service (COPFS) in Scotland). The offence is indictable only and as such will be heard in a Crown Court. Associated Health and Safety offences may also be brought in tandem by the HSE. On conviction, an organisation may be sentenced with an unlimited fine. In addition, the courts may impose a Remedial Order placing specific requirements for improvement on the organisation. A further sanction available to the judge is that of a Publicity Order whereby the organisation is required to explain the nature of their offence and sentence in an identified media source; this may be a newspaper, trade journal etc. as deemed appropriate by the sentencing judge.

Although no specific tariffs exist for sentencing of organisations found guilty of Corporate Manslaughter Sentencing Guidelines were introduced in February 2016. The guidelines themselves are tariff based and takes into account the potential harm (not the actual harm thus preventing organisations benefitting from luck that only one person was killed when realistically it could have been much worse) and the turnover of the organisation (not the profit - and will take into account the income of the organisation as a whole, not just the subsidiary in question). The recommended fines are a minimum of six figure sums with seven and even eight figure sums suggested for the more serious offences committed by organisations with an annual turnover in excess of £50 million. Fines will be lowered based on an early guilty-plea and/or mitigation and the effect on third parties will also be considered: fines imposed on charities or public bodies such as the NHS will often be lower to prevent the innocent suffering. However, in some cases, to fit the seriousness of the breach, the appropriate fine may cause the ultimate bankruptcy of the business, a result which though not necessarily the aim of the courts they will not deliberately avoid.

Relevant cases

In the first 8-years since the introduction of the Act, which was not applicable retrospectively, a total of 26-cases have been brought for prosecution, 21 of them in England, 5 in Northern Ireland but none in Scotland; at the time of writing, the verdict for one and the sentence for one remain outstanding. Although providing some useful information, 25 of the 26 cases have been brought against quite small organisations against whom a prosecution could potentially have been secured under the old common-law offence. As such, little clarification exists on the interpretation by the courts of the requirement for the activities of the management to be regarded as a substantial element of the breach.

Of the 26 cases, 14 organisations pleaded guilty to the alleged offence(s), six were convicted following a trial and three acquitted with two charges being held on the books following health and safety convictions. Many of the charges for corporate manslaughter cases were accompanied by charges for breaches of health and safety legislation (either against the company or individual liability of a senior office by virtue of Section 37) or against an individual officer for manslaughter by gross negligence. In some cases where the organisation pled guilty to the Corporate Manslaughter charges, associated charges against individuals were dropped in return.

The first case brought under the Act was against Cotswold Geotechnical (Holdings Ltd) following the death of an employee when the trench he was working in collapsed. Charges of gross manslaughter negligence and Section 37 offences against a director were dropped to his ill-health but the company was convicted following a trial of the Corporate Manslaughter offence. The fine of £385,000 plus costs was payable over ten years.

Princes Sporting Club pleaded guilty to the charge of Corporate Manslaughter following the death of an 11-year old girl who fell from a banana boat and was subsequently struck by the towing boat. Despite the guilty plea, the judge considered the offence so serious that they were fined a figure equivalent to the total value of all of their assets and also issued the first Publicity Order imposed under the Act which required them to publish details of the conviction in industry publications.

In comparison Mobile Sweepers were required to publicise their conviction in local newspapers. The fine imposed on Cavendish Masonry Limited was lower than originally planned. The defence argued that the fine of

£500,000 (which was considered appropriate by the judge for the severity of the offence) would put the company out of business resulting in none of the fine being paid; as such the fine was reduced to £150,000 in the hope that the business would continue to function and pay the fine over the five years allowed.

To date, the only publicity notice to be issued was imposed on Peter Mawson Limited. Following an incident where a man fell to his death through a fragile roof, the company was fined £200,000 and the owner, Peter Mawson, was given a suspended sentence of eight months. In addition, Mawson was ordered to advertise what happened on the company website and take out a half page spread in a local newspaper.

The first company to be acquitted was PS and JE Ward Ltd in 2014, though it was found guilty of breaching Section 2 of the Health and Safety at Work etc. Act 1974 leading to a fine of £50,000 plus costs.

The first large organisation to face charges under the Act was the Maidstone and Tunbridge Wells NHS Trust when a mother died shortly after giving birth by caesarean section when a heavy bleed led to a cardiac arrest.

Although two doctors attending at the birth were also charged with manslaughter by gross negligence, only one stood trial, the other having fled the country. Both the NHS Trust and the doctor who stood trial were acquitted after the judge presiding ordered the jury to do so. He criticised the prosecution describing the doctor's actions as being as far away from gross negligence as they could possibly be and describing some of the allegations made against the Trust as 'perverse'.

Although only 26 cases have been brought in eight years, ten of those came to a conclusion in 2015 indicating that the number of cases may now be increasing. Attention to the outcome and comments made by trial judges of future cases could be invaluable to help understand how the Act should be interpreted.

COMMON LAW MANSLAUGHTER

Homicide is the unlawful killing of a human being and is a common law criminal offence. Homicides include murder, manslaughter and causing death by dangerous driving. Manslaughter (culpable homicide in Scotland) is a crime where the offender has caused a death, but the act itself falls short of murder.

Manslaughter can be committed in one of three ways:

1) Killing with the intent for murder but where a partial defence applies, namely loss of control, diminished responsibility or killing pursuant to a suicide pact.
2) Conduct taking the form of an unlawful act involving a danger of some harm that resulted in death, is manslaughter ('unlawful and dangerous act manslaughter').
3) Conduct that was grossly negligent given the risk of death, and did kill, is manslaughter ('gross negligence manslaughter').

Although in reality a death may occur in the workplace as a result of any one of these, the issue most worthy of discussion in this context is manslaughter by gross negligence. Everyone owes a duty of care to those who will be affected by our acts and omissions. A significant deviation from the duty would normally be regarded as negligence which, if sufficiently serious to result in a death would be considered gross negligence.

"The ordinary principles of the law of negligence apply to ascertain whether or not the defendant has been in breach of a duty of care towards a victim who has died. If such a breach of duty is established the next question is whether the breach of duty caused the death of the victim. If so, the jury must go on to consider whether the breach of duty should be characterised as gross negligence and therefore is a crime. This will depend on the seriousness of the breach of duty committed by the defendant in all the circumstances in which the defendant was placed when it occurred. The jury will have to consider whether the extent to which the defendant's conduct departed from the proper standard of care incumbent upon him, involving as it must have done a risk of death to the [person] was such that it could be judged criminal."

Figure A3-10: Statement from R v Adomoko 1995 House of Lords Judgement. *Source: Lord Mackay.*

When people are killed by gross negligence arising out of work, those guilty of the gross negligence may be prosecuted for manslaughter. As well as individual workers, persons in authority may be called to account, particularly where a death has occurred that could have been avoided if the person had taken such action as lay within the normal routine of their job. Delegation of a job does not, in itself, absolve a person from responsibility. Cases of gross negligence manslaughter are initiated by the police and brought by the CPS in a criminal court.

The CPS taking a case against an individual for gross negligence manslaughter would have to show that the defendant had not fulfilled their common law duty of care, were grossly negligent and the death was due to this negligent behaviour. This would include showing that the defendant failed to take reasonable care and that their behaviour was reckless. The defence would need to show that there was reasonable doubt that the behaviour had departed from the standard of care incumbent upon the defendant and that the defendant had not acted in a reckless way or if they did that it did not lead to the death. The term 'involuntary manslaughter' is commonly used to describe a manslaughter falling within (2) and (3) while (1) is referred to as 'voluntary manslaughter'.

There are two types of involuntary manslaughter, where the accused had no intention of either seriously harming or killing the victim (i.e. there was no malice or aforethought).

1) Unlawful or dangerous act.
2) Gross negligence.

Unlawful or dangerous act manslaughter

Involuntary manslaughter caused by an unlawful or dangerous act is where the killing is the result of:

- The defendant's unlawful act (not omission).
- Where the unlawful act is one which all sober and reasonable people would realise would subject the victim to the risk of some physical harm resulting there from, albeit not serious harm.
- Whether or not the defendant realised this.

The act need not be directed against a person (for example, arson). The knowledge attributed to the 'sober and reasonable person' is that which such a person would acquire as an observer of the whole course of the defendant's conduct, throughout the unlawful act. In manslaughter arising from an unlawful and dangerous act, the state of mind of the person committing the act is relevant only to establish that the act was committed intentionally and that it was an unlawful act.

Gross negligence manslaughter

An important category of involuntary manslaughter is that of 'gross negligence' on the part of the accused. Where an individual causes death through gross negligence a common law criminal offence has been committed, which may be considered to be a criminal variation of the civil tort of negligence; negligence to such an extent that it results in a death. Lord Mackay stated in a 1995 House of Lords Judgement of ***R v Adomako [1995] 1 AC 171:***

"The ordinary principles of the law of negligence apply to ascertain whether or not the defendant has been in breach of a duty of care towards a victim who has died. If such a breach of duty is established the next question is whether the breach of duty caused the death of the victim. If so, the jury must go on to consider whether the breach of duty should be characterised as gross negligence and therefore is a crime. This will depend on the seriousness of the breach of duty committed by the defendant in all the circumstances in which the defendant was placed when it occurred. The jury will have to consider whether the extent to which the defendant's conduct departed from the proper standard of care incumbent upon him, involving as it must have done a risk of death to the [person] was such that it could be judged criminal."

Figure A3-11: Statement from R v Adomoko 1995 House of Lords Judgement. *Source: Lord Mackay.*

The case of *R v Adomako (1994) 3 All ER 79* introduced a four stage test of gross negligence manslaughter, known as the ***'Adomako Test'*** which involves:

1) The existence of a duty of care to the deceased.
2) A breach of that duty of care, which.
3) Causes (or significantly contributes) to the death of the victim.
4) The breach should be characterised as gross negligence, and therefore a crime.

"The facts must be such that, in the opinion of the jury, the negligence of the accused went beyond a mere matter of compensation between subjects and showed such disregard for the life and safety of others, as to amount to a crime against the State and conduct deserving punishment."

Figure A3-12: Civil liability becomes criminal liability. *Source: R V Bateman (1925) 19 Ct App R8.*

When people are killed by gross negligence arising out of work, those causing the gross negligence may be prosecuted for manslaughter. As well as individual workers, persons in authority may be called to account, particularly where a death has occurred that could have been avoided if the person had taken such action as lay within the normal routine of their job. Delegation of a job does not, in itself, absolve a person from responsibility; since they must satisfy themselves (ensure) that such delegation is effective. For example, in 1997 the Managing Director of a transport firm was imprisoned for 12 months for causing gross negligence manslaughter. Cases of gross negligence manslaughter are initiated by the police and brought by the CPS in a criminal court. The CPS taking a case against an individual for gross negligence manslaughter would have to show that the defendant had not fulfilled their common law duty of care, was grossly negligent and the death was due to this negligent behaviour. This would include showing that the defendant failed to take reasonable care and that their behaviour was reckless. The defence would need to show that there was reasonable doubt that the behaviour had departed from the standard of care incumbent upon the defendant and that the defendant had not acted in a reckless way or if they did that it did not lead to the death.

Examples of individual gross negligence manslaughter

Railway guard

A railway guard was jailed for five years for causing the death of a 16-year-old girl who went under a train at a Liverpool station. She died when she fell between the carriage and platform at James Street station in October 2011. A railway guard signalled for the driver to depart as the girl, who was drunk, stood on the platform leaning against the train. He denied gross negligence manslaughter. He told the jury he thought the sixth form student was moving away from the train when he gave the signal to depart. The prosecution said the railway guard should have been able to see that she was in an 'intoxicated state'. He was found guilty of manslaughter by gross negligence at Liverpool Crown Court and the jury was then discharged from reaching a verdict in an alternative, lesser charge under HASAWA 1974.

Construction company owner

A company co-owner and a health and safety advisor have been jailed following the death of a labourer in Fulham in December 2010. The labourer was crushed to death while working on a basement excavation in Ellerby Street, Fulham. The excavations were not properly supported and collapsed, crushing and trapping the labourer. The emergency services were called but he was pronounced dead at the scene.

Southwark Crown Court heard on 2 December 2014 how Mr Conrad Sidebottom of Siday Construction Ltd who was also the site manager, was aware of the dangerous state of the excavations, but took no steps to ensure it was safe. It was also heard that Mr Richard Golding, a qualified health and safety advisor employed by AllDay Safety Services Ltd, was also aware of the risks as he was responsible for drafting the method of work statement. This document was found to be inadequate and was not followed, despite him having the authority to stop dangerous works, which he failed to do.

The Crown Prosecution Service (CPS), said: *"The labourer's tragic death was preventable and inexcusable. While working on the excavation of the house, his safety was the responsibility of Conrad Sidebottom, the commercial director of Siday, and Richard Golding, the independent health and safety consultant contracted to provide advice to Siday."* The CPS added: *"These men did not fulfil their duties, with a haphazard attitude to the safety of the employees working in the house. Their failures led to these tragic consequences, which should never have occurred."*

Co-owner of Siday Construction Ltd, Conrad Sidebottom, was found guilty of gross negligence manslaughter on 2 December and was sentenced to three years and three months in jail. Richard Golding was found guilty of failing to take reasonable care for the health and safety of others (Section 7 HASAWA 1974) and was jailed for nine months.

Corporate liability for gross negligence manslaughter

Prior to the introduction of the Corporate Manslaughter and Corporate Homicide Act 2007, which came into force on 6 April 2008, under common law a corporate body could also be found guilty of gross negligence manslaughter. For example, OLL Ltd was found guilty after 4 teenagers drowned on an activity holiday in Lyme Bay, Dorset. Proof of corporate gross negligence manslaughter required establishing gross negligence on the part of the corporate body's 'controlling mind'. This required the prosecution to prove that individuals that comprised the controlling mind of corporate bodies had themselves committed gross negligence manslaughter.

This then enabled a separate prosecution against the corporate body for corporate gross negligence manslaughter. Proving the gross negligence of individuals at a senior enough level to represent the controlling mind, including that their behaviour led to someone's death, was a lot easier to do in smaller organisations. It proved very difficult to show causal link between the actions of the controlling mind of large corporate bodies and the death of those affected, ***see examples of successful and unsuccessful gross negligence corporate manslaughter cases later in this Element***.

These difficulties and the failure of cases taken led to the introduction of the Corporate Manslaughter and Corporate Homicide Act 2007, which is not reliant on prosecuting individuals as the controlling mind. Instead, it allows that an organisation be prosecuted for Corporate Manslaughter by looking at the wider corporate picture. ***This is explained in more detail later in this Element***.

SUCCESSFUL AND UNSUCCESSFUL CORPORATE GROSS NEGLIGENCE MANSLAUGHTER CASES

R V OLL Ltd (Peter Kite and Joseph Stoddart) 1994 - Lyme Bay [March 1993]

Four teenagers, taking part in a canoeing trip organised by OLL Ltd (formerly Active Learning and Leisure Ltd), died. They were in a group accompanied by a schoolteacher - all were novice canoeists - and two unqualified instructors who had only basic proficiency skills. The group was swept out to sea, capsizing frequently. The centre had neither provided distress flares, nor informed the coastguard of the expedition. Despite cold sea temperatures, the victims wore only swimming suits under their wetsuits. Although the rescue boat had set out at midday to look for the group when it did not return, the centre manager did not call the coastguard until 3pm. OLL Ltd's managing director, Peter Kite, and the centre's manager, Joseph Stoddart, were charged with gross negligence manslaughter.

The court heard that the children would have survived if their lifejackets had been fully inflated. Although the instructors had not told them to inflate the jackets, the prosecution alleged that the blame lay with the regime which allowed these instructors to be sent on such an expedition. Numerous criticisms were made, for example, there was no system for alerting the authorities within a given time of an expedition being overdue.

Kite argued that he delegated safety matters to the centre manager. But, the prosecution countered, this did not absolve him from his duty to put in place a system to protect those in his care from foreseeable risk. The judge suggested that the latter meant he was under an even stricter duty than he might otherwise have been to monitor and supervise safety.

The jury found the steps that Peter Kite had taken to discharge his responsibilities were sufficiently negligent as to amount to a criminal neglect of duty. He was sentenced to three years' imprisonment. This was reduced to two years on appeal. The centre manager, Joseph Stoddart, was acquitted.

Kite admitted that he was one of the company's 'driving forces'. He was clearly part of its controlling mind and, when he acted as managing director, he also acted as the company. Accordingly, the company was also found guilty of manslaughter, and fined £50,000.

R v Jackson Transport (Ossett) Ltd [1996]

James Hodgson died after cleaning out a tanker containing residues of a dangerous chemical (para-chloro-ortho-cresol), with a steam pressure cleaner while wearing only a boiler suit and a baseball cap. He had been injured three months earlier when similarly exposed to a dangerous tank cleaning operation. In the subsequent trial, the court heard that the director Alan Jackson had failed to provide first aid facilities, a trained first aider or protective clothing.

The jury decided that both the company and Mr Jackson were guilty of manslaughter. The company was fined £15,000 for manslaughter and £7,000 for breaches of HASAWA 1974. The director was fined a total of £1,500 and jailed for one year. Once again in this case the controlling mind of the company was easily identifiable.

Tebay Rail Deaths 2004

In February 2004, four men were working on a railway line at Tebay, Cumbria, when a wagon ran out of control. The brakes had failed and the rail workers had no warning that the wagon was running toward them out of control. The four men were all killed.

Mark Connolly, the boss of MAC Machinery Services, had deliberately disconnected the hydraulic brakes on two wagons because it was cheaper than repairing the wagons properly. The hydraulic systems were in such a bad way that they would not work properly in conjunction with a crane.

MAC Machinery had been contracted by Carillion to replace track on the West Coast Main Line. Connolly's employee, Roy Kennett, was using a crane to load steel rails onto two flat-bed wagons. Because the hydraulic brakes were in such a bad way with the hydraulic fluid having leaked out, Connolly disconnected them. He then filled the hydraulic brake cables with ball bearings, giving the impression that they were filled and in working order and that all was well. Kennett had placed pieces of wood under the wagons' wheels as a makeshift brake. One of the wagons rolled over the chocks acting as brakes, careered down a hill for three miles and collided with a gang of workers who were carrying out track maintenance. Four of the men were killed by the 19-tonne wagon and another three were injured.

Connolly was charged with health and safety offences and gross negligence (manslaughter) at Newcastle Crown Court. He was found guilty of four charges of gross negligence (manslaughter) and three counts of breaches of health and safety law. He was sentenced to nine years imprisonment for the manslaughter charges. Kennett, who had continued to use the wagons even though he knew they were faulty, was also imprisoned for two years for manslaughter charges.

The Prosecutor told the court that Connolly was 'grossly negligent' in his actions and had 'scant regard' for railway safety. It is interesting to note that TUC General Secretary Brendan Barber said that the case *"must be contrasted with the failure to prosecute the directors of large companies following other deaths."* This is indeed shown in the cases that follow.

P&O European Ferries (Dover) Ltd 1991- Herald of Free Enterprise [6 March 1987]

Less than half an hour after the ferry Herald of Free Enterprise sailed from Zeebrugge harbour for Dover, it capsized with the loss of 188 lives. The subsequent inquiry revealed that the Herald capsized because it went to sea with the bow doors open. The assistant boatswain had a duty to close the doors, but had fallen asleep and failed to attend for duty when the ship left harbour. The chief officer had been on deck at the time, but failed to ensure that the assistant boatswain had reported to work. His certificate of competence was later suspended as a result.

The inquiry report found severe fault with the system which allowed the capsizing to happen:

> *"The underlying or cardinal faults lay higher up in the company... All concerned in management, from the members of the board of directors down to the junior superintendents, were guilty of fault in that all must be regarded as sharing responsibility for the failure of management. From top to bottom the body corporate was infected with the disease of sloppiness."*

Figure A3-13: Fault with the system quote. *Source: Herald of Free Enterprise Inquiry.*

The tragedy was caused by:

> *"The faults of [the assistant boatswain, the chief officer, the captain and the company] at all levels from the board of directors through the managers of the marine department down to the junior superintendents."*

Figure A3-14: Cause of tragedy quote. *Source: Herald of Free Enterprise Inquiry.*

The inquest jury decided 188 victims had been unlawfully killed. Subsequently, the company and seven employees were charged with manslaughter. Those charged included the Assistant Boatswain and the Chief Officer. Five senior staff were also charged and, as one or all of the five were thought to represent the company's directing mind, the company was also charged.

Despite the damning conclusions of the inquiry and inquest, the prosecution failed to prove its case against any of the five senior staff and therefore had no case against the company. The judge discontinued their trials, and the prosecution then dropped charges against the chief officer and assistant boatswain.

The cases failed for lack of evidence that the conduct of the accused fell below the standard to be expected, i.e. what was common practice in the industry. The prosecution failed to show that a reasonably skilled person in the same line of work would have behaved differently to any of the senior officers. This meant that if no single senior officer was sufficiently at fault, and then neither was the company. This was because of an earlier court ruling that it was not possible to aggregate the faults of more than one individual and say that, together they amounted to such a high degree of fault as to be sufficient to convict the company, although not the individuals, of manslaughter.

R v Great Western Trains Company - Southall Train Collision (19 September 1997)

A high speed Intercity 125 train, operated by Great Western Trains (GWT), collided with an empty freight train killing seven and injuring a further 147 passengers.

The subsequent inquiry heard that the passenger train passed two amber warning signals 'at or very near its top speed of 125 mph' then ran a red light before the collision. The AWS (Automatic Warning System) system fitted in the driver's cab had been turned off because of a fault on the previous day. The AWS gives the driver an audible warning when a signal is passed. The train and the track were installed with a system of ATP (Automatic Train Protection) as part of a pilot scheme introduced by BR following the Hidden Report into the accident at Clapham Junction in 1988. However this ATP system had been switched off because the driver had not been trained in its use.

Manslaughter charges were made against both the driver of the train, Larry Harrison, and GWT. In dismissing the seven charges of manslaughter against GWT the judge (Mr Justice Scott-Baker) held that the prosecution must, in order to establish their case, identify GWT's negligent controlling mind. The Court of Appeal later upheld this principle, which is known as attribution or identification. The only candidate for this was the managing director Richard George who had ultimate responsibility for safety. Responsibility for safety is not however enough to charge a person with manslaughter. The judge concluded that if the Crown thought it could charge Mr George with manslaughter it would have done so.

Mr Justice Scott Baker said that:

> *"It is still necessary to look for ... a directing mind and identify whose gross negligence it is that fixes the company with criminal responsibility."*

Figure A3-15: R v Great Western Trains Company. *Source: Mr Justice Scott Baker.*

In making his decision the judge said that the lack of action since the publication of the Law Commission's report was regrettable but it was for Parliament to decide if it wished to change the law to remove the present obstacle to prosecuting large corporations for manslaughter.

Following this ruling, the prosecution dropped the charge of manslaughter against Driver Harrison. This was said to be in part because of the psychiatric reports that suggested he had been damaged by the accident.

Great Western Trains were fined a record £1.5 million under section 3 of HASAWA 1974. In doing so the judge followed the Howe sentencing criteria.

If the driver had been prosecuted for gross negligence (manslaughter), it would not have enabled the body corporate employer to be prosecuted for corporate gross negligence manslaughter as the driver was not the 'controlling mind' of the body corporate.

The law was changed in this respect with the introduction of the Corporate Manslaughter and Corporate Homicide Act (CMCHA) 2007, making it possible for the body corporate employer to be prosecuted for corporate manslaughter without having to find a person who is a 'controlling mind' to prosecute first. The change in the law required a statute to be made to do this, rather than rely on change through judicial precedent in common law.

CORPORATE MANSLAUGHTER AND CORPORATE HOMICIDE ACT (CMCHA) 2007

Previously, corporate liability for manslaughter was difficult to establish using common law principles, as the jury had to be satisfied that at least one individual who was part of the 'controlling mind' of the company caused the victim's death by gross negligence. The examples of successful and unsuccessful corporate gross negligence manslaughter cases explained earlier in this element show why it was necessary to introduce the Corporate Manslaughter and Corporate Homicide Act (CMCHA) 2007, as most of the major common law corporate gross negligence manslaughter cases failed to lead to successful prosecution. This was often because it was much easier to find the grossly negligent controlling mind in a small organisation than in a large one. The CMCHA 2007 allows for an organisation to be prosecuted by looking at the wider organisational issues.

The Corporate Manslaughter and Corporate Homicide Act (CMCHA) 2007 came into force on 6 April 2008 and established an offence enabling the conviction of an organisation where the gross failure in the way activities were managed or organised resulted in a person's death. The offence is not dependent on finding the individual controlling mind or designed to prosecute individuals, but is intended to be used when management failures lie across an organisation and it is the organisation itself that will face prosecution. Individuals will remain liable for the common law crime of gross negligence manslaughter as before.

The offence does not require organisations to comply with all new and current regulatory standards, but organisations should ensure they are taking proper steps to meet current legal duties. An organisation will be guilty of an offence under the CMCHA 2007 if the way in which its activities are managed or organised causes a death and amounts to a gross breach of a duty of care to the deceased. The organisation's conduct must have fallen far below what could have been reasonably expected, for example, serious health and safety breaches by the organisation. A substantial part of the failure within the organisation must have been at a senior level. Senior level means the people who make significant decisions about the organisation or substantial parts of it. This includes both centralised, headquarters functions as well as those in operational management roles.

Cases of corporate manslaughter and corporate homicide under the CMCHA 2007 are brought by the police. Penalties include an unlimited fine and the provision to require the organisation to publicise details of its conviction and fine. Courts may also require an organisation to take remedial steps to address the failures that led to the death. The offence does not apply to certain public and government functions whose management involves wider questions of public policy. Since the enactment of the CMCHA 2007 in 2008 there have been several prosecutions, some of them unsuccessful.

SUCCESSFUL CMCHA 2007 PROSECUTIONS

Cotswold Geotechnical Holdings Ltd

In April 2009 the CPS announced that it would bring the first prosecution under CMCHA 2007 against a Cotswolds firm and one of its directors. The CPS authorised a charge of corporate manslaughter against Cotswold Geotechnical Holdings Ltd, of Birdlip in Gloucestershire, in relation to the death of an employee on 5 September 2008. He was employed by the company as a junior geologist and was taking soil samples from inside a pit, which had been excavated as part of a survey of a site near Stroud, when the sides of the pit collapsed and crushed him.

Peter Eaton, a director of the company was charged with gross negligence manslaughter (charge was dropped) and under Section 37 of the Health and Safety at Work, etc. Act 1974 (charge was dropped). Cotswold Geotechnical Holdings Ltd was also charged with a breach of section 33 of the Health and Safety at Work, etc. Act 1974. Cotswold Geotechnical Holdings Ltd was found to be in breach of the Corporate Manslaughter and Corporate Homicide Act 2007 and was fined £385,000 payable over 10 years (annual turnover of company was £333,000).

JMW Farms

In May 2012 JMW farms was fined £187,500 after pleading guilty to corporate manslaughter in the first case of its kind in Northern Ireland. The company was given six months to pay the fine. The pig farm based in Co Armagh was handed the penalty at Belfast Crown Court over the death of employee on November 15, 2010.

The father-of-two suffered serious head injuries when a large metal bin, precariously positioned on the prongs of a forklift truck, toppled and fell on him as the vehicle was being reversed. Company director Mark Wright, a lifelong friend of the deceased, was driving the forklift.

A Police Service of Northern Ireland (PSNI) and Health and Safety Executive (NI) investigation found that the bin had not been attached or integrated with the forklift, a replacement vehicle on loan while the usual truck was being serviced.

Belfast Recorder Judge Tom Burgess said: *"The very definition of the offence of corporate manslaughter is an acceptance of a gross breach of duty. That is a high and totally unacceptable breach in circumstances where the risks involved were high, with the more than foreseeable likelihood of serious injury or death following if the proper steps were not taken."*

Lion Steel Manufacturing

In July 2012 Lion Steel Equipment Limited of Hyde, Greater Manchester, were successfully prosecuted under the Corporate Manslaughter and Corporate Homicide Act 2007 and ordered to pay a fine of £480,000 in four instalments: £100,000 by September 2012; £150,000 by the following September; £150,000 by September 2014; and £80,000 by September 2015. The prosecution arose from the death of an employee who fell through a plastic roof panel sustaining fatal injuries at one of Lion Steel's industrial sites in Hyde in May 2008.

The CPS also claimed approximately £140,000 in prosecution and investigation costs. However the judge ruled that Lion Steel should only be liable for 40% of these, in part because of what he called an 'unreasonable' three-year delay by the CPS in bringing charges after the death.

Three of the company's directors - Kevin Palliser, Richard Williams and Graham Coupe - originally faced charges of gross negligence manslaughter and failing to ensure the health and safety of their employees under Section 37 of the Health and Safety at Work Act 1974. However, at the close of the prosecution's case, the judge ruled there was no case to answer against Williams on both the gross negligence manslaughter and HASAWA 1974 charges and ordered the jury to return not guilty verdicts on these counts. He also instructed the jury to return a not guilty verdict on the manslaughter charge against Coupe, stating the prosecution had failed to prove their case against him. Negotiations took place between the CPS and the defence lawyers, and it was agreed the CPS would drop all the remaining charges against Palliser and Coupe in return for Lion Steel pleading guilty to a corporate manslaughter offence. A charge against Lion Steel of failing to ensure the safety of its employees under Section 2 of HASAWA 1974 was also dropped and will remain on file.

A number of complications affected the processing of the prosecution at the trial:

- Lion Steel had been charged with corporate manslaughter, but the judge ruled the charge could not be heard at the same time as the manslaughter case against the directors because the CPS intended to refer to alleged wrongdoing of the directors that supposedly occurred before the Corporate Manslaughter and Corporate Homicide Act 2007 came into force in April 2008. The CMCHA 2007 cannot be applied retrospectively.
- The judge accepted Williams had left the Hyde site in 1998 and had no involvement in its day-to-day running, meaning ***he had no duty of care to the deceased*** and therefore no case to answer.
- The judge also criticised the CPS for failing to look more closely at the legal test that should be applied before pursuing gross negligence manslaughter charges against individuals.

Princes Sporting Club

Princes Sporting Club were successfully prosecuted at Southwark Crown Court for corporate manslaughter and breach of Section 3(1) of HASAWA in November 2013.

The prosecution arose from the death of a girl, aged 11, who fell from a banana boat ride and was struck by the speed boat towing the ride. She sustaining fatal injuries during a birthday celebration at the Bedfont, Middlesex club on 11th September 2010. The court heard factors which contributed to the death included: lack of an observer on the boat at the time of the incident; speed of the boat; 'unnecessary' turning at tight angles; and the colour of the equipment the children were wearing making it difficult to spot them if they fell in the water.

Company director Frederick Glen Walker was charged with a breach of s.37 HASAWA, however the CPS dropped charges against him.

The company had ceased trading and was fined to the value of its assets - £134,579.69. Judge Alistair McCreath said: *"I propose to fine this company every penny that it has. I have no power to do anything other than impose a fine and I can impose no greater fine than all of its assets."*

Mobile Sweepers (Reading) Limited

Mobile Sweepers (Reading) Limited was charged with corporate manslaughter in March 2013 in relation to the death of an employee, who was working on a repair underneath the raised hopper of a road-sweeping truck. Investigation by Hampshire Police and HSE established that a prop designed to take the weight of the hopper when raised in 'tipping' position could not be used due to the poor condition of the vehicle. When loss of hydraulic pressure caused the hopper to fall back towards to the main chassis of the vehicle, there was nothing to stop it from crushing the employee as he worked underneath.

The company was fined £8,000 for the corporate manslaughter offence and ordered to pay a further £4,000 in costs at Winchester Crown Court in March 2014. The court heard that the fine imposed reflected the fact that the company ceased trading soon after the incident and had total assets of only £12,000. A CMCHA 2007 'Publicity Order' was also issued requiring an agreed account of the company wrongdoing to be publicised in the press. The court heard that the fine imposed reflected the fact that the company ceased trading soon after the incident and had total assets of only £12,000. Judge Boney commented that if the defendant company had been large and with considerable assets the penalty would have been at the upper end of a scale of between £500,000 and £1 million.

The company's sole director Mervyn Owens was also charged with failing to discharge his health and safety duties and was fined £183,000. Owens was also disqualified from being a company director for five years after the judge heard he had started an almost identical company after Mobile Sweepers (Reading) ceased trading on the day of the accident.

Cavendish Masonry Limited

Cavendish Masonry Limited was convicted on 22nd May 2014 of corporate manslaughter following the death of an employee in February 2010. A 23-year-old stone mason's mate, was erecting a large wall at the Well Barn Estate in Moulsford, Wallingford. A two tonne limestone block fell from a concrete lintel and crushed the employee. He was taken to the John Radcliffe Hospital by air ambulance and was pronounced dead later the same day. The company had pleaded guilty to breaching the Health and Safety at Work etc. Act 1974 at an earlier hearing. Cavendish Masonry Limited was found guilty of a gross breach of its duty of care in the management and organisation of its activities at The Well Barn estate, by failing to take reasonable care in the planning and execution of those activities, contrary to Section 1 (1) of the Corporate Manslaughter and Corporate Homicide Act 2007. The HSE Inspector who supported the police-led investigation, said: *"The employee's tragic death was completely avoidable had Cavendish Masonry Limited properly planned and managed the installation of the heavy limestone."*

Sterecycle (Rotherham) Limited

Sterecycle (Rotherham) Limited, which has since gone into liquidation, was found guilty on the unanimous verdict of a jury at Sheffield Crown Court on 7 November. It was fined £500,000, the highest penalty to that date under the Corporate Manslaughter and Corporate Homicide Act 2007.

On 11 January 2011 the door of the autoclave at the company's plant at Templeborough, Rotherham, which had for some time been recognised as a problem, failed and blew out. An autoclave operator suffered fatal head injuries and his colleague was left fighting for his life. The force of the blast was so great it blew a hole in the wall of the factory. Addressing the company in his sentencing remarks, Mr Justice Jay said the employee: *"Died as a result of systemic failings in the way the running of this autoclave was managed and operated by you. A substantial element in those failings was attributable to your senior management."*

The jury was presented with evidence outlining how Sterecycle, which had around 50 employees, cut corners to save money and keep production going until the autoclaves were retired from service. The company even removed safety devices because they slowed production. The court heard how there was not a proper system in place to ensure that the moving parts of the locking ring and the door were properly cleaned, greased and maintained. Instead, the maintenance system was essentially reactive. As a result, the court heard, wear built up in many critical areas of the mechanism, defects in the machine were not addressed and a screw that had been missing for months before the incident was not replaced. To make matters worse, the door on the autoclave was misaligned and was never properly repaired, but simply patched up. The door seals consistently failed and the locking ring was seen to move under pressure during operations.

Mr Justice Jay added: *"The overall impression given by the evidence taken as a whole is that you were way out of your depth, and overly driven by the desire to maintain production. It is an aggravating feature of this offence that the breaches of duty which occurred took place over a lengthy period of time measurable in years, not months."* However, as the company had gone into liquidation, the judge acknowledged that little, if any, of the fine would be paid. Instead, the judge said the fine *"will serve to mark society's condemnation of your behaviour, and may act as a deterrent to others."*

Pyranha Mouldings Limited

Pyranha Mouldings Ltd became the tenth (seven in England and three in Northern Ireland) company to be convicted in the UK under the Corporate Manslaughter and Corporate Homicide Act 2007, following the tragic death of an employee who became trapped in an industrial oven.

The employee was carrying out cleaning work on the oven at the Runcorn-based firm in December 2010. It had been shut down for maintenance and once the work had been completed, the oven was switched back on by a colleague, who was engaged to deceased's daughter, and put into warm-up mode. A few minutes later, smoke was seen coming from the oven and when operators opened the door to find the source, they discovered his body. He suffered severe burns and died of shock after becoming trapped in the large bespoke oven, which was designed to accommodate the moulds for two kayaks. He was pronounced dead at the scene.

Martin Heywood, investigating inspector at the Health and Safety Executive, said the employee: *"Tragically lost his life because the way in which maintenance work was carried out on the moulding ovens at the factory was fundamentally unsafe."*

Mr Justice MacDuff, hearing the case, said there were *"a number of serious failings in the design of this and other moulding ovens"*, which had been designed and commissioned in-house. *"The doors were set to automatically close whenever the electrical supply was switched back on, which meant there was a high risk of someone being trapped inside. There had been no risk assessments and staff had not received suitable training on how to use the new ovens and there were no written instructions on cleaning and maintenance."*

If Pyranha Mouldings and the individual prosecuted over the death had properly considered the risks to employees when they designed, installed and operated the ovens then he would still be here today.

At the sentencing hearing, the trial judge, sitting at the Royal Courts of Justice in London, imposed the £200,000 fine on the company. Pyranha's refusal to own up to its failures together with the fact that in September 2013, it was fined £50,000 plus costs when it pleaded guilty to breaches of the Health and Safety at Work Act 1974 following the injury of an employee as the result of unsafe systems of work, lack of risk assessments and failure to train employees was taken into account.

Peter Mackereth, the company's technical director who designed the oven which was used to mould kayaks, was also convicted at Liverpool Crown Court of two charges of breaching the Health and Safety at Work Act 1974. He was given a prison sentence of nine months, suspended for two years and a fine of £25,000. Mr Justice MacDuff also awarded costs against the two defendants jointly in the total sum of £90,000.

UNSUCCESSFUL CMCHA 2007 PROSECUTIONS

MNS Mining Limited

MNS Mining Limited was charged with four counts of corporate manslaughter in January 2013 following the deaths of four of its colliery workers in a flooding incident at its Gleision Colliery in Cilybebyll on 15 September 2011. The mine manager Malcolm Fyfield also faced four counts of gross-negligence manslaughter.

Prosecutors said Mr Fyfield was negligent by allowing the men to dig towards an area where underground water was present. The prosecution said Mr Fyfield took staggering risks in allowing the work to go ahead. Two witnesses for the crown - outgoing mine manager Ray Thomas and Neath Port Talbot Council planning chief Alun Rees - both said they warned Mr Fyfield about underground water weeks before the massive flood.

But Mr Fyfield insisted he carried out safety inspections on the eve of the tragedy, which showed no water was present. The prosecution said this did not happen as it would have been 'impossible' to inspect the Old Central Workings because the area was covered in water. However, expert witness and mine hydrology academic Dr Alan Cobb - who was called as part of MNS Mining's defence - said he believed it 'was more likely than not' the water came from another part of the mine.

Both MNS Mining and Malcolm Fyfield were cleared of charges.

PS & JE Ward Ltd

The CPS charged a Norfolk garden nursery with corporate manslaughter and with failing to discharge a duty imposed by Section 2(1) of HASAWA 1974 in relation to the death of an employee. The employee died on 15 July 2010 from an electric shock, caused when the metal hydraulic-lift trailer he was towing came into contact with an overhead power line. The incident occurred in Terrington St Clement, at Belmont Nursery, which was run by PS & JE Ward Ltd, based in King's Lynn. The HSE investigated the incident on the same day and immediately served the company with two Prohibition Notices, relating to: the operation of vehicles of a certain height from coming into contact with or closely approaching the overhead power lines in the field adjacent to the nursery buildings; and the movement of metal irrigation pipework under the overhead lines without undertaking a suitable risk assessment and implementing a safe system of work.

A month later, a further Prohibition Notice was served to prevent the use of the trailer after the Vehicle and Operator Services Agency (VOSA) found that the brakes did not work. An Improvement Notice was also issued to the company to provide information, instruction and training for employees relating to transport and deliveries close to overhead power lines. The firm complied with all the notices.

The jury at Norwich Crown Court cleared the nursery's operator PS & JE Ward of corporate manslaughter, but found it guilty of a breach of Section 2 of HASAWA.

Relevant decided cases

R V ASSOCIATED OCTEL CO LTD [1996] 4 ALL ER 846 [HASAWA 1974 SECTION 2AND3]

Liability for the safety of persons affected by the undertaking

Facts

In June 1990 Associated Octel had closed down its production processes at its Ellesmere Port chemical plant for pre-planned annual maintenance. The repair of the lining of a tank in the chlorine plant was carried out by a specialist contractor, Resin Glass Products Ltd (RGP). During the maintenance, one of RGP's employees was badly burned when the bulb of the lamp he was holding shattered, igniting vapour from the acetone - a highly flammable solvent - that he was using to clean the surface. The contractor RGP was convicted under Section 2 of HASAWA 1974 and Associated Octel was prosecuted under Section 3.

Decision

Associated Octel was fined £25,000, with £60,000 costs. They appealed to the Court of Appeal, asserting that it was not responsible under HASAWA 1974 for the work of the independent contractor, RGP, the appeal failed. Octel subsequently appealed to the House of Lords (now the Supreme Court), who also rejected the appeal.

Relevance

Section 3(1) stipulates that: *"It shall be the duty of every employer to conduct his undertaking in such a way as to ensure, so far as is reasonably practicable, that persons not in his employment who may be affected thereby are not exposed to risks to their health and safety."*

The duty to prevent exposure to risks arising from the conduct of the undertaking includes the activities of independent contractors carrying out the works of cleaning, repair and maintenance, which were necessary for the conduct of the employer's business or enterprise.

R V BRITISH STEEL PLC [1995] IRLR 310 [HASAWA 1974, SECTION 3 (1)]

Reasonable care - criminal controlling mind of a company

Facts

In 1990, at its Sheffield plant, British Steel planned to reposition a section of steel platform. This involved cutting the platform free and using a crane to move it to its new position.

The repositioning work was carried out by a contractor, on a labour only basis, with British Steel providing supervision and equipment. The two men supplied by the contractor cut the platform free, but they neglected to secure it. The platform collapsed and killed one of the men.

Decision

British Steel was convicted of an offence under Section 3(1) for failing to discharge its duty under that Section.

The company appealed to the Court of Appeal on the grounds that, it argued, Section 3(1) allowed a company to avoid liability if, at the level of directing mind it had taken reasonable care. The Court of Appeal was asked to infer that the words 'through senior management' appeared in Section 3(1) immediately after the word 'employer'. That is: *"It shall be the duty of every employer 'through senior management' to conduct his undertaking in such a way as to ensure, so far as is reasonably practicable, that persons not in his employment who may be affected thereby are not thereby exposed to risks to their health and safety."*

The Appeal Court decided that a corporate employer was not able to avoid liability for an offence under Section 3(1) on the basis that the company at 'directing mind' or senior management level was not involved, having taken all reasonable care to delegate supervision. On summing up the Judge said: *"If the appellant's submissions were accepted, it would be particularly easy for large companies, engaged in multifarious hazardous operations to escape liability on the basis that the company, through its directing mind or senior management was not involved. That would emasculate the legislation."*

The Court ruled that the appeal be dismissed.

Relevance

It was accepted that Parliament considered it necessary for the protection of public health and safety to impose absolute criminal liability, subject to the defence of reasonable practicability. It was said that it would 'drive a juggernaut' through the legislative scheme if companies could avoid criminal liability where the potentially harmful event was committed by someone who was not the directing mind of the company.

R V CHARGOT LTD [2008] UKHL 73 [HASAWA 1974 SECTION 2]

Facts

Chargot Ltd was prosecuted for a breach of Section 2 following the death of an employee caused by the dumper truck he was driving overturning. The details of how the accident happened were not clear as no-one witnessed the event. The employee driving the truck had been asked to replace the usual driver. He was required to move truck-loads of top-soil but on his third trip, apparently whilst descending a ramp in the defined route, the truck overturned and he was buried under the soil. As well as his employer, Chargot, prosecutions were also brought against Ruttle Contracting (the principal contractor on site) and its managing director.

Although it was clear that the stand-in driver had been killed whilst carrying out his duties as required by his employer, the exact course of events which led to the accident were unclear. While the prosecution alleged this was a prima facie case that the standards required by his employer had not been achieved, the defence argued that before Section 40 applies, the prosecution must define the exact failure.

Although the argument by the defence was rejected by both the Crown Court and the Court of Appeal it was allowed to be referred to the, then, House of Lords (now the Supreme Court) who also rejected the argument.

Decision

The decision of the final appeal was explained by Lord Hope who agreed it was not enough for the prosecution to make a bald assertion there had been a breach. However, in this case the basis for the allegation were explained more specifically and it was not necessary for the prosecution to define the specific actions which should have been taken. It was evident that the case arose from the fatal accident which resulted from the appellants' failure to ensure this safety and absence of risk.

The argument that the imposition of Section 40 is incompatible with principles of innocent until proven guilty as endorsed by the Human Rights convention was also rejected.

Relevance

The House of Lords decision shows that where the prosecution can demonstrate that an individual has been exposed to a risk of injury, the employer has failed to achieve the result expected by Sections 2 (and 3) of HASAWA74. As such the prosecution have a prima facie case and the onus falls on the defence to demonstrate that they had done everything reasonably practicable.

R V HTM [2006] EWCA CRIM 1156 [HASAWA 1974 SECTION 2 AND MHSWR REG 21]

Unforeseen act of employee and Reg 21 MHSWR as a criminal defence

Facts

HTM Limited was providing traffic management services to a contractor who was resurfacing part of the A66. The contra-flow signs were lit by mobile telescopic towers, provided by HTM Limited, and which extended to a height of 9.1m. Two employees of HTM Limited were instructed to move the towers and did so without first lowering them, as it was said they had been trained to do. A tower struck an overhead power line and both employees were killed.

HTM Limited was prosecuted under Section 2 of HASAWA 1974 for failure to provide and maintain a system for the movement and erection of mobile lighting towers which was, so far as was reasonably practicable, safe and without risks to health; and failure to provide such information, instruction or supervision as was necessary to ensure, so far as was reasonably practicable, the health and safety of its employees.

HTM Limited denied liability, stating that they had taken all reasonable steps in the training they had provided to their employees and it could not have foreseen that they would act as they did.

The HSE contested the HTM Limited defence in that:

- Foreseeability was irrelevant to whether there had been a breach. The HSE argued that it would find it difficult to implement improvement and prohibition notices if the somewhat subjective concept of foreseeability was relevant.
- Regulation 21 of the Management of Health and Safety at Work Regulations (MHSWR) 1999 which provides 'nothing in the relevant statutory provisions shall operate so as to afford an employer a defence in any criminal proceedings for a contravention of these provisions by reason of any act or default of an employee' precluded such a defence.

Decision

The trial judge rejected the HSE's arguments and, upon appeal, the Court of Appeal held:

- The concept of foreseeability is 'merely a tool' with which to assess the likelihood of the risk arising.
- Concepts of fault should not be imported from the civil law and that the concept of reasonable foreseeability did not provide a defence in itself and they had no intention to reduce the protection offered by HASAWA 1974 to the equivalent of liability under common law.
- Ruled that the legal burden on a defendant to show that he had reduced the risk, so far as was reasonably practicable, should be analysed as a ***qualification*** to the offence rather than a ***defence*** in law.

Relevance

Following the HTM Ltd case, in determining compliance with HASAWA 1974, the employer must still demonstrate that everything reasonably practicable had been done to ensure that the employee had the appropriate skill and instruction, that there were safe systems of work, adequate supervision and the employee has been provided with safe plant and equipment for the proper performance of the work.

Regulation 21 of MHSWR 1999 does not prevent employers from pointing to the act of an employee as part of their argument that they had taken all reasonably practicable steps to control that risk.

R V NELSON GROUP SERVICES (MAINTENANCE) LTD [1998] ALL ER 420 [HASAWA 1974 SECTION 3(1)]

Isolated act of employee as a criminal defence

Facts

Nelson Group Services (Maintenance) Ltd (N Ltd) installed, maintained and serviced gas appliances and employed a large number of gas fitters. A fitter employed by the company had been properly trained to enable him to carry out his work competently and safely. While removing a gas fire from a house, the fitter left the gas fittings in a condition which exposed the occupier of the house to health and safety risks.

N Ltd was prosecuted under Section 3(1) of HASAWA 1974. At first instance, N Ltd was convicted. At the Crown Court trial the judge directed the jury on the issue of reasonable practicability, saying that in the circumstances the defence of reasonable practicability was not available.

The company appealed to the Court of Appeal on the basis that:

- It was accepted that gas fittings had been left in such a state that the occupants of houses had been exposed to risk to their health and safety.
- The question was whether this alone made N Ltd guilty of an offence under Section 3, or whether.
- The negligent act by the fitter was not the employer conducting its undertaking for the purposes of Section 3.
- The fitter's negligent act or omission did not prevent N Ltd from showing that it had done all that was reasonably practicable to ensure that the occupants were not exposed to risk.

Decision

The fact that an employee has done work carelessly does not preclude an employer from establishing a defence of reasonable practicability.

The question of what is reasonably practicable is a question of fact for the jury depending on the circumstances of the case, for example, that an employee had safe systems of work, the appropriate skill and instruction, and had been provided with safe plant and equipment.

The appeal succeeded.

Relevance

The 'isolated act' of a negligent employee, in circumstances where a properly trained and instructed employee does something carelessly or contrary to orders, should not give rise to liability in the employer. The implication was that systematic negligence or systematic disobeying of orders would be needed as evidence that the employer had not done enough.

R V PORT OF TILBURY DOCKS AND GLADDING [2007] EWCA CRIM 1937 [HASAWA 1974 SECTION 37]

Criminal neglect - elements to prove

Facts

The proceedings of this case arose out of a tragic accident at the docks, when a six year old boy was thrown from a fork-lift truck. The boy was being carried on the fork-lift truck as a passenger when it collided with a second truck which was 'riding reel', which is the description given to a practice of carrying a third and unclamped newspaper reel on the top of two clamped reels. This practice was alleged to be obviously dangerous.

The prosecution in the trial alleged that the practice of 'riding reel' was commonplace at the docks. The prosecution asserted that Mr Gladding, as managing director of the company and chair of their Strategic Health and Safety Management Committee, had ultimate responsibility for ensuring that the company employed safe systems of work. As such, he was accordingly guilty on the basis that the breaches by the company of its statutory duties had arisen due to his consent or connivance, or was attributable to his neglect.

This is a rather unusual case: the defendant argued that he was not aware of the practice and as such was not guilty of an offence by virtue of Section 37. The judge accepted the argument and this charge was dropped. As this was not itself an acquittal, the prosecution was able to appeal the basis of the decision under section 35(1) of the Criminal Procedure and Investigations Act 1996 with leave of the trial judge. The appeal constituted a ruling at a preparatory hearing given by the Judge in relation to what the prosecution would have to prove in relation to neglect.

Decision

In the Court of Appeal, Lord Justice Latham (the Vice President of the Court of Appeal Criminal Division) disagreed with this approach. He said the argument presented by the defence placed too high a burden on the prosecution. Citing with approval the earlier case of ***Wotherspoon v HM Advocate (1978)***, Lord Justice Latham concluded that, in those cases where there is no actual knowledge of the conduct giving rise to the offence by the company, the test should always be whether: *"The officer in question of the company should have, by reason of the surrounding circumstances, been put on enquiry so as to require him to have taken steps to determine whether or not the appropriate safety procedures were in place."*

Relevance

The case of ***R v P Limited*** makes it clear that it is possible to bring a successful prosecution against a director of a company in respect of one of the 'consent, connivance or neglect' offences without necessarily proving that the director concerned knew of the practice and/or of its dangers.

If the evidence is sufficiently compelling that a director ought to have been put on enquiry as to the relevant practice, a successful prosecution for 'neglect' is possible. It is not necessary to show 'guilty mind' (knowledge of the material facts) in relation to neglect under Section 37 of HASAWA 1974; it is enough to show that there was a duty and the accused was neglectful in their duty.

R V PORTER [2008] EWC CRIM 1271 [HASAWA 1974 SECTION 3]

Facts

Porter was the headmaster of a small private primary school. The school playground was split over two levels connected by a small number of well-maintained steps. The case was brought when a young boy jumped down the steps, landed awkwardly and bumped his head. Porter was prosecuted for a breach of Section 3 of the 1974 Act alleging he had failed to ensure the safety of his pupils (those affected by his undertaking other than his employees) due to inadequate supervision.

In his defence he was able to demonstrate a good safety record as well as suitable and sufficient risk assessments.

Decision

In the first instance, the Crown Court jury found Porter guilty of the offence; Porter appealed. The Court of Appeal accepted his appeal and the conviction was overturned. Lord Justice Moses explained that 'what is important is that the risk which the prosecution must prove should be real as opposed to fanciful or hypothetical'. As there was nothing inherently wrong with the steps, and as such the risk was nothing more than the child would encounter within everyday life, the incident could not have been identified by a risk assessment.

Relevance

This case helps to define the circumstances when a risk must be assessed. The wording from Lord Justice Moss refers to 'real risks' and indicates that fanciful or hypothetical risks need not be considered. Trivial risks, such as those encountered in normal day to day living need not be assessed - unless circumstances, including those introduced by the undertaking, increase those risks to a higher level.

R V SWAN HUNTER SHIPBUILDERS LTD AND ANOTHER [1982] 1 ALL ER 264 [HASAWA 1974 SECTION 2]

Section 2(2)c HASAWA provision of information to third parties

Facts

A fire broke out during the building of HMS Glasgow in 1976. The fire was caused by sub-contractors (Telemeter Installations) who failed to disconnect a hose from the oxygen supply at the end of the shift. This resulted in an overnight build-up of oxygen. Eight workers were killed and two firemen hospitalised. The subsequent investigation showed that although Swan Hunter's employees had been given information (a 'blue book' prepared by their health and safety officer) and training about the danger of oxygen enrichment, Telemeter's employees had not.

Decision

Swan Hunter was found guilty under Section 2(2)c of HASAWA 1974 and fined £3,000. The Court of Appeal upheld the conviction on the basis that Swan Hunter had endangered their own employees because they had failed to provide information and instruction to their contractors (persons other than their employees). The Court of Appeal confirmed: *"If the ignorance of another company's employees places its own employees at risk, then it is the company's duty for the protection of its own employees to inform the employees of another, of any special risks within its knowledge."*

Relevance

The duty of the employer (Swan Hunter) under Section 2(2)c was to provide information 'as necessary' to ensure, as far as is reasonably practicable, the health and safety at work of employees. It was held that, providing information to a third party (Telemeter Installations) that was readily available to the employer (Swan Hunter) was reasonably practicable.

Contractors who are ignorant of the dangers involved in work they carry out for an employer can pose a risk to the employer's employees. Employers have a duty to give information to contractors in order to ensure the health and safety of their own employees.

The conviction underlined the need for an employer to communicate and adequately inform non-employees of special risks that may be created, in order to protect the employer's own employees, because pertinent information, effectively imparted, would diminish such risks.

Civil law

Learning outcomes

On completion of this element, candidates should be able to demonstrate understanding of the content through the application of knowledge to familiar and unfamiliar situations. In particular, they should be able to:

A4.1 Explain the principles of common law.

A4.2 Explain the criteria required to establish a successful civil action for breach of statutory duty and negligence, the main defences available and the procedure for assessment of damages under civil law.

A4.3 Outline the main civil law statutory duties owed by the occupiers of premises to lawful and unlawful visitors.

Contents

Relevant statutory provisions

Enterprise and Regulatory Reform Act (ERRA) 2013 (Section 69)

Health and Safety at Work etc. Act 1974 (Civil Liability) (Exceptions) Regulations (HSWCER) 2013

Social Action, Responsibility and Heroism Act (SARAH) 2015

Occupiers' Liability Act (OLA) 1957 and 1984

Occupiers' Liability Act (Northern Ireland) 1957

Damages (Scotland) Act 2011

Occupiers' Liability (Scotland) Act 1960

Sources of reference

N/A

A4.1 - Common law duties

Principles of tort (delict - in Scottish law) of negligence

THE TERM NEGLIGENCE

The tort of negligence (delict in Scotland) is the most common cause of civil action for work related injury. Negligence was defined in the case of Blyth v Birmingham Waterworks 1856. Here the claimant alleged that the water company had failed in their duty of care by laying a water pipe at too shallow a depth in the ground, which, as a result, became frozen in winter.

"Negligence is the omission to do something which a reasonable man guided upon those considerations which ordinarily regulate the conduct of human affairs would do, or doing something which a prudent and reasonable man would not do."

Figure A4-1: Definition of negligence. *Source: Baron Alderson, Blyth v Birmingham Waterworks [1856].*

Thus, negligence is failing to do something a reasonable person would do or doing something that a reasonable person would not do (i.e. an act or omission) in the same circumstances. The "man on the Clapham omnibus" is often cited as representing the reasonable person. Ultimately it is the Judge who will decide what is reasonable, based on the facts of the case. The Judge will take into account precedents established in prior cases. Thus the law of torts relating to negligence is being continually developed with old cases being applied to new situations.

"The categories of negligence are never closed." In effect the standard of 'reasonableness' changes with the changing morals and attitudes of society as a whole."

Figure A4-2: Development of the term negligence. *Source: Lord Macmillan said in Donoghue v Stevenson [1932].*

In order to prove negligence, the claimant must show that:

- The defendant owed the claimant a duty of care.
- The defendant was in breach of that duty.
- That the claimant had suffered an actionable loss as a result of the breach.

DUTY OF CARE OWED

Thus, before negligence is proved, it must first be established that a duty of care is owed by the defendant to the claimant. Donoghue v Stevenson (1932) helped establish a principle of care that was owed, called the 'neighbour principle'.

"You must take reasonable care to avoid acts or omissions which you can reasonably foresee would be likely to injure your neighbour. Who then, in law, is my neighbour? The answer seems to be - persons who are so closely and directly affected by my act that I ought reasonably to have them in contemplation as being so affected when I am directing my mind to the acts or omissions which are called into question."

Figure A4-3: The neighbour principle. *Source: Lord Atkin.*

BREACH OF THE DUTY OF CARE THROUGH NEGLIGENCE

A breach of the duty of care is required when proving negligence. This depends on the standards of care that may be expected from the 'reasonable person'.

CAUSAL LINK BETWEEN THE BREACH AND THE LOSS SUFFERED

The injury (loss) was a result of the breach. The injury or loss must be as a consequence of the breach. Generally, the harm must be 'reasonably foreseeable'. The breach of duty must be connected to the loss, for example, a claimant may claim that the noise level in a workplace caused them to conduct a manual handling task badly and injure their back.

This may be considered by the court hearing the case to be too remote a link. In addition, it is necessary to prove that the loss suffered was not too remote from the breach of duty. This is particularly applicable to situations where a claimant claims for what may be seen as consequential loss. Whilst a court can confirm compensation for damages directly arising from the breach of duty this does not generally extend to more remote consequential loss, for example, missing a wedding because of injury caused.

See also - 'Decided cases' for 'Fairchild v Glenhaven Funeral Services Ltd and Others [2002] UKHL 22'.

Concept of res ipsa loquitur

Normally the burden of establishing that the defendant has broken the duty of care and there was a link to the loss suffered rests with the claimant. In some instances the facts that the claimant establishes are such that 'the thing may speak for itself' - ***res ipsa loquitur***.

In this situation, it is made obvious to the court that there is negligence by the expression of the facts of the case alone. In this case, the claimant does not have to offer further evidence and it then remains for the defendant to establish that the damage occurred without their negligence. In ***Cassidy v Ministry of Health***

(1951), for example, the Court of Appeal was of the view that the very fact that the claimant emerged from treatment in a hospital with a paralysed hand raised a presumption that the defendant hospital had been negligent. The presumption may arise when the claimant has satisfied the court as to what has occurred, but has not been able to show exactly how it occurred. For the presumption to arise the situation must have been under the control of the defendant; thus it is particularly likely to arise where an accident has occurred on premises under the control of the defendant.

The principle of ***res ipsa loquitur*** was established in ***Scott v London and St Katherine Docks Co (1865)***. Here the defendants dropped a bag of sugar from a crane onto the claimant, causing injury. The claimant satisfied the court on this fact. The Court of Exchequer (now the High Court) was then prepared to find that the claimant had raised a presumption that the defendant had been negligent. In the judgement the rule is stated as:

> *"There must be reasonable evidence of negligence, but where the thing is shown to be under the management of the defendant, or his servants, and the accident is such as, in the ordinary course of things, does not happen if those who have the management of the machinery use proper care, it affords reasonable evidence, in the absence of explanation by the defendant, that the accident arose from want of care."*

Figure A4-4: Res ipsa loquitur. *Source: Erle Court Judge (CJ), Scott v London and St Katherine Docks Co [1865].*

It should be noted that ***res ipsa loquitur*** is a rule of evidence rather than a defence.

FORESEEABILITY OF THE TYPE OF DAMAGE

The damage arising from alleged negligence must be foreseeable. The court considering a new matter, with no established case law showing foreseeability of harm would have to consider whether there was a theoretical possibility of harm or that there was a reasonable foreseeability of harm. It is not enough to show theoretical harm - the harm must be foreseeable to a reasonable person. In the past it was necessary to prove a clear causal link with exposure to noise and noise induced deafness; it was also necessary to show that exposure to mineral oils could lead to scrotal cancer.

Concept of 'duty of care'

TO WHOM A DUTY IS OWED (THE 'NEIGHBOUR TEST')

In the case of Donoghue v Stevenson [1932], the claimant, Ms Donoghue, became ill when she drank a bottle of ginger beer which contained a decomposed snail. The bottle was opaque and she had no reason to believe that it contained anything other than ginger beer. She sued the manufacturer.

As it was Ms Donoghue's friend who had purchased the ginger beer, the manufacturer claimed that there was no contract between them.

Damages were awarded to the claimant. The House of Lords held that a duty of care was owed and determined the extent of that duty known as the 'neighbour principle'.

The question was asked, 'To whom is this duty of care owed?' said:

> *"You must take reasonable care to avoid acts or omissions which you can reasonably foresee would be likely to injure your neighbour."*

Figure A4-5: The neighbour principle. *Source: Lord Atkin, Donoghue v Stevenson [1932].*

Who is my neighbour, in the legal sense?

> *"Persons who are so closely and directly affected by my act that I ought reasonably to have them in contemplation as being so affected when I am directing my mind to the acts or omissions which are called in question."*

Figure A4-6: Clarification of term neighbour. *Source: Lord Atkin, Donoghue v Stevenson [1932].*

In other words, my neighbour is someone who is affected by what I do.

THE DUTY OF CARE OF VARIOUS CATEGORIES OF PERSON

Designers, manufacturers and suppliers to customers/users

The duty of care owed by designers, manufacturers and suppliers to customers/users follows on from the Donoghue v Stevenson case.

At common law a manufacturer has a duty of care to both industrial and domestic users of his product (including servicemen and repairers) in respect of defects. He has a duty to the ***"end user"***.

In order for an action for negligence to succeed, the claimant must show that the defect:

- Was caused by lack of reasonable care on the manufacturer's part.
- Was foreseeably likely to cause injury, and loss occurred.
- Existed at the time that the product left the manufacturer.
- Was not one which a legitimate intermediary (for example, the retailer) or the user could be expected to discover and modify.

Manufacturers and suppliers are not automatically liable for injuries caused by a product. They will be liable if they fail to take reasonable care. For example by:

- Failing to consult available technical information.
- Failing to give suitable warnings or clear instructions.
- Poor ergonomic details in, for example, instrumentation leading to user error.
- Defective insulation in an electrical appliance.

The seller as well as, or instead of, the manufacturer may be liable for negligence if they failed to make reasonable enquiries about the use and purpose of the product in relation to its safety or suitability.

The supplier's liability can be offset or reduced if, for example:

- The user (for example, employer) is expected to test the goods himself.
- Where the goods are put to a purpose for which they are not intended.
- They are not maintained in accordance with the manufacturer's instructions.
- The user uses the equipment knowing it is defective.

In recent years product liability has moved from the need to establish negligence towards strict liability. The sale and supply of goods is regulated by statutes with both civil and criminal liability under the Consumer Protection Act (CPA) 1987 and the Sale and Supply of Goods Act (SSGA) 1994.

Occupiers of premises to those using or visiting the premises

Under the Occupiers' Liability Act (OLA) 1957, an occupier owes a 'common law duty of care' to all visitors. The duty is to take such care as is reasonable (given the circumstances) to see that visitors are reasonably safe in using the premises for the purposes for which they are invited or permitted to be there.

A visitor is usually extended an 'area of invitation' in which they are expected to stay and to which the duty applies. If a visitor strays outside this area then they become uninvited. A visitor going off to search for toilets, however, is likely to receive special sympathy from the courts.

The OLA 1957 reminds the employer (or occupier) that they must be prepared for children to be less careful than adults. They must also take care that 'allurements' such as attractive poisonous berries or machines are considered. The occupier may be liable even if the child is a trespasser as far as the allurement itself is concerned.

Under Section 2(1) of the OLA 1957 the occupier may exclude or restrict liability by giving adequate warning or by contract. The Unfair Contract Terms Act (UCTA) 1977 states that the common law duty of care for death or personal injury cannot be excluded in relation to business premises. In addition loss or damage can only be excluded when it is reasonable to do so.

In determining whether the occupier of premises has discharged the common duty of care to a visitor, regard is to be had to all the circumstances, so that (for example):

- Where damage is caused to a visitor by a danger of which he had been warned by the occupier, the warning is not to be treated as an excuse absolving the occupier from liability, unless in all the circumstances it was enough to enable the visitor to be reasonably safe.
- Where an injury arises through the defective work of a contractor, the occupier can avoid liability provided that the contractor was selected with reasonable care.

The common law duty of care does not impose on an occupier any obligation to a visitor in respect of risks willingly accepted by that visitor. The question whether the visitor accepted a risk is a matter for the court to decide using the same principles as in other cases in which one person owes a duty of care to another.

However, an occupier may expect that persons, in the exercise of their calling, will appreciate and guard against any special risks ordinarily incident to that calling. An employer (or occupier) cannot be prosecuted for a breach of their duties under the OLA 1957, but if a visitor is injured, the occupier may be liable to pay compensation for the injury.

Persons who enter premises for any purpose in the exercise of a right conferred by law (for example, HSE inspector, Customs and Excise Officer) are to be treated as permitted by the occupier to be there for that purpose, whether they have his permission or not.

The position towards non-visitors is governed by the Occupiers' Liability Act (OLA) 1984. Prior to the OLA 1984 certain duties of common humanity were owed under common law by an occupier to those other than lawful visitors - for example, to trespassers or ramblers. The OLA 1984 extended the rules of common law (see British Railways Board v Herrington (1972)) and defines under what circumstances a duty is owed. In particular, occupiers owe a duty if:

- They are aware of a danger.
- They know (or should know) that a person may put themselves at risk.
- The risk is one which the occupier might reasonably be expected to do something about.

The extent of the duty is as follows:

- The occupier must take such care as is reasonable in the circumstances of the case, to see that the other person does not suffer injury.

The Act provides that the duty may be discharged by:

- Giving warning of the danger.
- Discouraging people from putting themselves at risk in the first place (for example, by making it more difficult for trespassers, etc. to enter the premises).

As with the OLA 1957, the duty is greater towards children. What constitutes a reasonable warning for an adult may not do so for a child. If the occupier, knowing that children trespass onto his land, fails to take action (for example, by repairing fences) this might be taken as evidence of implied permission and an injured child would therefore be considered a visitor.

Contractors to clients and vice versa

Under the neighbour principle espoused in Donoghue v Stevenson, a reciprocal duty of care exists between clients and contractors. The duty owed by a client to visiting contractors is that laid down by the Occupiers Liability Act (OLA) 1957 ***(see above)***. That is, the client must have safe premises and take reasonable care of the contractor's safety. In return, the client may assume that the contractor will take all precautions reasonable to his trade and the client is not expected to tell a highly skilled workman how to do his job. If the client (occupier) is involved in supervising operations, however, then they have to ensure that adequate safety arrangements have been made.

Clients are, however, liable for risks that are peculiar to the premises but are not occupational hazards. For example, a client could not expect a contractor such as a telephone engineer to deal with working on an unusually fragile roof as part of his normal expertise.

The client would not be liable under English civil law for the dangers created by contractors on or near his premises - provided they have taken reasonable care in selecting them. The exception to this is if the injured person is the client's own employee (a strict liability exists for the reasonable care of employees which cannot be delegated to a contractor). It should be noted however that there might be criminal liability under the HASAWA 1974 for exposure to risks arising from the employers' undertaking ***(see also - R v Swan Hunter and R v Associated Octel).***

EXTENT OF DUTY

Remoteness

This is the principle that the damage or loss is not sufficiently connected to the defendant's breach of duty of care. The principle of a new act intervening (novus actus interveniens) is an extension of the defence of remoteness of damage. To succeed it must be shown that the chain of causation is broken by an extraneous act. This principle was the defence in Scott v Shepherd [1773]. Here the defendant lit a squib at a market fair and threw it onto a stall (A). The stallholder A threw it onto another stall (B) where, in turn, it was thrown onto another stall (C) where it exploded and injured C. This defence was unsuccessful and the defendant was held liable because there was a direct chain of causation that was not broken by the acts of A and B.

The chain of causation can also be broken if the claimant acts unreasonably by, for example, taking an avoidable and foreseeable risk of injury himself. If this act can be seen to break the chain then this might be enough to establish a complete defence for the defendant (if this action is not seen to break the chain the defence will fail; however the claim may be reduced due to contributory negligence).

In some cases, the loss may be too remote where there is a sequence of physical cause and effect without human intervention. This is illustrated by The Wagon Mound [1961] case that involved the negligent spilling of oil into the sea by a ship in Sydney Harbour. The oil drifted to a wharf 200 yards away where a ship was being repaired. The owner of the wharf on seeing the oil ordered welding to cease due to the fire risk. He was later advised that the risk of fire from sparks was unlikely and accordingly allowed welding to continue albeit with additional safety precautions. A spark fell onto a piece of floating cotton waste and ignited the oil causing damage to the wharf. The charterers of the Wagon Mound were sued by the owners of the wharf. The claim failed based on the test of reasonable foresight in that pollution was a foreseeable risk; however the fire was too remote from the cause.

Note: cases involving ships are usually referred to by the ship's name. The Wagon Mound case was brought on appeal to the Judicial Committee of the Privy Council from Australia. It has persuasive precedent in a court in England and Wales.

Reasonableness

What is meant by reasonable care will be decided by the judge in the light of contemporary knowledge and thinking. For example, what was considered reasonable conduct in the 1960s will not necessarily be seen as reasonable many years later. As in the case of Blyth v Birmingham Waterworks [1856], what is reasonable is doing something a reasonable person would do in the same circumstances. Who then is the reasonable man? The 'man on the Clapham omnibus' is often cited as representing the reasonable person. This does not mean that the average prudent person has a low standard of care. Most people behave unreasonably from time to time and it is no defence to say that we are usually reasonable people. Ultimately it is the Judge who will decide what is reasonable based on the facts of the case.

Reasonableness could include the making of decisions for controlling a risk being based on the foreseeability and remoteness of a risk causing loss and the level of risk involved.

Foreseeability

The more likely or probable an accident is, the greater the duty to guard against it. If there is only a remote possibility of loss, the need for precautions is usually much reduced. Indeed it may not be reasonable to take any precautions at all against very unlikely events. In Latimer v AEC [1953], a flash flood was unforeseeable and the company could not have been expected to guard against it.

A car driver knows that even though he exercises due care while driving on a normal road surface, his tyres may conceivably throw a stone against the windscreen of the car behind and so cause a serious accident. The only way he could ensure that it did not happen would be to drive at 10 mph or not drive at all.

The law does not of course require such drastic precautions in normal road conditions and so it would not make him liable in these circumstances. If the road had recently been re-surfaced and stones were all over the road surface, it is foreseeable that a stone be thrown up and loss result. It then becomes reasonable to drive at a reduced speed - particularly when warning signs have been displayed.

In Millard v Serck Tubes Ltd [1969] a worker was using a power drill with his hand resting on the guard. A piece of swarf thrown out from the work-piece wound around the worker's hand and drew it into the drill causing injury. The Court of Appeal rejected the employer's defence that, although the drill was inadequately guarded, the accident itself was unforeseeable.

They could not escape liability for injury on the grounds of unforeseeability when there was a duty under the Factories Act to fence the dangerous parts of machinery

In Doughty v Turner Manufacturing Co [1964] an asbestos cement lid accidentally fell into a vessel containing sodium cyanide at 800°C. There was an explosion due to a chemical reaction which was previously unknown to science.

The defendant was injured by the molten liquid ejected by the explosion rather than the splash caused by the lid falling into the cauldron. His claim failed because splashing by sodium cyanide was foreseeable but the explosion was not.

Thus in summary the defendant is only liable for damages in respect of both events and loss that can be reasonably foreseen.

GREATER DUTY OF CARE TO MORE VULNERABLE INDIVIDUALS

Where vulnerable individuals may reasonably be foreseen to be affected by the acts or omissions of someone, they owe them a greater duty of care. This extends particularly to children, who may be vulnerable due to their early stage of development or their lack of knowledge, skill or perspective of risk. This greater duty of care may be extended to the disabled. For example, if the disability constituted poor eye sight of a worker an employer would be under a greater duty of care to safeguard the limited eyesight the person had than to a person with normal eyesight. This principle can be applied to a situation where workers have been exposed to noise and have suffered hearing loss, for which an employer has been sued and provided compensation. The workers will now be vulnerable individuals with regard to their remaining hearing. Continued exposure would cause them progressively more harm and the employer would be under a greater duty to protect it.

See also - 'Paris v Stepney Borough Council [1951] 1 ALL ER 42, HL' in 'Relevant decided cases' - later in this Element.

Common law duties owed by employers to employees

The common law duty of care owed by an employer may be summed up as:

- The employer must take reasonable care not to subject his employees to unnecessary risk.

The elements of this duty were established by the case of Wilsons & Clyde Coal Company v. English in 1938. Here the House of Lords held that the employer must provide:

- Proper and safe plant and appliances.
- Safe systems of work (with adequate supervision and instruction).
- Safe premises (this includes safe access and egress).
- Competent fellow employees.

It was also held that the duty owed was a personal one, which could not be delegated to an agent. Employers have no liability for risks that a reasonable employer could not have foreseen. Reasonable steps for employees' health and safety can only be taken if the employer knows of the danger. Employers have a duty to keep themselves abreast of the latest developments within their sphere of activity. Thus ignorance of a danger can be reasonable provided that a reasonable effort has been made to keep up to date with developments.

A SAFE PLACE OF WORK AND SAFE ACCESS AND EGRESS

An employer has a duty to ensure that the place of work is safe together with safe access and egress. The extent of the duty is to guard against foreseeable risks, not for those which are transient and exceptional ***(see***

also - Latimer v AEC [1953]). The duty to ensure safe access and egress extends beyond the duty to clear snow and ice from around the workplace but also to ensure, for example, that someone working at height can reach and leave their place of work safely.

As with all elements of the employers' duty, the duty is to take reasonable steps. In ***Markwell v Suffolk Coastal District Council*** an employee slipped on a patch of ice 18 inches in diameter. His claim failed because the judge considered that, when taken as a whole, the yard was reasonably safe. In contrast in ***Bath v BTC [1954]*** a man worked on a narrow ledge at the top of a dock, without any edge or fall protection to prevent death or serious injury if he slipped. He did so and the employer was held liable for failing to provide a rail or a net in a dangerous place. The judge held that the fact that the fall may have been the employee's fault was unimportant. The required precautions would undoubtedly be expensive and have ramifications throughout the industry; however these were justified due to the risk.

SAFE SYSTEMS OF WORK

The employers' duty to take reasonable steps to prevent injury to employees reaches every aspect of the employer's undertaking. The provision of a safe system of work encompasses the working environment, the work equipment, the way in which the job is carried out and the provision of the colleagues that the employer gives employees to work with. Thus it relates not so much to the place or the work equipment but to the way in which they are used. The law does not require employees to be treated as though they are children. The employer should be able to assume that employees can deal with routine and straightforward operations.

"An experienced workman must know the ordinary risks of the work he is employed to do. In doing that work he is expected to take ordinary routine precautions common to it and should not expect to be told by his employers of every danger which might arise and of every step that should be taken to counteract that danger."

Figure A4-7: Ordinary risks and precautions. *Source: The judge, Ferner v Kemp [1960].*

It is the employer's duty to reduce risks as far as is reasonably practicable. If the job cannot be made absolutely safe, the employer must do all he can to make it as safe as can be reasonably expected in the circumstances. The extent of the duty to provide a safe system of work has progressed over the years to more than just the prevention of physical injury.

The duty also extends to the provision of reasonable care to take steps to ensure that employees' mental health is not adversely affected by overwork or stress at work (as was the case in Walker v Northumberland County Council [1994], where damages were awarded to a social worker for the mental stress caused by overwork).

SAFE PLANT, EQUIPMENT AND MATERIALS

The employer has a duty to provide safe and proper plant and equipment that must be suitable for the task being undertaken. This applies to machinery, apparatus, goods, vehicles and the like used by employees. As previously stated the precautions taken by the industry or in the trade provides important evidence that the appropriate standard of care has been taken in any particular case.

A general industry practice which ignored hazards that are inherent to the job, or treated an occupational disease as something about which very little can be done (when precautions are appropriate) would be rejected by the court.

In Bradford v Robinson Rentals [1967], the employer failed to provide suitable equipment when a van without a heater had to be used for travel on behalf of the company, resulting in the driver contracting frostbite. Frostbite is usually associated with mountain climbing or suffering extreme cold if lost in the open countryside in the depths of winter, not driving a van. Looking at the foreseeable risks associated with driving a van would not normally include the driver getting frostbite.

However, this was not deemed to be the point. The fact was that the employer had not provided suitable equipment for use at work, i.e. a van without a heater for use in winter, and this resulted in the claimant's loss. The nature of the loss was not relevant. The only time the nature of the loss would be relevant would be when an amount of compensation was awarded for that loss. The nature of the loss had nothing to do with the point of law, i.e. the employer failed to provide suitable equipment and this caused the claimant's loss.

There must also be a suitable system for the testing, inspection and maintenance of equipment. A suitable system should take reasonable steps to detect both obvious and non-obvious causes of defects. In Bell v Arnott and Harrison (1967) the employer bought a new Black and Decker electric hand drill. After a year without inspection a screw came loose and a bush flew out injuring an employee.

The employer argued that they had a reasonable right to assume that equipment supplied by a reputable manufacturer was safe. This argument failed as the judge held that the equipment should have been subjected to regular checks.

INSTRUCTION, TRAINING AND SUPERVISION

The employer must take reasonable steps to ensure that adequate instruction, training and supervision is provided to employees in order to take reasonable steps to ensure their safety. Training is relevant to all levels and may take many different forms.

To quote some previous judges' remarks:

> *"The employer must remember that men doing routine tasks are often heedless of their own safety, and may become careless about precautions. He must by his foreman keep them up to the mark and not tolerate any slackness."*
> *"The defendants having given these lectures and prescribed the right method did what reasonable and careful employers could do in order to institute a safe and effective system of work."*
> *"If an employer ... put a young girl in charge of such a machine, they are under a duty of giving special instructions to her."*

Figure A4-8: Comments on instruction, training and supervision. *Source: Various judges.*

The level of duty is dependent on the person and the circumstances. Young people and inexperienced workers for example require more training and supervision than that of an adult, experienced, employee. Clearly one-to-one supervision cannot be provided at all times and the law does not require it particularly where experienced employees are concerned.

COMPETENT FELLOW EMPLOYEES

The provision of competent fellow employees applies to all levels within the hierarchy of an organisation. In simple terms this means that the employer should select employees who have both the mental and physical capabilities required to perform their duties in a manner which is safe to both themselves and others. The term employee usually encompasses everyone from the chief executive, managing director etc. through to newly recruited trainees. An incompetent manager can have a significant effect on those who are affected by his or her decision-making - even though they may be remote from that manager's direct influence. It should be noted, that while the duty to provide competent employees rests with the employer as a corporate entity, there can be a personal liability. A manager assigned with responsibility for the safety of their department is not personally liable for an accident unless there is a personal failure to take reasonable care.

This duty extends beyond competence through to behaviour in the workplace. An employee who is known to engage in practical jokes and horseplay could constitute a hazard, which the employer would be expected to deal with. In Hudson v Ridge Manufacturing [1957] an employee was a known practical joker who had made a nuisance of himself over the previous 5 years. Examples of his behaviour included tripping people from behind.

Although this was done for fun, and without malice or intent to bully, injury was inevitable sooner or later. In one incident an employee broke his wrist in an effort to break his fall after being gripped around the neck and pulled back by the practical joker. The employer was held liable in the resulting case on the basis that they had failed to take steps to curb his horseplay. It was held that the duty of the employer is to provide safe colleagues to work with. In this case the employer's duty was to take steps by reprimand, discipline and, if necessary, dismiss the practical joker in order to remove the risk of injury to other employees.

Damage for which tortfeasor is liable

RELEVANCE OF DAMAGE OF FORESEEABLE TYPE

A tortfeasor is defined as someone who has carried out a tort. Tortfeasors are liable for damages that occurred while they were in breach of duty. In ***Thomson and others v Smiths Ship Repairers (North Shields Ltd [1984]),*** relating to a claim for damages for occupational deafness, it was held that damages should be apportioned to all previous employers. It was, however recognised that most of the damage from noise occurs in the early years of exposure, and that subsequent employers may be liable for a smaller portion of the claim.

See also 'Element A4.2 - Concept of joint tortfeasors'.

Damage might also be for psychological injury. Employers are normally entitled to assume that employees can withstand the normal pressures of a job. However, there have been successful claims for psychological injury, in particular occupational stress. In the case of ***Sutherland v Hatton and others [2002]*** the Court set out a number of practical propositions for future claims concerning workplace stress. This precedent was set to 'close the floodgates' on claims for stress.

DATE OF KNOWLEDGE OF RISK

As previously stated employers have a duty to keep themselves abreast of the latest developments within their sphere of activity. Thus ignorance of a danger can be reasonable provided that a reasonable effort has been made to keep up to date with developments. The concept of date of knowledge of risk is illustrated by the ***Thomson and others v Smiths Ship Repairers (North Shields Ltd)*** case outlined above. Here it was held that employers were deemed to have first known about the risk of noise induced hearing loss when the publication 'Health and Safety booklet 25 'Noise and the Worker'' was first published in 1963 by the Department of Employment. Similarly, it has been held that industry should have been aware of the risk of vibration white finger in relation to certain processes involving the use of vibration inducing machinery since 1976. Employers are not, however, expected to change overnight and, in one case, the judge allowed three years from this date of knowledge for employers to make modifications to reduce vibration (Shepherd v Firth Brown [1985] unreported).

Concept of vicarious liability

CIRCUMSTANCES IN WHICH EMPLOYER IS VICARIOUSLY LIABLE FOR NEGLIGENCE OF EMPLOYEE

The two major principles regarding liability at tort are that:

- A person is only responsible for the loss or damage caused by his or her own acts and omissions.
- A person is only liable if he or she is at fault.

The doctrine of the employer's vicarious liability in respect of employees appears to go against these principles and may be stated as:

> *"The employer is vicariously liable for the torts of the employee provided that they are committed during the course of employment and where a third party is hurt."*

Figure A4-9: Vicarious liability. *Source: Ambiguous.*

It has been argued that justice is served because, if the employer chooses to employ a careless employee, then that employer should be liable for the losses caused by that employee's torts. More usually the case given is that the employee is acting on the employer's behalf. Effectively this liability is paid for by all employers, all of whom are obliged to take out insurance under the Employers' Liability (Compulsory Insurance) Act (ELCIA) 1969. The employer is only liable if there has been fault. A driver who has a sudden and unexpected heart attack cannot be said to be to blame for a resulting accident and therefore the employer would not be vicariously liable. The second proviso is that the action must be within the course of employment. If the wrongful act is done purely for the employee's own benefit, then he alone is responsible for it. Acting for his own benefit, the negligent employee is said to be 'off on a frolic of his own'.

Acts within the scope of employment

As has been stated, an employer can be liable for the torts of the employee where that employee has acted improperly or has broken the rules laid down by the employer. This is logical and will often be the case. Negligent acts do not usually occur when employees are doing only the things that they have been told to do - or are supposed to do. In Century Insurance Co Ltd v Northern Ireland Transport Board (1942) a tanker driver lit a cigarette and threw away a match whilst delivering petrol. An explosion resulted and the employer was found to be liable for the negligence of the driver. In this case the driver was said to have acted improperly in the course of employment. The employee was still basically engaged in the work of unloading petrol. In other words he was acting within his employment.

Acts beyond the scope of employment

As indicated, the employer is not liable for acts that are personal to the employee. For example, if an employee working in a shop has a fight with a customer, then that would be regarded as being beyond the scope of his employment unless the employee is required to use force as part of his job (for example, the employer could be liable if a door supervisor at a night club used excessive force to eject a troublemaker).

In Twine v Bean's Express Ltd (1946) the driver of a van gave a lift to a third person against the specific instructions that were displayed on a sign in the van. When the third person was killed due to the driver's negligent driving, the employer was not held vicariously liable because the driver acted beyond the scope of his employment. The passenger was a trespasser and, in offering a lift, the driver was not acting in the course of his employment.

Similarly, if an employee is on a jaunt of his own then the employer is not liable for any negligence. Thus when a brewer's deliveryman took a van without permission in order to deliver a neighbour's coffin, the brewers were not held liable for a person who was injured on the return journey.

Similarly, if a vehicle driver took a new and independent journey for his own purposes then the employer is unlikely to be liable. If, however, it was a minor deviation (for example, a slightly longer road) then this would not be considered to be outside the scope of employment. If an employer lends a vehicle to an employee entirely for his own use then the employer would not be vicariously liable.

VICARIOUS LIABILITY FOR ACTIONS OF CONTRACTORS

As a general rule, a client is not liable in civil law for the torts of an independent contractor. This is provided that the client has fulfilled the common law duty to select his contractor with reasonable care. There are some exceptions, however, where the law imposes strict liability on someone in the client's position. This is based on the idea that the client is in breach of a primary duty. These include:

- Where the client is negligent in that he has failed to select his independent contractor with reasonable care or has given the contractor imperfect instructions or information.
- Where the client authorises or ratifies the tort of the contractor (for example, the employer turns a blind eye or condones an illegal activity such as fly-tipping).
- The client's duty to his employees under health and safety legislation (for example, the duty to prevent access to dangerous parts of machinery could not be delegated to a contractor).

- Where the work requires extra precautions, for example.
- Where the work is on, or adjacent to, the public highway.
- Where the work is particularly hazardous.
- Where a person brings and keeps anything on the land likely to do mischief, he is liable if it escapes (the Rylands v Fletcher rule). The liability cannot be avoided by employing an independent contractor.
- Where the liability for the tort is strict, responsibility cannot be delegated. The duty of an employer to prevent access to the dangerous part of a machine is laid down by statute and cannot be delegated to an independent contractor.

See also - 'Decided cases' for 'Viasystems (Tyneside) Ltd v Thermal Transfer (Northern) Ltd (2005) CA'.

LIABILITY FOR EMPLOYEES OF ANOTHER COMPANY

The general principle established by Wilsons and Clyde Coal Company v English is that the employer's liability in respect of injury cannot be delegated. This is a well-developed point of common law. In McDermid v Nash Dredging and Reclamation Company Ltd [1987] Nash entered into a joint venture with a Dutch company, Sevin, to dredge a Swedish fjord. Mr McDermid, a deckhand, was employed by Nash but seconded to a vessel belonging to Sevin under the control of a Dutch captain. Nash was seriously injured when his leg became entangled in a mooring rope. The House of Lords, in upholding the award of damages against Nash, ruled that it was the employer's responsibility to devise and operate a safe system of work. They could not delegate this responsibility to the Dutch Captain.

Another key principle is that the employer is vicariously liable for the torts committed by the employee during the course of employment. This liability may still exist even where the contract of employment is temporarily transferred. The case of Mersey Docks and Harbour Board v Coggins and Griffiths (Liverpool) Ltd [1946] illustrates the point. Coggins, a firm of stevedores, hired a crane and driver (Mr Newall) from Mersey Docks and Harbour Board. Mr Newall negligently operated the crane thereby injuring a third party, Mr McFarlane.

The hire contract stipulated that Mr Newall was to be a temporary employee of Coggins for the duration of the agreement. The House of Lords held that Coggins were not liable, as they, as stevedores, did not have the skill or experience to direct how the crane was to be operated. Mersey Harbour Board were therefore held vicariously liable even though Newall was a temporary employee of Coggins.

LIABILITY FOR THE LOSSES CAUSED BY CRIMES AND FOR CRIMINAL ACTS

It should be noted that the principle of vicarious liability can be extended to include civil liability in respect of a crime - provided that it was committed during the course of employment. In Nahhas v Pier House [1984] an employer was held liable for the losses caused by thefts committed by a resident porter who had been entrusted with their tenant's keys. The employer is not however liable for criminal acts merely because the work enabled the act to be committed (for example, a cleaner making unauthorised telephone calls or a docker stealing goods whilst unloading a ship).

Vicarious liability can also extend to an employer being prosecuted and convicted for the acts of employees under criminal law. This is designed to ensure that the organisation does not routinely delegate their criminal responsibilities to individual employees. For example, food shops are required to sell food that is fit for human consumption. If they were able to delegate this responsibility to individual shop assistants, they would have no incentive to ensure that only wholesome food was sold. In order to defend such a case, the employer would have to demonstrate that they had done everything appropriate to ensure that the employee was both fully instructed and that arrangements were in place to see that these instructions were carried out. In other words, they had acted with 'due diligence'.

RECOVERY OF COSTS

The employer can claim redress against the negligent employee. This happened in Lister v Romford Ice and Cold Storage (1957). Here one employee injured another through careless driving. The employer, acting on behalf of the insurance company, successfully claimed indemnity against the negligent driver. The modern practice, however, is only to make such a claim where there has been wilful misconduct or there is evidence of collusion. The negligent employee also risks dismissal where there has been a breach of contract of employment (subject to the provisions of employment law).

A4.2 - Breach of statutory duty and negligence

Principle that a breach of a statutory duty may give rise to civil liability

A breach of statutory duty (failing to comply with the obligations imposed by statute law) is a recognised tort and may also be a cause for a civil claim for compensation in cases that result in personal injury or loss.

In practice, there is nothing to prevent a claimant basing their case on both breach of statutory duty and negligence, the former having a limited interpretation but affording ready proof of liability, and the latter being capable of wide interpretation but being more difficult to prove. Pursuing a claim on the basis of negligence and breach of statutory duty is called a 'double barrelled action'.

This principle is best illustrated in the case of Stark v Post Office (2000). Mr Stark, a postman, was injured at work when his cycle front brake snapped in two, the front wheel locked and he was thrown over the handlebars. The bicycle was supplied by the Post Office.

The brake had snapped because of metal fatigue or manufacturing defect and the defect would not and could not have been discoverable on any routine or rigorous inspection. The court found the Post Office had done their best to maintain the bike and had done everything they could to check for faults.

The question for the court was whether the Post Office had breached their statutory duty under Regulations 6 of the Provision and Use of Work Equipment Regulations (PUWER) 1992 that says that *"Every employer shall ensure that work equipment is maintained in an efficient state, in efficient working order and in good repair."* Was that an absolute duty, or a duty just to take reasonable steps to keep the equipment in good order?

The Court said the duty was not breached as it required a reasonable level of maintenance. But the Court of Appeal have overturned the decision and ruled that the Regulation imposes an absolute duty, and, since the bike broke, the employers must have been in breach.

However, the introduction of the Enterprise and Regulatory Reform Act 2013 (ERRA) means that in most work related cases an injured party cannot simply base their claim on a breach of statutory duty. Instead claimants will need to show that there has been negligence on the part of the employer. In the past, some Regulations imposed a 'strict liability' on the employer. This meant that the employer could have been liable for damages even though there was nothing the employer could have done to prevent the incident.

Criteria for a successful action

In order to establish a claim for breach of statutory duty, the claimant must establish the following:

- That the statutory provision, properly construed, was intended to protect an ascertainable class of persons of whom the plaintiff was one.
- The defendant was in breach of the statute.
- The plaintiff suffered the type of injury that the provision was designed to prevent.
- That the injury was caused by the breach.

Breach of statutory duty including the criteria for a successful action

Section 47 of the HASAWA Act 1974 and the Enterprise and Regulatory Reform Act 2013

HASAWA 1974 is essentially a criminal statute enforced by criminal sanctions. Section 47 specifically provided that nothing in Part I (which contains the relevant provisions so far as this statute is concerned) shall be construed as conferring any right of action in any civil proceedings in respect of a failure to comply with any duty imposed by Sections 2-7 or a contravention of Section 8.

The Enterprise and Regulatory Reform Act 2013 (ERRA), Section 69, amended the HASAWA 1974 in order to provide that there should be no civil right of action for breach of a duty imposed by certain health and safety legislation, other than where such a right is specifically provided for.

The Explanatory Notes to ERRA 2013 confirm the intended consequence of this reform as follows: "The amendment ... reverses the present position on civil liability, with the effect, unless any exceptions apply, that it will only be possible to claim for compensation in relation to breaches of affected health and safety legislation where it can be proved that the duty holder (usually the employer) has been negligent. This means that in future, for all relevant claims, duty-holders will only have to defend themselves against negligence ...... there should be no civil right of action for breach of a duty imposed by certain health and safety legislation, other than where such a right is specifically provided for."

Previously, Section 47(2) of the HASAWA 1974 provided a right of action for breach of a duty contained in a health and safety Regulation (a Regulation made under Section 15 of the HASAWA 1974) if that breach caused damage. The Section also contained a power to make exceptions to this rule in the Regulations that imposed the duty.

Health and Safety at Work etc. Act 1974 (Civil Liability) (Exceptions) Regulations 2013

The amendment of Section 47(2) of the HASAWA 1974 by ERRA 2013 meant the right to any civil action for breach of statutory duty related to a Regulation made under the HASAWA 1974 was removed and could only exist where a specific exemption had been made by Government Regulations.

A specific exemption was created for new or expectant mothers under the ***Health and Safety at Work etc. Act 1974 (Civil Liability) (Exceptions) Regulations 2013*** to protect the rights of new or expectant mothers derived from the Pregnant Workers Directive 92/85/EEC, which required member states to make provisions to enable workers who believe they have been wronged by a failure to comply with the duties in the Directive to enforce their rights by judicial process.

In the UK the rights of new and expectant mothers to meet the directive were established in the Employment Rights Act (ERA) 1996 and the Management of Health and Safety at Work Regulations. The Health and Safety at Work etc. Act 1974 (Civil Liability) (Exceptions) Regulations 2013 amended the two laws to maintain the judicial process for new and pregnant mothers at work and enable them to take civil action for breach of statutory duty. The parts of these laws that were affected are outlined below.

Section 72(1) of the Employment Rights Act (ERA) 1996 requires employers to prohibit a worker working within the compulsory maternity leave period (2 weeks from the date of giving birth). A right of action is created to the extent that a breach of this Section of ERA 1996 causes damage.

The Management of Health and Safety at Work Regulations (MHSWR) 1999 provide that new and expectant mothers (including agency workers) will continue to have a right of action in civil proceedings in relation to breaches of Regulations 16 - 17A of MHSWR 1999, to the extent that the breach causes damage.

These Regulations relate to requirements for the employer to carry out risk assessments and make particular arrangements for new and expectant mothers to protect their health and safety.

The main defences to the tort (delict) of breach of statutory duty

The first defence to a claim of breach of statutory duty (BOSD) is that a specific exemption has not been made as required by the ERRA 2013 and that therefore, a civil right of action is not provided for.

STATUTORY DUTY NOT ON THE DEFENDANT

One of the first defences available to the defendant in an action for breach of statutory duty would be that there was no duty on the defendant to protect the claimant. Some statutory duties expressly relate to protection of employees and a claimant who is the employee of a contractor working on another employer's (client) premises may wrongly claim breach of statutory duty of the client rather than their employer.

This defence may also be applicable to shared premises where another person than a claimant's employer is the person in control of common parts of the premises and is the duty holder.

NO BREACH OF STATUTORY DUTY

Another defence to a breach of statutory duty claim could be that there was no breach of duty under the statute. The defence would show that everything that must be done was done or was done so far as is practicable or reasonably practicable, or was suitable and sufficient, whatever the statute required.

INJURED PARTY NOT WITHIN THE CLASS OF PERSONS PROTECTED BY THE STATUTE

The defendant might defend a claim for breach of statutory duty by establishing that the claimant was not one of the classes of persons that the statute was there to protect.

For example, if a female who was not a new or expectant mother was the claimant in a situation where the claim was related to Regulations 16 – 17A of MHSWR 1999 then the claim would fail.

HARM NOT OF THE TYPE THAT THE STATUTE WAS DESIGNED TO PREVENT

If the harm done to the claimant were not the type the statute was there to prevent, the claim would fail. For example, Regulations 16-17A of MHSWR 1999 establishes responsibilities of employers to prevent harm of a type that relates to new or expectant mothers only.

If a claim of breach of statutory duty was made related to these Regulations that was for harm not relating to the claimant being a new or expectant mother it would provide an opportunity for the defendant to use this as a defence.

NO CAUSAL CONNECTION BETWEEN THE BREACH AND THE LOSS SUFFERED

The defence of no causal connection between the breach and the loss suffered is applicable where the breach of statute is admitted, but the breach did not lead to the loss.

In the case of McWilliams v Sir William Arrol & Co Ltd [1962] the breach was under the Factories Act and stated as failing to provide safety harnesses for the steel erectors and thus Mr McWilliams fell to his death.

The defendant agreed they were in breach, but claimed that it did not lead to the loss, as even if they had provided harnesses, the employees (including Mr McWilliams) would not have worn them.

The defendant showed that they had once provided harnesses, but as nobody would wear them, they took them away again. The claimant failed, as the judge decided that the breach did not lead to the loss.

See also - 'Corn v Weirs Glass (Hanley) Ltd [1960]' for a further example of the breach not leading to the loss.

CONTRIBUTORY NEGLIGENCE

Where the loss or damage is partly the fault of the claimant, and if that claimant may be said to have contributed to the loss or damage, then any compensation they may recover will be reduced in proportion to their blameworthiness.

Main defences to claims of negligence

DENIAL

The first defence is one of a denial of liability. This may be based on a variety of grounds:

NO DUTY OWED

The defendant did not owe the claimant a duty of care. This is unlikely to apply in an action for damages between an employer and an employee as the duty of care relationship is well established. The employer could however claim that the claimant was not an employee and was working for another company ***(see also - Mersey Docks and Harbour Board v Coggins & Griffiths [1946]).***

NO BREACH OF DUTY

Foreseeability

The risk and loss were not foreseeable, as in Latimer v AEC [1953] where there was a flash flood. No reasonable person could have foreseen this occurrence.

If the risk of harm arising from the breach of common law duty is not foreseeable then it may be considered that the defendant had taken reasonable care. As our knowledge of health and safety risks grows then more risks may be considered to be foreseeable. However, until a specific risk of harm can be determined, to the extent that a reasonable person would have foreseen it, the defendant may assert that harm of the type being claimed by the defendant was not foreseeable from the circumstances of the breach of duty.

It is, however, no defence of foreseeability to show that an injured person was unforeseeably weak ('the thin skull' rule).

Reasonableness

Everything that could reasonably have been done was done.

BREACH DID NOT LEAD TO DAMAGE

There may have been a failure to fulfil the common law duty of care, but that was not what led to the loss. The employer may have failed to provide training on manual handling, but the claimant hurt his back reaching over to pick up his pen.

In some cases, the loss may be too remote where there is a sequence of physical cause and effect without human intervention. ***This is illustrated by the Wagon Mound [1961] case.***

This is the principle that the damage or loss is not sufficiently connected to the defendant's breach of duty of care. The principle of a new act intervening (novus actus interveniens) is an extension of the defence of remoteness of damage ***(see also - Scot v Shepherd [1773]).***

The chain of causation can also be broken if the claimant acts unreasonably by, for example, taking an avoidable and foreseeable risk of injury, that breaks the chain (if it does not, then the claim may be reduced due to contributory negligence).

TYPE OF DAMAGE NOT FORESEEABLE

In addition, the type of damage must not be too remote from the breach and there must be a direct causal link established by the facts of the case. For example, it may be possible to show causal link to physical harm to a claimant, caused as a result of an accident at work. In the past, it has been more difficult to persuade a court of close causal link where the harm is mental rather than physical. In more recent times, persuading the courts has been less difficult, but it remains a matter of legal perspective as to whether causal link can be established for mental harm experienced by someone who witnessed an accident. Moreover, harder still for someone to prove causal link of mental harm caused when the claimant heard about harm caused to a person in an accident.

VOLENTI NON FIT INJURIA

Where a person has agreed either expressly or by implication to accept the risk of injury, he cannot then recover damages for loss caused to him by that risk.

Volenti non-fit injuria - translated from Latin this means 'to one who is willing no harm is done'. This is a complete defence to negligence by the defendant, and is used where the claimant agreed to run the risk of accidental harm.

In the case of ***Smith v Baker & Sons Ltd***, the claimant, who was a railway navvy, was working in a cutting. A crane, used for removing stones, passed its load directly over a workman's head and he had made several complaints. When stones fell on him he claimed damages. The employer claimed the defence of volenti non-fit injuria as they deemed that the navy had accepted the danger by continuing to work.

The House of Lords found that even though he knew of the danger; and still continued to do the work, there was no evidence that he had volunteered for the risk of injury.

> *"...The duty of taking reasonable care to provide proper appliances and maintain them in a proper condition and so to carry on his operations as to not subject those employed by him to unnecessary risk."*

Figure A4-10: Duty of care to employees. *Source: Lord Hershell, Lord Atkin, Donoghue v Stevenson [1932].*

An allowance should be made for the fact that most workers cannot afford to terminate their employment. After this case, the courts have rarely found that an employee has accepted the risk of injury at work.

This true consent must be freely given and has been used in connection with spectators injured in the course of events with inherent risks, for example, motor racing. In Hall v Brooklands Auto Racing [1933] a spectator paid for a ticket to watch car races. During a race two cars collided which resulted in one of them spinning off the track, through the barriers and into the stands injuring Mr Hall. He lost his subsequent claim for negligence as the precautions were deemed to be adequate (this was the first time such an incident had happened). By purchasing his ticket, the claimant was held to have assumed the risk from such an accident.

This principle can also apply to minors. In Murray v Harringay Arena Ltd (1951), a 6 year old boy was taken by his parents to watch an ice hockey match. He was sitting in the front row when he was hit in the eye by the ice hockey puck. He was unsuccessful in his claim for damages as it was held that the claimants had voluntarily undertaken the risk. The defendants had provided protection by means of a wooden barrier and netting. Had the protection been more substantial then it would have seriously interfered with the spectator's view.

An employee, however, does not consent to abnormal or unnecessary risks merely by accepting the job or continuing to do it ***(see also - Smith v Baker & Sons [1891]).*** The 'volenti' defence may, however, be successful where employees conspire to breach statutory duty against the employer's specific instructions.

In ICI Ltd v Shatwell [1964] two brothers, who were certified shot blasters, were injured whilst testing electrical circuitry used to detonate explosives. They did so in the open, contrary to the Quarries (Explosives) Regulations 1959 which required them to be in a position of safety. The resulting claim for damages against the employer was unsuccessful on the basis that they had voluntarily accepted the risk of each other's negligence and breach of statutory duty by defying the employer's instructions and breaching the Regulations which were well known to them. The volenti defence, however, will not normally succeed in cases where there is a direct breach of common law or statutory duty by the employer.

This defence is unlikely to succeed when defending claims made by rescuers who volunteered for the risk in order to safeguard others, whether or not the rescuers are members of the public or public service employees.

In Haynes v Harwood [1935], for example, it was held that the defences of volenti and contributory negligence did not apply when a policeman was injured by a runaway horse and van in a crowded street. Similarly, it was held that the defence of volenti did not apply in the case of Baker v T E Hopkins and Sons Ltd [1959] where a local doctor died trying to rescue two men who had been overcome by a lethal concentration of carbon monoxide fumes when working in a well.

CONTRIBUTORY NEGLIGENCE

Where the loss or damage is partly the fault of the claimant, and if that claimant may be said to have contributed to the loss or damage, then any compensation they may recover will be reduced in proportion to their blameworthiness.

TIME LIMITATION

The Limitations Act (LA) 1980 places a time limit on actions in respect of tort of six years. Damages for personal injuries and death in respect of negligence, nuisance, and breach of statutory duty must be started within three years from the date of the claimant becoming aware of the injury, or three years from the medical diagnosis of occupational ill-health. At common law personal actions did not survive the person: if either claimant or defendant died the right to litigate was lost. This principle was changed with the introduction to the Fatal Accidents Act 1846.

This allowed the wrongdoer to be sued to maintain the income for the deceased's dependants at the level which the deceased had been providing. The Fatal Accident Act (FAA) 1976 gives rights to certain persons to sue for negligence that has resulted in death. The objective of the FAA 1976 is to provide compensation for relatives who have been deprived of maintenance due to death. The claimants must have suffered financial loss because of the death and must have been financially dependent on him. Classes of persons include husband, wife, children, grandchildren, parents, grandparents, brothers, sisters, aunts, and uncles, and their issue. The relationship may be traced through step-relatives, adoption or illegitimacy, and relatives through marriage have the same rights as the deceased's own relatives. Any person who was living with the deceased for two or more years prior to death may also claim (for example, an unmarried co-habitant).

A single action must be brought on behalf of all eligible dependants and damages apportioned according to dependency. The action may be bought by a personal representative(s) of the deceased. However, if there are none or if they fail to bring an action within six months of the death, the dependants may bring it themselves.

The three year rule under the 1980 Limitations Act still applies; however the court has discretionary powers. If the claimant dies before the limitation period has expired, the three years starts from the date of death or the date of the personal representative's knowledge of the cause of action (whichever is the later).

The implications of the Social Action, Responsibility and Heroism Act 2015

The Social Action, Responsibility and Heroism Act (SARHA) 2015 can be used as a defence against negligence and breach of statutory duty (where necessary) and is intended to provide a greater degree of reassurance and protection to 'good Samaritans', volunteers and those who may be deterred from participating in socially useful activities due to worries about risk or liability.

SARHA 2015 requires a court, when determining whether someone has behaved negligently or in breach of a statutory duty, to have regard to each of three additional factors:

1) Social action - whether the alleged negligence or breach of duty (where relevant) occurred when the person was acting for the benefit of society or any of its members.
2) Responsibility - whether, in carrying out the activity in which the alleged negligence or breach of duty occurred, the person demonstrated a predominantly responsible approach towards protecting the safety or interests of others.
3) Heroism - whether the alleged negligence or breach of duty occurred when the person was acting heroically by intervening in an emergency to assist an individual in danger.

This legislation applies to compensation claims made after 13 April 2015 and applies only in England and Wales.

Factors to be considered in the assessment of damages

In deciding the amount of damages awarded the Judge will take many factors into account including:

- Loss of actual and future earnings.
- Pain and suffering.
- Loss of amenities (i.e. enjoyment of life).

Damages are categorised as follows:

SPECIAL DAMAGES

Special damages represent the losses incurred by the plaintiff that can be proved to have occurred up to the date of the hearing of the claim, i.e. the tangible losses, (for example, loss of earnings prior to the trial). For example:

- Itemised expenses, for example, legal costs.
- Repair or replacement of damaged property.
- Lost earnings (historical).
- Loss of irreplaceable items.
- Modifications to the home.

Special damages can include direct losses (such as amounts the claimant had to spend to try to mitigate problems) and consequential or economic losses resulting from lost profits in a business. Special damages basically include the compensatory and punitive damages for the tort committed in lieu of the injury or harm to the plaintiff.

GENERAL (OR ORDINARY) DAMAGES

General damages compensate the claimant for the non-monetary aspects of the specific harm suffered. This is usually termed 'pain, suffering and loss of amenity'. Examples of this include:

- Physical or emotional pain and suffering.
- Loss of companionship.
- Disfigurement.
- Loss of reputation.
- Loss or impairment of mental or physical capacity.
- Loss of enjoyment of life etc.
- Loss of future earnings.

These are not easily quantifiable, and depend on the individual circumstances of the claimant. Judges in the United Kingdom base the award on damages awarded in similar previous cases. For example, an accident in which the claimant has suffered the total effective loss of both hands and for which the defendant was legally responsible, will typically attract general damages (at 2016) in the region of £210,000 to £350,000.

OTHER DAMAGES

Exemplary or aggravated damages: intended to punish the defendant and to deter others. The House of Lords ruled that these could only be awarded where the defendant is calculated to make more money from his tort than he would have to pay in damages; where a government official acts oppressively, arbitrarily or unconstitutionally, or where statute allows this.

Nominal damages: where the claimant has suffered injury but no real loss, a small sum (i.e. a few pounds) may be awarded. For example, the defendant has trespassed onto the claimant's land without causing any damage.

Contemptuous damages: where the court has no sympathy for the action (case) that the defendant has brought (i.e. damages of one penny).

Liquidated and unliquidated damages: liquidated damages are those which result from breach of contract which are fixed by the terms of the contract. Unliquidated damages occur when a contract has been breached; however damages are not fixed by the contract. These are awarded by the court based on the evidence relating to the loss that the claimant has suffered as a result of the breach. Liquidated damages do not normally apply in the law of tort.

From the above list it can be seen that special and general damages are the most significant factors in many cases.

In general terms the aim of the award is to place the injured person back to where he or she was prior to injury - as far as this can be achieved with money.

There is no exact table of awards to be paid according to the type of injuries; however certain kinds of injury should fall within a 'bracket' of awards.

The judge will normally explain the factors he or she considered when making the award. The claimant should however not be materially better off because of the injury. ***Some factors are given in the following points***.

For example, factors relating to loss of earnings include:

- Estimated annual earnings.
- Length of working life.
- How much the injury has affected earning/working capacity?
- Life expectancy.

In making the award the judge has to consider some quite complex factors that can overlap and contradict each other.

In one case of injury to 'a strong, confident mentally stable and physically fit man of 40...reduced to a shadow of himself, a pathetic piece of human wreckage' by an accident on the docks; the judge considered that:

"Whilst the claimant is denied the satisfaction of a docker's life, he is also spared its rigours. In view also of his acute awareness of his inability to look after his wife and his family the very substantial sum coming to the claimant by way of damages for loss of earnings past and future, will be a great source of consolation for the claimant...because he will realise that the sum, plus the additional figure to be awarded to him for general damages, really makes his family's economic future perhaps more secure than it has ever been. It is also necessary of course to have due regards to the interests and well-being of the other side's insurance company."

Figure A4-11: Damages. *Source: Ambiguous.*

Fletcher v Autocar [1968]

In recent times there has been increasing concern about the perceived low levels of damages awarded. In 1999 the Law Commission recommended that damages be increased by between 50% and 100%. The move towards higher awards has however been resisted by both the insurance industry and the NHS (which self-insures). In March 2000 five judges of the Court of Appeal (including the Master of the Rolls Lord Woolf) heard submissions relating to eight test cases. The claimant's lawyers argued that the current level of damages was too low in respect of the amounts awarded for pain, suffering and loss of amenity (PSLA), and general damages. The Court ruled that such damages should be increased by as much as one third. The judges did not want to see the 'daunting' levels found in the US but thought that a 'modest increase' that recognised factors such as increased life expectancy and increases in standards of living. They held that, as damages were the only available remedy, they should be fair, even though they cannot compensate for injury. There should be a consistency so that everyone knows what to expect and so as to facilitate settlements.

The Court ruled that:

- Damages at the top end of the scale should be increased by one-third.
- That there should be no increase in awards below £10,000.
- There should be a tapering increase between the above points so that the largest increases occurred at the most serious end.

As a result of one of the above test cases the Court of Appeal increased the damages awarded to:

- A woman and a man who died of asbestos-induced mesothelioma - from £45,000 to £50,000 and from £40,000 to £44,000 respectively.
- A woman, who suffered severe multiple injuries and brain damage in a road traffic accident, from £110,000 to £138,000.

The Court of Appeal said that this process of review should not be repeated unless there was real reason to suspect that damage awards had fallen behind the times.

SOLATIUM AND PATRIMONIAL LOSS

Solatium is a form of compensation for emotional rather than physical or financial harm. It is used in Scots law mainly to denote reparation for pain and suffering in personal injury cases (although it can also be awarded in other types of cases). In this respect it is similar, but not identical, to the English law concept of general damages; Scots law damages are divided into pecuniary and non-pecuniary losses, rather than general and special damages.

Under the Damages (Scotland) Act 1976 where a person dies as a result of personal injury the relatives may be entitled to claim damages for patrimonial (for example, financial) loss and/or non-patrimonial loss. Patrimonial damages are awarded for loss of financial support, while non-patrimonial damages are awarded in respect of:

- Distress because of the suffering of the injured person before death.
- Grief and sorrow at the death of injured person.
- Loss of deceased's society and guidance.

Contributory negligence and its effects

This was a complete defence until the law was reformed by the Law Reform (Contributory Negligence) Act (LRCN) 1945. This Act established the following position when dealing with civil cases:

Apportionment of liability in case of contributory negligence:

1) Where any person suffers damage as the result partly of his own fault and partly of the fault of any other person or persons, a claim in respect of that damage shall not be defeated by reason of the fault of the person suffering the damage, but the damages recoverable in respect thereof shall be reduced to such extent as the court thinks just and equitable having regard to the claimant's share in the responsibility for the damage.

4) Interpretation.

'Fault' means negligence, breach of statutory duty or other act or omission which gives rise to a liability in tort or would, apart from this Act, give rise to the defence of contributory negligence.

5) Application to Scotland.

"... The expression 'fault' means wrongful act, breach of statutory duty or negligent act or omission which gives rise to liability in damages, or would, apart from this Act, give rise to the defence of contributory negligence."

Now contributory negligence is a partial defence whereby claimants are held partly responsible for their own injury or loss. Any damages awarded are reduced by a corresponding percentage.

In Uddin v Associated Portland Cement Mfrs Ltd [1965], Mr Uddin was injured by falling in to the top of a machine, which had guards round the perimeter, but not across the top. He was in an area of the factory he should not have been, chasing a pigeon, which was not his job. Technically, the employer was in breach of his statutory duty because he had not 'securely fenced all dangerous parts of all machinery', but the claimant had contributed to his own accident. The judge said 'he was off on a frolic of his own' and reduced his compensation by 80%.

See also - 'Decided cases' for 'Jones v Livox Quarries Ltd [1952] 2 QB 608'.

Concept of joint tortfeasors

A tortfeasor is defined as someone who has carried out a tort. Where two (or more) people acting together commit a tort then they are joint tortfeasors.

MEANING OF JOINT AND SEVERAL LIABILITIES

Each individual has liability for his own torts, but in some cases he may be liable for his own and others. The liability of joint tortfeasors is said to be joint and several. This means that the claimant can sue them both together or may recover the full amount from just one of them. Where one joint tortfeasor is successfully sued, he can, under the Civil Liability (Contribution) Act (CLCA) 1978, claim a contribution from the other to the extent of his liability. Where joint tortfeasors are sued together, damages are divided according to the degree of responsibility that the court considers each party to have.

RECOVERY OF DAMAGES FROM JOINT TORTFEASOR

As previously stated, a claimant can recover damages in full from just one of the joint tortfeasors. Prior to the law being reformed, there was no right for one joint tortfeasor to claim contribution from another. Now, where a joint torfeasor is sued as an individual, he may claim a contribution from the fellow tortfeasor(s). The amount recoverable is left to the discretion of the court on the principle of a just and equitable division of responsibility.

The law was further developed by the Law Reform (Husband and Wife) Act (LRHWA) 1962. This treats spouses as separate parties when considering the liability for damages of joint tortfeasors. For example a wife is injured in a car accident due to the joint negligence of her husband and a third party. If the wife then successfully gains damages from the third party, then that third party can claim a contribution from the negligent husband.

A joint tortfeasor must bring the action for recovery of contribution within two years from the date that that liability was admitted or a judgement against him was made.

Personal injury action protocol under the civil procedures rules (England and Wales only)

CIVIL PROCEDURES RULES

The procedures for personal injury cases were radically overhauled as a result of a review by Lord Woolf (the Woolf Report). This was done in an attempt to speed up and simplify the process as well as keeping costs within reasonable limits. The resulting Civil Procedures Rules were introduced on 26 April 1999 and have the following aims:

- To ensure that all parties are dealt with on an equal footing so that wealthier parties do not have an unfair advantage.
- To save expense.
- To deal with cases in a way that is proportionate to the sums involved, the complexity and importance of the case and the financial position of the parties.
- To ensure that cases are dealt with expeditiously and fairly in ways which promote an open and full exchange of information.
- To allot an appropriate share of the Court's resources, while taking into account the needs of other cases.

The rules include a pre-action protocol for personal injury claims, which is designed to make the processing of these claims more efficient and effective.

See also 'Element A2 - Principles of health and safety law' for 'Basic procedures for pursuing civil actions'.

PRE-ACTION PROTOCOL

The pre-action protocol for personal injury claims applies to those claims that relate to such things as road traffic accidents, a member of the public tripping and slipping on a pavement and accidents at work involving personal injury where the value claimed is less than the fast track limit. An efficient and effective process is essential for these cases because they need to be ready for trial within 30 weeks.

The process of the claim begins with a pre-action protocol that aims to reduce the need for litigation and encourages contact between the parties prior to the action starting. The pre-action protocol suggests that the claimant's legal representative may wish to notify the defendant that a claim may be made at the earliest opportunity, this is before the formal 'timetable' of actions starts. An on-line portal has been established to enable the electronic submission of smaller claims (most personal injury claims of the value £1,000 to £25,000) as part of the Low Value Injury Scheme.

The pre-action protocol for personal injury involves a 3 stage process:

1) The letter of claim.
2) The reply.
3) The investigation.

The letter of claim

The action is started on a claim form. This contains a summary of the facts of the cases together with the details of the loss (for example, financial; injuries suffered). The claimant must make an effort to identify the defendant's insurer (by use of a database) and send the claim form directly to the insurer if there is one. The claim letter enables the defendant/insurer to have sufficient information to assess liability and to estimate the likely size of the claim.

The reply

The insurer must acknowledge the claim within 24 hours of receipt. Letters of claim and response are not intended to have the same status as a statement of case in proceedings. Matters may come to light as a result of investigation after the letter of claim has been sent, or after the defendant has responded. These circumstances could mean that the 'pleaded' case of one or both parties is presented slightly differently than in the letter of claim and response. It would not be consistent with the spirit of the pre-action protocol for a party to 'take a point' on this in the proceedings, provided that there was no obvious intention by the party who changed their position to mislead the other party.

The investigation

The defendant has a maximum of 30 days for Employer's Liability claims and 40 days for Public Liability claims to investigate the claim. A reply must be then be sent to the claimant (or the claimant's solicitor) which:

- Admits liability.
- Denies liability in full.
- Gives a partial admission of liability.

Reasons for full or partial admission of liability must be given together with any relevant documents. These are the same documents that the defendant would have to disclose anyway under court proceedings. Thus the defendant will not be able to issue a denial of responsibility without giving reasons or to delay the decision as to whether to settle or fight. The claimant will be able to make a decision, based on the defendant's reply, about whether or not to proceed with the claim. If the defendant fails to make a proper reply, then the claimant will be able to start proceedings without being liable for costs.

Documents

The pre-action protocol has a standard list of documents for different types of cases; these must be disclosed as a minimum. These include the following for workplace claims:

(i) Accident book entry.

(ii) First aider report.

(iii) Surgery record.

(iv) Foreman/supervisor accident report.

(v) Safety representative's accident report.

(vi) Reporting of Injuries, Diseases and Dangerous Occurrences Regulations (RIDDOR) 2013 report to the enforcing authority.

(vii) Other communications between defendants and HSE.

(viii) Minutes of health and safety committee meeting(s) where accident/matter considered.

(ix) Report to the Department of Social Security.

(x) Documents listed above relative to any previous accident/matter identified by the claimant and relied upon as proof of negligence.

(xi) Earnings information where defendant is employer.

Documents produced to comply with requirements of the Management of Health and Safety at Work Regulations (MHSWR) 1999.

(i) Pre-accident risk assessment.

(ii) Post-accident re-assessment.

(iii) Accident investigation report.

(iv) Health surveillance records.

(v) Information provided to employees.

(vi) Documents relating to the employees' health and safety training.

Further documents are listed in the protocol that relate to situations where other, more specific, Regulations apply.

Special damages

The claimant must send to the defendant as soon as is practicable a schedule of special damages, with supporting documents, particularly where liability has been admitted.

Experts

Details of proposed experts must be sent to the other party before instructing them to make a report. The other party may object to one or more experts on the list, the first party would then instruct one of those not objected to. If all are objected to, the party proposing them may instruct an expert of their choice.

Either party may send questions to the nominated experts, costs related to the expert answering questions go to those that posed the questions. The protocol promotes the practice of the claimant obtaining a medical report, disclosing it to the defendant who then asks questions and/or agrees it and does not obtain his own report.

A4.3 - Occupiers' liability

The Occupiers' Liability Acts (OLA) 1957 and 1984

Under the ***Occupiers' Liability Act (OLA) 1957***, and ***OLA (NI) 1957*** an occupier owes a 'common law duty of care' to all visitors, i.e. a duty to take such care as is reasonable (given the circumstances) to see that visitors will be reasonably safe in using the premises for the purposes for which they are invited or permitted to be there. The Act reminds the employer (or occupier) that they must be prepared for children to be less careful than adults. A higher duty of care exists for visitors who are children, which includes the need to foresee the attractiveness of everyday work things that may be considered to be allurements.

An occupier may expect that persons, in the exercise of their calling, will appreciate and guard against any special risks ordinarily incident to that calling, for example, the contractor electrician should be aware of electrical risks as part of his work.

An employer (or occupier) cannot be prosecuted for a breach of their duties under the OLA 1957, but if a visitor is injured, the occupier may be liable to pay compensation for the injury if it can be proved that the occupier had been negligent.

The position is slightly changed by the ***Occupiers' Liability Act (OLA) 1984***. Prior to this Act, minimal duties were owed under common law by an occupier to those other than lawful visitors - for example, to trespassers. The common law duty to trespassers was considered common humanity. In other words, don't deliberately set a trap for them. The situation was changed by the case of ***Herrington v British Railways Board 1972.*** A six year old boy wandered onto a railway line and was seriously injured when he came into contact with a live electrified rail.

The defendants knew that the fence alongside the line was in a state of disrepair and had taken no remedial action - despite previous reports of children being seen on the line. There was clear evidence of the presence of trespassers (likely to be children) since the path was worn from the broken fencing to the railway line. The OLA 1984 followed that case and extends the rules of common law to unwelcome visitors (for example, trespassers and ramblers), and defines under what circumstances a duty is owed.

In particular, occupiers owe a duty if:

- They are aware of a danger.
- They know (or should know) that a person may put themselves at risk.
- The risk is one which the occupier might reasonably be expected to do something about.

The extent of the duty is to take such care as is reasonable in the circumstances of the case and to see that the other person does not suffer injury. The Act provides that the duty may be discharged by:

- Giving warning of the danger.
- Discouraging people from putting themselves at risk in the first place (for example, by making it more difficult for trespassers, etc. to enter the premises).

In summary, OLA 1957 is the common law duty of care to welcome visitors and OLA 1984 the common law duty of care to unwelcome visitors.

OCCUPIERS LIABILITY (SCOTLAND) ACT 1960

An 'occupier' of land has a duty to show care towards people on that land. The level of this duty of care is that which is 'reasonably foreseeable' to ensure that people do not suffer injury. The occupier must consider the nature of any dangers, such as the condition of the property, or of anything done or omitted to be done by the occupier which is his legal responsibility.

Occupier - the occupier is defined as the person (or body) 'occupying or having control of land or other premises'. Where premises are leased, then the respective duties of the landlord and tenant will depend upon the terms of the lease.

Nature of danger - the hazard must be one which is due to the state of the premises (which might include natural dangers as well as man–made ones) or to anything done or omitted from being done on the premises for which the occupier is legally responsible (negligence). An occupier is not normally expected to guard against dangers which are obvious ***(see also - Relevant Decided Cases)***, not required by the current syllabus). For example, fencing a lake to prevent someone falling in or fencing a cliff to prevent someone falling off.

Persons entering onto the premises - the occupier owes an equal duty of care to all persons entering onto his/her premises. This is irrespective of whether they have permission to be there or not, for example, trespassers.

Willingly-accepted risks - the 1960 Act explicitly excludes any obligation of occupier liability over risks willingly accepted by the visitor, for example, mishaps inherent in quite ordinary activities like walking, swimming rock climbing, they will be taken to have accepted the risk of injury if they have an accident (volenti non-fit injuria).

Case title and date - Duff v East Dunbartonshire Council and others, 1999, and appeal to House of Lords, 2002.

Case issue - duty to fence an obvious and natural hazard.

Case outline - a person was injured when he slipped and fell down an embankment, and the case was brought that the slope should have been effectively fenced off from the nearby car park.

Case judgement - the case hinged around the distinction between what is obvious and what is concealed as a matter of fact. The court took the view that the exit from the car park itself could not be said to be dangerous, because of the area of level ground beyond it before the ground started to slope, and the sudden sharp slope of the embankment itself was an obvious and natural hazard against which the pursuer would have been expected to protect himself, and therefore dismissed the case. On appeal the Lords agreed with the principles applied by the lower court.

Relevant decided cases

CAPARO INDUSTRIES PLC V DICKMAN [1990] 2 AC 605

[Duty of care in negligence]

[Three-fold test]

Facts - a company called Fidelity plc was the target of a takeover by Caparo Industries plc (Caparo). In March 1984 Fidelity had issued a profit warning, which had halved its share price. In May 1984 Fidelity's directors made a preliminary announcement in its annual profits for the year up to March. This confirmed the position was bad, the share price fell again.

At this point Caparo had begun buying up shares in large numbers. In June 1984 the annual accounts, which were completed with the help of the accountant Dickman, were issued to the shareholders, which now included Caparo. Caparo continued to buy shares and reached a shareholding of 29.9% of the company, at which point it made a general offer to take over the remaining shares.

It was discovered, upon completion of the takeover, that the accounts of Fidelity plc were in a worse state than Dickman had divulged and Caparo sought damages from Dickman, in relation to having purchased the shares of Fidelity plc at an inflated value.

When the case was first heard Caparo failed to show that a duty of care existed in the circumstances, but was successful in establishing a duty of care might exist in the Court of Appeal. It was held that as a small shareholder, Caparo was entitled to rely on the accounts. Had Caparo been a simple outside investor, with no stake in the company, it would have had no claim.

Decision - the decision of the Court of Appeal was overturned by the House of Lords, which unanimously held there was no duty of care. The House of Lords held that the issuing of company accounts (by Dickman) was intended to be for the use of company shareholders for the purpose of exercising their class rights in general meetings etc. The accounts were not published for the intention of illustrating the potential of future investment by shareholders or other interested parties.

The House of Lords established what is known as the 'three-fold test' (a series of three factors), which required for one party to owe a duty of care to another, the following must be established:

- Harm must be a 'reasonably foreseeable' result of the defendant's conduct.
- A relationship of 'proximity' between the defendant and the claimant.
- It must be 'fair, just and reasonable' to impose liability.

The main turning point of the decision was that it was held that Dickman had no knowledge as to the use that the published accounts would be put to by Caparo, and thus there could be no proximity of relationship as the accounts were published and issued to shareholders as a whole.

CORR (ADMINISTRATRIX OF CORR, DECD) V I B C VEHICLES LTD [2008] HL

[A defendant may be held liable for loss attributable to suicide where the deceased has suffered an injury in consequence of the defendant's negligence]

Facts - the deceased, an employee of the defendant company, was injured in an accident at work in June 1996. Following the accident, in which most of his right ear was severed, he suffered from post-traumatic stress disorder, severe headaches and difficulty sleeping. He also developed depression. His depression worsened until, in May 2002, while suffering from an episode of severe depression, he committed suicide.

The deceased had begun proceedings in June 1999 claiming damage for his physical and psychiatric injuries. After his death, proceedings were amended, his widow becoming the claimant. She claimed for the benefit of his estate pursuant to the Law Reform (Miscellaneous Provisions) Act 1934 and for herself as a dependant under the Fatal Accidents Act 1976.

The parties agreed that:

- The defendant owed the deceased a duty to take reasonable care to avoid causing him personal injury. Following Page v Smith [1996] AC 155, personal injury encompasses psychological as well as physical injury. According to Page v Smith, a claimant may recover in relation to a psychological injury suffered where a physical injury was reasonably foreseeable (whether or not physical injury occurred). Psychological injury need not have been reasonably foreseeable.
- The employer's breach of duty had caused the accident at work.
- The deceased suffered physical and psychological injuries as a result of this accident.
- The deceased's depressive illness was caused by the accident. He had not suffered from depression prior to this.
- The deceased's suicide was the result of his depressive illness.

The issue in contention was whether (as had been found by the Court of Appeal (CA)) damages could be recovered in respect of the financial loss attributable to the deceased's suicide. Were the damages claimed too remote?

The defendant submitted that the suicide:

- Fell outside the duty of care owed to him by the employer.
- Was not reasonably foreseeable and therefore the employer should not be liable for it.
- Broke the claim of causation and constituted a novus actus interveniens.
- Was an unreasonable act which broke the chain of causation.
- Was a voluntary act of the deceased and so precluded by the volenti non-fit injuria principle.
- Amounted to contributory negligence.

Decision - the House of Lords dismissed the defendant's appeal and held:

- The employer owed the deceased a duty to avoid causing him both psychological and physical injury. Breach of this duty caused him injury of both kinds. Whilst the deceased was not insane at the time of his death, he was not fully responsible. He acted in a way which he would not have done but for the injury. His suicide was not outside the scope of the employer's duty.
- Depression, possibly severe, was a foreseeable consequence of the employer's breach. It was not incumbent on the claimant to show that suicide itself was foreseeable. A tortfeasor who reasonably foresees the occurrence of some damage need not foresee the precise form which that damage may take. Suicide is a not uncommon consequence of severe depression. The suicide was thus reasonably foreseeable.
- The chain of causation could be broken where a person of sound mind voluntarily committed suicide. However, here the suicide was not voluntary. It was the response of a man suffering from a severely depressive illness which impaired his capacity to make reasoned judgments. It was fair to hold the defendant liable for this consequence of its breach. Whilst it is true that the more unsound the mind of the victim, the less likely it is that his suicide will break the chain of causation, it would be a retrograde step to bar recovery because the deceased was not insane. The suicide was not a novus actus interveniens.
- In light of the reasons given for holding that the suicide was not a novus actus, it would be impossible to hold that the damages attributable to the death were rendered too remote because the deceased's conduct was unreasonable.
- The deceased did not consent to the accident at work and he did not voluntarily consent to the suicide. The suicide occurred because of the psychological condition resulting from the defendant's breach.
- No deduction was made for contributory negligence.

In principle, a deduction in damages could be made for contributory negligence where a person suffering from depression committed suicide. However, in the present case, no finding in relation to this point was made in the courts below and the question was hardly touched on in argument in the CA.

Lord Bingham absolved the deceased from any causal responsibility and assessed his contributory negligence at 0%. Lord Walker agreed with this assessment. Lord Scott, however, assessed the deceased's contributory negligence at 20%, noting that had he injured someone else by his suicide (jumping from the top of a multi-storey car park), the injured person would have had a claim against him in negligence; the deceased's conduct would have constituted fault in tort. Lord Mance and Lord Neuberger considered that a deduction could be made for contributory fault, recognising the element of choice which may be present even where the deceased was suffering from an impairment. A deduction was not appropriate here given that, amongst other things, no satisfactory material was available to their Lordships to enable whether a deduction could be made.

The defendant's appeal was thus dismissed.

DONOGHUE V STEVENSON [1932] AC 562

[Reasonable care and the neighbour principle; duty of manufacturer to end user]

Facts - the claimant (Ms Donoghue) became ill after she drank a bottle of ginger beer which contained a decomposed snail. The bottle was opaque and she had no reason to believe that it contained anything other than ginger beer. She sued the manufacturer.

As it was Ms Donoghue's friend who had purchased the ginger beer, the manufacturer claimed that there was no contract between them.

Decision - damages were awarded to the claimant. The House of Lords held that a duty of care was owed and determined the extent of that duty ('the neighbour principle').

Relevance - extent of duty of care: to whom is this duty owed?

As per:

> *"You must take reasonable care to avoid acts or omissions which you can reasonably foresee would be likely to injure your neighbour."*
> *Who is my neighbour in the legal sense?*
> *"Persons who are so closely and directly affected by my act that I ought reasonably to have them in contemplation as being so affected when I am directing my mind to the acts or omissions which are called in question."*

Figure A4-12: The neighbour principle. *Source: Lord Atkin, Donoghue v Stevenson [1932].*

FAIRCHILD V GLENHAVEN FUNERAL SERVICES LTD AND OTHERS [2002] UKHL 22

[Causation, employer liability, material increase in risk]

Facts - Mr Fairchild had worked for a number of different employers, as a subcontractor for Leeds City Council, all of whom had negligently exposed him to asbestos. Mr Fairchild contracted pleural mesothelioma. He died, and his wife was suing the employers on his behalf for negligence. A number of other claimants were in similar situations, and joined in on the appeal. The problem was, a single asbestos fibre, inhaled at any time, can trigger mesothelioma. The risk of contracting an asbestos related disease increases depending on the amount of exposure to it. However, because of long latency periods (it takes 25 to 50 years before symptoms of disease become evident) it is impossible to know when the crucial moment was. It was impossible therefore for Mr Fairchild to point to any single employer and say 'it was him'. Moreover, because the traditional test of causation is to show that 'on the balance of probabilities' X has caused Y harm, it was impossible to say that any single employer was the cause at all. While it was possible to say 'it was one of them' it was impossible to say which. Under the normal causation test, none of them would be found, on the balance of probabilities to have caused the harm.

Decision - the House of Lords held that, following ***McGhee v National Coal Board*** the appropriate test in this situation, was whether the defendant had materially increased the risk of harm toward the plaintiff. The employers were joint and severally liable against the plaintiff (though amongst themselves they could sue one another for different contributions). Lord Bingham, in particular, noted that in this case it was not possible to speak of 'probabilities' in a simple way, because, 'it is on this rock of uncertainty, reflecting the point to which medical science has so far advanced, that the three claims were rejected by the Court of Appeal and by two of the three trial judges'. Moreover, 'the overall object of tort law is to define cases in which the law may justly hold one party liable to compensate another'. It was wrong to deny the claimants any remedy at all. Therefore the appropriate test of causation is whether the employers had materially increased the risk of harm to the claimants.

The House of Lords overturned the Court of Appeal and ruled in favour of the Claimants.

The cost of this ruling was enormous. It is estimated that this single judgment was worth £6.8 billion. Approximately 13 Britons die every day from asbestos related diseases, and the rate of deaths are increasing.

JONES V LIVOX QUARRIES LTD [1952] 2 QB 608

[Contributory Negligence]

Facts - the claimant worked in the defendant's quarry. One lunch break he hitched a lift back to the canteen by standing on the tow bar of a traxcavator. The driver of the traxcavator was unaware that the claimant had jumped on the back and it was against company rules to stand on the back of the traxcavators. Unfortunately a dumper truck, driven recklessly by another employee, crashed into the back of the traxcavator crushing the claimant's legs. Consequently the claimant had to have his legs amputated.

Decision - the defendant was liable but the claimant was held to be 1/5 to blame under the Law Reform (Contributory Negligence) Act 1945. He had acted against orders and exposed himself to danger.

Lord Denning MR: "Just as actionable negligence requires the foreseeability of harm to others, so contributory negligence requires the foreseeability of harm to oneself. A person is guilty of contributory negligence if he ought reasonably to have foreseen that, if he did not act as a reasonable, prudent man, he might be hurt himself: and in his reckonings he must take into account the possibility of others being careless. Once negligence is proved, then no matter whether it is actionable negligence or contributory negligence, the person who is guilty of it must bear his proper share of responsibility for the consequences. The consequences do not depend on foreseeability, but on causation. The question in every case is: What faults were there which caused the damage? Was his fault one of them?"

Singleton LJ: "The plaintiff, in riding on the traxcavator, was disobeying the orders of his employers. In so doing he was exposing himself to danger. It may well be that the chief danger was that he might fall off, or be thrown off, or that he might become entangled in some part of the machine on which he was riding; but those were not the only risks to which he subjected himself. He had put himself in a dangerous position which, in fact, exposed him to the particular danger which came upon him. He ought not to have been there. The fact that he was in that particular position meant that he exposed himself, or some part of his body, to another risk, the risk that some driver following might not be able to pull up in time - it may be because that driver was certainly at fault. That is the view which the trial judge took of this case, and I do not see that that is a wrong view. It is not so much a question of was the plaintiff's conduct the cause of the accident as did it contribute to the accident on the assumption that it was something of a kind which a reasonably careful man so placed would not have done. If he unreasonably, or improperly, exposed himself to this particular risk, I do not think that he ought to be allowed to say that it was not a cause operating to produce the damage, even though one may think that the prohibition against riding on the vehicle was not made with that particular risk in mind."

LATIMER V AEC LTD [1953] 2 ALL ER 449, HL

[Discharge of duty to take reasonable care]

Statute - Factories Act 1937

Facts - an exceptionally severe thunderstorm flooded the factory. The water mixed with the cooling oil used on lathes which caused the floor to become slippery. The company kept a quantity of sawdust but this was insufficient to cover all of the slippery area.

Mr Latimer hurt himself by slipping while trying to lift a 2 cwt bin. He sought damages under the Factories Act which required that floors be 'properly maintained'.

Decision - Mr Latimer lost his case. The House of Lords held that a transient and exceptional condition of an otherwise sound floor did not cause it to be improperly maintained. The employer had taken reasonable steps to deal with the danger to his employee.

Relevance - maintained means 'maintained in an efficient state, in an efficient working order and in good repair'. The term within the Act was clearly directed to the state of construction of the floor and not to a temporary and unexpected condition or obstruction on its surface.

MERSEY DOCKS AND HARBOUR BOARD V COGGINS & GRIFFITH (LIVERPOOL) LTD [1946] 2 ALL ER 345

[Extent to which employer may be held vicariously liable for negligence of contractor]

Facts - Coggins, a firm of stevedores, hired a crane and driver (Mr Newall) from Mersey Docks and Harbour Board. Mr Newall negligently operated the crane thereby injuring a third party. The hire contract stipulated that Mr Newall was to be a temporary employee of Coggins for the duration of the agreement.

Key points

- The authority supplied the driver with the crane.
- The driver was hired, paid and liable to be dismissed by the authority.
- The contract between the authority and the stevedores stipulated that the driver supplied would be deemed to be the servant of the stevedores while carrying out the work under the contract.
- The hirers had power to control what Mr Newall lifted but not how it should be lifted.

Decision - it was accepted that the driver was a servant, not a self-employed person and it was also accepted that the accident was due to his negligence. The only question was which of the two organisations should be held vicariously liable to the injured person. The House of Lords had to consider an appeal by the general employer from a decision against them by the Court of Appeal. It was not suggested that there was any personal fault on the part of either of the organisations.

It was held that control over Mr Newall's work had not passed to the hirers. Only if there is control over what work the person does and how he does it can control be held to have been transferred.

Relevance - vicarious liability cannot be easily passed to a 'quasi-employer'. There are two tests which must be passed in order for this to succeed. The new 'employer' must be in a position to dictate:

- What work is to be done.
- How it is to be done.

PARIS V STEPNEY BOROUGH COUNCIL [1951] 1 ALL ER 42, HL

[Higher duty of care owed to vulnerable employees]

Facts - Mr Paris worked in the Borough Council's trucks maintenance garage. He was blind in one eye, but this was not known by his employer until he was examined by a doctor for the Council's superannuation scheme. He was given two weeks' notice of dismissal. Two days before his leaving date, he was working underneath a truck when a piece of metal flew off into his good eye blinding him.

He claimed damages for negligence saying that he, as an individual with extra susceptibility of serious injury, should have been provided with goggles.

Decision - the House of Lords upheld his claim. The duty of the employer to take reasonable steps to prevent injury to employees is owed to each employee individually. If the employer knows of a condition which makes the employee more susceptible to injury, or could make the injury more severe than usual, he must take extra precautions. The provision of goggles to Mr Paris would have been reasonable even if goggles were not provided to others doing the same job.

SUTHERLAND V HATTON (2002) [REASONABLENESS IN RELATION TO HARM FROM STRESS AT WORK]

Sutherland v Hatton and others [2002] EWCA Civ 76

Statute ref - these claims were brought in common law negligence

Facts - the Court of Appeal heard four appeals by employers against compensation awards to employees who had suffered stress-induced psychiatric illness.

Decision - three of the appeals succeeded. The Court ruled that the general principle was that employers should not have to pay compensation for stress-induced illness unless such illness was reasonably foreseeable. Employers are normally entitled to assume that employees can withstand the normal pressures of a job. The Court set out a number of practical propositions for future claims concerning workplace stress.

These are as follows:

- Employers do not have a duty to make searching inquiries about employees' mental health. They are entitled to take what they are told by employees at face value unless they have good reason to disbelieve the employees' statements.
- Where an employee wishes to remain in a stressful job and the only alternative is demotion or dismissal, the employer is not in breach of duty in allowing the employee to continue.
- Indications of impending harm to health at work must be clear enough to show an employer that action should be taken, in order for a duty on an employer to take action to arise.
- An employer is in breach of duty where he fails to take reasonable steps bearing in mind the following: the size of the risk; the gravity of the harm; the cost of preventing the harm; any justification for taking the risk.
- No type of work may be regarded as intrinsically dangerous to mental health.
- Employers, who offer confidential counselling advice services, with access to treatment, are unlikely to be found in breach of their duty of care in relation to workplace stress.
- Employees must show that their illness has been caused by a breach of duty and not merely by occupational stress.
- The amount of compensation will be reduced to take account of pre-existing conditions or the chance that the employee would have become ill in any event.

The Court of Appeal dealt with the following cases:

- Penelope Hatton, a schoolteacher, had been awarded £90,000 compensation for depression and debility. Her employer's appeal was allowed on the grounds that her workload was no greater than her colleagues' and her absences could be put down to reasons other than workplace stress.
- Olwen Jones, a local authority employee, had suffered from depression and anxiety as a result of overwork. It was foreseeable that her workplace conditions would cause harm; therefore the employer's appeal against an award of £150,000 damages was dismissed.
- Leon Barber, a teacher, developed symptoms of depression. He was awarded £100,000 compensation. The employer's appeal was allowed on the grounds that the claimant had not told the employers about his illness until he suffered a breakdown.
- Melvyn Bishop, a factory worker awarded £7,000 compensation following a nervous breakdown and attempted suicide. The employer's appeal was allowed because the Court ruled that the demands of his work had not been excessive.

(Source: Croner's Health and Safety Case Law 2003).

VIASYSTEMS (TYNESIDE) LTD V THERMAL TRANSFER (NORTHERN) LTD (2005) CA [Vicarious Liability]

Facts - Viasystems Ltd sued for damages for a flood, caused while Thermal Transfer Ltd had installed an air conditioning in their factory on Eldon Street, South Shields. Thermal Transfer Ltd subcontracted to S&P Darwell Ltd, which subcontracted to CAT Metalwork Services, which hired young Darren Strang. Darren made 'a foolish mistake on the spur of the moment' by climbing down from a roof, where he was sent by his workmate to get fittings, through a duct. He was meant to use crawling boards and roof purlins. The duct collapsed, hit a sprinkler, and flooded the site. Viasystems sued Thermal, S&P Darwell and CAT Metalwork, while Thermal brought an action for contribution against S&P Darwell and CAT Metalwork.

Decision - the decision at first instance was based on the assumption that vicarious liability is an entire liability, i.e. two distinct legal entities could not be vicariously liable for the same act. This assumption was based on a case dating back as far as 1826.

The assumption has implicitly been applied in numerous cases in which an employee of a company (Company A) has been hired or seconded to work for another company (Company B). In such cases, although there is no legal transfer of the individual's employment, the law has in some instances been prepared to treat the individual as an employee of Company B such that Company B has been found vicariously liable for the individual's negligence. However, the assumption was that liability had to attach to either Company A or

Company B, it could not attach to both. Against this assumption, it is not difficult to see why the question of who should be held vicariously liable was so contentious. Mr Strang was working closely with Mr Megson, and did as he was instructed by Mr Megson (except when it came to taking the fateful shortcut), so there were strong arguments that CAT should be liable. Equally, the pair were working under the overall supervision of Mr Horsley who checked on them from time to time and who gave them general instructions on what they were to do, and who was responsible for health and safety on the site. The judge at first instance found that CAT were liable.

The Court of Appeal approached the question from the standpoint that the long standing assumption that dual vicarious liability could not exist was not binding in the Court of Appeal. Much of the judgment is taken up by an examination of previous authorities, the extent to which they were founded on the assumption and the extent to which any authorities would bind the Court of Appeal. May LJ and Rix LJ concluded that they were entitled to find that dual vicarious liability could indeed exist. The proper approach is to look at the negligent act and to ask who was entitled (and, possibly, theoretically obliged) to give orders about how the work should or should not be done? The alternative approach of establishing which party was in 'entire and absolute control' was inappropriate. In many instances a person will not be in the 'entire and absolute control' of one single party. A person may perform his functions at the direction of more than one party.

On this basis the Court of Appeal found that both Mr Megson and Mr Horsley were entitled to give Mr Strang instructions about what he did and how he did it. Weighing their respective 'shares' of responsibility, the Court found that each should contribute equally, i.e. 50%.

Relevance - a case which establishes (contrary to accepted opinion since the 1800s) that dual vicarious liability can apply in an employment situation. This decision is potentially significant in outsourcing contracts.

WILSONS AND CLYDE COAL CO V ENGLISH [1938] 3 ALL ER 628 [COMMON LAW DUTY OF CARE OF EMPLOYER AND NON-DELEGABLE NATURE OF DUTY OWED]

The nature of employers' duties at common law

Circumstances - Mr English was a miner employed by Clyde Coal Company. He was crushed by haulage equipment. His employers argued that they had discharged their duty by appointing agents who managed the mine.

Decision - the House of Lords held that the employer can delegate the performance of the duty but not the responsibility which attaches to it. The employer was therefore negligent and that negligence led to Mr English's injuries. The employer's duty of care to his employees was determined to extend to the provision of:

- Safe place of work including access and egress.
- Safe procedures.
- Safe plant and appliances.
- Competent fellow employees.

Relevance - the duty is therefore 'personal', such that the employer must take reasonable care for the safety of his employees whether or not the employer takes any share in the conduct of the operation.

OTHER RELEVANT DECIDED CASES

Adsett v K & L Steelfounders and Engineers Ltd (1953) 1 ALL ER 97 and 2 ALL ER 320

[Meaning of practicable]

Statute Factories Act (FA) 1937, s.47; later Factories Act (FA) 1961, s.63

Facts - Mr Adsett contracted pneumoconiosis in a foundry. The main cause was inhalation of silica dust, which became airborne as he shovelled various casting sands and compounds through a grate onto a conveyor below. To control this, the employers installed a dust extractor near the conveyor. This was done as soon as they thought of it, but it was well after the onset of Mr Adsett's disease and too late to save him from disablement.

Decision - the question concerned the meaning of the word 'practicable' in Section 47 of the Factories Act (FA) 1937, which required employers to take all practicable measures to protect employees from inhalation of dust, etc. Dictionary definitions said, 'possible to be accomplished with known means or resources', 'capable of being carried out in action' and 'feasible'. 'Practicable' is a stricter standard than 'reasonably practicable' so the question of cost should be eliminated. No measure could be 'practicable' if it was not within current knowledge and invention. The employers were held not to have been in breach of statutory duty.

Note - there was an appeal, as it was argued, on behalf of Mr Adsett, that the technology to install an extractor did exist, but that the employers (and everyone else in the industry) had not thought of this particular application for an extractor. It was said that the employers could not escape liability just because they did not think of this particular use for an extractor. The appeal was rejected on the grounds that 'practicable' meant that a measure had to be known for its application by people in the industry and especially by the experts.

Armour v Skeen (1977) IRLR 310

[Responsibilities of senior management Section 37 HASAWA 1974]

Statute - HASAWA 1974 Section 37 and Section 2 Personal Liability - Criminal

Facts - a workman fell to his death whilst repairing a road bridge over the River Clyde. The Strathclyde Regional Council and its Director of Roads (Mr Armour) were both prosecuted. The basis of Mr Armour's prosecution under Section 37 of the HASAWA 1974 was that the duty of supervising the safety of council workmen on the roads was his. He had failed to formulate a sound safety policy.

Mr Armour claimed in his defence to have no personal duty to carry out the council's statutory obligations (one of which was the formulation of a detailed safety policy for the roads department).

Decision - he was convicted of the offences and the conviction was upheld on appeal. The fact that Section 2 of HASAWA 1974 imposes a duty on the employers to provide a safe system of work did not mean that there was no duty on his part to carry out that duty. Section 37(1) refers to 'any neglect', not to the neglect of a duty imposed. The offences were committed by the body corporate, but were due to his neglect. Further, although his title as 'Director of Roads' did not mean he was a 'director' within the meaning of Section 37 (1), he was within the meaning of the words 'manager or similar officer'.

Relevance - the case gives guidance as to those who may be considered as the 'directing mind' of the organisation who therefore are liable under Section 37 of the HASAWA 1974. This includes persons who purport to act as directors; company secretaries or other similar officers are equally liable. Anyone who acts in a managerial capacity must be held liable under Section 37 (1) whatever the title he or she may have. If the affairs of the body corporate are being managed by its members (for example, a workers' co-operative), then the acts of the member which are in connection with his managerial functions are within the meaning of this Section (Section 37 (2)).

Note - for a person to be convicted under Section 37, it must be shown that he or she has some responsibility for the making of management decisions, and be in a position of responsibility. In R v Boal, the accused was the assistant general manager of Foyle's Bookshop. He had been given no management training, in particular, none in health and safety matters, or fire precautions. He was, however, in charge of the shop while the general manager was away on a week's holiday. During this period, the premises were visited by officers from the local fire authority, which discovered there were serious breaches of the fire certificate which had been issued. Foyle's and the accused were charged with a number of offences under the Fire Precautions Act (FPA) 1971 - Foyle's as the 'body corporate' and the accused because the 1971 Act provides that:

> *"... Where an offence committed by a body corporate is proved ... to be attributable to any neglect on the part of any director, manager, secretary or other similar officer of the body corporate ... he as well as the body corporate shall be guilty of that offence ..."*

Figure A4-13: Liability as well as the body corporate. *Source: Section 37 of HASAWA 1974.*

Foyle's were convicted on 11 counts, and were fined. The accused pleaded guilty to 3 counts, and was found guilty of 7 others. He was sentenced to 3 months' imprisonment, suspended for 12 months. He then appealed against the conviction on the ground that he was not '... a manager or similar officer ...' within the meaning of the Act. The Court of Appeal (Criminal Division) allowed his appeal. A person was 'a manager' if he had the power of '... the management of the whole of the affairs of the company ...' or was, '... entrusted with power to transact the whole of the affairs of the company ...' or was '... managing in a governing role the affairs of the company itself ...'. The Court further thought that the intended scope of Section 23 of the Fire Precautions Act was '... to fix with criminal liability only those who are in a position of real authority, the decision makers within the company who have both the power and responsibility to decide corporate policy and strategy...'. It is to catch those responsible for putting proper procedures in places; it is not meant to strike at underlings. The case was overturned at appeal.

Baker v T E Hopkins and Sons Ltd [1959] 1 WRL 966

[Non-application of Volenti non-fit injuria where aid given to co-employee or third party]

Facts - it was held that the defence of volenti did not apply in the case of Baker v T E Hopkins and Sons Ltd [1959] where a local doctor died trying to rescue two men who had been overcome by a lethal concentration of carbon monoxide fumes when working in a well. The defendants used a petrol engine to clean out a well. Two employees were told not to start the engine until the manager arrived. They ignored the instruction and were overcome by carbon monoxide fumes. The claimant, who was a doctor, attempted to rescue them, but all three died.

Decision - the defendant Company was negligent in not warning the employees of the danger.

It was foreseeable that a rescue attempt would be made if employees were in difficulties. The claimant was not reckless in attempting to save life and was therefore owed a duty of care.

Barkway v South Wales Transport Co Ltd [1950] 1 ALL ER 392 HL

[Employers duty to maintain work equipment]

Facts - the defendant's bus mounted the pavement and then went down an embankment. The claimant's husband, a passenger was killed. Evidence showed a defective tyre had burst.

Decision - the principle of res ipsa loquitur did not apply and the claimant had to establish negligence. She did this by showing the system of inspecting the tyres was inadequate, being only twice a week, when the tyres suffered unduly harsh treatment from being driven over kerbstones and other obstacles. There should have been a reporting system to record incidents of heavy blows to the tyres. The defendant did not take these reasonable steps and therefore was negligent.

Cambridge Water Co v Eastern Counties Leather plc [1994] 2 WLR 53

[Further to decision in Rylands v Fletcher (later in this element), liability does not apply retrospectively in absence of forseeability of harm]

Facts - Cambridge Water Company owned a borehole at Sawston from which it pumped water into the public supply. The water company bought the land in 1976 for the purposes of constructing the borehole to supply water to its customers. The pumping house was commissioned in 1979.

As part of its operations in the 'industrial village' of Sawston, Eastern Counties Leather plc used organochlorines, one of which was perchloroethylene (PCE), a solvent used for degreasing skins. Until about 1976, the chemical was delivered to the site in drums, kept in storage there and when needed, tipped into the reservoir supplying the degreasing machine. This transfer led inevitably to spillage, which over the years filtered through the ground into the aquifer. The degreaser is volatile and easily evaporated, and Eastern Counties might have considered that it evaporated before it could pass through the ground. In 1980 the EC drinking water directive was issued and became effective in 1985. This Directive included maximum permissible concentrations of organochlorine compounds. Tests by Cambridge Water showed that supplies from the borehole were in excess of those limits and as a result Cambridge Water had to stop pumping and find an alternative supply. This cost something in excess of £900,000. Cambridge Water Company sued Eastern Counties Leather under the Rylands v Fletcher Rule.

Decision - the High Court held that, as Sawston is an industrial village, the creation of employment by the defendants was clearly for the benefit of the community. The storage of chemicals on the land was therefore a natural use of the land.

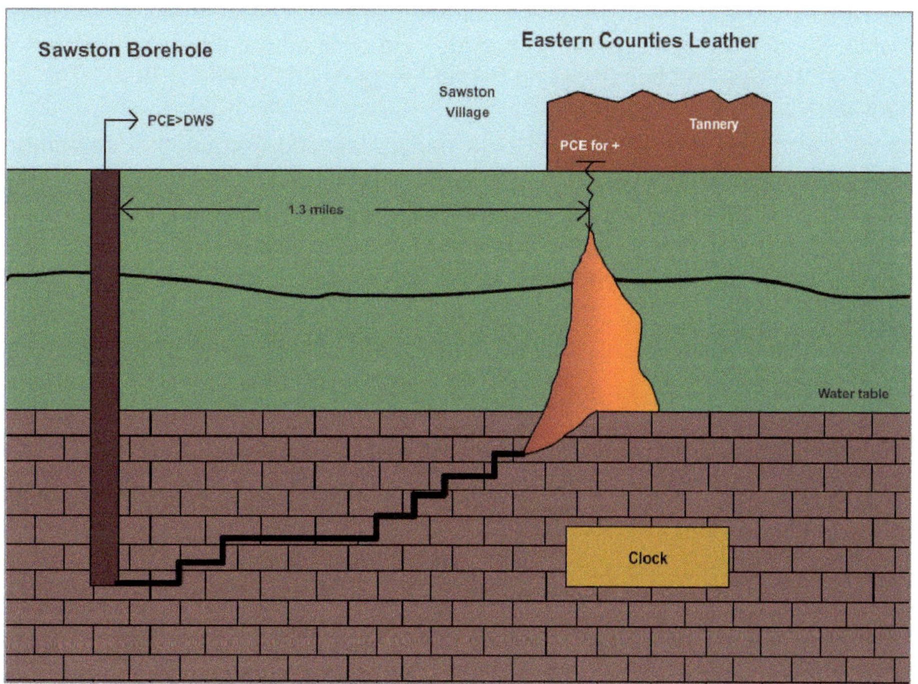

Figure A4-14: Sawston borehole. *Source: RMS.*

The Court of Appeal reversed this decision and held that Easter Counties Leather were liable under Rylands v Fletcher. The storage of chemicals on the land could not be regarded as a natural use.

The House of Lords found that the defendants were not liable for the damages of £2 million awarded to Cambridge Water. Lord Goff explained that although the fact that a defendant had taken all reasonable care will not exonerate him or her from liability in nuisance, 'it by no means follows that the defendant should be held liable for damage of a type which he could not reasonably foresee'. This means that strict liability would only apply if damage could be reasonably foreseen. Lord Goff also commented that the storage of large amounts of chemicals should be considered a non-natural use of land.

Thus the House of Lords agreed that this was an almost classic case of non-natural use; however they reversed the Court of Appeal decision based on the principle of foreseeability of consequences. Eastern Counties could not have foreseen the consequences of the spillage or the maximum permissible levels in the EC directive, therefore they were not liable.

Note - this case illustrates the development of common law as well as extending the defences to strict liability imposed by the Rylands v Fletcher rule. The defence of the consequences of an escape being unforeseeable is now available to defendants.

It should be noted that the House of Lords doubted whether the fact that Eastern Counties gave employment could lead to the conclusion that keeping chemicals on land was a natural use.

Edwards v National Coal Board [1949] 1 ALL ER 743

[Meaning of reasonably practicable]

Facts - Mr Edwards was killed when a Section of underground roadway collapsed. The NCB argued that the cost of shoring up would have been prohibitive and was not justified.

Decision - the NCB were found liable on the basis that the cost of making this particular road Section safe was not great when compared to the risk.

Relevance - the 'Quantum of Risk' test should be applied in determining whether or not the steps that were taken in the circumstances of the incident were sufficient to satisfy the requirements of the term 'reasonably practicable' where such an obligatory term is stipulated.

This 'Quantum of Risk' test is the analogy of a scale balance whereupon the degree of risk is placed on one pan and the sacrifice involved in the measures necessary for averting the risk (whether in money, time or trouble) is placed on the other; and that, if it be shown that there is a gross disproportion between them - the risk being insignificant in relation to the sacrifice - the defendants discharge the onus upon them. (Lord Asquith).

Ferguson v Dawson and Partners Ltd [1976] IRLR 376, CA

[Self-employed workers may be regarded as employees]

Facts - Ferguson, a labourer, joined the company 'on the lump', which was the basis they hired labourers. To avoid tax, Ferguson gave a false name. He and the company argued that he was self-employed.

He fell off a roof while removing some scaffolding boards. He claimed damages against Dawson for breach of statutory duty relying on the Construction (Working Places) Regulations (CWPR) 1966. The duty would only be owed if the claimant were an employee of Dawson.

Decision - it was held that, despite the fact that both parties labelled Ferguson a 'self-employed labour only contractor', the relationship was in fact that of employer-employee.

Relevance - a contract of employment was said to exist because: the employer directed the manner in which the work was to be done; the worker was not allowed to use his discretion; he was an integral part of the employer's business; the worker was not allowed to employ others to assist him; and the worker was not in business on his own account.

General Cleaning Contractors v Christmas [1953] 2 ALL ER 1110

[Employer's duty to provide a safe system of work]

Facts - Mr Christmas was a window cleaner. To save time, a method had been adopted in the trade of climbing onto the window sills instead of using ladders and holding onto the window frames. Although there was some danger, it was accepted that if a proper routine were followed, it could be done safely. It was also accepted that the use of ladders was unduly restrictive and impracticable.

Mr Christmas pulled the upper half of a sash window down a little so he could hold on to the top bar of the bottom frame. The top frame slipped down and trapped his fingers. He let go and fell to the ground suffering injury. He claimed damages, alleging negligence against his employer in the way the method of work was organised.

Decision - the House of Lords held the employer liable. It was held that it was the employer's duty to go to the site, assess the risks and work out a safe method of doing the work. In this case it would have been relatively easy to devise wedges or some other system to prevent the window frames slipping. This measure would have made the work reasonably safe.

Note: *"Where a practice of ignoring an obvious danger has grown up I do not think that it is reasonable to expect an individual workman to take the initiative in devising and using precautions. It is the duty of the employer to consider the situation, to devise a suitable system, to instruct his men what they must do, and to supply any implement that may be required such as in this case wedges or objects to be out on the window sill to prevent the window from closing."* (Lord Reid).

Hudson v Ridge Manufacturing Co. Ltd [1957] 2 ALL ER 229

[Employer's duty to provide competent fellow employees; issue of horseplay at work]

Facts - in Hudson v Ridge Manufacturing [1957] an employee was a known practical joker who had made a nuisance of himself over the previous 5 years. Examples of his behaviour included tripping people from behind. Although this was done for fun and without malice or intent to bully, injury was inevitable sooner or later.

In one incident an employee broke his wrist in an effort to break his fall after being gripped around the neck and pulled back by the practical joker. This duty extends beyond competence through to behaviour in the workplace. An employee who is known to engage in practical jokes and horseplay could constitute a hazard which the employer would be expected to deal with.

Decision - the employer was held liable in the resulting case on the basis that they had failed to take steps to curb his horseplay.

It was held that the duty of the employer is to provide safe colleagues to work with. In this case the employer's duty was to take steps by reprimand, discipline and, if necessary, dismiss the practical joker in order to remove the risk of injury to other employees.

ICI v Shatwell [1965] AC 656

[Volenti non-fit injuria and sole fault of claimant as defences in employers' liability cases]

Facts - in ICI Ltd v Shatwell [1964] two brothers, who were certified shot blasters, were injured whilst testing electrical circuitry used to detonate explosives. They did so in the open, contrary to the Quarries (Explosives) Regulations 1959 which required them to be in a position of safety.

Decision - the resulting claim for damages against the employer was unsuccessful on the basis that they had voluntarily accepted the risk of each other's negligence and breach of statutory duty by defying the employer's instructions and breaching the Regulations which were well known to them. The 'volenti' defence can be successful where employees conspire to breach statutory duty against the employer's specific instructions.

Note - the volenti defence, however, will not normally succeed in cases where there is a direct breach of common law or statutory duty by the employer.

Knowles v Liverpool County Council [1993] 1 WLR 1428

[Employers' Liability (Defective Equipment) Act 1969 extends to materials used at work]

The word equipment includes any plant and machinery, vehicle, aircraft and clothing. This is a wide ranging definition of the word 'equipment' as illustrated by such cases as:

Knowles v Liverpool City Council (1992) where the employer was held liable for injury caused by a 'green' concrete flagstone.

Mr Knowles was employed by the appellant Council as a labourer flagger. Whilst repairing a pavement a flagstone he was handling broke causing injury to his finger. The breakage occurred because the manufacturer had not cured the flagstone properly.

The House of Lords held that the flagstone which the respondent was handling in the course of his employment was equipment provided by his employer for the purpose of the employers' business within the meaning of the Employers' Liability (Defective Equipment) Act (ELDEA) 1969 Section 1 (1) (a) and consequently his employer was liable to him under this Act. The flagstone had broken due to a defect in manufacture.

Lister v Romford Ice and Cold Storage Co. Ltd [1957] AC 555 [Negligence of fellow employees re: vicarious liability, joint tortfeasers and subrogation]

Facts - Mr Lister (junior), a lorry driver employed by Romford Ice and Cold Storage Co (RI&CS), took his father with him on a job to act as mate. While reversing the lorry, he injured his father. The father claimed damages against RI&CS.

Decision - the father successfully recovered damages from RI&CS for the negligence of Lister (junior) as their employee.

The company then claimed a contribution from Lister (junior) on the basis that he had broken an implied term in his contract of employment, which was that he would use reasonable skill and care in driving.

Subrogation generally occurs when an insurance company pays damages on behalf of its insured client and then sues the party, which the claimant contends caused the injury/loss. Lister (junior) had broken an implied term in his contract of employment that he would take reasonable skill and care in driving.

Relevance - the employer is vicariously liable for the negligent acts of his employees during the course of work if a third party suffers loss. RI&CS was liable for Lister's (junior) actions.

Lister (junior) had a duty of care to his employer in the performance of his duty as a driver. The employer has a right to claim damages from a negligent employee.

Note - generally, there is an agreement between insurance companies that they will not pursue claims against employees for their negligent acts except if the circumstances seem suspicious.

Machray v Stewart and Lloyds Ltd 1964 [Employers duty to provide sufficient plant]

Facts - the claimant, an experienced rigger was asked to get some piping from the ground to seventy feet up on a steel structure in the course of construction. A crane that would normally be used was not available. The rigger decided he could use a chain block and tackle, but could not find one on the site. As the foreman was pressing him to do the job, he used a rope block and tackle, which was not suitable. As the pipe was being lifted, it swung out of control and hit the rigger. He claimed damages for the company's failure to provide suitable equipment.

Decision - the Company was held liable. It was the company's duty to take reasonable steps to prevent injury to employees by providing equipment that was suitable for the job in hand.

The Company claimed contributory negligence in their defence, but this was completely rejected as Mr Machray had been prevented from adopting a safe system of work for the job because of the pressure put on him by the foreman.

Marshall v Gotham and Co. Ltd (1954) AC 360

[Difference between practicable and reasonably practicable]

Facts - Mr Marshall was killed by a roof collapse whilst working in a gypsum mine. The roof had been tested but the roof fall had been caused by an undetectable geological fault known as 'slickenside'. His wife claimed compensation for breach of statutory duty.

Decision - the employer was not liable as they had taken reasonable steps to secure the roof.

Relevance - the term 'secure' means safe against normal or likely hazards. Thus totally unexpected or abnormal movements or falls do not prove a failure of duty. In this case the movement was caused by an unforeseeable geological fault.

In this case, a comparison between the levels of duty required by the terms 'practicable' and 'reasonably practicable' was made.

> *"If a precaution is practicable, it must be taken unless in the whole circumstances that would be unreasonable and as men's lives may be at stake it should not lightly be held that to take a practicable precaution is unreasonable."*

Figure A4-15: Practicable. *Source: Lord Reid.*

Practicable denotes a stricter standard than 'reasonably practicable'. It means 'possible to be accomplished with known means or resources' or 'feasible', with cost probably not a factor.

Paine v Colne Valley Electricity Supply Company [1938] 4 ALL ER 803

[Employer's duty to provide a safe place of work]

Facts/decision - an employer was held liable for injuries to his employee caused by the failure of contractors to install sufficient insulation in an electrical kiosk.

Relevance - the employer's duty is a personal one so that he remains liable even though he has delegated the performance of the duty to a competent independent contractor.

Qualcast v Haynes 1959 [Higher duty of care expected of an experienced employee]

Facts - Mr Haynes was an experienced foundry man. He splashed molten metal on his foot while pouring from a hand-held ladle. He was not wearing foundry boots or leather spats even though they were readily available from the stores. He claimed damages for the employer's failure to urge him to wear spats.

Decision - the judge found for Qualcast. The duty of care remains constant to employees, but may be discharged in varying degrees according to the gravity of risk and the experience of individual employees. In this case, the employee had sufficient experience to know the risks well and the employer's duty was discharged by making the safety equipment available and leaving it to the discretion of the experienced men to use it.

Rylands v Fletcher 1865 [Tort relative to escape of stored materials]

Rylands v Fletcher (1868) LR 3HL 330

Facts - Mr Rylands was a mill owner who employed independent contractors to construct a reservoir on his land in order to provide water power for his mill. During the work, the contractors discovered an old mine whose shafts and passages connected with another mine on neighbouring land owned by Fletcher.

The contractors did not inform Fletcher and did not block up the shafts. When the reservoir was filled with water, the water seeped through some disused shafts and passages and flooded a mine belonging to Mr Fletcher. Mr Fletcher sued Mr Rylands for the resulting damages. Fletcher himself had not been negligent as he had no knowledge of the existence of the shafts. He was not vicariously liable for the actions of the contractors as they were not his employees.

The key points arising in the Rylands case were that:

- The contractor was competent.
- There was no negligence on the part of the client.
- There was a 'non-natural' use of land.
- There was an escape of matter.
- The escape resulted in damage.

Decision - Mr Rylands was found to be liable in tort, both by the Court of Exchequer and on appeal to the House of Lords. The Rylands v Fletcher rule of strict liability was established in this decision.

The Rylands v Fletcher rule is stated as:

> *"Where a person for his own purposes brings and keeps on land in his occupation anything likely to do mischief if it escapes, he must keep it at his peril, and if he fails to do so, he is liable for all damage which is a natural consequence of the escape."*

Figure A4-16: Strict liability - Rylands and Fletcher rule. *Source: Rylands and Fletcher [1868].*

During the appeal Lord Cairns, in agreeing with the above statement, added the qualification that the rule only applied to a 'non-natural' use of the land and not to circumstances where a substance accumulated naturally on land. The word 'natural' has since been extended to mean 'ordinary'.

The rule has been held to apply to beasts, water, filth and stenches and fire. This 'strict liability' rule can apply to all persons (corporate or individual) and a claimant does not have to prove negligence.

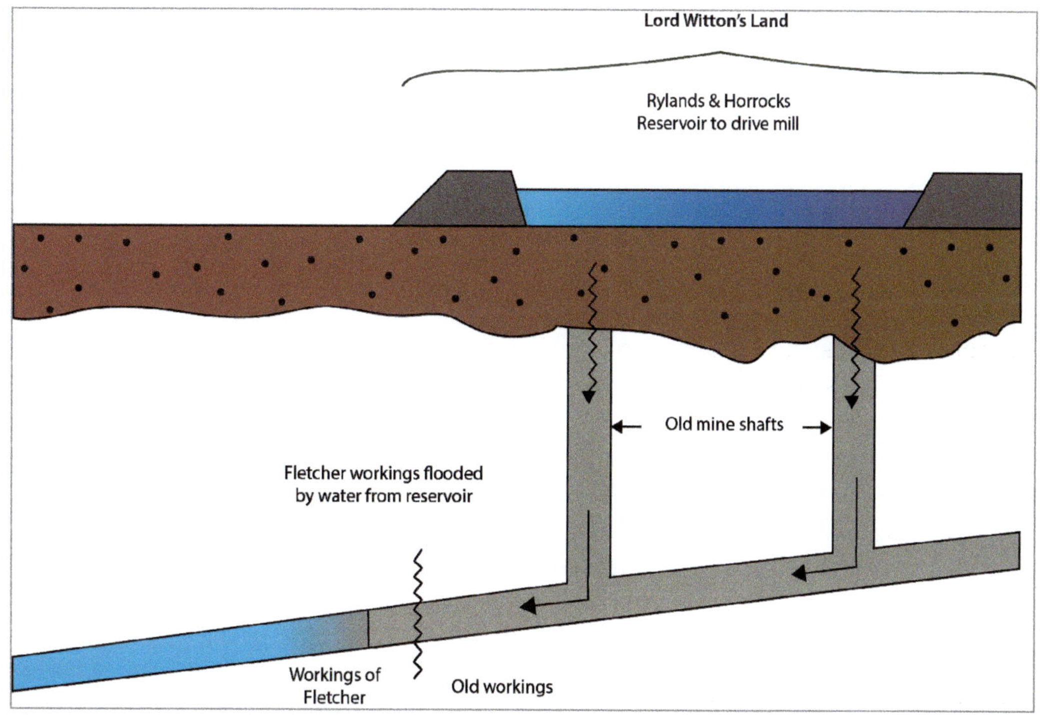

Figure A4-17: Rylands and Fletcher. *Source: RMS.*

Smith v Crossley Brothers 1951 [Vicarious Liability relative to an extreme act of horseplay]

Circumstances - the claimant, an apprentice employed in the defendant's apprentice training school, was seriously injured by a practical joke played on him by two fellow apprentices.

Decision - the defendants were found to be not liable because the claimant's injury had occurred through an act of wilful misbehaviour, which the defendants could not have reasonably foreseen.

The employer was not liable when he did not know and had no reason to suspect that there was a risk of an apprentice being injured by being subjected to 'initiation rites' by his fellows.

Speed v Swift (Thomas) and Co. Ltd (1943) [Employer duty to provide, an element of, a safe system of work]

Facts - a hook on a winch on board a ship caught on the perimeter railing of the ship. The railing broke and fell into a barge below injuring a workman. The winch was known to be unsafe and the railing was known to be broken.

Decision - the employer was found to have failed to take reasonable steps to prevent injury to employees. As part of a safe system of work, he should have ensured that the fencing was covered or protected in some way so that hooks or other objects could not hit it or get caught up in it.

Summers (John) and Sons v Frost (1955) AC 740 [Nature of absolute duty]

John Summers and Sons v Frost 1955 All ER 870 AC.740

F.A.1937 S14 (1), F.A.1961 S14 (1)

Securely fenced (absolute duty)

Circumstances - Mr Frost was injured whilst using a grinding wheel. His thumb was caught between the small (1cm) gap between the guard and the grinding wheel. Although the wheel was partially guarded, s.14 of the Factories Act imposed an absolute duty to guard the dangerous parts of machinery.

Decision - the House of Lords ruled that there had been a breach of statutory duty.

Relevance - the obligation to guard securely those parts of a machine which are dangerous must be fulfilled in such a manner, as to afford 'complete protection' to the 'careless or inattentive worker as well as the meticulous, careful one'.

Moreover, the 'dangerous part is securely fenced if and only if the presence of the fence makes it no longer dangerous'.

The duty was to provide fencing which absolutely prevented any part of any person working on the premises from coming into contact with any dangerous part of the machine.

This illustrates the literal interpretation of statutes that the courts are required to make - if the effect of this is to make the machine unusable then so be it.

Uddin v Associated Portland Cement Manufacturers Ltd. (1965) 2 ALL ER 213 [Contributory negligence]

F.A. s.14 Breach of Statutory Duty

Circumstances - Mr Uddin went into an unauthorised part of the factory in an attempt to catch a pigeon. He climbed up onto some machinery housing. As he reached over for the pigeon, he fell onto a rotating shaft which was only accessible from above.

Decision - even though the claimant was in a part of the factory where he should not have been and not there in relation to his normal work, the injury he sustained as a result of a revolving steel shaft being left unfenced, constituted a breach of the provisions of the Factories Act. Mr Uddin was however held to be 80% responsible for his own action (contributory negligence) and was awarded 20% of damages.

Relevance - it was enough to show that an employee suffered injury as a result of a breach of statutory duty for the case to succeed.

Note - it should be noted that a worker who was authorised to be in the roof space (for example, maintenance personnel) could equally have fallen onto the unguarded shaft.

Walker v Northumberland County Council 1995 [Employers' duty of care in relation to mental ill-health arising from excessive workload]

Facts - Mr Walker was employed as a social worker dealing with cases of child abuse. His workload increased steadily over the years and in 1986 he had a nervous breakdown. When he recovered and returned to work, he was promised additional resources to help him with his workload, but they failed to materialise. He had a second breakdown six months later. Mr Walker sued the council claiming they were in breach of their duty of care to provide a safe working environment.

Decision - the council were not held liable for the first breakdown as they could not reasonably have foreseen Mr Walker was exposed to a significant risk of mental illness through his job. They were, however, liable for the second breakdown, given that the same circumstances were there that caused the first. The second breakdown was a reasonably foreseeable risk.

The council were found to have failed in their duty of care by not providing effective support to alleviate Mr Walker's suffering.

Element

A5

Loss causation and incident investigation

Learning outcomes

On completion of this element, candidates should be able to demonstrate understanding of the content through the application of knowledge to familiar and unfamiliar situations. In particular, they should be able to:

A5.1 Outline theories/models and use of loss causation techniques.

A5.2 Explain the use of quantitative methods in analysing loss data.

A5.3 Explain the significance and use of statutory and internal reporting of loss events.

A5.4 Explain the reasons for loss and near miss investigations and the procedures to be followed.

Contents

Relevant statutory provisions

Reporting of Injuries, Diseases and Dangerous Occurrences Regulations (RIDDOR) 2013

Reporting of Injuries, Diseases and Dangerous Occurrences Regulations (Northern Ireland) 1997 (as amended)

Sources of reference

Reference information provided, in particular web links, was correct at time of publication, but may have changed.

Investigating accidents and incidents - a workbook for employers, unions, safety representatives and safety professionals (HSG245), ISBN 9780717628278, HSE Books

Reporting accidents and incidents at work, INDG453, http://www.hse.gov.uk/pubns/indg453.htm

The above web links along with additional sources of reference, which are additional to the NEBOSH syllabus, are provided on the RMS Publishing website for ease of use - www.rmspublishing.co.uk.

A5.1 - Theories/models and use of loss causation techniques

Losses result from lack of control and are revealed by loss causing events. These events may be known by a variety of names, the most common of which is 'accident'. There are many different definitions for the term accident, ranging from simple to complex. A useful definition of an accident is:

An unplanned, uncontrolled event which led to, or could have led to injury to persons, damage to plant or some other loss to the company.

Figure A5-1: Definition of an accident. *Source: RMS.*

This definition encompasses events that result in a wide range of losses and has, for a long time, helped to provide a good perspective of events with different outcomes. This has assisted greatly in encouraging people to learn from events and the subsequent prevention of accidents. The definition therefore, includes 'near misses', i.e. where no injury or damage etc. occurs. It is important not to think of injuries, damage and other losses as accidents, but rather as the results of accidents.

The following accident model is offered to illustrate the above statement: a brick falls from a height. The following consequences could result:

1) The brick falls into a pile of sand and there is no damage or injury. The brick hits an item of equipment, resulting in damage, but no injury.
2) The brick strikes a person causing a cut and bruising to the hand, this is an injury accident.
3) The brick strikes a person working directly underneath causing a fatality.

Figure A5-2: Accident. *Source: HSG245.* Figure A5-3: Near miss. *Source: HSG245.*

The difference between a near miss and a fatal accident in terms of time and distance can be very small indeed. It is therefore clear that the damage to persons or property is not the accident, but part of the effects of the accidents (i.e. the result or consequences).

An old adage says never waste an accident. Apart from being unpleasant and perhaps very costly, every accident constitutes an opportunity to correct some problem. For this purpose, a near miss which has the potential to cause loss is just as important as a serious injury/damage, in fact even more important if we are to avoid a future loss incident, a golden opportunity not to be missed.

In the HSE Guidance Document HSG245 'Investigating accidents and incidents', the HSE refers to an 'Adverse Event'. An adverse event includes:

- Accidents.
- Incidents.

The term 'adverse event' used by the HSE is similar to the term 'accident' used in ***figure ref A5-1*** above and encompasses events that have a wide range of outcomes. The HSE reserves the term 'accident' for events that involve harm to people.

The HSE define an Accident as:

An event that results in injury or ill-health.

Figure A5-4: Definition of an accident. *Source: HSE, HSG245.*

The HSE states that an incident includes a:

- Near miss.
- Undesired circumstance.

The HSE considers a 'near miss' to be: an event that, while not causing harm, has the potential to cause injury or ill-health. They take the term near miss to include dangerous occurrences specified in the Reporting of Injuries, Diseases and Dangerous Occurrences Regulations (RIDDOR) 2013.

In the view of the HSE the term incident includes 'undesired circumstance', which they take to be: a set of conditions or circumstances that have the potential to cause injury or ill-health, for example, untrained nurses handling heavy patients i.e. the potential for back injury.

HSG245 sets out an interesting perspective, which has generated some debate. Some practitioners may feel a definition of accident that limits itself to outcomes that harm people to be too narrow, others may find it useful to have a focused definition, for reporting and analysis purposes. It is important to remember that HSG245 is a guidance document and carries no direct legal duty to follow or use the definitions contained in it.

Understand the following theories/models

ACCIDENT/INCIDENT RATIO STUDIES AND THEIR USE AND LIMITATIONS

Some years ago, a study of 1,750,000 accidents, in 21 industry sectors, led by Frank Bird, showed that there is a fixed ratio between losses of different severity (and accidents where no loss occurred, i.e. near misses).

This can be demonstrated with a pyramid model:

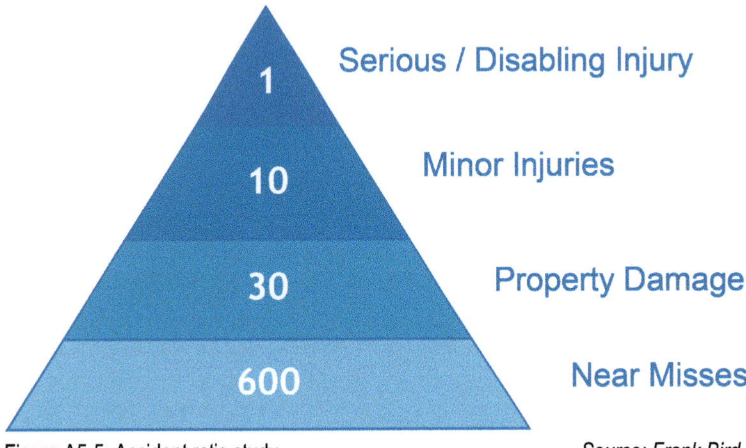

Figure A5-5: Accident ratio study. *Source: Frank Bird.*

There have been several versions of the accident pyramid, with some in HSE publications, for example, HSG 96 'The Costs of Accidents at Work'.

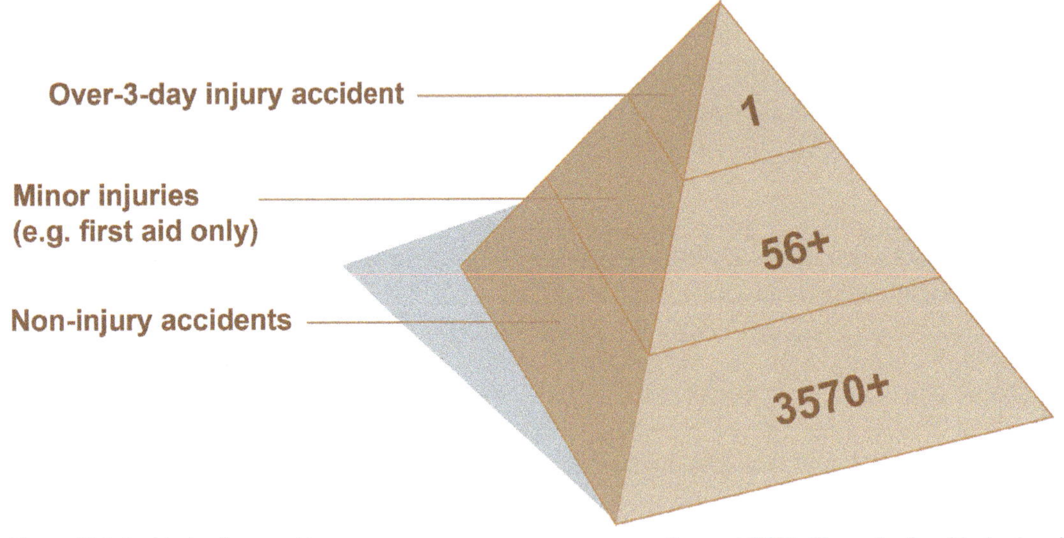

Figure A5-6: Accident ratio pyramid. *Source: HSG96: The costs of accidents at work.*

Use

The models illustrate that if limited interest is taken in the full range of events that occur, such that only those resulting in injury are considered, many opportunities to learn about what goes wrong are being missed. If near misses are also studied they can provide more opportunities to learn and possibly prevent some of the events that result in injury.

The Bird model includes property damage in addition to near misses, and again if measured, analysed and acted on this will help to prevent the injuries events.

The models help to convince people of the value of reporting a wide range of events and show that there are usually more near misses than injury events, which provide more opportunities to learn and improve health and safety.

Limitations

These results show that Bird's findings are not uniform throughout industry and in fact differ from industry to industry. This may be mostly due to the range of risk involved, but there could also be cultural differences, and variances in the level of reporting.

The accident ratio studies may not necessarily show the extent of the loss to the organisation. For example, the 'property damage' category may include extensive damage to large plant and equipment. Also, shown from Bird's and HSE's examples, there are no universally agreed definitions of each subset of accident type.

Statistical analysis of loss events relies on large numbers, comparable work and worker skills over the measured time-frame, to be effective. In smaller organisations, the first recorded accident may be the top event, i.e. fatality or major injury. Near miss reporting may mean different things to different people, for example, an office employee who enters a production plant will have a higher perception of danger than a worker who has been at work there for several years.

DOMINO AND MULTI-CAUSALITY THEORIES

The Domino Theory

HW Heinrich, an American safety engineer, proposed one of the first coherent theories of accident causation in the mid-1920s. He suggested that accidents were not 'acts of God' but were caused by the failures of people. His domino theory suggested that the series of events, which led to an injury or some other loss, were a succession of events which followed a logical pattern.

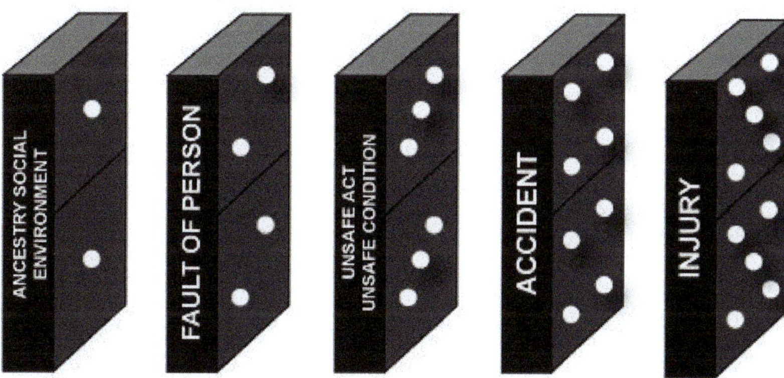

Figure A5-7: Accident causation domino. *Source: H.W. Heinrich.*

Further research by the International Loss Control Institute (ILCI) into accident causation led them to put forward a modified domino theory.

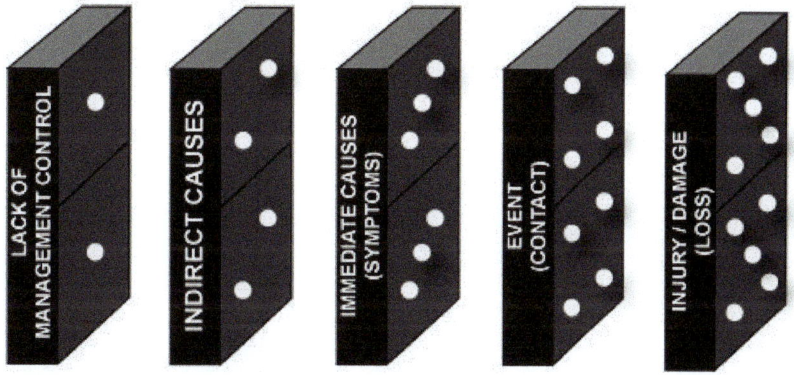

Figure A5-8: Accident causation domino. *Source: Frank Bird - ILCI.*

Considering each stage of the Frank Bird - ILCI domino separately:

'Loss'

This is the consequence of the accident and can be measured in terms of people (injuries), property (damage) or loss to the process (failed telecommunication) or the potential for any of these (near miss) and hence loss of profit.

'Event (accident or incident)'

The event producing the loss involving contact with a substance or source of energy above the threshold limits of the body or structure.

'Immediate (direct) causes'

These are the substandard (unsafe) ***acts*** (for example, using tools and equipment for tasks they were not designed to do) and substandard (unsafe) ***conditions*** (for example, a trailing telephone cable in an office) which gives rise to an accident. These are symptoms, which can be observed. Whilst these symptoms should not be ignored, action solely at this level will not, by itself, ensure that recurrence is prevented. Unsafe acts and conditions may be considered as workplace hazards.

Underlying (indirect or root) causes

These are the underlying or root causes of accidents. Identifying these causes will explain why the substandard act happened or the condition arose. They are not always easy to identify. Underlying causes fall into three major categories:

- Organisational factors - (Procedural).
- Job factors - (Technical).
- Personal factors - (Behavioural).

Organisational factors

- Work standards and procedures.
- Communication.
- Co-ordination.
- Supervision.

Job factors

- Design of equipment and layouts.
- Maintenance.
- Purchase of materials and equipment.

Personal factors include

- Physical capability.
- Mental capability.
- Physical stress.
- Mental stress.
- Information and knowledge.
- Skill.
- Motivation.

'Lack of management control'

This is the initial stage centred on the management functions of:

- Policy.
- Planning.
- Organising.
- Controlling.
- Monitor.
- Review.

HSG245 sets out the HSE's perspective of the domino theory. 'Each domino represents a failing or error which combines with other failings and errors to cause an adverse event. Dealing with the immediate cause (B) will only prevent this sequence. Dealing with all causes, especially root causes (A) can prevent a whole series of adverse events'. This is an important aspect of loss causation/prevention which is why thorough investigation to determine root causes is necessary.

Figure A5-9: Sequence of dominoes. *Source: HSG245.*

In this guidance document the HSE explains its view on what is an immediate, underlying and root cause, HSG245 should be read with care as some of the data on what is meant by these different causes appears in more than one place and is contradictory.

The meanings that follow good practice conventions have been selected from the document and are set out below:

"Immediate cause: the most obvious reason why an adverse event happens, for example, the guard is missing."

"Underlying cause: the less obvious 'system' or 'organisational' reason for an adverse event happening, for example, pre-start-up machinery checks are not carried out by supervisors."

"Root cause: an initiating event or failing from which all causes or failings spring. Root causes are generally management, planning or organisational failings."

Multiple accident causes

It should be remembered that accident investigation experience confirms that there is usually more than one causative factor, therefore each of the multiple causation factors may be seen as one domino in its own line of dominoes (just as the roots of a tree branch out).

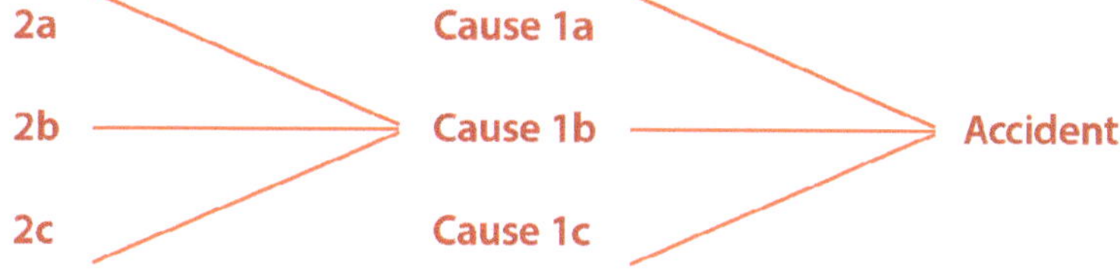

Figure A5-10: Multiple accident causes. *Source: RMS.*

For example, consider a tired gas fitter; he does not check his work equipment before leaving home and is delayed by heavy traffic on his way to work. On arrival at work, he finds a co-worker has not arrived on time, so starts a two-man job on his own. The work is to be carried out at height. On checking, he finds that his ladder is not on his van, so uses a drum for access to a high level pipe, slips and injures his leg.

LATENT AND ACTIVE FAILURES - REASON'S MODEL OF ACCIDENT CAUSATION (SWISS CHEESE MODEL)

Reason's original work (1990) involved operators of a nuclear power plant; he was particularly concerned with latent (hidden) human failures which could jeopardise the overall safety of the system at some future time. Reason proposed four levels of human failure, each influencing the next. Using an example of an accident involving a train crash, each of these levels of human failure can be considered. Working backwards from the accident the first level that Reason identifies is unsafe acts of persons, which ultimately lead to the accident. These unsafe acts were considered to relate to the errors, such as the skill-based slips/lapses and mistakes identified in Rasmussen's work in 1987. These were called active failures. Reason argued that there was little advantage in blaming the operator for errors when there were many latent failures that led to the error. In the case of a train driver, it may be too easy to blame the cause of the accident on 'driver error' in circumstances where there were latent failures that lead to the error.

Reason identified three further levels of human failure that comprised latent failures. The second level of human failure involves preconditions that lead to unsafe acts taking place. This level involves preconditions such as mental fatigue, poor communication and poor work practices, such as a train driver passing a faulty signal when it shows danger. In many instances, these preconditions can be traced back to instances of unsafe supervision, the third level of human failure identified by Reason. In the example of the train crash, this may relate to a driver being asked to do extra driving time to cover for the absence of others or an acceptance of the practice of passing signals at danger if the drivers believe it to be faulty.

Importantly, Reason identified that causation did not stop at the supervisory level. He recognised that the fourth level, the organisation itself, can impact on performance at all levels. The fallible decisions of designers, higher level managers and others can have adverse consequences. For example, in the case of the train crash, if supervisory level line management are provided with limited resources for driver training or signal maintenance and they are under pressure to maintain train service this can lead to active failures at the lower level of human failure. The schematic in ***figure ref A5-11*** represents barriers at each level, with circles of potential failure which may vary and change over time, but do not result in an active failure. However, when all the circles of failure in each barrier are in alignment then the path to active failure will exist and the accident will occur. The structure of this model is likened to Swiss cheese because of the holes - hence its nickname the 'Swiss Cheese Model'.

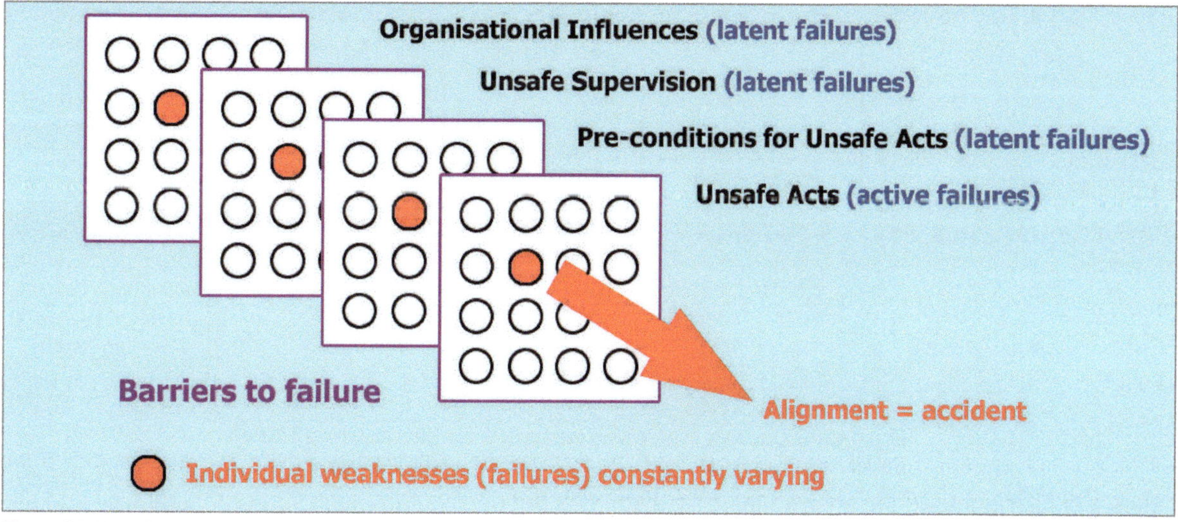

Figure A5-11: J. Reason's accident model. *Source: Reason/RMS.*

Fault tree

In many accidents and incidents there are multiple causes for the loss-making event. Fault tree analysis is an analytical technique for tracing the events which could contribute. It can be used in accident investigation and in a detailed hazard assessment. The fault tree is a logic diagram based on the principle of multi-causality, which traces all branches of events which could contribute to an accident or failure. It uses sets of symbols, labels and identifiers.

The starting point for the fault tree is the undesired event (known as the 'top event') because it is considered first. The immediate contributory fault conditions leading to that event are then considered. These may each in turn be caused by other faults. ***See also 'A7.5 - Failure tracing methodologies' where a fault tree is discussed in more detail.***

Event tree

An event tree differs from a fault tree in that it is a forward thinking approach and the analyst begins with a projection of an initiating event and then goes on to consider the potential for accidents/incidents arising out of that event taking into account successful event management and unsuccessful event management. ***See also 'A7.5 - Failure tracing methodologies' where an event tree is discussed in more detail.***

Bow tie model

The bow tie methodology is a risk evaluation method used to analyse and demonstrate causal relationships in high risk operations. The method takes its name from the shape of the diagram created which resembles a gentleman's bow tie.

The bow tie approach (combining fault trees and event trees, centred on an undesired event) was utilised to present the major hazards of the facility in such a way as to facilitate workforce understanding of hazard management and their role in it. In this approach the hazard is represented by a top event (realisation of hazard) which can be triggered by one or several threats. ***See also 'A7.5 - Failure tracing methodologies' for a diagram of a bow tie analysis.*** The bow tie method then goes further to consider the possible failure of one of the control methods, such failures are known as 'escalating failures'. By considering safety measures for escalating factors the bow tie analysis provides a more complete view of a safe system of work.

Behavioural root cause analysis

Root cause analysis methods are designed to assist investigation processes to move from considering only the immediate causes to establishing the causes that lead to the immediate causes and the root (underlying) causes. Therefore, any technique that encourages this in a systematic way is an acceptable method. Two well established methods are the 'Five Whys' and Fault Tree Analysis (FTA).

The 'Five Whys' is a method where the investigator moves from the point of the problem, for example, a person slipping while walking through a warehouse and injuring their back.

The process might follow this sequence:

1) Why did the worker slip? - Because of oil on the floor.
2) Why was there oil on the floor? - It had leaked from a fork lift truck.
3) Why did it leak from the fork lift truck? - The oil sump was loose.
4) Why was the oil sump loose? - The fork lift truck had not been maintained.
5) Why was it not maintained? - No planned maintenance programme was in place.

An example of how the 'Five Whys' can be depicted graphically is shown in ***figure ref A5-12*** in the example of two control valves 'tripping'.

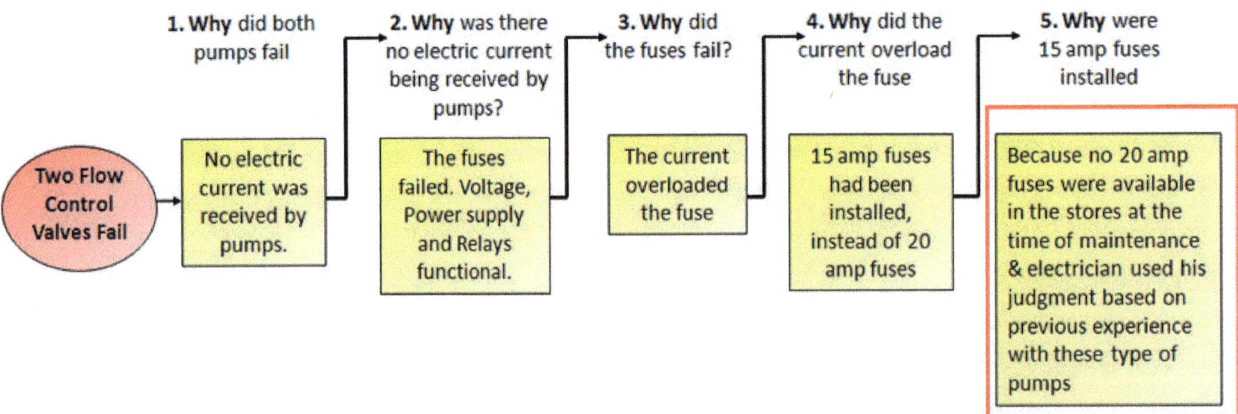

Figure A5-12: Discovery of underlying causes by 'Fault Tree Analysis' (FTA).
Source: Behaviour Based Safety, Behavioural Safety and Safety Leadership.

FTA uses a similar approach to analyse the circumstances but depicts the result in a graphical way and allows for more than one branch of investigation. ***See figure ref A5-13*** for an example of an FTA.

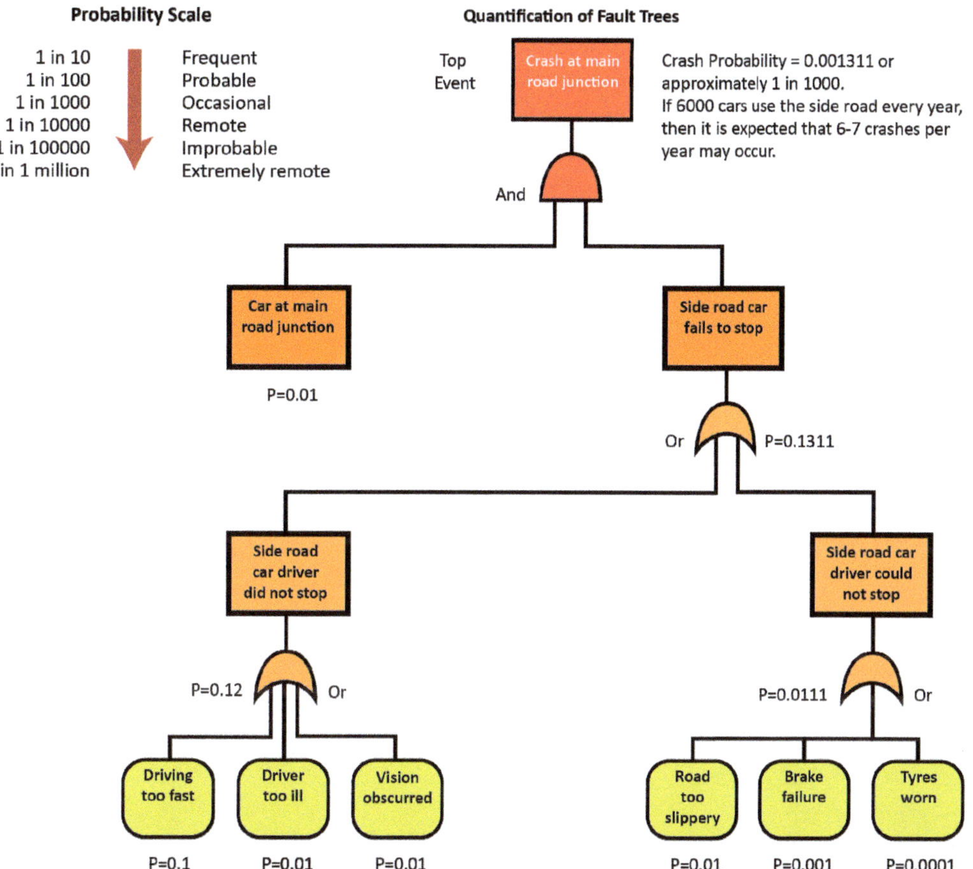

Figure A5-13: Discovery of underlying causes by 'Fault Tree Analysis' (FTA). *Source: HSE HSG65 Successful Health and Safety Management.*

A5.2 - Quantitative analysis of accident and ill-health data

Methods of calculating injury rates from raw data

ACCIDENT FREQUENCY RATE

$$\text{Frequency rate (accidents compared with time)} = \frac{\text{Number of accidents in the period}}{\text{Total hours worked during the period}} \times 1{,}000{,}000$$

This uses the '1,000,000' multiplier that the HSE and International Labour Office (ILO) use. In the UK, organisations frequently use multipliers of 100,000 or less, in order to bring the numbers to a manageable size, and the USA use 200,000. Care should be taken that the multiplier is specified when making comparisons.

ACCIDENT INCIDENCE RATE

Incidence rate (accidents compared with number of people) =

$$\frac{\text{Number of accidents in the period}}{\text{Average number employed during the period}} \times 10{,}000$$

ACCIDENT SEVERITY RATE

$$\text{Severity rate (average number of days lost compared with hours worked)} = \frac{\text{Total no. of days lost}}{\text{Total no. of hours worked}} \times 1{,}000$$

ILL-HEALTH PREVALENCE RATE

Prevalence rate (similar to incidence rate but is an index of ill-health rather than accidents) =

$$\frac{\text{Number of ill - health conditions observed over the period}}{\text{Number of people in the population exposed to agent over period}} \times 1{,}000$$

These calculations are made from the actual numbers of accidents, hours worked, numbers employed and days lost. In order for the result to be a meaningful number, a large multiplier is included in the equation. The injury rates themselves only mean something if they are compared to other injury rates and are to be used to show trends. They may be compared to rates from previous years, to rates from other departments in the company, to other companies or to rates in a particular industry.

The definition of all of these factors must be agreed in order to make comparisons. It is only meaningful to compare like with like. For example, accident may mean any period of 'lost time' to some organisations, but to others a lost time accident is a 'RIDDOR reportable' accident which is greater than seven days. The number of people employed in a given period may or may not include part-time workers and/or contractors, agency staff, other temporary workers. Calculating the hours worked may be inaccurate depending on where the figures come from. The finance department may give the paid hours rather than the actual hours. If workers get double time for overtime, this can skew the figures.

If a fatality occurs, it may not show up in the severity rate, as it will not be possible to say how many days have been lost. With the severity rate being low, it will appear that the accidents that have occurred have been minor.

If all rates are used together and compared with rates using the same definitions they will be valid. If the rates rise, it does not necessarily mean that there has been no improvement. It could mean that the reporting procedures are better and an improved health and safety culture means more people are willing to report. As with all statistical data, injury rates must be carefully analysed.

Presenting and interpreting loss event data

GRAPHICAL AND NUMERICAL FORMAT

Histogram

Histogram

Accidents

200
150
100
50
0

Year 1
Year 2

Manufacturing
Maintenance
Despatch
Administration
Sales

Department

Figure A5-14: Histogram. *Source: RMS.*

The above diagram is a histogram. In this particular histogram, the scale on the y-axis (vertical) shows the number of recorded accidents and the x-axis shows a comparison between two years in each department.

Where the scale is on the x-axis this type of histogram is known as a bar chart.

Pie chart

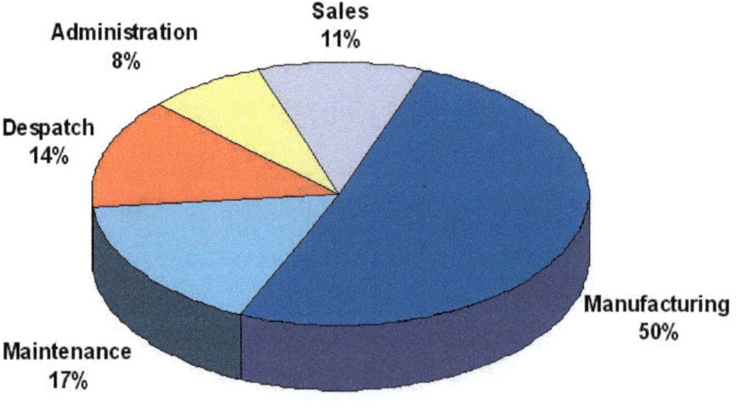

Figure A5-15: Pie chart. *Source: RMS.*

The pie chart shows the pie as the whole amount being considered and the slices as ratio parts of that whole. The above pie representation could be representative of accidents in the different departments. It is easy to see at a glance where most of the accidents are happening.

Line graphs

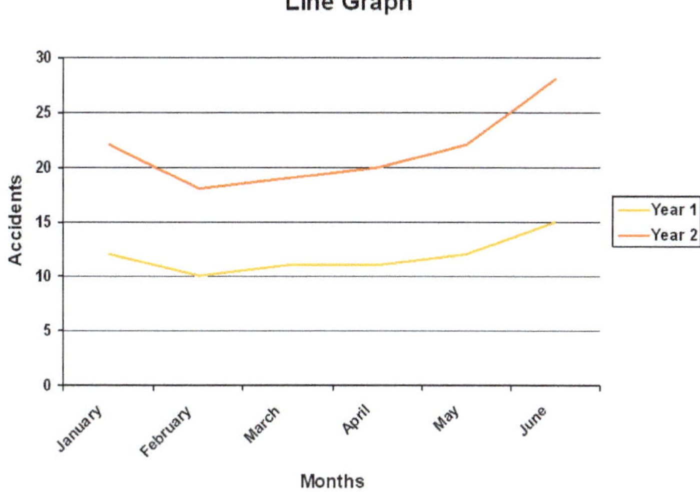

Figure A5-16: Line graph. *Source: RMS.*

Line diagrams, such as the simple one above can show at a glance, accidents over a period, but not in much detail.

Principles of statistical variability, validity and the use of distributions

SAMPLE POPULATION

The sample population is the entire group under analysis; the group about which the analyser wishes to draw conclusions.

SAMPLING VARIABILITY

Sampling variability is concerned with the different values obtained for the same property under consideration when compared with several samples taken from the same sample population, for example, the weight of males of the same age or the health and safety condition of fork lift trucks of the same type.

SAMPLING VALIDITY

Independent samples are those samples selected from the same sample population, which have no influence on each another, for example, the individual eyes colour of each person taken from a random sample of 100 non-related individuals.

RANDOM SAMPLING

Random sampling is a technique where each individual is chosen entirely by chance and each member of the population has a known, but possibly non-equal, chance of being included in the sample. The difference between the estimates derived from the sample and the true population values is referred to as the sampling error. The amount of variation and the likelihood of bias can be reduced by increasing the size of the random sample.

SYSTEMATIC ERROR

Systematic error occurs when the data is biased. Bias is a term which refers to how far the average statistic lies from the parameter it is estimating, that is, the error which arises when estimating a quantity.

Errors from chance will cancel each other out in the long term, but those from bias will not.

The following illustrates bias and precision, where the target value is the bulls-eye.

For example, the police decide to estimate the average speed of drivers using lane three of the motorway and consider how it can be done. One method suggested is to tail cars using police patrol cars and record their speeds as being the same as that of the police car.

This is likely to produce a ***biased*** result as any driver exceeding the speed limit will slow down on seeing a police car behind them.

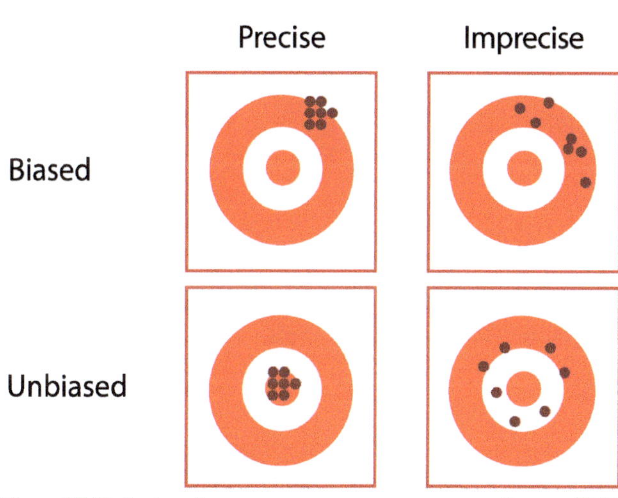

Figure A5-17: Systematic error. *Source: RMS.*

The police then decide to use an unmarked car for their investigation using a speed gun operated by a constable. This is an ***unbiased*** method of measuring speed, but is ***imprecise*** compared to using a calibrated speedometer to take the measurement, which would be ***unbiased*** and ***precise***.

Precision

Precision is a measure of how close the sample result (measurement technique for the speed of drivers) is expected to be to the true value of a parameter (the actual speed of the drivers).

AVERAGES, MEDIANS, MODES AND MEANS

In presenting material your main aim is to ensure the reader can understand what is being presented and that they draw the correct interpretation of the statistics. You should not distort the results by presenting the material in a way that over emphasises mere small shifts in accidents or obscures the true facts by statistical manipulation.

Averages are a common way of presenting accident statistics, but can be misleading especially when there are wide variations. For example, if the number of accidents in several departments were 10, 5, 5, 5 and 100 the average would be 25.

The word 'average' can also be misleading. Average most often refers to the arithmetic mean, but may be used to also refer to the mode, median, or midrange. The Arithmetic Mean is obtained by summing all elements of the data set and dividing by the number of elements.

You should always clarify which average is being used:

- ***Mode*** is the data element which occurs most frequently.
- The ***median*** is the middle element when the data set is arranged in order of magnitude.
- The ***midrange*** is the average of the highest and lowest numbers.

USE OF DISTRIBUTIONS

Confidence intervals

The estimate produced from a sample survey will rarely be identical to the population value, but statistical theory allows us to measure its accuracy to a level of confidence.

The confidence interval can be calculated around the estimated value, which gives a range in which the true value for the population is likely to fall. The standard error measures the precision with which the estimates from the sample approximate to the true population values and is used to construct the confidence interval for each survey estimate.

The 95% confidence intervals can be calculated for each estimated value presented. These are known as such, because if it were possible to repeat the survey under the same conditions a number of times, the expectation would be that 95% of the confidence intervals calculated in this way to contain the true population value for that estimate. A normal distribution of data means that most of the examples in a set of data are close to the 'average', while relatively few examples tend to one extreme or the other.

The standard deviation is a statistic that tells you how tightly all the various examples are clustered around the mean in a set of data. When the examples are pretty tightly bunched together and the bell-shaped curve is steep, the standard deviation is small. When the examples are spread apart and the bell curve is relatively flat, that tells you that you have a relatively large standard deviation.

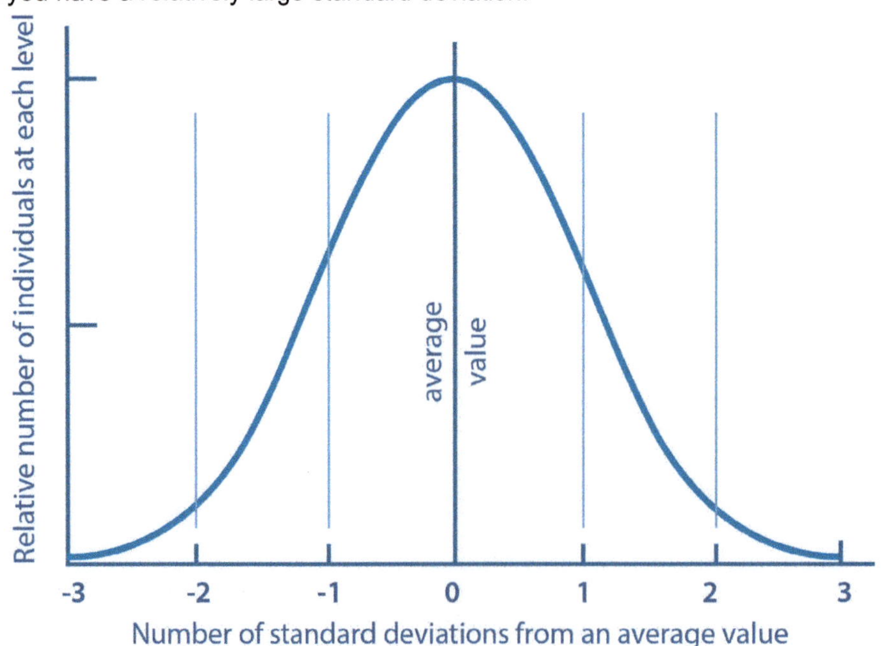

Figure A5-18: Normal distribution and standard deviation. *Source: RMS.*

A5.3 - Reporting and recording of loss events (injuries, ill-health, dangerous occurrences) and near-misses

Statutory reporting requirements and procedures

The legal requirements for recording and reporting accidents, diseases and dangerous incidents at work are laid down in the Social Security Act (Claims and Payments) Regulations (SSACPR) 1979, the Social Security Administration Act (SSAA) 1992 and the Reporting of Injuries, Diseases and Dangerous Occurrences Regulations (RIDDOR) 2013. RIDDOR reports are sent to the relevant authority and are used to compile statistics to show trends and to highlight problem areas in particular industries or companies.

The social security legislation referred to above sets out requirements on employers to confirm whether a claim for 'industrial injury benefit' made to the social security department by an employed earner relates to an accident at work. The request for confirmation of circumstances is initiated by the social security department, following a claim by an employed earner. The legislation places duties on the employer and the employed earner, to enable this process to take place.

The employed earner who suffers personal injury for which benefit may be payable must give 'notice' to the employer in writing or orally as soon as practical after it happens. Notice can be given to the employer directly or to someone supervising the work of the employed earner at the time of the injury. This can be done by someone acting on the employed earner's behalf.

If the employer occupies specified premises then an accident book (BI 510), or other means of holding a record of the same data, must be kept at the location.

Specified premises are:

- Every factory, mine and quarry.
- Works or premises where the Factories Act 1961 applies.
- Any premises where or about ten or more people are employed at the same time.

Particulars to be given of accidents are:

- Full name, address and occupation of injured person.
- Date and time of accident.
- Place where accident happened.
- Cause and nature of injury.
- Name, address and occupation of person giving the notice, if other than the injured person.

The means of recording, for example, the accident book, must be easily available at all reasonable times.

An entry of the appropriate particulars of the accident in a book or similar, if made as soon as practicable after the accident, will be considered sufficient notice.

When notified, the employer has a duty to investigate the cause of the injury. If the employer finds anything different from what was recorded in the accident book this must be noted. Records must be kept for three years.

The Social Security (Industrial Injuries) (Prescribed Diseases) Regulations (SSIIPDR) 1985 list those diseases that are prescribed for the purpose of payment of disablement benefit.

A prescribed disease is defined in the Social Security Act (SSA) 1975 as:

- A disease that ought to be treated, with regard to its causes, incidence and other relevant considerations, as a risk of occupation and not a risk common to everyone.
- Such that, in the absence of special circumstances, the attribution of particular cases to the nature of the employment can be established with reasonable certainty.

Schedule 1 to the regulations classifies prescribed injuries or diseases as:

- Conditions due to physical agents - such as cataract from an occupation involving frequent or prolonged exposure to radiation from red-hot or white-hot material.
- Conditions due to biological agents - infection by leptospira from an occupation involving work in places liable to be infested with rats, field mice, or voles; work at dog kennels or handling of dogs; contact with bovine animals or pigs or their meat products.
- Conditions due to chemical agents - anaemia with a haemoglobin concentration of 9g/dL or less, and a blood film showing punctuate basophilia from an occupation involving the use or handling of, or exposure to the fumes, dust or vapour of, lead.
- Miscellaneous conditions - diffuse mesothelioma (primary neoplasm of the mesothelium of the pleura or of the pericardium or of the peritoneum) from an occupation involving exposure to asbestos at a level above that commonly found in the environment at large.

Some of the prescribed diseases carry a specification of how much harm has occurred, for example, a 50dB reduction in hearing in both ears, this level of harm is required before benefit is payable.

The Reporting of Injuries, Diseases and Dangerous Occurrences Regulations (RIDDOR) 2013

RIDDOR 2013 covers the requirement to report certain categories of injury and disease sustained at work, along with specified dangerous occurrences and gas incidents, to the relevant enforcing authority. These reports are used to compile statistics to show trends and to highlight problem areas, in particular industries or organisations.

THE MAIN POINTS OF RIDDOR 2013

Reporting

1) When a person ***dies or suffers any specified injury*** listed in regulation 4 (non-fatal injuries to workers) as a result of a work-related accident or dies as a result of ***occupational exposure to a biological agent*** or an incident occurs of the type listed as a ***dangerous occurrences*** in Schedule 2 the responsible person must notify the relevant enforcing authority by the quickest practicable means (usually by telephone) without delay and must send them a report in an approved manner (on-line) within ***10 days***. This therefore includes accidents connected with work where:
    - An employee or a self-employed person at work is killed or suffers a specified injury (including as a result of physical violence).
    - A member of the public is killed or taken to hospital.
    - A work-related ***'accident'*** in the context of RIDDOR 2013 includes an act of non-consensual physical violence done to a person at work.
2) In cases of work-related diseases that are listed in regulation 8 and 10 the responsible person must send a report of the diagnosis in an approved manner (online) to the relevant enforcing authorities without delay. In cases of diseases related to carcinogens, mutagens and biological agents that are listed in regulation 9, the responsible person must notify the relevant enforcing authority in an approved manner.
3) If personal injury results in ***more than 7 days (excluding the day of the accident) incapacity*** for routine work, but is not one of the specified non-fatal injuries, the responsible person must send a report to the relevant enforcing authority in an approved manner (online) as soon as is practicable and in any event within ***15 days*** of the accident. The day of the accident is not counted, but any days which would not have been working days are included.
4) If there is an accident connected with work (including an act of physical violence) and an employee, or a self-employed person at work, suffers an over-three-day injury it must be recorded by the employer.
5) The enforcing authority for most workplaces is either the Health and Safety Executive or the Local Authority, for railway operations it is the Office of Rail Regulation (ORR).

Responsible person

Reportable event	***To***	***Responsible person***
Death, specified non-fatal injury, over 7 day injury, disease.	Employee.	Employer.
	Self-employed person working in someone else's premises.	Person in control of the premises: ■ At the time of the event. ■ In connection with trade, business or undertaking.
Specified injury, over 7 day injury, disease.	Self-employed in own premises.	Self-employed person or someone acting for them.
Death, being taken to hospital or a specified non-fatal injury on hospital premises.	A person not at work.	Person in control of the premises: ■ At the time of the event. ■ In connection with trade, business or undertaking.
Dangerous occurrences - general.		Person in control of the premises where, or in connection with the work going on at which, the dangerous occurrence happened: ■ At the time of the event. ■ In connection with trade, business or undertaking.

Road traffic accidents

Road traffic accidents only have to be reported if:

- Death or injury results from an accident involving a train.
- Death or injury results from exposure to a substance being conveyed by a vehicle.
- Death or injury results from the person being engaged in work connected with loading or unloading of any article/substance or results from another person engaged in these activities.

- Death or injury results from the person being engaged in work on or alongside a road or results from another person engaged in these activities.

Work on or alongside a road means work concerned with the construction, demolition, alteration, repair or maintenance of:

- The road or the markings or equipment on the road.
- The verges, fences, hedges or other boundaries of the road.
- Pipes or cables on, under, over or adjacent to the road.
- Buildings or structures adjacent to or over the road.

Non-employee

The responsible person must not only report non-employee deaths, but also cases that involve major injury or being taken to hospital if caused by an accident out of or in connection with their work.

Employee death

Where an employee dies within one year of the date of an accident, as a result of a reportable injury, as soon as the employer knows they must inform the enforcing authority in writing of the death.

Gas incidents

Specified gas incidents are notified without delay and reported within 14 days to the Health and Safety Executive.

Injury under medical supervision

Reporting and recording requirements do not apply in situations where the injury or death of a person arises out of the conduct of an operation, examination or other medical treatment of that person whilst under the supervision of a registered medical practitioner or dentist.

Self-employed people

If a self-employed person suffers a specified non-fatal injury while working at premises that are owned or occupied by themselves they do not need to notify the enforcing authority immediately. However, they or someone acting for them must report the injury within 10 days.

Where an injury is not a specified non-fatal injury, but causes a self-employed person to be incapacitated from routine work for more than seven consecutive days, the self-employed person or someone acting for them must report it within 15 days of the accident.

There is no reporting requirement for situations where a self-employed person suffers a fatal accident or fatal exposure on premises controlled by that self-employed person.

Recording

In the case of an accident at work, the following details must be recorded:

- Date and time.
- Name.
- Occupation.
- Nature of injury.
- Place of accident.
- Brief description of the circumstances in which the accident happened.
- In the case of a person not a work, instead of occupation a record of their status should be made (for example, passenger, customer, visitor or bystander).
- The date on which the accident was first notified or reported to the relevant enforcing authority and the method used.

For non-reportable injuries that incapacitate for more than 3 days a record of notification or reporting is not relevant, but a record of the accident must be maintained.

Similar information should be recorded for dangerous occurrences, except that details of injured persons will not be relevant.

In the case of a diagnosis of a reportable disease, the following details must be recorded:

- The date of diagnosis of the disease.
- The name of the person affected.
- The occupation of the person affected.
- The name or nature of the disease.
- The date on which the disease was first reported to the relevant enforcing authority.
- The method by which the disease was reported.

Records must be kept for at least 3 years and kept at the place where the work it relates to is carried out or at the usual place of business of the responsible person.

Defences

A person must prove that they were not aware of the event requiring reporting and that they had taken all reasonable steps to be made aware, in sufficient time.

SPECIFIED INJURIES (RIDDOR 2013 - REGULATION 4)

The list of specified non-fatal injuries is:

- Any bone fracture diagnosed by a registered medical practitioner, other than to a finger, thumb or toe.
- Amputation of an arm, hand, finger, thumb, leg, foot or toe.
- Any injury diagnosed by a registered medical practitioner as being likely to cause permanent blinding or reduction in sight in one or both eyes.
- Any crush injury to the head or torso causing damage to the brain or internal organs in the chest or abdomen.
- Any burn injury (including scalding) which:
    - Covers more than 10% of the whole body's total surface area.
    - Causes significant damage to the eyes, respiratory system or other vital organs.
- Any degree of scalping requiring hospital treatment.
- Loss of consciousness caused by head injury or asphyxia.
- Any other injury arising from working in an enclosed space which:
    - Leads to hypothermia or heat-induced illness.
    - Requires resuscitation or admittance to hospital for more than 24 hours.

DISEASES (RIDDOR 2013 - REGULATIONS 8, 9 AND 10)

Regulation 8 - Occupational diseases

- Carpal Tunnel Syndrome (CTS), where the person's work involves regular use of percussive or vibrating tools.
- Cramp in the hand or forearm, where the person's work involves prolonged periods of repetitive movement of the fingers, hand or arm.
- Occupational dermatitis, where the person's work involves significant or regular exposure to a known skin sensitizer or irritant.
- Hand Arm Vibration Syndrome (HAVS), where the person's work involves regular use of percussive or vibrating tools, or the holding of materials which are subject to percussive processes, or processes causing vibration.
- Occupational asthma, where the person's work involves significant or regular exposure to a known respiratory sensitizer.
- Tendonitis or tenosynovitis in the hand or forearm, where the person's work is physically demanding and involves frequent, repetitive movements.

Regulation 9 - Exposure to carcinogens, mutagens and biological agents

- Any cancer attributed to an occupational exposure to a known human carcinogen or mutagen (including ionising radiation).
- Any disease attributed to an occupational exposure to a biological agent.

Regulation 10 and Schedule 3 - Diseases offshore

Examples of diseases listed in this Schedule are:

- Chickenpox.
- Cholera.
- Diphtheria.
- Dysentery (amoebic or bacillary).
- Mumps.
- Food poisoning.
- Legionellosis.
- Malaria.
- Measles.
- Meningitis.

DANGEROUS OCCURRENCES (RIDDOR 2013 - SCHEDULE 2)

Dangerous occurrences are events that have the potential to cause death or serious injury and so must be reported whether anyone is injured or not.

Examples of dangerous occurrences that must be reported are:

- The failure of any load bearing part of any lifting equipment, other than an accessory for lifting.
- The failure of any pressurised closed vessel or any associated pipework.
- Any unintentional incident in which plant or equipment either:
    - Comes into contact with an uninsulated overhead electric line in which the voltage exceeds 200 volts.
    - Causes an electrical discharge from such an electric line by coming into close proximity to it.
- Electrical short-circuit or overload attended by fire or explosion which results in the stoppage of the plant involved for more than 24 hours.
- The schedule also identifies Dangerous Occurrences that are specific to mines, quarries, transport systems, and offshore workplaces.

Note: This information is a summary of requirements under RIDDOR 2013; the regulations should be consulted for further information.

The significance of internal reporting and recording systems

All accidents to employees, however minor, should be recorded. This is a requirement under social security legislation. As a result of a workplace injury an employee may need to claim for benefits in the future, and the relevant checks will be made to confirm that the accident occurred at work.

RELEVANT REPORT FORMS

A number of report forms are utilised to identify and inform that accidents and ill-health have occurred. These include:

- Accident book, in the form of BI 510. This accident book has been revised to take into account the Data Protection Act (DPA) 1998. Completed accident records should be detached from the book, passed to the nominated person and stored securely, for example, in a lockable cabinet.
- First aid treatment reports.
- Medical treatment reports.
- Medical (doctor) reports of ill-health.
- Sickness absence reports.
- Event (accident) reports.
- Event (near miss) reports.
- Maintenance/repair reports.
- Insurance reports.
- Reporting Injuries, Diseases and Dangerous Occurrences Regulations (RIDDOR) 2013 reports.

INVESTIGATION REPORT FORMS

Investigation report forms vary in design, layout and content. Many organisations recognise that a different report form may be necessary for first line managers' initial investigations (a level 1 report) and those done by other managers and health and safety practitioners (a level 2 report). The main difference is the section relating to causes of the accident. The version used by other managers and professionals often has more analysis in this area and causes greater investigation of underlying causes. In the same way, reports prepared by an investigation team would not tend to be on a pre-printed format, but would be designed around agreed headings and the content/extent of the report would depend on the matter being investigated and findings (a level 3 report).

Common structure of a report tends to determine:

- What happened - the loss.
- How it happened - the event.
- Why it happened - the causes.
- Recommendations - remedial (and preventive) action.

The report is usually supported by drawings, photographs and statements as appendices.

REPORTING ROUTES

Reporting of an accident or ill-health may be by a number of means and includes:

- Person suffering harm.
- Person causing loss.
- Person discovering loss.

PERSON SUFFERING HARM

This person is often the source of first reporting of less serious events. The reporting system must make available to them the means to make a report. They have a right to report in an 'accident book' BI 510 (or equivalent) any event that may cause them to claim Social Security benefit. This might be fulfilled by using: a copy of the BI 510 book or first aid/medical treatment documents/event report forms that are adapted to contain the same data. These reports should be under the control of a responsible person who would then initiate an investigation which would usually require the completion of an event (for example, accident) report.

PERSON CAUSING LOSS

This person would be expected to bring the loss to the attention of a line manager who would fill in the appropriate event (for example, accident report) and initiate an investigation to complete the remainder of the report that the person reporting the loss may not be able to do.

If this person is not the manager responsible for the location in which the loss took place they would have to bring the loss to the attention of a line manager, as above. If the person is the line manager they would initiate an investigation and report on the appropriate event form.

COPIES AND DISTRIBUTION OF REPORTS

Reports from first line managers may be copied to the next line manager (middle manager), health and safety practitioner and employee representative. It is important that the originator retains a copy till action to prevent re-occurrence is complete. This will help to encourage ownership and continued involvement. The copy passed to the next line manager is usually seen as the primary document. The manager confirms/adds to the investigation, retains a copy and passes the report to a central record point. Clearly this may be done in part or whole as a computerised or paper system. Records held by the line manager/health and safety professional may be held for varying periods depending on their role. Central records should be under the control of the responsible person and are usually maintained in accordance with the organisation's own practices. A minimum period is usually 3 years for accident (in order to respond to civil claims) and 40 years for events resulting in ill-health (in order to deal with the long-lived nature of the problem).

FOLLOW-UP

This virtually finishes the work of the investigator, but management is still responsible for seeing that the necessary remedial actions are implemented, and monitored to ensure that the causes are satisfactorily controlled. The line manager, health and safety practitioner and health and safety committee/members will monitor these actions.

ACTION FOLLOWING A FATAL ACCIDENT

Reporting a death at work following an accident should include informing:

- Enforcing authority.
- The senior manager.
- Health and safety practitioner.
- Coroner.
- Next of kin.
- Employee representatives.
- Other employees.
- Insurance company.

A5.4 - Loss and near miss investigations

Implied legal requirements

It is not overtly stated in law that accidents must be investigated. However, the implication is there in statute. The Health and Safety at Work etc. Act (HASAWA) 1974 states that the employer must ensure so far as is reasonably practicable the health, safety and welfare of employees at work. This implies that should an accident occur then the employer should find a way of preventing it happening again. In order to do this they need to investigate the accident, find out what caused it and put further measures in place to prevent a recurrence. The Management of Health and Safety at Work Regulations (MHSWR) 1999, Regulation 5, requires the employer to put arrangements in place for the preventive and protective measures, with one of the arrangements being for monitoring. The approved code of practice (ACOP) to the Regulations considers monitoring as including 'adequately investigating the immediate and underlying causes of incidents and accidents to ensure that remedial action is taken'. Therefore, although not stated in the legislation itself, in order to comply with legislation, accident investigation should be carried out. Also, following the Woolf Report on civil action (which gave rise to the Civil Procedures Rules); full disclosure of the circumstances surrounding the accident to the person(s) considering legal action is required. Conducting a prompt investigation will enable this obligation to be met. In addition, carrying out a full investigation and taking remedial action will show the court that the employer has a positive attitude to health and safety. The results of the investigation will also be essential information for the insurers in the event of a claim.

HSE GUIDANCE

'Successful health and safety management', HSG65 is a practical guide for those with responsibility for health and safety to improve health and safety within their organisations. It was first prepared by HSE's Accident Prevention Advisory Unit. In the section on 'monitoring', there is information on reactive monitoring, which includes accident investigation. There is advice on the level and nature of investigation, collecting the data from immediate and underlying causes and analysing it. This should be followed by recommendations and actions to prevent recurrence. Following HSG65 will ensure compliance with the legal requirements. Further guidance is provided by HSE in the document HSG245 'Investigating accidents and incidents'. It takes a step-by-step approach to investigations:

Step 1: Gathering the information - the 'where, when and who' of the adverse event. The information gathered will include results of interviews, photographs of the equipment involved and the area in which it was positioned at the time, sketches of the workplace layout, weather conditions, etc.

Step 2: Analysing the information - the 'what happened and why' stage. Analysing the information to find the immediate, underlying and root causes. At this stage it should be considered if human error is a contributory factor. This can be cross-referenced with HSG48 'Reducing error and influencing behaviour'. Job factors, human factors and organisational factors can all influence human behaviour and will all need to be considered in the analysis.

Step 3: Identifying suitable risk control measures. Possible solutions can be identified. This will involve looking at the technical, procedural and behavioural controls, with the technical or engineering risk control measures being more reliable than those that rely on human behaviour.

Step 4: The action plan and its implementation - which risk control measures should be implemented in the short and long term? This is the risk control action plan, which should have smart objectives, i.e. specific, measurable, agreed and realistic, with timescales. This will also state which risk assessments need to be reviewed and which procedures need to be updated; any trends that need further investigation; and what the adverse event cost. HSG245 contains some useful investigation forms that could be used and developed for individual companies. It also has examples of simple, but effective investigation tools such as the accident/incident investigation tree.

The reasons for carrying out investigations

The reasons for carrying out investigations include:

- Establish legal liability.
- Information/data gathering.
- Discovery of underlying causes.

The purpose of an investigation of this nature should never seek to blame any individual or group of individuals. If human error is believed to be a significant cause, the reasons for this must be investigated. Lack of knowledge, training or unsuitability for the job may be the causes of this error. These are management and not operator failings. Only when these have been evaluated can the conclusion of wilful and intentional acts or omissions be considered.

LEGAL REASONS

Part of investigation is to highlight where there are any breaches of the law and make sure they are dealt with. If there is any legal action, the accident investigation report may be asked for and a prompt, thorough report could help to show the organisation has a good safety culture and is committed to improving standards. It may be used in a criminal court as mitigation as part of a defence if a guilty plea has been lodged. It may be used as a defence in civil court if someone is claiming compensation. Investigation of an accident does not indicate that the employer accepts legal liability for the accident; since it might have been unforeseeable. Investigation shows commitment to managing health and safety.

INFORMATION/DATA GATHERING

The investigation will enable data to be collected. These data may be numbers of incidents involving a certain type of work equipment, injuries occurring in a certain department and/or problems during a certain time of the shift. Records of interviews will be compiled together with related procedures and instructions and relevant training. This is important to determine any flaws in the system of work, rather than to focus on individual failure alone. At first the data may be too insignificant to do anything with, but it will be available should there be other investigations etc.

ESTABLISHING THE ROOT, UNDERLYING AND IMMEDIATE CAUSES

The 'root cause' of an accident is the initiating event or failing from which all other causes or failings spring. Root causes are generally management, planning or organisational failings. Investigation should identify both immediate and underlying causes, including the design of appropriate risk control systems (RCSs).

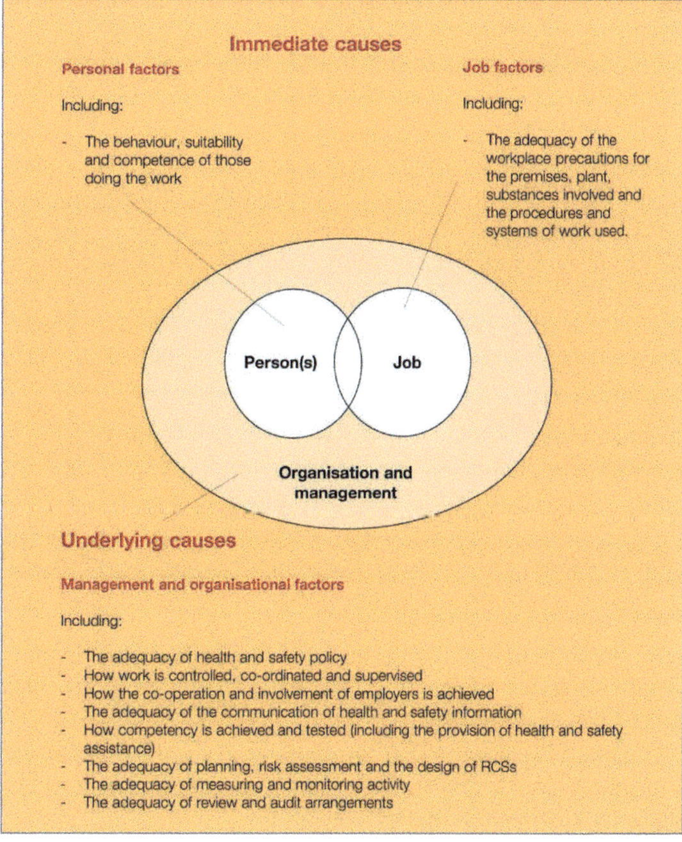

Figure A5-19: Discovery of underlying causes. *Source: HSG65 Successful Health and Safety Management.*

The benefits of carrying out an investigation

The benefits of carrying out an investigation include:

- Prevention of recurrence.
- Improved employee morale.
- Management skills development.

PREVENTION OF RECURRENCE

The main purpose of investigation is to find out what has gone wrong and put further controls in place to prevent that accident happening again. It is not to find someone to blame, although the person's actions may well need to be changed as part of the range of control measures. The further controls therefore will be taken from technical, procedural and behavioural issues.

IMPROVED EMPLOYEE MORALE

Employees who see the organisation's reaction to incidents will respond in two ways:

- If the organisation restricts its investigation to only establishing the immediate causes and disciplining workers who were responsible, the workers will conclude that the management is 'not concerned about safety'.
- Employees who see the management taking a sympathetic and thorough approach will conclude that the management are determined to find and eliminate threats to their safety. Workers will feel that the organisation is concerned for their safety and welfare.

DEVELOPING MANAGERIAL SKILLS

The process of investigating accidents in order to discover root causes will teach managers to identify and eliminate future weaknesses in safe systems of work.

Investigating an accident is also similar to investigating the reasons for a budget overspend or establishing why a production target may have been missed. The skills involved in investigating a safety accident are transferrable when investigating performance generally.

Investigation procedure with reference to HSG245

INITIAL REPORT (PRESERVE THE SCENE, NOTE PEOPLE AND EQUIPMENT INVOLVED, REPORT EVENT)

After an accident has occurred, it is important that it is dealt with quickly. In the first instance it will be necessary to preserve the scene. If evidence is removed or tampered with it can cause problems for the investigator. The resting place of machinery, debris and, in worst case scenario, people can tell a lot about what happened. Any equipment that might have caused or contributed to the cause of the incident should be identified and its condition noted.

The identity of those involved should be established; the identity of any injures parties and anyone who might be a witness should be established. Depending on the severity of the incident it may be necessary to ascertain the names off first aiders and supervisors.

In most cases, accidents only need to be reported internally, i.e. kept within the company or organisation. However, there are certain types of accident that need to be reported to the Health and Safety Executive (HSE), or the Local Authority.

These 'statutory reportable cases' are found in a piece of health and safety law known as the Reporting of Injury Disease and Dangerous Occurrences Regulations (RIDDOR). This law requires employers to report certain accidents by telephone (or via the internet) within 24 hours of the accident occurring. Also, the accident should be reported to the authority on a special form and submitted within 10 days of the accident.

DECIDE WHETHER FURTHER INVESTIGATION IS REQUIRED

When deciding whether to investigate and determining the level of investigation, factors to consider are the likelihood of a recurrence and the potential worst consequence should the incident happen again.

Likelihood can be measured in terms such as certain, possible and rare whilst consequence might be categorised as fatal, specified or seven-day injury (as defined in RIDDOR 2013), minor and damage only. ***Figure ref A5-20*** will help the investigator to settle on the level of investigation that is required.

First line manager (supervisory) investigations

As the person in immediate operating control of an area or activity, it is logical to expect the first line manager to gather information on all accidents that happen in their sphere of responsibility. This investigation is normally all that is necessary for the majority of accidents.

It should result in swift remedial actions being implemented, and underlines the first line manager's responsibility for health and safety, on a day-to-day basis. In some cases, this is seen as only the initial investigation, further investigation taking place by another party.

Likelihood of recurrence	*Potential worst consequence of adverse event*			
	Minor	***Serious***	***Major***	***Fatal***
Certain				
Likely				
Possible				
Unlikely				
Rare				

Risk		Minimal		Low		Medium		High
Investigation level		Minimal level		Low level		Medium level		High level

Figure A5-20: Investigation level risk rating table. *Source: HSE, HSG245.*

Employee involvement in investigations

Involvement can bring a useful perspective and go some way towards acceptance of the changes that may have to be put into place following the investigation. Asking the union appointed safety representative or the non-union representative of employee safety to work with the group or individuals conducting the investigation can do this. The other forum for employee involvement may be through health and safety committee meetings or briefing groups that might be asked to consider a particular accident and its underlying causes.

Health and safety practitioner investigations

The health and safety practitioner will observe and evaluate the events reported and decide what is considered to be noteworthy, but not warranting a formal investigation team. These events will then be investigated by the health and safety professional to a greater depth. Ill-health events may require involvement of other specialists with a medical background or environmental workplace assessment.

Manager investigations

Organisations that encourage management involvement identify events that warrant further investigation by a nominated manager. A manager, independent of the line managers where the event occurred, may be required by the senior manager of an organisation to conduct a more in depth (detailed) investigation. They would take advice, including that from the health and safety practitioner, as appropriate.

In some cases, middle managers who receive reports of initial investigations conducted by first line managers may decide that a more in-depth investigation is required. This may be in order to gain a better understanding of the underlying causes of the accident, something that may be outside the perspective of the first line manager.

Formal investigation team

In some cases a formal investigation team will be convened to take the information gained by the first line manager, in the initial stages, and conduct an investigation which is independent from line managers.

This is often conducted in situations where a significant loss has resulted (for example, those that result in a major injury). Organisations with a more developed health and safety programme will also use a team to investigate events that have a high potential for loss, for example, a significant near miss.

The team size should be kept small in number and could include the following people:

- A senior manager to sit as chair person.
- A manager at a lower level than the chairperson, for example, supervisor/team leader.
- A health and safety practitioner.
- An employee representative.
- As necessary, a person competent to give technical advice, for example, an electrical engineer in the case of electrocution.

Any person whose responsibilities or actions may have been involved in the incident being investigated should be excluded from sitting on the investigation team, but would, of course, be valuable witnesses.

Accident/Incident report forms

Investigation report forms vary in design, layout and content. Many organisations recognise that a different report form may be necessary for first line managers' initial investigations (a level 1 report) and those done by other managers and health and safety professionals (a level 2 report).

GATHERING INFORMATION

This stage of the investigation is about gathering information about the incident and the events that led up to the incident. The information that is collected should allow the investigator to create a mental picture of the situation immediately prior to the incident. Information can be gathered from photographing or sketching the scene of the accident, eye witness accounts, a variety of documents: risk assessments, safe systems of work, maintenance records, training records, inspection reports, technical files; previous investigation reports.

Consider the people, equipment (including any PPE that was or should have been worn), materials, environment, task, position (location), time, as well as paper evidence.

Interviewing witnesses

- Interview witnesses promptly after the event, to avoid lapse of memory or confusion through witnesses discussing with each other.
- Conduct the interview in private with no interruptions.
- Do not interview more than one person at a time.
- Protect the reputation of the people you interview.
- Set a casual, informal tone during the interview to put the individual at ease. The witnesses may be very defensive, feeling that they may be blamed, it is important to state that the purpose of the interview is to determine facts to prevent a re-occurrence.
- Ask probing questions (these should not put words in the witnesses' mouth).
- Avoid jumping to conclusions.
- Approach the witness with an open mind.
- Notes should be taken, so that the investigator is not relying on memory.
- Record details: names of the interviewers and interviewee; place, data and time of the interview; and any significant comments or actions during the interview.
- Summarise your understanding of the matter.
- Express appreciation for the witnesses' information.

ANALYSE THE INFORMATION

It is at this stage that the investigator can start to piece together the evidence and start to develop a true picture of the events that led to the incident. There are a number of computerised analysis programmes available, from health and safety publications or exhibitions, which can help to analyse the collected data to highlight common features and underlying causes. Many investigators use Fault Tree or Event Tree analysis techniques to identify the root cause of the accident.

IDENTIFYING RISK CONTROL MEASURES

After the identification of the root cause, risk control measures that effectively cut off the root can be identified and implemented. Each measure has to be evaluated in its own right to assess its effectiveness in preventing a recurrence.

The fact that some measures might be more difficult and/or expensive to implement should not be a barrier to them being considered. Where possible, measures which eliminate risk should be given priority followed by measures that combat the risk at source. Generally, measures that rely on human intervention are less effective than engineering controls that require no human input.

PRODUCE AND IMPLEMENT AN ACTION PLAN

Senior managers who have the authority to influence performance must be responsible for creating an action plan based on the SMART (specific, measurable, agreed, and realistic with timescales) principles. Recommend corrective action, including both short and long-term controls, to prevent the same thing from happening again. Short-term controls should prevent a recurrence until longer-term controls can remove the fundamental causes. Actions should be prioritised and a senior manager appointed to be responsible for its implementation.

Sharing of information/lessons learned to prevent recurrence

It is important to complete the accident investigation promptly and to issue and implement any recommendations to prevent a reoccurrence as soon as is practical.

VERBAL REPORTING

It is often necessary to announce investigation findings ahead of the issue of a written report, particularly where a serious or life threatening incident has occurred. This will be the expectation of those who may be directly involved, such as the injured parties and their relatives.

It will also assist by providing assurance to others, for example, employees or fellow workers or perhaps the news media, that management are taking the matter seriously and are looking to prevent a reoccurrence.

Care should be taken to ***only state the facts*** and where there is still doubt or uncertainty in the findings, at the time of giving the verbal report, to make this clear.

WRITTEN REPORT

Executive summary

An executive summary should provide sufficient information to enable the management of the organisation to understand the significance of the event investigated, to establish the findings and the main recommendations for implementation.

Main body of report

The main body of the report should establish the findings of the investigation in terms of the event, direct causes and underlying causes. The report must convince management to take action on the issues identified from the accident investigation.

This should include clear legal, moral and financial arguments. The possible costs of not taking action should also be included. The report should be well structured and concise.

Conclusions and recommendations

The report should include the lessons learnt and provide balanced arguments on remedial action required. It should explain the effect recommendations would have on the standards of health and safety at the workplace and the possible effects on the business overall.

Remedial actions

The report should specify which issues require action by management, identify the line manager responsible for completion, prioritise each action in order of criticality, and provide agreed time frames for completion, with estimated costs.

LESSONS LEARNT

It is important that lessons learnt from incidents are shared with as many people who would benefit from it as is possible. As a minimum, this must include different departments within an organisation. It would be regrettable if an accident happening in one department was seen as of no relevance to others, without considering the causes.

The more that the root causes are examined the more likely the lessons are to be relevant to other departments. An incident in one department may reveal a need to improve job induction processes, which may affect all departments of an organisation.

In the same way, it is important to share lessons learned with other locations of the same organisation. Some organisations contribute to an industry group network, where organisations issue a briefing note to others so that they can learn from the experience of others.

BN0901A1649 - Transport of Zip-Up (Mobile Tower) Scaffolding

Background

- A task had been carried out in the HAST 18 area of B215 which required the use of a zip-up scaffold.
- The type of scaffold used was a Planet Platforms Protec scaffold (GRP tower).
- The scaffold had been dismantled and the base frames (1.35m wide / 15-20 kg weight) were being carried up the HAST 15 stair case which is of an open nature.
- One of the adjustable legs with affixed castor (approx 5-6 kg) became detached from the bottom of one of the frames and fell approximately 9m, glanced off some lagging and came to rest on the landing of a lower floor.

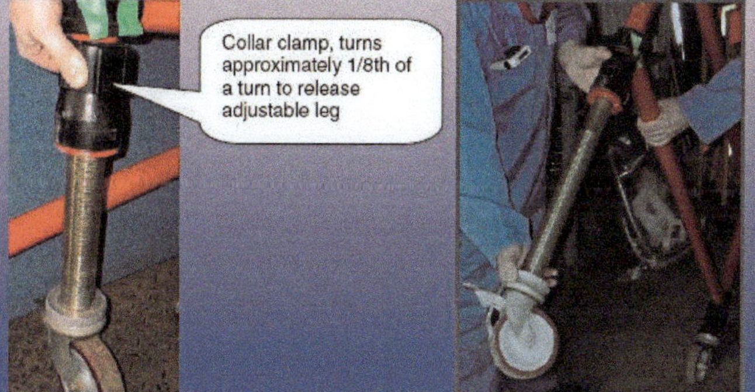

Details

- The adjustable leg / castor assembly is inserted into the base frame which has an adjustment collar that is designed to be used to adjust the legs to level up a scaffold on an uneven surface.
- The adjustment collar has only a small range of movement between 'lock' and 'unlock' (approx one eight of a turn). When in the unlock position, the adjustable leg can be fully released in an uncontrolled manner if the castor is not resting on something.

Key Learning

- <u>Do not use this type of scaffold with the legs extended until investigations have been concluded.</u>
    - If the adjustable legs are used to raise the height of a scaffold there is the potential that if not properly locked or if the collar is in a poor state of repair, if knocked the collar could release resulting in the scaffold tipping.
- Remove the leg / castor assemblies from the base frame before transporting zip-up scaffolding.
- Do not carry zip-up scaffold on stair cases if possible – look for alternative routes where a lift can be used instead.
- If equipment has to be carried around Plant / on stair cases, ensure areas where personnel are at risk of falling items have barriers and signs are erected.

Figure A5-21: Incident lessons learned alert. *Source: Jamie Lockie Sellafield Fellside site.*

This page is intentionally blank

Element

A6

Measuring and reviewing health and safety performance

Learning outcomes

On completion of this element, candidates should be able to demonstrate understanding of the content through the application of knowledge to familiar and unfamiliar situations. In particular, they should be able to:

A6.1 Explain the purpose and use of performance measurement in relation to health and safety objectives and arrangements.

A6.2 Explain the need for, and the objectives and limitations of, health and safety monitoring.

A6.3 Describe the variety of health and safety monitoring and measurement techniques.

A6.4 Explain the need for and process of reviewing health and safety performance.

Contents

Relevant statutory provisions

N/A

Sources of reference

Reference information provided, in particular web links, was correct at time of publication, but may have changed.

Guidelines for auditing management systems, BS EN ISO 19011:2011, ISBN: 978-0-5807-7524-6

The above web links along with additional sources of reference, which are additional to the NEBOSH syllabus, are provided on the RMS Publishing website for ease of use - www.rmspublishing.co.uk.

A6.1 - Purpose and use of health and safety performance measurement

The meaning of health and safety performance measurement

Measurement is a key step in any management process and forms the basis of continual improvement. Organisations need to monitor and measure their health and safety performance to find out the degree to which they are being successful, just like finance, production or sales. Monitoring is an essential component of good management. An old maxim is 'what gets measured tends to get done'. If measurement is not carried out correctly, the effectiveness of the health and safety management system is undermined and there is no reliable information to inform managers how well the health and safety risks are controlled. Effective performance measurement provides information on both the level of performance and why the performance level is as it is.

Managers and directors usually measure health and safety performance in the negative while measuring other things in the positive. Health and safety differs because success results in the absence of an outcome (injuries or ill health) rather than a presence. Low injury or ill-health rate, however, is no guarantee that risks are being controlled and will not lead to injuries or ill health in the future.

Even where managers do utilise positive - proactive - measures of performance there is often no systematic approach to deriving these measures and how they link to the risk control process. Measurements are taken in a haphazard unscientific way with not enough linkage with the management system. If the information derived from measurement cannot be used as a means to understand the basis of performance then it is of little use.

WHY MEASURE PERFORMANCE?

In modern, active and changing organisations it is essential to identify and confirm what is working and why in order to prevent these successful management actions being 'lost' at a time when resources are being reviewed. Measurement is an accepted part of the 'plan-do-check-act' management process.

Key purposes of performance measurement are to:

- Determine whether safety plans have been implemented and objectives achieved.
- Comparison of actual performance with targets.
- The identification of trends.
- Check that risk controls have been implemented and are effective.
- The identification of substandard practices and or conditions.
- Learn from safety management system failures, including hazardous events (accidents and incidents).
- Promote implementation of plans and risk controls by providing feedback to all parties.
- Allow for benchmarking with similar organisations.
- Provide information that can be used to review and, where necessary, improve aspects of a current safety management system via an action plan.

WHO SHOULD MEASURE PERFORMANCE?

Health and safety performance needs to be measured at each management level in an organisation, starting with the most senior management. Senior managers must guard against a culture of management, or measurement of health and safety, by exception. This means that unless a problem or deficiency is brought to their attention they presume that everything is working as intended and do not inquire any further.

The dangers of this approach have been highlighted in several reports of official inquiries into major incidents in the UK and abroad. Senior managers must satisfy themselves that appropriate arrangements to control health and safety risks are:

- In place.
- Complied with.
- And effective.

Need for a range of both active and reactive measures

Organisations need to use a range of both active (proactive) and reactive measures to determine whether objectives have been met. A balanced approach to monitoring seeks to learn from all available sources. Hence two forms of monitoring system are required:

- ***Proactive monitoring***, before the event, involves identification through regular, planned observations of workplace conditions, systems and the actions of people to ensure that standards are being implemented and management controls are working. Are you achieving the objectives and standards you set yourself and are they effective?
- ***Reactive monitoring***, happens after things go wrong; involves learning from your mistakes, whether they have resulted in injuries and illness, property damage or near misses. Investigating injuries, cases of illness, property damage and near misses - identifying in each case ***why*** performance was substandard.

Using a range of measures will provide correlation of facts; enable a more accurate analysis and a better basis for conclusions to be drawn. If an organisation only used reactive monitoring to measure success it would be a limited source of data, as it would not identify a situation where objectives had not been met and the circumstances had not yet resulted in an accident or illness. Active measures, such as audit and inspection, provide data on completion of objectives without the need for an accident or illness to have occurred.

Organisations need to ensure that information from both active and reactive monitoring is used to identify situations that create risks, and to do something about them. Priority should be given where risks are greatest. Look closely at serious events and those with potential for serious harm. Both monitoring methods require an understanding of the immediate and the underlying causes of events.

Investigate and record what happened - find out why. Refer the information to the people with authority to take remedial action, including organisational and policy changes.

The following are examples of methods that can be used to measure safety performance:

- Systematic workplace inspections using checklists.
- Safety tours - for example, on a 'walk through' basis.
- Accident analysis.
- Safety sampling - examining specific aspects of the system.
- Safety survey - a detailed examination of a narrow field of activity.
- Behaviour sampling - assessing workers' behaviour to identify unsafe work practices that might require correction, for example, by work design improvements or through training.
- Attitude surveys of personnel at all levels.
- Analysis of documentation and records.
- Benchmarking against good safety practices in other organisations.
- Environmental sampling - measuring exposure to substances or energies and comparing with recognised standards.
- Inspections of specific machinery and plant to check that safety-related parts are fitted and in good condition.
- Audits - how well health and safety is being managed.

The meaning of key performance indicators and their role in setting business objectives

A key performance indicator (KPI) is a measurement used to evaluate factors that are important to the success of an organisation. A lagging indicator measures a company's past incidents, accidents, non-conformances, unsafe behaviour etc. A leading indicator measures activities carried out to prevent and control incidents and safety failures. Both types of indicator are needed by an organisation to set objectives and targets which will ensure continual improvement in health and safety performance.

By measuring something it is possible to establish whether you are on target to meet a goal within a given period or not. This is what a KPI should do.

In numerous companies, KPIs are only used if they are considered to be SMART.

If a KPI is said to be SMART, then it must be:

- ***Specific*** - it has to be unambiguous and not overlap with other KPIs.
- ***Measurable*** - it must be consistently quantifiable whichever method is used.
- ***Achievable*** - it has to be seen by all parties concerned as a fair indicator.
- ***Realistic*** - it cannot set impossible-to-achieve targets or goals.
- ***Time related*** - it has to have a start and end date which are agreed in advance.

Managers and employees often set SMART KPIs at an annual performance review, but they can be used at any time by mutual agreement. Understanding your KPIs and where you are with them can help you to focus on important factors relating to health and safety performance in the workplace.

Key performance indicators are also used by senior executives to set measurable goals for departments and divisions within larger organisations. Sometimes individual KPIs will contribute to team ones which can then feed into departmental ones. Remember that your own measurable contribution is part of how a larger organisation ultimately performs.

The types, benefits and limitations and leading and lagging indicators

Some examples of KPI's an organisation might use:

KPI	***Suggested measurement***
Compliance (leading and lagging)	
Compliance with regulations and guidance.	Review and scoring of legislation register.
Permit-to-work infringements.	PTW infringements per shift or week/month.
Number of risk assessments completed.	Comparison with objectives and targets.

KPI	***Suggested measurement***
Implementation of risk controls from findings of risk assessment.	Comparison with objectives and targets.
Incidents and accidents (lagging)	
Injury rate compared with industry average.	Data from reporting system.
Accident costs.	Using HSE's costs model.
Management performance (leading and lagging)	
Management safety tours.	From management review proceedings.
Management attitude to health and safety.	From opinion surveys (workforce and management).
Health and safety resources (leading)	
Time and finance allocated to health and safety management.	Data from budget allocation.
Competence (leading)	
Training and competence of management and workers.	From training records and training budget.

The limitation of lagging indicators is that they tell you how many people got hurt and how badly, but not how well the organisation is preventing incidents and accidents.

Leading indicators are only effective if they are used by the management team to identify and introduce improvements to the health and safety management system, by means of an action plan and by allocating sufficient resources.

The organisation should arrange a regular independent audit of its safety management performance, with specific recommendations which will achieve continual improvements in health and safety standards.

Assessment of the effectiveness and appropriateness of health and safety objectives and arrangements

The Management of Health and Safety at Work Regulations (MHSWR) 1999, Regulation 5, requires employers employing five or more employees to record their arrangements for the effective monitoring of preventive and protective measures.

This must include the effectiveness and appropriateness of objectives and arrangements. Monitoring and measurement can enable the organisation to determine if arrangements are appropriate, for example, control measures to safeguard machinery. In a situation where arrangements are made for certain equipment to have a two handed control fitted as a safeguard, monitoring methods like near miss reporting and audits may identify that a more appropriate safeguard would be a fixed guard.

The effectiveness of arrangements for control of chemicals may be assessed by measuring the amount of the chemical in the atmosphere over a period. This can confirm if the controls in place to limit exposure to the chemical are effective. In a similar way, objectives can be assessed by active monitoring of how they are being implemented and considering the feedback from managers and workers. It is important that organisations use a wide range of monitoring methods to measure health and safety performance in order to maximise on opportunities to assess objectives and arrangements.

Review of current health and safety management systems

It is essential for organisations to learn from their experiences and take the opportunity to decide how to improve performance. One of the purposes of measuring performance is to enable recommendations to be made for the review of the current health and safety systems.

Audits by an organisation's own staff or people from outside the organisation complement other monitoring activities by assessing if the policy, organisation and arrangements (systems) are actually achieving the right results.

Consulting employees or their representatives about matters to do with their health and safety is good management practice, as well as being a requirement under health and safety law. Employees are a valuable source of information and can provide feedback about the effectiveness of health and safety management arrangements and control measures. Where safety representatives exist, they can act as an effective channel for employees' views. Safety representatives' experience of workplace conditions and their commitment to health and safety means they often identify potential problems, allowing the employer to take prompt action. They can also have an important part to play in explaining safety measures to the workforce and gaining commitment.

It is important that reviews are conducted in a systematic way and that recommendations are made after consideration of a range of active and reactive measures. Review involves establishing priorities for necessary remedial action that were discovered as a result of monitoring to ensure that suitable action is taken in good time and is completed

Periodically the whole of the health and safety management system should be reviewed including the elements of planning, organisation, control and monitoring to ensure that the whole system remains effective.

A6.2 - Health and safety monitoring

The objectives of active monitoring

The objectives of active (proactive) monitoring are to:

- Check that health and safety plans have been implemented.
- Monitor the extent of compliance with the organisation's systems/procedures against legislative/technical standards.

Active monitoring includes the assessment of the appropriateness, implementation and effectiveness of health and safety standards (in the form of objectives and arrangements, including Risk Control Systems (RCSs)). This includes the operation of the current management systems. Organisations need to know:

- Where they are.
- Where they want to be.
- What is the difference - and why.

Active monitoring will tell the organisation about the reliability and effectiveness of its systems. This provides a good basis from which decisions and recommendations for maintenance and improvement may be made.

Active monitoring also provides an opportunity for management to confirm commitment to health and safety objectives. In addition, it also reinforces a positive health and safety culture by recognising success and positive actions, instead of 'punishing' failure after an undesired event.

Organisations must see active monitoring as an integral (and normal) part of the management (work) function. As such it must take place at all levels and opportunities in the organisation. Managers should be given responsibility for the monitoring of objectives and compliance with standards for which they and their subordinates are responsible. The actual method of monitoring will depend on the situation and the position held by the person monitoring.

The various forms and levels of active monitoring include (source HSG65):

- Routine procedures to monitor specific objectives, for example, quarterly or monthly reports or returns.
- Periodic examination of documents to check that systems relating to the promotion of the health and safety culture are complied with, for example, the way objectives for managers are established or appraised, assessment of records of training needs and delivery of training.
- Systematic inspection of premises, plant, and equipment by supervisors, maintenance staff, management, safety representatives and other employees to ensure continued effective operation of workplace precautions.
- Environmental monitoring and health surveillance to check on the effectiveness of health control measures and to detect early signs of harm to health.
- Systematic direct observation of work and behaviour by first-line supervisors to assess compliance with risk control systems (RCSs) and associated procedures and rules, particularly those concerned with risk control.
- The operation of audit systems.
- Consideration of regular reports on health and safety performance by the board of directors.

Active monitoring effort should be applied on a risk basis. Monitoring of workplace precautions would typically be more detailed and frequent than management system activities that carry a low risk if misapplied.

The objectives of reactive measures

The primary objectives of reactive monitoring are to analyse data relating to:

- Accidents.
- Ill-health.
- Other loss causing events, such as property damage.

Reactive monitoring activities are deemed to be after the event and include:

- Identification.
- Reporting.
- Investigation.
- Collation.
- Analysis of data and statistics on the events.

The events monitored include those resulting in:

- Injuries.
- Cases of illness (including sickness absence).
- Property damage/environmental damage.
- Near misses.

It is important to identify, in each case, why performance was sub-standard. Trends and common features may be identified, such as when, where and how these events occur. This provides an opportunity to learn and put into place improvements to the overall management system and to specific risk controls.

In order to carry out reactive monitoring effectively systems must be in place to identify the event for what it is, record it and report it. Without this nothing may be learnt. Indeed, what little data that is communicated might serve to reinforce that there is no need to put in a great deal of health and safety effort. If reporting etc. is planned and encouraged it is not uncommon to find a large increase in recorded events. This does not necessarily mean an increase in events, merely an increase in reporting.

Events contribute to the 'corporate memory', helping to prevent a repeat in another part of the organisation or at a later time. Though it should be remembered that the 'corporate memory' is said to be short, in the average organisation (one undergoing some change) it is said to be 4 years. Data may be gained from other organisations to reinforce or extend experience of events and the hazards involved.

The limitations of reliance on accident/incident and ill-health data

Many organisations spend considerable time developing data on their health and safety performance based on the accidents and ill-health they have experienced. Whilst there is value in doing so, it has the limitation of being after the event, after the harm has been done. Accidents must occur to get the data, thus tending to reflect what ***was*** done to prevent them rather than what ***is*** being done. A more complete approach to monitoring will tend to include 'before the event' actions, like audits and inspections, to indicate what is currently being done to prevent accidents and ill-health. A low injury accident rate is not a guarantee that risks are being effectively controlled. In some cases this might be a matter of good fortune, rather than effective management.

If organisations wait until an event happens to determine where health and safety effort is required then some sort of loss is likely to have occurred. In some particularly reactive organisations, in order to gain sufficient management attention this may have to be an event resulting in serious personal injury to someone. Clearly this is an undesirable way of learning, particularly as, with an amount of effort, planning and thought, the event could have been foreseen and prevented. The more mature, less reactive, organisation seeks to learn most from activities (for example, risk assessment) before the event or, at the very least, learn from those events that result in no personal injury, for example, near misses.

Distinction between performance measures

ACTIVE/REACTIVE

Reactive performance measures are those that measure performance by analysis of data obtained after loss or potential loss events have occurred. They may focus on such things as the number, type, timing, and outcome of these events. Whereas, active performance measures are those that measure performance by analysis of the functions of the management system that can prevent loss or potential loss events. They may focus on the actions of management and workers to prevent harm, such as the condition of the workplace or the amount of training conducted.

OBJECTIVE/SUBJECTIVE

Objective performance measures are those that are detached from an assessor's personal judgement. For example, reading a calibrated noise meter, counting the number of workers using hearing protection or confirming whether an occupational health and safety practitioner is in post. However, things that can easily be measured objectively may not be the most important things to measure. Subjective performance measures are those that may be influenced by the person doing the measuring. Examples are measures of adequacy of risk control measures that are often difficult to measure objectively, for example, safe systems of work where no defined standard has been laid down. These measures can be very useful, but need to be treated with caution as two people may report different findings from the subjective consideration of the same thing.

QUALITATIVE/QUANTITATIVE

Qualitative performance measures are those that are descriptions of conditions or situations that cannot be recorded numerically, for example, a commentary on the deliberations of an occupational health and safety committee. While qualitative performance measures may be very important they may be difficult to relate to or compare with other performance measures. Quantitative performance measures are those that can be described in terms of numbers and/or recorded on a scale. Where possible, it is desirable to quantify performance measures so that comparisons can be made over time. However, efforts to derive quantitative measures from some aspects of health and safety may give an unjustified impression of precision.

A6.3 - Health and safety monitoring and measurement techniques

Evaluation of the health and safety performance of an organisation

RANGE OF MEASURES AVAILABLE

An organisation's performance measurement system should incorporate both reactive and proactive monitoring. The data gained from reactive monitoring will allow an organisation to see if they are achieving the proactive objectives that are contained in the health and safety policy. Health and safety performance in organisations that manage it successfully is measured against predetermined standards, which have been set from the findings of the risk assessments throughout the organisation.

The success of action taken to control risks is assessed through self-monitoring involving a range of techniques. This includes both examination of both hardware (premises, plant and substances) and software (people, procedures and systems), including individual behaviour. Failures of control are assessed through reactive monitoring that requires the thorough investigation of any accidents, ill health or incidents with the potential to cause harm or loss. In both active and reactive monitoring, the objective is to determine the immediate, underlying and root causes of sub-standard performance.

Active monitoring data

The range of active measures include:

- The extent to which plans and objectives have been set and achieved.
- Staff perceptions of management commitment to health and safety.
- Whether a director for health and safety has been appointed.
- Whether health and safety specialist staff has been appointed.
- The extent of the influence of health and safety specialists.
- Whether a safety policy has been published.
- Whether a safety policy has been adequately communicated.
- The numbers trained in health and safety.
- Effectiveness of health and safety training.
- Number of risk assessments completed as a proportion of those required.
- Extent of compliance with risk controls.
- Extent of compliance with statutory requirements.
- Number and effectiveness of senior managers' health and safety tours.
- Number of staff suggestions for health and safety improvements.
- Staff attitudes to risks and risk controls.
- Staff understanding of risks and risk controls.
- Frequency of health and safety audits.
- Time to implement health and safety audit recommendations.
- Frequency and effectiveness of health and safety committee meetings.
- Frequency and effectiveness of staff health and safety briefings.
- Health and safety specialist reports.
- Time to implement action on complaints or suggestions.
- Health surveillance reports.
- Personal exposure sampling reports.
- Workplace exposure levels (for example, noise, dust, fumes).
- Personal protective equipment use.

Reactive monitoring data

The range of reactive monitoring measures include:

- Unsafe acts.
- Unsafe conditions.
- Near misses.
- Damage only accidents.
- Reportable dangerous occurrences.
- Lost-time accidents - when at least one work shift (or other time period) is lost by a person as a result of an accident injury.
- Reportable accidents involving absence from work for more than seven days.
- Reportable major injuries.
- Sickness absences - employee absences due to illness (occupationally-related or non-occupationally-related).
- Complaints made, for example, by members of the public.
- Criticisms made by regulatory agency staff.
- Regulatory enforcement action.

HOW THE RANGE OF MEASURES CAN BE USED

Organisations should adopt a well-formulated combination of both active and reactive types of measure in an occupational health and safety programme. This allows a much better overall assessment of occupational health and safety performance than reliance on any single measure. Attention should be given to the level of competence required of those responsible for devising, carrying out and analysing data from all performance monitoring measures.

As can be seen by the range of active data to be measured, it can provide an early insight into the extent that the health and safety management system is operating in an effective manner. If measurements of this type are made, it can quickly indicate a decline or improvement in the management system.

For example, if the numbers of people that have received health and safety training decline this could quickly cause the management system to become ineffective.

A great deal of effort may go into managing health and safety in an organisation; however, reactive monitoring might show it is not effective as, despite the effort, accidents are occurring. This provides a clear indication that effectiveness needs to be improved and may show some gaps or poor performance not identified by active monitoring.

Reactive data can be translated into performance indices, such as frequency rates, and used to track trends in performance of the health and safety management system.

Collecting and using sickness absence and ill-health data

Keeping accurate sickness absence records can help employers comply with their duty to risk assess any hazards in the workplace.

These records can provide information that identifies:

- Who has been off sick.
- Why they have been off sick.
- Where the high levels of absence occur.
- Whether there are any underlying workplace-related health issues.

From these records, the organisation can identify how to improve working conditions so that their workers are not exposed to harmful environments or stressful conditions.

For example, 'absenteeism' is a term used for workers taking a day or two off, because they are suffering from work-related stress. By collecting absenteeism data, the employer can provide early intervention to support workers who are showing symptoms of stress. This requires that the senior staff such as supervisors and managers have the necessary training and awareness to identify unacceptable stress and manage the factors which produce it.

Role and purpose of main monitoring techniques

HEALTH AND SAFETY AUDITS

The role of an audit is to determine the level of compliance of something to a set of agreed standards. In the field of health and safety, minimum standards may be derived from relevant legislation. Therefore, an audit may determine the extent to which an organisation is compliant with legislation.

A health and safety management system audit is that part of the management system that subjects the other components of the management system to a rigorous, systematic examination.

> *"An audit is a systematic, critical examination of an organisation's systems to determine the extent of compliance with a set of agreed standards."*

Figure A6-1: Definition of term audit. *Source: RMS.*

As an audit will assess how well health and safety is being managed compared with expectations (standards), the audit process will identify strengths, weaknesses and areas of vulnerability. The outcome from the audit is a report to management and an action plan, which will allow health and safety to be managed successfully.

The entire health and safety management system should be subjected to a comprehensive audit from time to time. Individual elements of the health and safety programme can, of course, be subjected to individual audits, for example:

- Evaluation of compliance with set occupational health standards.
- Evaluation of compliance with physical safeguards (health and safety hardware).
- Evaluation of compliance with fire prevention/control standards.

Senior management

Before starting ensure that the senior management are on board and are prepared to implement the findings. Senior management should communicate to all staff to be audited their commitment and ensure that other staff co-operate. A senior person within the organisation should be responsible for the audit, prepare and oversee the process and ensure sufficient resources are in place.

Audits can prove stressful both to those being audited, who often see them as a threat and critical of their efforts, and to the auditors, who may be confronted by people who may be unreceptive, uncooperative, obstructive or even aggressive. Every effort should be made by the auditors to defuse such situations by emphasising that the audit has the full support and commitment of senior management. They should explain that it will benefit both the organisation and the individual by identifying potential problems and weaknesses in the health and safety management system, and that the audit is not concerned with recrimination or the allocation of blame.

Pre-audit preparations

What to audit

The nature and extent of the audit to be undertaken should be determined. This involves answering such questions as:

- Will the audit look at the whole or just part of the organisation, or focus on a specific activity, location or issue?
- Will the audit look solely at the safety management system or will it involve technical matters concerning plant, equipment and processes?
- Is the audit intended to establish the effectiveness or otherwise of the safety management system (validation audit), or to verify whether the organisation is complying with its own standards and procedures (compliance audit), or both?
- Should the audit be carried out by internal or external auditors?
- Will the audit as proposed require any special skills of the auditors?

Once decided the terms of reference of the audit should be agreed and made known to both auditors and the manager of the activities to be audited.

Information gathering

The audit must be structured and co-ordinated in its assessment of the systems. This is best achieved by utilising audit checklists developed or obtained before the audit.

The audit involves interviewing people, observations in the workplace and assessment of documents such as:

- Health and safety policy.
- Health surveillance records.
- Accident/incident reports.
- Records of statutory inspections such as lifting equipment and portable appliances.
- Maintenance procedures.
- Procedures for method statements and permits to work.
- Risk assessments.
- Insurance documents.
- Training records, etc.

Whilst some of this information may be assessed in the workplace, it is useful to gather information through analysis of documents that set out the expectation of the organisation before conducting the audit.

Information should be obtained on the results of prior audits, the organisational structure, policy and objectives of the organisation and other important information that will enable the audit to be effective, such as shift patterns, where work is taking place off site and critical activities that may be of interest to the audit.

Notifications and interviews

In order to prepare for an audit it is necessary to decide who needs to be interviewed and organise a timetable in order to meet them on a planned and organised basis. Suitable start and completion dates for the audit should be agreed with the auditors and those to be audited and the date by which the audit report is to be completed. It is important that people involved in the audit are notified of arrangements for interview and informed what type of documents they need to make available. In this way, there is a better chance that they will be provided at the time of audit.

Interviews should be structured and provide the interviewee with opportunity to express what they are doing to meet the requirements being audited. It may be necessary to utilise pro-forma questionnaires to ensure that interviews are carried out in a structured manner and that all the information required is obtained efficiently and with the minimum of inconvenience to the parties involved.

Selection of staff

Audits should be conducted by people that are both independent and competent. Health and safety practitioners that have received specific training in health and safety auditing techniques would usually be able to carry out this function. Audits can be carried out by the management of the organisation, provided that the managers do not audit their own efforts directly (bias must be eliminated) and that the managers concerned have been trained in audit techniques. Often a small team will be commissioned to conduct the exercise, in order to widen the experience base and establish some degree of independence.

An internal audit team may comprise three main groups of people:

1) A manager.
2) A representative from the workforce.
3) A health and safety practitioner.

Extra individuals may be co-opted to the team when specific topics are under assessment. If audits are to be carried out internally, organisations should be mindful of the fact that, very often, the result of the audit may appear to be critical of the organisations and/or individual performance. This can lead to feelings of resentment or even violence towards the auditor. When selecting the people to carry out the audits, employers should be sure that the auditor is sufficiently strong of mind to deal with these situations.

It is important when carrying out internal audits that the auditor is impartial, this will usually mean that the auditor does not audit a function of the organisation for which they have responsibility.

Therefore, where an audit is to be conducted of the whole health and safety management system, a more independent approach would be to conduct an audit using auditors from outside the organisation or location (external auditors).

Advantages and disadvantages of external and internal audits

	Advantages	***Disadvantages***
Internal audits	■ Internal audits ensure local acceptance to implement recommendations and actions. ■ The auditor often has intimate knowledge of the hazards and existing work practices. ■ An awareness of what might be appropriate for the industry. ■ Familiarity with the workforce including their strengths and weaknesses. ■ Relatively low cost and easier to arrange.	■ May not possess auditing skills. ■ May not be up to date with current legislation and best practice. ■ The auditor may also be responsible for implementation of any proposed changes and this might inhibit recommendations because of the effect on workload. ■ May be subject to pressure from management and time constraints.
External audits	■ External audits are usually impartial; auditors will have a range of experience of different types of work practices. ■ May be able to offer solutions to what might be considered unsolvable problems within. ■ Not inhibited by criticism. ■ Will assess the organisation's performance without prior bias.	■ Need to plan well to identify nature and scope of the organisation. ■ Individuals may not be forthcoming, be nervous or resistant to discussing their workplace with an outsider. ■ May seek unrealistic targets.

Figure A6-2: Advantages and disadvantages of external and internal audits. *Source: RMS.*

Competence of auditors

The auditor must be familiar with audit techniques, familiar with work practices, have the ability to interpret standards and be in a position to be able to keep up to date with new information and standards. Consideration should be given to the auditor receiving formal, generic audit skills training to enable them to conduct audits efficiently and effectively.

Time and resources

Audits are an in-depth analysis of compliance with standards and must not be treated lightly. The planning of the audit alone can be very time consuming. Evidence gathering and verification can also take a long time, depending on the scope of the audit. Employers should not apply pressure on the auditor to complete audits in less time than is appropriate and must be prepared to allocate sufficient time to the task.

Similarly, the auditor might need other resources than time, such as access to documentation, measuring equipment, electronic storage facilities, and research facilities (internet, library, etc.) in order to do a thorough job.

Assessments processes during audit

The audit process

An audit of the health and safety management system will determine:

- If a comprehensive system exists.
- If employees and those working on behalf of the organisation, are fully aware of the requirements and their duties with respect health and safety.
- If the documentation system reflects the practices.
- If the procedures, work instructions etc. are being worked to and satisfy those they are supposed to protect.
- If there areas that are deficient and are non-conformities.
- If there areas where improvements can be made.

The audit process will involve interviewing people, evaluation of documents and observation of the workplace.

The auditor will assess sufficient evidence to enable them to confirm the level of compliance for what they are auditing.

"Records, statements of fact or other information which are relevant to the audit criteria and verifiable."

Figure A6-3: Definition of term audit evidence. *Source: ISO 19011.*

Assessment decisions

Objective measurements are those that are detached from an auditor's personal judgement. This compares with subjective measurements, which are influenced by the judgement of the auditor. Quantitative measurements are those that can be described in terms of numbers and recorded on a scale, whereas qualitative measurements are descriptions of conditions or situations that cannot be recorded numerically.

Implicit in any health and safety audit is some form of quantitative and qualitative measuring/evaluation process, usually in the form of a scoring system. This allows the organisation to see how it is performing in comparison to an absolute standard and compared with other organisations using the same measurements. There are many types of scoring system available to the organisation. These can be qualitative or quantitative. The examples below illustrate the principles involved.

Sampling based scoring system (quantitative)

In this process an appropriate sample size is selected which relates to the number of items to be audited and a risk factor. The risk factor is determined by the auditor and can be based on consequence, likelihood and number of people exposed. For example:

400 items with a risk factor of 15%

Total affected x $\frac{15}{100}$ = sample quantity, therefore: 400 x $\frac{15}{100}$ = 60 items to be sampled.

22 items of the 60 sampled conform to requirements, therefore: $\frac{22}{60}$ x 100 = 36.7%.

All or nothing points scoring system

Some items to be audited can best be assessed by awarding all the points - 100%, or no points at all - 0%. In some instances a number of detailed sub-questions combine to give a maximum of 100%.

Judgmental scoring system (subjective)

Some questions can only be effectively assessed by the auditor making a judgement of the level of compliance, based on auditor's guidance. This is fundamentally a qualitative measure which can be made semi-quantitative by, for example, selecting a figure between 0% and 100% based on guidance on what performance warrants what percentage.

The value of an audit is dependent upon the experience, training and knowledge of the auditors and their ability to interpret and use the findings. It is also dependent on the integrity of all the parties involved. A system of checks and balances should be built into the system in order to avoid misrepresentation.

Actions taken following audit

Initial feedback

At the end of the data collection phase the auditor or audit team should summarise and feedback their initial findings to local senior management and, in particular, draw attention to any issues that are of such significance as to necessitate immediate attention.

Draft report

The audit report should be submitted to local management in draft form to enable factual accuracy to be checked and to ensure that the report is understood.

Final report

The outcome from an audit should be a detailed report of findings and recommendations to improve or maintain the health and safety management system. A structure and approach to the report should be agreed at the pre-audit stage.

The final report should give a clear assessment of the overall performance of the organisation. It should identify deficiencies and make recommendations for improvement. It should also identify the observed strengths and suggest how they can be built upon. All audit reports need to be accurately and clearly communicated. In addition to the provision of a detailed written report, a verbal presentation of the report may be provided soon after the close of the audit, in order to give an early opportunity for management to learn and take action.

Recommendations should flow logically from the main body of the report and should be clearly connected to the findings set out in the report. Recommendations should consist of a plain statement of action without repeating the arguments of the preceding findings section of the report. They should include justification for the recommendation by referring to information summarised from findings.

Correcting non-conformities

Responsibility for carrying out actions for correcting non-conformities should be assigned to those in line management responsible for them, together with target completion dates. It is essential that management take ownership of both the audit and the subsequent action plans.

Progress on correcting non-conformities should be monitored; this can be through reports or feedback at meetings.

WORKPLACE INSPECTIONS

The role of health and safety inspections is to identify the health and safety status of what is being inspected and what improvements are needed. They are particularly well suited to identifying workplace hazards and determining if they are under satisfactory control or not.

In addition to general workplace inspections, inspections are used to confirm the safe condition of work equipment.

Regulation 6 of Provision and Use of Work Equipment Regulations (PUWER) 1998 lays down requirements for inspecting work equipment to ensure that health and safety conditions are maintained and that any deterioration can be detected and remedied in good time.

Inspections involve examination of the workplace, or items of equipment, in order to identify hazards and determine if they are effectively controlled.

Four different types of inspections are common:

1) General workplace inspections - carried out by local, first line managers and employee representatives.
2) Statutory inspections (thorough examination) of equipment, for example, boilers, lifting equipment - carried out by specialist competent persons.
3) Preventive maintenance inspections of specific (critical) items - carried out by maintenance staff.
4) Pre use 'checks' of equipment, for example, vehicles, forklift trucks, access equipment - carried out by the user.

Planned inspections help to:

- Identify potential problems before they cause injuries or illnesses.
- Identify equipment problems resulting from such things as wear and tear or improper use.
- Identify improper work practices.
- Draw attention to good safety and health practices.
- Identify new hazards resulting from changes in the workplace.
- Identify inadequacies in corrective action that has been taken.

Your inspection schedule must:

- Identify what will be inspected, (work areas, equipment, tools, procedures, practices, etc.), by whom, and how frequently.
- Include inspections of work procedures and production processes.
- State what inspection records must be produced to ensure accountability.
- Provide for correcting defects found during each inspection.
- State what training those who carry out inspections require.

The frequency of inspections should be planned to take place at regular intervals. The time between most inspections is often at the employer's discretion. Factors that will influence the timescales are the changing nature of the workplace, manufacturer's recommendations, type and frequency of use, environmental conditions, severity of failure and previous history of failure. Principally this establishes a frequency on the basis of risk. Occasionally, minimum intervals are set by legislation such as that referring to local exhaust ventilation, lifting equipment, scaffolds and excavations.

General workplace inspections may be carried out in lower risk organisations that are not affected by change on a frequency so that every workplace is inspected once in a three month period - this may suit an office environment. In organisations that have higher risks, such as manufacturing, this may be increased to once in a month. Where risks are higher and changes to the workplace are more likely, such as in maintenance or construction operations, this may be increased to weekly or daily.

Competence of inspector

Inspections normally involve physical examination of the workplace, or equipment, with a view to identifying hazards and determining if they are effectively controlled. They are usually carried out by a manager, worker health and safety representative, equipment user or technical specialist.

It is important that the inspector be competent to inspect what they are inspecting. This will include techniques of conducting the inspection, recognising conditions that are of a good standard, recognising conditions that are, or will become, substandard and the acceptable response to what they find.

One of the most important competencies an inspector must have is the ability to know their own limitations of competency and what action they should take when they identify something that falls beyond their limitations.

Use of checklists

In many cases inspections are based on checklists. If this is the case, then managers relying on inspections must be mindful that the system of using a checklist has advantages and disadvantages.

Advantages

- Enables prior preparation and planning.
- Quick and easy to arrange.
- Brings a consistent approach.
- Clearly identifies standards.
- Thorough.
- Provides a ready-made basis for an inspection report.
- Provides evidence for audits.

Disadvantages

- Does not encourage the inspector to think beyond the scope of the checklist.
- Items not on checklist are not inspected.
- May tempt people who are not authorised/competent to carry out the inspection.
- Can be out of date if standards change.
- Inspectors might be tempted to fill in the checklist without checking the work area/equipment.

An inspection checklist is typically a list of 'the way things should be'; when a work area or item of equipment fails this test it is considered substandard and represents a hazard. Each substandard condition should be assessed and corrective action identified and details recorded.

Checklists relating to work equipment tend to reflect the specific, critical health and safety conditions of the equipment.

Factors to be considered when creating a checklist for a general workplace inspection should include:

- Substances or materials being used.
- Condition of traffic routes and means of access and egress.
- Work equipment.
- Work practices (manual handling, etc.).
- Work environment.
- Workstation ergonomics.
- Electricity.
- Fire precautions.
- Physical agents hazardous to health, including noise.
- Welfare provision, including first aid arrangements.

Allocation of responsibilities and priorities for action

Inspections will not improve health and safety performance unless, where a deficiency is found, corrective and preventive actions are identified, responsibilities are allocated and they are fully implemented. Standards will continue to deteriorate where workers/managers find their inspection efforts, which identify improvement needs, are not followed through and actions are not implemented.

They may quickly regard their efforts as 'a waste of time'. It is, however, equally important to avoid putting in a substandard or inadequate solution as this is bound to result in a waste of the organisation's resources (time, equipment, money etc.).

To get the right solution implemented will usually take a quantity of management time and effort and if the risk is significant this effort is warranted.

Whenever a substandard (at risk) situation is identified two important steps must be taken:

1) Each situation must be evaluated as to its risk potential.
2) The underlying cause(s) of the situation must be identified.

The first step ensures that resources are prioritised and allocated on a worst first basis, and the second step ensures that the appropriate corrective action will reduce the risk and prevent its return to the same level.

When situations that require most urgent attention are identified and corrective actions have been determined, the maximum time period by which the action is to be completed can be agreed.

A structured approach is necessary to ensure that the actions needed are initiated and followed up (monitored) in such a way that they are not simply forgotten. Responsibilities for action should be clearly allocated to those that have responsibility to ensure the action takes place.

HEALTH AND SAFETY TOURS

These present an opportunity for management to explore the effectiveness of risk control measures through planned visits to the workplace to observe and discuss the controls in use. It is important, when developing a positive health and safety culture, that management commitment is visible.

The conducting of planned tours to workplaces to meet work groups is one effective way of achieving this. As such, it is a monitoring method that senior and middle managers would find useful. It has the advantage of enabling direct contact and communication between employees and senior management.

This gives an accurate picture of work conditions and the understanding of employees. It can indicate deficiencies or success in managers carrying the organisation's objectives through to action. It also provides a forum for gaining the viewpoint of employees directly, without the translation that takes place through formal management channels.

In order to be planned there must be an intended outcome, for example, to communicate or review a topic, even if this includes some free time for other comment. Details of the tour and outcomes, including improvement actions, must be recorded to be effective.

HEALTH AND SAFETY SAMPLING

Sampling is where only a partial amount of a potential group/area/equipment is examined to establish facts that can indicate the standard of compliance of the whole.

A very small sample, such as the examination of three pieces of lifting tackle, may only give a rough, but acceptable, indication of the situation relating to lifting tackle as a whole.

When a representative sample is taken this may be considered to reasonably represent the situation for the whole group.

Sampling is conducted relating to the following:

- ***Specific hazards*** - such as noise or dust - typically conducted by staff trained in appropriate hygiene techniques.
- ***Good practice*** - such as the wearing of personal protective equipment - typically conducted by first line managers.
- ***General workplace hazards*** - such as those identified during a defined walk through a work area - typically conducted by first line managers, employee representatives and employees.

OBSERVATIONS OF PHYSICAL CONDITIONS

Category	Number Checked (C)	Number Standard (S)	% Meeting Standard S/C x 100	Comments
EQUIPMENT *For Guidance Notes consider Audit 123 Level 1 - Sections 4 and 7*				
Guarding				
Hand tools				
Power tools				
Electrics (visual)				
Electrics (technical inspection)				
Pressure systems				
Ladders and mobile towers				
Personal protective equipment				
Total Score				Total % Compliance Total S / Total C x 100

Figure A6-4: Sampling general workplace conditions - extract from Audit 123 Level 3 Section 1 Workbook [ISBN 978-1-900420-50-1]. *Source: RMS.*

HEALTH AND SAFETY SURVEYS

It may be described as an examination of a narrow field of the health and safety programme on an exploratory basis, with no fixed expectation of findings.

The term survey is usually applied to an exercise that involves a limited number of critical aspects, for example:

- Noise survey - usually with the aid of noise measuring equipment.
- Lighting survey - usually with the aid of a light meter.
- Temperature survey (to measure both high and low temperature levels) - usually with the aid of a thermometer.
- Personal protective equipment needs survey - usually involving review of standards and workplace conditions/activities.

The term survey is also used to define an exercise in which managers/employees are interviewed in order to identify knowledge, understanding, and details of specific needs within the working environment.

Examples of these might be:

- Training needs usually involving written questionnaires to managers and employees.
- Attitudes to health and safety, usually involving written questionnaires to managers and employees.
- The need for specific health and safety rules for specific tasks - usually involving review of standards and workplace conditions/activities.

Should these exercises seek to identify details from only a small number of people within a subject group or small geographical area it would cease to be a survey and would become a sample?

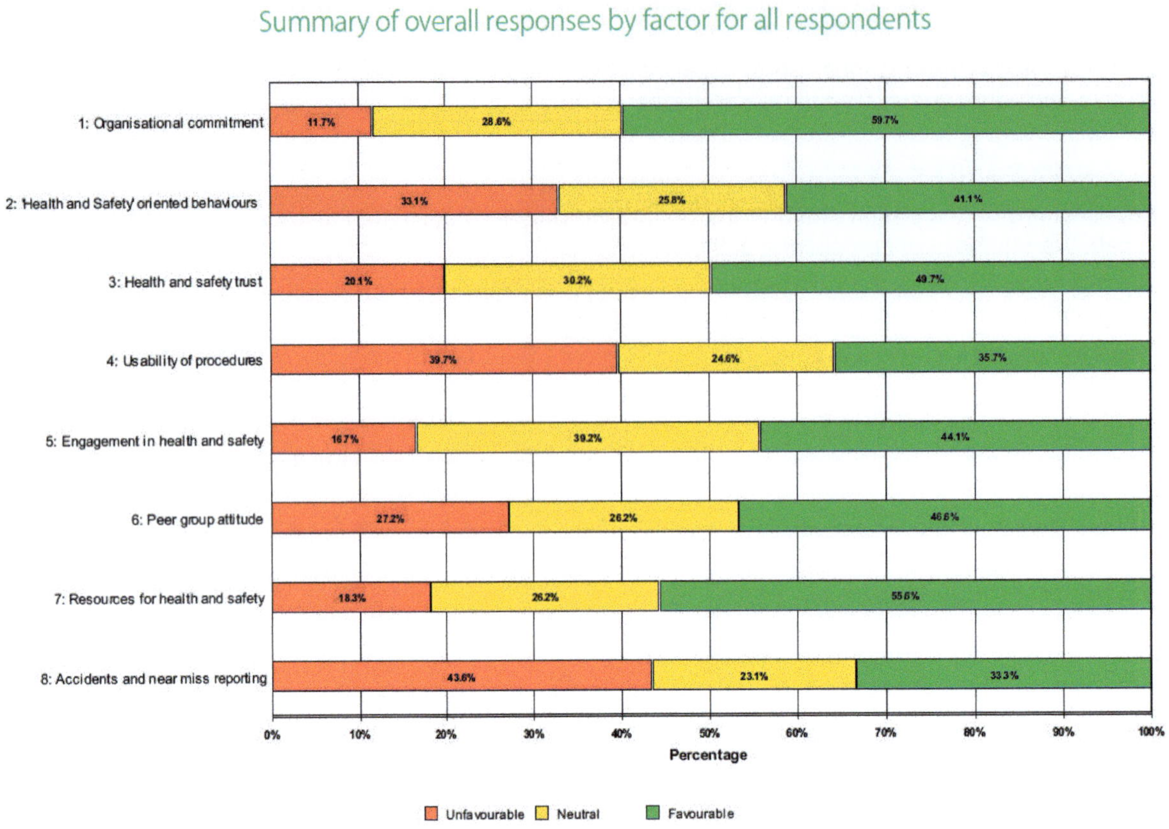

Summary of overall responses by factor for all respondents

1: Organisational commitment 11.7% 28.6% 59.7%
2: Health and Safety oriented behaviours 33.1% 25.8% 41.1%
3: Health and safety trust 20.1% 30.2% 49.7%
4: Usability of procedures 39.7% 24.6% 35.7%
5: Engagement in health and safety 16.7% 39.2% 44.1%
6: Peer group attitude 27.2% 26.2% 46.6%
7: Resources for health and safety 18.3% 26.2% 55.6%
8: Accidents and near miss reporting 43.6% 23.1% 33.3%
0% 10% 20% 30% 40% 50% 60% 70% 80% 90% 100%
Percentage
Unfavourable Neutral Favourable

Figure A6-5: Results of a climate survey. *Source: Health and Safety Laboratory.*

SAFETY CONVERSATIONS AND BEHAVIOURAL OBSERVATIONS

Safety observations and conversations are an important part of a health safety culture. They are used in the offshore oil industry to promote observation and discussion of workers safety behaviour. British Petroleum and others use the acronym A.W.A.R.E.

Announce

The first step in the process is to let the person you intend to observe know you are there.

Watch

After the individual is aware, you should spend a designated time watching the job task.

Ask

If you see a safe precaution being taken, or an exposure to risk, ask the most appropriate questions: Why did you do it that way? Is that the way you always do it? Do you feel safe doing it that way? Is there a safer way to do it? Were you trained to do it that way?

Reinforce

Emphasis should be placed on reinforcing what the worker is doing right. Ask how they felt. Did they feel safer?

Express concern

If a worker behaves in an unsafe manner during an observation, the language chosen to provide feedback is critical. Expressing concern is a preferred approach instead of stating someone is 'at risk' and 'unsafe'.

In-house health and safety practitioner's role in audits carried out by external/third parties

It is likely that in-house OHS practitioners will be appointed custodian of the safety management system on behalf of the organisation, with a key role in its effective implementation and regular review. They should also be able to make key contributions to audit processes, both internal and external.

They may be responsible for monitoring the progress of internal audits and ensuring that non-conformances are put right before the visit of independent external auditors. It is important that the OHS practitioner has the authority and support of the Chief Executive of an organisation.

This will ensure that the practitioner does not meet unnecessary resistance or delays in dealing with non-conformances.

Comparison of performance data and potential benefits of benchmarking

When comparing statistical data on health and safety performance, the larger the sample analysed the more indicative it will be of future potential incidents, if controls remain the same.

Multi-sited organisations, such as a distributors, banks and retail should not only measure performance data at the local level, but regionally, as well as gather statistics for the organisation as a whole. Each sample becomes larger from the single site to regional area and finally the whole organisation. Outcomes can then be compared, location to location, region to region and as the total business.

This data can be particularly important because it may show a particular area of concern in one sector (for example, a near miss, injury or significant incident) which has the potential to occur elsewhere at some time in the future if corrective action is not taken, thus turning reactive monitoring into a potentially predictive tool.

The HSE reports annually and publishes data on health and safety performance of all business sectors, utilising data collected from reports under the Reporting of Injuries, Disease and Dangerous Occurrence Regulations (RIDDOR) 2013. Organisations may usefully compare this with their own performance over the same period, to give an indication of their performance against the national results for their organisational sector.

Many organisations belong to trade associations or other such bodies which will often ask for performance data and create reports which, not only show how the sector is changing with time, but may give some idea how organisations are performing compared to each other (bench marking). Those organisations which are subject to ombudsman or regulatory control, such as rail companies and utility companies, will also collect and publish comparative data on performance for their specific sectors.

A6.4 - Reviewing health and safety performance

Need for formal and informal reviews of performance

The purpose of the health and safety performance review is to evaluate the effectiveness of what is being done, compared with expectation, experience and foreseeable factors that might affect future performance.

The Health and Safety Executive states that:

> *"Reviewing is the process of making judgements about the adequacy of performance and taking decisions about the nature and timing of the actions necessary to remedy deficiencies. [The purpose is that] the organisation learns from all relevant experience and applies the lessons."*

Figure A6-6: Reviewing performance. *Source: HSE.*

Regulation 5 of the Management of Health and Safety at Work (MHSWR) 1999 requires:

> *"(1) Every employer shall make and give effect to such arrangements as are appropriate, having regard to the nature of his activities and the size of his undertaking, for the effective planning, organisation, control, monitoring and review of the preventive and protective measures."*

Figure A6-7: Regulation 5 of MHSWR 1999. *Source: Managing of Health and Safety at Work Regulations (MHSWR) 1999.*

Learning from all relevant experience (including that of the organisation and of other organisations) needs to be done systematically, through regular reviews of performance. Reviews draw on sources of data from active and reactive monitoring activities. It is important that a formal review of performance be conducted at least annually, preferably involving senior management, in order to consider performance against objectives and assist in setting new objectives. In addition, informal reviews may be conducted by management, health and safety practitioners and employee health and safety representatives on a needs basis, as part of a reflective approach. These form the basis of continuous improvement, necessary to maintain compliance and effectiveness. This helps to maintain a management system that is fresh, dynamic, appropriate and effective.

The review process

Health and safety performance needs to be reviewed at each management level in an organisation, starting with the most senior management. Senior managers must guard against a culture of management, or measurement of health and safety, by exception. Management by exception would mean that unless a problem or deficiency is brought to senior management attention they presume that everything is working as intended, and do not inquire any further.

It is important that the most senior management of an organisation are involved in the process of reviewing health and safety performance as it may have an influence on policy, objectives, the management system and resources. It may be that the review is done by a working party drawn from the senior management team, but all senior management should be involved with deciding the parameters of the review and considering the outcomes of the review process.

A more local level review of performance may take place for part of an organisation or location, involving a number of people in a review team. This would usually involve the manager of that part of the organisation, another line manager who reports to the manager, someone who represents workers and someone who can give specific health and safety input. By using a team of this nature a wide range of experiences and perspectives can be taken into account and it will enable them to review the overall performance or focus on specific aspects of health and safety, such as transport accidents or manual handling.

Reviewing health and safety performance is an ongoing activity, so in one sense the review process is continuous. But, like any other activity, review should be both efficient and effective, so the frequency with which it takes place needs to be planned appropriately. This will ensure it happens on a timely basis, particularly where no monitoring activities have stimulated it to happen.

When considering when reviews should take place the following factors should be taken into consideration:

- Suitable intervals to ensure that specific planned objectives are achieved.
- The level of risk relating to the organisation and the consequences of ineffective performance.
- Where there are indicators of non-compliance or compliance.
- The potential for change, resources, external factors, legislation and standards.
- Where intervals for review are prescribed by standards, such as BS OHSAS 18001.
- If the review relates to a particular activity, the frequency at which the activity takes place.

Planned reviews may take place on the following basis:

- Monthly - for individuals, first line managers or sections.
- Quarterly - for middle managers or departments.
- Annual - for senior managers, sites or the whole organisation.

The results of the reviews can be communicated to the next level of the organisation for consideration in the review at that level. If critical health and safety issues are identified during monthly reviews it is not necessary to wait to pass what has been learnt to the next levels, if they require senior management input this can obtained outside the review structure.

Inputs to a review process

INTERNAL PERFORMANCE DATA

A thorough review should take account of internal performance data gathered from both active and reactive measures.

Whilst review of accident and incident data is important it should be remembered it is after the event. Any analysis should consider the potential for each occurrence to have resulted in a more significant outcome. Two elements to consider when reviewing accident and ill-health data are:

1) The effectiveness of planning, before the accident or ill-health.
2) Any failure of controls, this will often include the review of current risk assessments.

When carrying out the review, it is important to consider the effectiveness of corrective and preventative actions taken following the accident/ill-health. The review should also consider the timeliness of fulfilment of the corrective and preventive actions and if there was delay what caused it. The review of corrective and preventive actions applied following accidents/ill-health may indicate a need for further strategic level action. For example, a number of accidents may have led to action to provide refresher training for the specific risks associated with the hazards causing the accidents and this may indicate a need for a more organised, strategic approach to refresher training.

The active internal performance data considered during reviews should include the full range available, audits, inspections, tours surveys and sampling. Part of the review should be whether they were conducted as expected and their effectiveness.

The review should particularly consider information from audits because the audit process should provide a strong source of independent, verified data on the management system, implementation of objectives or specific risk being audited. This may provide a different perspective to that delivered by data from managers, as it should be free from bias.

HEALTH AND SAFETY OBJECTIVES

It is essential that an organisation review progress against its objectives in order that they can be successfully fulfilled. It could be that objectives set, when implemented, meet resource problems or are not as effective as planned. By identifying this through a review process, as they are implemented, they are more likely to be met and be effective.

In addition, objectives have to be reviewed to ensure objectives completed are removed, new ones are added, amendments are made to those that are determined to be inappropriate or unclear and ones that continue for a period are retained. This process will ensure that objectives are established with careful, thorough consideration and they remain valid.

ORGANISATIONAL ARRANGEMENTS AND CHANGE

It is a well established principle that things like health and safety policy and risk assessments are reviewed. This is done when it is believed that something affecting them has changed. Legislation may change and cause a review to determine if the arrangements and controls in place are relevant to the proposed change; for example, signage and work practices had to be changed when the exposure limits and action values were reduced by the Control of Noise at Work Regulations (CNWR) 2005.

Changes to the organisational structure can have a significant effect on the effectiveness of the health and safety management system. The review should, therefore, take account of changes to the organisational structure and its effects. It is important that where responsibility for the fulfilment of a set of health and safety objectives become redistributed or shared that this is done in an assured way. This will involve all objectives being formally assigned to someone and that those with responsibility can fulfil the objectives within their remit.

EXTERNAL STANDARDS AND EXPECTATIONS

Innovation in health and safety is happening all the time and even well intending organisations can find themselves out of step with legislation and good practice. It is essential that formal reviews of changes take place and improvement plans are put into place to take account of them. Information on legislative changes is often communicated well in advance of the actual change being made. This provides an early opportunity for the possible impacts to be reviewed and plans put in place to deal with it on a timely basis.

Feedback on levels of performance should be encouraged from customers and neighbours; this will help ensure that any health and safety issues are addressed before an incident occurs. Similarly, any communication or complaints from the enforcing authorities will need to be taken into account. In addition, suppliers may provide information on the performance of things they supply, gathered from other users. All of this data informs the review process of the wider perspective of others that might be affected by the organisation or may have an opinion on how it is performing.

Outputs from a review process

The results of the review of health and safety performance should be reported at the most senior (top) management level. This is particularly important in situations where the review has been conducted by a work group drawn from the top management team as this will enable all of top management to understand and accept the implications of the review.

The outputs required should be agreed at top management level, and signed by the most senior (top) manager to demonstrate commitment and authority from the most senior level of the organisation.

ACTIONS AND IMPROVEMENT PLANS

Outputs to the review process will include what has been achieved and what is next to be done. The review process will provide formal health and safety objectives and the establishment of actions and improvement plans.

The actions and improvements will need to be appropriately resourced, in terms of time and money to ensure they are achieved. This will often require a degree of prioritisation to enable the most beneficial action to be completed; the development of a clear action and improvement plan from the review will assist this.

The results of the review should be communicated widely in the organisation and in particular to those managers that have responsibility for responding to the actions and improvements arising from the review.

STAKEHOLDER REPORTS

It is a formal requirement of some organisations that they conduct an annual review, which includes health and safety, and report this to stakeholders. This is a specific requirement of the Corporate Governance Code and affects stock market listed companies.

> *"C.2.1 The board should, at least annually, conduct a review of the effectiveness of the company's risk management and internal control systems and should report to shareholders that they have done so15. The review should cover all material controls, including financial, operational and compliance controls."*

Figure A6-8: Reviewing performance. *Source: FRC, Corporate Governance Code.*

For organisations that do not have to meet this requirement this is a highly recommended practice and illustrates good governance. Such reports should be available to all employees and other stakeholders.

PERFORMANCE TARGETS

From the review process objectives and actions to improve health and safety may be identified and fed into development and improvement plans. The strategic level plans enable the production of local level plans through information cascade.

These plans should be used to develop clear performance targets related to the completion of the objectives and actions.

Performance targets should follow the SMART principle and be specific, measurable, achievable, relevant and timely. In this way health and safety in an organisation is maintained dynamically, leading to continuous improvement.

This page is intentionally blank

Element

A7

The assessment and evaluation of risk

Learning outcomes

On completion of this element, candidates should be able to demonstrate understanding of the content through the application of knowledge to familiar and unfamiliar situations. In particular, they should be able to:

A7.1 Explain how to use internal and external information sources in identifying hazards and the assessing of risk.

A7.2 Outline the use of a range of hazard identification techniques.

A7.3 Explain how to assess and evaluate risk and to implement a risk assessment programme.

A7.4 Explain the analysis, assessment and improvement of system failures and system reliability with the use of calculations

A7.5 Explain the principles and techniques of failure tracing methodologies with the use of calculations.

Contents

Relevant statutory provisions

Management of Health and Safety at Work Regulations (MHSWR) 1999 - Regulation 3

Management of Health and Safety at Work Regulations (Northern Ireland) 2000 (as amended)

Sources of reference

Reference information provided, in particular web links, was correct at time of publication, but may have changed.

Controlling the risks in the workplace, HSE, http://www.hse.gov.uk/risk/controlling-risks.htm

The health and safety toolbox: How to control risks at work, HSE, http://www.hse.gov.uk/toolbox/index.htm

Managing for health and safety, HSG65 (third edition), HSE Books ISBN: 978-0-717664-56-6, http://www.hse.gov.uk/pubns/priced/hsg65.pdf

Reducing risks, protecting people, R2P2, HSE books, ISBN: 978-0-7176-2151-0, http://www.hse.gov.uk/risk/theory/r2p2.htm

Risk assessment, A brief guide to controlling risks in the workplace, INDG163, HSE Books, http://www.hse.gov.uk/pubns/indg163.htm

The above web links along with additional sources of reference, which are additional to the NEBOSH syllabus, are provided on the RMS Publishing website for ease of use - www.rmspublishing.co.uk.

A7.1 - Sources of information used in identifying hazards and assessing risk

Accident/incident data and rates

INCIDENCE, FREQUENCY, SEVERITY, PREVALENCE

Information used in the identification of hazards and assessment of risk may be reactive, after the event, and include accident/incident data and rates. The value of this data to risk assessment is that it is information on experience of actual harm. This may indicate the presence of hazards, potential for future harm and may provide an insight into the effectiveness of current controls related to the risk.

Accident/incident data can indicate hazards that might not have been identified; either missed or was not foreseen. Accident/incident rates are often too general to do other than indicate likelihood of harm. Unless the rates are derived from events related to specific risks they can only be used in a general way. In order for them to be even more meaningful they need to be derived for the specific work activity or location being considered in the risk assessment process.

Presuming the rates derived are relevant to specific risks, the incidence rate will indicate how likely harm is going to occur to the workforce from that risk. The frequency rates can show how often harm is likely to occur and the severity rate can show the organisation's current experience in the level of harm arising from the risk. These rates can help evaluate the risk by establishing the likelihood that they will occur and the severity of the loss.

Prevalence rates are often produced related to specific ill-health conditions, for example, asbestosis, pneumoconiosis, vibration white finger, and provide an indication of risk of harm. Ill-health data is also often produced for specific work groups, which will particularly aid the risk assessment process.

External information sources

Information that can assist in risk evaluation can be sourced from external bodies. These might include trade associations, Health and Safety Executive (HSE) publications and reference to BS EN standards, for example, guarding and maintenance.

HSE

The HSE produces a range of documents including the legal series, guidance notes (HSG, INDG and HSR), information sheets and leaflets. The range available is printed in a free catalogue and may be ordered by phone, fax or e-mail. Information is also available from the web site, www.hse.gov.uk, where many of these documents are available to download for free as pdfs and provide a ready source of information.

The website contains comprehensive information on risks and their control in specific industries. Following the HSE's simplification initiative, the website also contains a good number of sample risk assessments, including those relating to shops, motor vehicle work, offices, call centres, factory maintenance work and warehousing.

OTHER RELEVANT GOVERNMENTAL AGENCIES

Information can be sourced from the Office for Nuclear Regulation, Office of Rail and Road (formerly Office of Rail Regulation), Local Authorities, Environment Agency, the Fire Authority, the Health and Safety Laboratory and Public Health England.

EUROPEAN SAFETY AGENCY

The European Agency for Safety and Health at Work was set up by the European Union (EU) to bring together the vast reservoir of knowledge and information on occupational health and safety and health related issues and preventive measures.

Its website, www.osha.europa.eu/en, includes a European Risk Observatory, which aims to identify emerging health and safety risks. The website also provides access to a database of risk assessment tools and fact sheets on health and safety risks.

ILO

The International Labour Organisation (ILO) deals with labour standards, fundamental principles and rights at work. They have a website and a database of all types of work-related information.

WHO

The World Health Organisation (WHO) is the United Nation's specialised agency for health and was established in 1948. It has information on health, where health is defined as 'a state of complete physical, mental and social well-being and not merely the absence of disease or infirmity.'

PROFESSIONAL BODIES AND TRADE BODIES

There is a range of professional and trade bodies that can provide information to help with risk evaluation: the TUC, the British Safety Council (BSC), the Institution of Civil Engineers (ICE), the Institution of Occupational

Safety and Health (IOSH), The Royal Society for the Prevention of Accidents (RoSPA), Faculty of Occupational Medicine, the Institute of Electrical Engineers, The Confederation of Business Industry (CBI), and many more.

Internal information sources

COLLECTION, PROVISION, ANALYSIS AND USE

Damage, injury and ill-health data

The information that is gathered during accident investigation, health surveillance and absence records can provide useful data when evaluating risk. The existence of several cases of dermatitis could show a problem with a particular cleaning fluid, whereas the existence of one case, when analysed, may show the problem could have been caused by the person's own individual sensitivity or possibly pursuit of a leisure time hobby.

Near-miss information

Near miss reporting is an important way of finding out where the potential for harm is. The reports can be analysed and show where further investigations are needed.

For example, a box falling off the racking and almost hitting an employee passing by may show that the racking is damaged or tilting, or the boxes have been badly stacked. This highlights areas of risk that might be thought to have been dealt with.

Maintenance records

These records show where and when equipment used by workers or provided as a health and safety precaution breaks down and therefore can be a source of failure rates and reliability data.

Uses and limitations of external and internal information sources

EXTERNAL INFORMATION SOURCES

- Information from external sources is beneficial in as much as it utilises the experience of others.
- One organisation's incident data can enable another organisation to evaluate their risk.
- The wealth of information could not possibly be amassed by just one organisation.
- It is time consuming and expensive to collect data, so having access to it externally is cost effective.

There are ***limitations*** to the use of external data:

- There is so much of it; it may be difficult to decide what is actually relevant to a particular situation.
- You have to know what you are looking for or you may become lost within the plethora of detail.
- Although it saves resources in collecting your own data, it may require extensive resources to reach a conclusion from it.

INTERNAL INFORMATION SOURCES

- Internal information will be relevant to the issues being considered.
- It will be easy to access and easy to ask questions should clarification be required.

There are also ***limitations*** to the use of internal data:

- Often data held by the organisation may be statistically too small to be of any use. For example, one lost time accident in a year is not going to provide enough information to be all that useful.
- Collecting data is time consuming and can easily get mislaid.
- Organising a near miss reporting procedure is often difficult to do as employees sometimes do not see (or have not been trained in) the usefulness of reporting something that did not cause any damage.

A7.2 - Hazard identification techniques

Regulation 3 of the Management of Health and Safety at Work Regulations (MHSWR) 1999 places an absolute duty on the employer to make a suitable and sufficient assessment of the risks to health and safety of his employees at work and to non-employees who are affected by his undertaking. The self-employed similarly have an absolute duty to assess the risks to them and to others they may affect.

The first step in assessing risks is to identify the hazards that can create a risk. A hazard may be considered to be:

"Something that has the potential to cause harm (loss)"

> *"The potential to cause harm, including ill-health and injury, damage to property, plant, products or the environment, production losses or increased liabilities."*

Figure A7-1: Definition of hazard. *Source: Managing for health and safety, HSG65 (3rd Edition), HSE.*

For example, hazards may be electricity, trailing cables and working at height. A substance would be a hazard in a particular way depending on the properties of the substance; a strong acid would present a corrosive hazard.

There are many ways hazards can be identified. The methods used will need to be reviewed in order that the risk assessor uses the most suitable ones that ensure all the hazards associated with a task can be identified.

For complex activities it may be necessary to break the activity down into its component parts by, for example, task analysis or hazard and operability study methods.

Observation

Hazard identification can be carried out by observation of the activity and noting the hazards as they occur in the actual work setting. Observation methods have advantages over carrying out a desktop exercise, for example using the health and safety manual for a machine for the assessment of its use.

This could be because operators may have developed their own method of working, contrary to the manual's instructions and training. It is important that work conditions and practices are observed to identify the actual hazards present in the workplace.

Task analysis

Task analysis is a methodology used in the assessment and reduction of human error. There are two main forms of analysis that can be used, hierarchical and step analysis.

The methods generally take into account the structure of the task and the observable operator behaviour, in various levels of detail.

Analysing the tasks can highlight where there could be errors and eliminate them before they can occur. They can be used at the design stage and when a task is already being conducted.

They can also be used in accident investigation, where how the task was carried out can be compared to how the task was supposed to be carried out, and differences identified. This can assist with the identification of immediate causes of loss.

Initially it is necessary to identify tasks for analysis. When considering tasks that relate to a complex process or large machine, this could mean the analysis of a number of tasks, for example:

- Installation.
- Normal operation.
- Breakdown.
- Cleaning.
- Adjustment.
- Dismantling.

By breaking down the item being analysed into tasks, the hazards associated with each task could then be identified more easily and considered more thoroughly. This analysis can help to identify contingent hazards which can arise from the system, components, checking or maintenance failures, as well as continuing hazards, i.e. those present continuously.

Examples of hazards include:

- Mechanical hazards.
- Electrical hazards.
- Thermal hazards.
- Noise and vibration.
- Radiation.
- Toxic materials.
- Ergonomic design.
- Psychosocial

HIERARCHICAL TASK ANALYSIS

This method considers the overall objective of the task and then how it is organised to achieve that objective. It is a 'top down' approach that involves looking at all the sub-tasks that need to be carried out in order to achieve the top objective.

At each stage it may be necessary to break the sub-task down even further. At each stage the potential for error can be noted. This breakdown of the task can help the assessor decide on the control measures needed to reduce the risk of loss and can then be written into a safe system of work or safe procedure.

Example of application:

Task: preparing a soft boiled egg

1) Boil water.
2) Place egg in water.
3) Boil for 3/4 minutes.
4) Remove egg from water.

A decision would have to be made as to whether any of the stages need to be expanded, for example, does there need to be more information at stage one about the size of the pan and the amount of water to be added.

Similarly, stage three may need to be expanded, details of the equipment for timing may have to be provided.

See figure ref A7-2 which shows an expanded view of the stages involved in the task. At each of the stages the likelihood of errors that can lead to a hazard can be noted and this will lead to a decision on the appropriate control measures. For example, what would happen if at stage 1.3 the gas did not stay lit? There is also a potential for scalding at stages 2.2 and 4.1.

As well as identifying hazards, the analysis could be used for the development of a written safe procedure on the preparation of a soft boiled egg. Of course there are far more onerous tasks than boiling an egg, but the methodology remains the same.

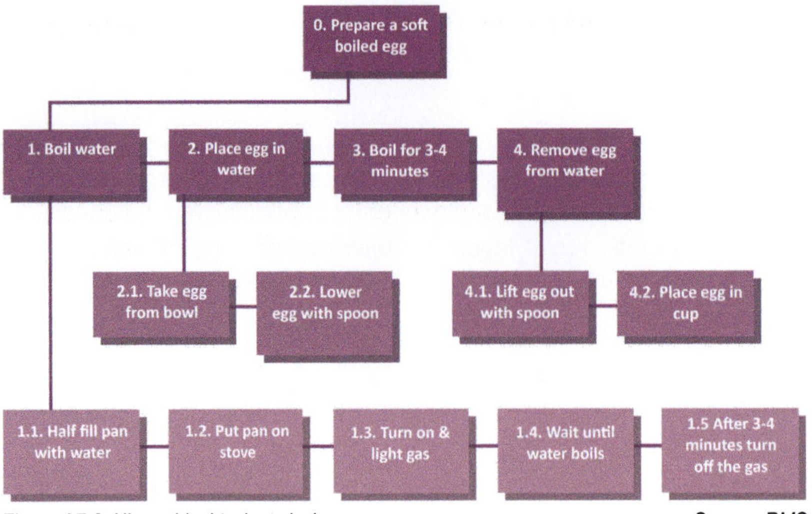

Figure A7-2: Hierarchical task analysis. *Source: RMS.*

STEP TASK ANALYSIS

This is a variation on the method described above; it has been used for some years in the field of health and safety and is often called 'job safety analysis'. Step task analysis consists of a formal step-by-step review of the work to be carried out. All aspects of the task should be considered and recorded in writing to ensure that nothing is overlooked. When used as part of the hazard identification stage of the risk assessment process, the objective is to establish the hazards/risk factors and controls at each stage of the procedure to ensure a safe and healthy result.

The process can be considered by looking at the acronym SREDIM:

- Select the task to be analysed.
- Reduce the task to steps.
- Evaluate the risk at each of the steps by identifying the hazards, take into account the existing controls and add in further control measures to reduce the risk.
- Develop a safe system of work (SSW).
- Implement the SSW.
- Monitor and review.

Example of the application of step task analysis

Consider dispensing sulphuric acid from a 200 litre drum:

Job step	***Risk factor***	***Control action***
Select drum.	Error in selection.	Check drum identification and hazard label.
Fit eye, hand, body and foot protection.	Splash of acid to eyes/face. Hands, forearms, body.	Select and fit suitable personal protective equipment (PPE), for example, chemical goggles/visor, protective suit, long gloves and boots.
Loosen and remove cover from drum discharging hole.	Acid vapour on ventilation.	Ensure local ventilation operating. Controlled release of cover.
Insert drum discharging tap in discharging hole.	Tap is in open position when fitted or poor seal, leading to leaks and spills of drum contents.	Ensure tap is closed before inserting, align carefully, and tighten securely.
Locate horizontal drum holder.	Unstable load.	Ensure drum holder on level and even floor.
Position drum horizontally.	Unstable load.	Use mechanical aid and kinetic technique. Drum in correct contact with drum holder.
Place drip tray beneath drum discharging tap.	None if discharge tap closed and securely fitted.	Selection of tray size and careful positioning to ensure tray will catch drips from tap and containers being filled.

Job step	***Risk factor***	***Control action***
Release cover from drum ventilation hole.	Leakage of acid from ventilation hole if drum has been overfilled or contents expanded. Breathing acid vapour released during ventilation.	Controlled release of ventilation cover.
Position container to receive discharge of acid.	Unsuitable container.	Acid resistant container. Suitable opening to receive acid. Correct identification/hazard label.
Discharge acid.	Risk of splash of acid/overfilling container.	Full attention, regulate discharge.
Close discharge tap.	Small quantity of drips may come from tap after it is closed. Contact with tray or container when reaching over.	Position of feet and hands. Clockwise direction. Allow last drips to fall from tap into container.
Replace ventilation hole cover.	Breathing acid vapour released during ventilation. Contact with tray or container when reaching over.	Position of feet and hands. Clockwise direction, close securely but not over tighten.
Check PPE for signs of acid contamination.	Risk of acid burns when removing PPE.	Wash off any acid contamination with water at emergency shower. Remove PPE to storage.

Figure A7-3: Example of the application of step task analysis. *Source: RMS.*

Step task analysis is also used to analyse tasks for the purpose of establishing procedures and developing training requirements. In this case, the factors being analysed may not only be hazards and controls, but might include knowledge requirements, which indicate good skill, influences on behaviour and learning methods. The findings from conducting a step task analysis may be used to develop a safe system of work, which can then be used in training and as a checklist during the task.

Checklists

Checklists are useful in the hazard identification process when used in conjunction with carrying out an inspection or observation of work activities. They can list the variety of hazards that are possible in an area. They provide a means to remind the assessor of the range of hazards that might be found and help prevent a competent assessor failing to perceive hazards in a workplace they are familiar with. A range of checklists may be produced for different work situations to make them more relevant to what is being assessed. The checklists may have a space on the form where hazards not listed can be added for future reference.

The checklist format provided by the HSE in their guidance to the Manual Handling Operations Regulations (MHOR) 1992, to help when conducting a manual handling risk assessment, helps the assessor to identify the features of work that present a manual handling hazard. This is a very useful way of ensuring that the assessor considers and comments on each of the features.

Failure tracing techniques

Failure tracing techniques are useful for detailed analysis of hazards; they provide data on how hazards may arise and their possible effects. Failure tracing techniques include hazard and operability studies, fault tree analysis and event tree analysis.

See also - Failure tracing methodologies - later in this element.

Importance of employee input

Employee input into the identification of hazards and assessing risks, for the work practices they are employed to carry out, is essential if all the hazards and risks to them are to be identified and assessed. Experienced workers will often be able to bring issues such as near miss events, which may never have been reported, into the analysis. They will also have a good understanding of how work is done in practice, rather than how it should be done in theory. They will have experience of how work hazards and risks change in different circumstances, for example, out of hours, in the dark, hot/wet weather, work at height and below ground. They will also be able to comment on the effectiveness of existing risk controls systems, such as permits to work, guarding and personal protective equipment.

Employee input should be gained whenever a new process, process change or the introduction of new plant or equipment is to be considered. They will often be able to provide valuable practical contributions to the analysis, identify the implications of the change and predict hazards that might arise. Hazard identification and risk assessment should not be delegated to employees, but their contribution should be sought and recognised. This will ensure that any changes in risk controls are more readily accepted and used by the employees.

Finally, and just as importantly, it will promote employee involvement and participation, leading to an improving health and safety culture.

A7.3 - Assessment and evaluation of risk

Key steps in a risk assessment process

Risk may be considered to be:

"The risk is the chance, high or low, that somebody could be harmed by these and other hazards, together with an indication of how serious the harm could be."

Figure A7-4: Definition of risk. *Source: HSE.*

There are a number of definitions for the term risk assessment, including one used in the NEBOSH glossaries:

"The process of evaluating the risks to safety and health arising from hazards at work."

Figure A7-5: Definition of risk assessment. *Source: NEBOSH, Technical Glossary.*

Risk assessment is the process to evaluate the risk(s) arising from the hazard(s), identifying preventive and protective measures, taking into account the adequacy of any existing controls, and deciding whether or not the risk is acceptable.

Risk assessment is therefore an analytical process that identifies hazards, who may be harmed and in what way they may be harmed, it also takes into account factors that make the risk more likely and those that make it less likely. The HSE gives guidance on assessment and evaluation of risk in its publication 'Risk assessment - A brief guide to controlling risks in the workplace' INDG163 the five steps are:

Step 1 Identify the hazards.

Step 2 Decide who might be harmed.

Step 3 Evaluate the risks.

Step 4 Record the significant findings.

Step 5 Regularly review the assessment.

STEP 1 IDENTIFY THE HAZARDS

It is important not to overcomplicate the process, observe the workplace and determine what could reasonably be expected to cause harm. Employees and their representatives should be consulted to see if they have any particular concerns regarding current work practices and the hazards they present. Additional information on likely hazards can be obtained from the HSE website and trade associations. Check manufacturer's instructions for both substances and work equipment in use, these will often give information on the less obvious hazards, for example, the longer term health effects of a sensitiser substance, and sources of noise and vibration.

All aspects of the work activity must be considered, routine and non-routine activities, for example, maintenance, cleaning, loading or unloading vehicles, changes in production cycles, response to emergencies and special activities like 'motivational days'. Care should be taken to include hazards to those that work offsite and in other people's premises, for example as contractors, home workers, mobile employees and those that visit members of the public.

STEP 2 DECIDE WHO MIGHT BE HARMED

The identification of the ***persons at risk*** is often best done by grouping individuals into work categories or areas, rather than on an individual basis, for example, warehouse order pickers, store room workers and despatch workers. Consider those people who may not be in the workplace all the time, for example, visitors, maintenance workers or contractors. Some workers may not work core hours, such as cleaners and shift workers. Also consider remote working individuals, such as welfare workers, street cleaners, delivery workers, consultants and sales people. For some workers, who work in a shared workplace, it will be necessary to consult with another employer to determine any risks associated with joint occupancy. Particular care should be taken to identify where members of the public may be harmed.

The Management of Health and Safety at Work Regulations (MHSWR) 1999 place a statutory duty to take account of certain categories of worker who may be at particular risk, for example, new and young workers, new or expectant mothers and people with disabilities.

STEP 3 EVALUATE THE RISKS AND DECIDE ON PRECAUTIONS

Factors affecting probability (likelihood) and severity of risk

There are many factors affecting the level of risk that someone may be exposed to. They all need to be taken into account to arrive at a risk rating or an evaluation of the risk.

Hazard exposure outcomes

The effect of a hazard will depend on its ability to cause a degree of harm in the circumstances in which it is encountered. A simple example would be to compare the slip or trip risk associated with:

1) A spillage of water onto a carpet on a level surface in a quiet office.
2) An equivalent spillage of water onto vinyl flooring on a level surface in a frequently used corridor.
3) The same quantity of water spilt onto a steep flight of steps, also frequently used.

The probable outcomes of these three scenarios would be, respectively, (1) typically a possible slip, but no fall and no injury, (2) a likely slip with possible fall and minor injury and finally (3) a high likelihood of slip and major injury. Taking another example, being exposed to a defective electrical cable can result in contact with a live conductor and electric shock, with varying degrees of harm, including death. The severity of harm depends on a number of factors: the current passing through the body, the length of time the current passes through the body, which part of the body is affected, etc.

Skin contact with a strong acid will destroy the tissue it comes into contact with and all people will be equally affected if exposed to the same substance. However, exposure to flour dust will cause a sensitised reaction (asthma) only in certain people. Therefore, it is difficult to predict who might be affected in a work group exposed to flour dust.

Persons and numbers exposed

The more people exposed to a hazard, for example, to flour dust, the greater the probability that someone will be affected. If it were the case that one in fifty people exposed to flour dust developed asthma and one hundred were similarly exposed, then it can be postulated that two people could develop asthma. The level of awareness, training and competence of the person exposed to hazards can greatly influence the probability of being affected by a hazard. New employees may be more likely to make mistakes that could lead to contact with a hazard and the general public, particularly vulnerable people with characteristics that may limit their perception, may make contact with hazards that workers would observe and respond to.

Duration of exposure

The longer someone spends exposed to a hazard, the greater the chance (the risk) that they will make contact with it and be harmed. For example, consider a damaged electrical conductor to an office lamp. If the circuit is protected by a fuse and someone makes contact with the exposed conductor there is a high risk of fatal electric shock. However, if the circuit to the lamp is protected with a residual current circuit breaker, the moment that the circuit is discharged by touch the device will interrupt the power within 30ms, reducing the risk of a fatal shock significantly compared with that of the much slower fuse. Some exposures, such as inhalation of an airborne substance and noise, will present an increasing risk of harm from initial exposure until the risk is removed. The level of harm will be dependent not only on the quantity of the substance or noise level, but the duration of exposure.

Frequency of exposure

The likelihood of contact with a hazard increases the more a person is exposed to it. The more frequent the exposure, the more likely other variables come into play. Variables like distractions and fatigue can increase the risk, for example, on one of the many times a person is exposed to the hazard of pipework set at head height they may be distracted and forget to avoid the pipework, contact with the hazard is made and harm is done to the person.

Current controls

The current controls can greatly influence the probability (likelihood) and severity (consequences) of risks and their consideration is essential in risk evaluation. Deciding on how well hazards are controlled may involve comparing the hazard controls to ***required risk control standards*** and good practice, much of which is readily available on the HSE web site. Risk control standards may be influenced by specific legal requirements to take certain action or to meet certain standards. The more technical the matter being considered the more likely it is that the risk assessor will have to consider specific information contained in documents like British Standards, industry standards and codes of practice. The type of control used may greatly affect the risk; some controls will provide general protection from contact with a hazard for many classes of person, whereas, others may be limited to the worker and be less reliable. For example, in response to a noise hazard an organisation may provide acoustic sound absorption around equipment or provide a worker with personal hearing protection.

Risk evaluation

Once the hazards have been identified, the associated risks can be evaluated. This is done by considering the probability (likelihood) of the hazard being realised and severity (consequences) of the loss, taking account of the people who may be affected and the efficiency of the existing controls.

The type and depth of evaluation used will depend on the level of complexity involved in the task or area under consideration. Risk evaluation usually includes a form of risk rating in order to determine how significant the risk is and how this compares with other risks. This allows priorities to be set and can be used to help justify the cost of control measures. Once controls have been implemented, the risk can be re-evaluated and re-rated to confirm the effect of the change.

Relativistic methods of risk rating

A relativistic rating system is a qualitative way of evaluating risks in order to allow comparisons between the various risks, thereby indicating priorities for management action. Relative rating systems can take the form of a simple category system of high/medium/low or take the form of a risk rating system based on a numerical matrix. In this method, the factors affecting probability and severity are considered and the resultant level of risk is classed as high medium or low or similar. This method of rating is often used in qualitative and semi-qualitative risk assessment processes.

Probabilistic methods of risk rating

The probabilistic method of risk rating considers the chance of the harm arising in a quantitative way. This uses mathematical and statistical techniques in order to aid the decision making process. The level of risk may be arrived at purely by failure rates, for example, one component in a system of two components may fail with the probability of 1 in 10 or 0.1 and the next with a probability of 1 in 50 or 0.02. Therefore, the probability that the whole system will fail will be 0.1 X 0.02, which is 0.002 or 1 in 500.

Techniques such as Event Tree Analysis and Fault Tree Analysis can be used to establish a probability figure or failure rate of an undesired event occurring, for example, a 1 in 10,000 chance of there being an undetected fire in a laboratory in a given year. This has the advantage of being far less subjective than a relativistic method. Probabilistic methods also allow the risk assessor to be more authoritative when justifying the expenditure on controls and can be used to satisfy the enforcing authorities with regard to the control of complex risks. Although statistics and probabilistic methods can be invaluable as an aid to improved decision making they should be used with great care. There are several possible drawbacks involved in using them, including the following:

- Statistical probability information may be difficult or impossible to obtain.
- Probability statistics may contain bias, be interpreted to produce a desired result or can be misinterpreted through limited knowledge of the subject.
- The probability statistics may not encompass all the information and variables that affect the specific risk assessment being conducted. They should not be used as a substitute for executive evaluation and decision making. The most that this method can achieve is the narrowing of the range in which subjective judgement must operate.

In summary, when conducting risk evaluation, account is taken of the factors that influence the probability (likelihood) and severity (consequence). Once the risk has been evaluated it is necessary to decide on suitable precautions for the risk. If existing measures are not adequate consideration should be given to what measures would reduce the risk sufficiently. In deciding appropriate measures, it may be necessary to take account of specific legal requirements. Not all legal requirements are 'so far as is reasonably practical', which allows some balance and judgement on the part of the employer, some may carry 'absolute' and 'practicable' duties.

STEP 4 RECORD THE FINDINGS AND IMPLEMENT THEM

Those employers with five or more employees must ***record*** the results of their risk assessment, identifying the controls in place and any limitations, and this should be brought to the attention of those who might be affected and their representatives.

"3(6) Where the employer employs five or more employees, he shall record:
(a) The significant findings of the assessment.
(b) Any group of his employees identified by it as being especially at risk."

Figure A7-6: Requirement to record risk assessments. *Source: MHSWR 1999, Regulation 3.*

The record should represent an effective statement of hazards and risks and that it should lead management to take action. The record does not have to be in writing, it may be recorded by other means, such as electronically. It must be retrievable for use by the organisation and available to enforcing authority officers. It is essential that it remain retrievable, even when the chosen technology of recording changes. Significant findings should include:

- Preventive and protective measures in place.
- What further action is necessary to reduce risk sufficiently?
- Proof that a suitable and sufficient risk assessment has been made.

The findings should clearly identify where measures in place are not sufficient to control risk; this should be used to ***formulate actions*** necessary to improve risk control. It may be that a series of actions need to be considered in order to establish sufficient control. When implementing the findings, the most significant risks should be addressed first. Priority of risk will depend on a number of factors. Risks may be prioritised as needing attention in the "immediate, medium or long term", the higher the risk rating the higher the priority. Priority may also depend on legal compliance, for example, the risk of injury may be low, but the risk of enforcement action might be high. Prioritisation may also depend on public acceptability and perception of risk. The risk may be placed in a low category, but the public's view that it is high may give it high priority for example, the siting of microwave telecommunication repeater aerials on school buildings.

It is also important to ***prioritise actions***; this will ensure the actions that can have the most immediate effect on reducing the risk are taken first. For example, where there are risks of tripping from a trailing cable in a walkway the actions may be - removal of the cable or use a cable cover (immediate), provide re-routing for cables or extra electrical sockets (medium term), review equipment needs and socket locations in the department (long term).

STEP 5 REVIEW THE ASSESSMENT AND UPDATE IF NECESSARY

The risk assessment should be periodically reviewed and updated In addition, a review of risk assessments should be carried out following any significant changes to a workplace.

> *"(3) Any assessment such as is referred to in paragraph (1) or (2) shall be reviewed by the employer or self-employed person who made it if:*
> *(a) There is reason to suspect that it is no longer valid.*
> *(b) There has been a significant change in the matters to which it relates; and where as a result of any such review changes to an assessment are required, the employer or self-employed person concerned shall make them."*

Figure A7-7: Requirement to review risk assessments. *Source: MHSWR 1999, Regulation 3.*

Examples of circumstances that would require the review of the validity of a risk assessment are:

- When the results of monitoring (accidents, ill-health effects, environmental) are adversely not as expected.
- A change in process, work methods (introduction of shifts) or materials.
- Changes in personnel.
- Changes in legislation.
- The introduction of new plant or technology.
- New information becoming available.
- As time passes - the risk assessment should be periodically reviewed and updated. A common approach would be no longer than five years.

Use and limitations of generic, specific and dynamic risk assessment

GENERIC RISK ASSESSMENT

Generic risk assessment formalises thinking and gives consistency to the way an organisation works or an event or activity might take place. Such risk assessments can help ensure consistency when the same activity is carried out regularly or by different people. They offer basic core principles, which can then be applied and adapted. They can encourage a more proactive approach to preparation, through training and equipment that might be carried.

The disadvantage of a generic risk assessment, written for everyone, is that it can become static and forgotten in practice. In a constantly changing workplace, for example, construction sites, the hazards will change as the construction project moves through its different phases. The risk is that these may not be properly identified and therefore not adequately evaluated, managed and controlled.

SPECIFIC RISK ASSESSMENT

Specific risk assessment focusses on a defined and restricted hazard. Some examples would be:

- Risk assessment of isocyanate spray painting: Breathing in isocyanate fumes can cause respiratory sensitisation. Additionally, the worker may not know of their sensitivity, and because the harm is irreversible.
- Risk assessment of a confined space: Most legislative systems have a clear definition of a confined space, which combines the nature of the 'space' with the presence of certain hazards in it.
- Permit to Work: where there is non-routine work such as maintenance or repairs, which expose the workers to extra risk from hazards such as electricity or movement of dangerous machine parts, risk assessment must be carried out in order to control the specific extra risks before the work begins. The benefit of specific risk assessment is that it uses specialised knowledge to identify the unique hazards present and defines the necessary risk controls.

The obvious disadvantage is that its narrow focus does not identify other hazards. For example - confined space risk assessment does not include other hazards such as exposure to excessive noise or vibration.

DYNAMIC RISK ASSESSMENT

A growing number of workers have to make quick decisions regarding risk and identify control measures in high pressure, potentially stressful environments. Dynamic risk assessments are designed to deal with situations where conditions (and therefore risk) change rapidly.

For example recovery operatives generally working alone by the side of the road, often near moving traffic or alternatively in remote locations, and frequently in the dark or in poor weather, it's imperative that they are trained to continually assess risks so that they can make judgements about the safest way to work. Following the drowning of Max Palmer at Glenridding Beck 26 May 2002 the HSE recommended that activity leaders be *"competent in dynamic risk assessment for the activity, so that they can continuously evaluate the implications of changing conditions."*

A definition of a dynamic risk assessment used by emergency responders is: 'The continuous process of identifying hazards, assessing risk, taking action to eliminate or reduce risk, monitoring and reviewing, in the rapidly changing circumstances of an operational incident'.

The emergency services use dynamic risk assessment, because in their work they meet high risk situations which may change frequently and quickly. Their approaches may differ.

Fire services - recognising immediate hazards they expect to be present at some stage in any fire. Modifying their reactions to meet the change in conditions. Also, using rehearsals and constant practice to maintain the competence of their personnel.

Police services - applying judgment to choose a tactic which has previously been rehearsed. Using previous rehearsals and drills to decide their response.

The advantage of this approach is that the emergency services are able to respond safely and quickly to dangers. But dynamic risk assessment depends heavily on the competence of their personnel.

The drawback is the method's heavy dependence on the competence of the responders.

It is very important to learn from the experience of new or unexpected events with which the responding personnel were unfamiliar. The review then leads to further training and practice so that similar future circumstances can be safely handled.

Meaning of 'suitable and sufficient'

To be suitable and sufficient a risk assessment has to identify all the potential causes of harm in the workplace. This includes not only dangerous substances and equipment, but also work processes and work organisation. It must cover not only the risk of immediate injury but also the long-term risks to health. The risk assessment must be systematic, comprehensive and address what actually happens in the workplace, not what the employer says should happen. It should also cover non-routine activities such as spillages and cleaning, what happens during breakdowns, loading and unloading and maintenance operations. The assessment must cover everyone in the workplace, including groups like maintenance staff, security staff, cleaners, and visitors, and it should take account of preventative or precautionary measures that have already been introduced and check that these are effective in reducing risk.

Good practice, originally established in the withdrawn Approved Code of Practice to Regulation 3 of the Management of Health and Safety at Work Regulations (MHSWR) 1999, suggests that in order to be suitable and sufficient a risk assessment should:

a) 1) Identify the risks arising from or in connection with work.

   2) The level of detail should be appropriate to the risk.

   3) Once risks are assessed, insignificant risks can be ignored.

   4) Risks arising from routine activities associated with life in general, unless the work activity compounds or significantly alters the risk, may be ignored.

   5) The level of risk arising from a work activity should determine the degree of sophistication of the risk assessment.

   - For small businesses presenting few or simple hazards, a simple approach based on informed judgement and reference to appropriate guidance will be enough.
   - In many intermediate cases the risk assessment will need to be more sophisticated. Some areas of the risk assessment may require specialist advice, for example, relating to complex processes or where specialist analytical techniques are required to measure worker exposure to the hazard and to assess its impact.
   - Large and hazardous sites will require the most developed and sophisticated risk assessments, particularly where there are complex or novel processes.
   - Risk assessments must consider all those who might be affected, for example, contractors and members of the public.

b) Employers and the self-employed are expected to take reasonable steps to help themselves identify risks, for example, by looking at appropriate sources of information or seeking advice from competent sources.

c) The risk assessment should be appropriate to the nature of the work and should identify the period of time for which it is likely to remain valid.

Although not part of the definition of the term 'suitable and sufficient', if those conducting a risk assessment cannot confirm competence this could call into question whether a risk assessment is in fact suitable and sufficient. Risk assessments should be carried out by competent person(s) who have the necessary experience or training in hazard identification and carrying out risk assessments, knowledge of the process or activity, and good communication and reporting skills. The individual(s) should have the right attitude to carry out the task, the ability to interpret legislation and guidance, and technical knowledge of the plant or equipment involved.

Limitations of risk assessment processes

There are a number of limitations associated with the traditional risk assessment process:

Uncertainties permeate the whole process. These may include indecision regarding the particular technique to be used and as a consequence using an inappropriate model which lacks the detail required for the task or takes too much time and effort for low risk subjects.

Uncertainty due to lack of knowledge at individual, corporate or national level. If data regarding the toxicity of a substance was gathered from testing on animals it might be difficult to apply the information to assess the effects on humans.

Historic accident and ill health statistics can accurately predict the likelihood of a future occurrence if circumstances do not change. The absence of this information or the infrequent nature of past events will inevitably lead to subjective decisions being made.

Erring on the safe side. Adopting the view that it is better to be safe than sorry leads to an overcautious approach to risk. Indeed, many risk assessment systems lend themselves to this approach and it is easy to take the view that every incident could be fatal. This approach can lead to results that are highly distorted.

Individual differences in risk perception. This can be a barrier to good decision making because there is a tendency to see risk where there is none. Research suggests that people are more likely to accept risk if it is linked to benefits; driving a car from A to B is a much greater risk than living near to a nuclear power station. However people are far more willing to use a car than live next to a nuclear facility. Perception is also linked to control; passengers in cars perceive a much higher risk than drivers.

Differences of opinion regarding tolerability of risk. Everyone has their own view on risks that they would tolerate. Without realising it, many people base their opinions on a risk v benefit analysis. For example, nuclear power v increased utility bills, the risks associated with skiing off piste v the thrill of doing something different. In the workplace, risk acceptance is very controversial and leads to difficulties in decision making.

Not acting on the findings or carrying out a risk assessment to justify a decision that has already been made.

Temporary and non-routine situations

For activities where the nature of the work may change fairly frequently or the workplace itself changes and develops (such as a construction site), or where workers move from site to site, the risk assessment might have to concentrate more on the broad range of risks that can be foreseen.

When other less common risks arise, specific risk assessments will be required to enable planning and worker training to take account of those risks and enable them to be controlled.

Employers who control a number of similar workplaces containing similar activities may produce a 'generic' risk assessment reflecting the core hazards and risks associated with these activities. 'Generic' risk assessments may also be developed by trade associations, employers' bodies or other organisations concerned with a particular activity. Such 'generic' assessments may be applied by employers or managers at each workplace, but only if they:

- Satisfy themselves that the 'generic' assessment is appropriate to their type of work.
- Adapt the 'generic' risk assessment to the detail of their own actual work situations, including any extension necessary to cover hazards and risks not referred to in the 'generic' assessment.

Consideration of long-term hazards to health

Risk assessment should not only include immediate harm such as injury or acute response to a harmful substance.

A suitable and sufficient risk assessment must therefore include the possibility of the following:

Chronic health effects, which usually occur after repeated exposure over days, weeks and months must also be considered. Examples of these conditions would be noise-induced hearing loss and hand arm vibration syndrome.

Many occupationally acquired diseases produce symptoms of the condition many years after the exposure that caused the disease ('Long Latency Effects'). Examples include the development of mesothelioma (following asbestos exposure), other lung cancers and pneumoconiosis (silicosis, coal worker's pneumoconiosis, and asbestosis, etc.) which can occur decades after exposure has ceased.

Principles of and differences between qualitative, semi-quantitative and quantitative assessments

QUALITATIVE

The qualitative method of risk assessment can generally be considered to not involve equations based on numbers to evaluate risk. Using a qualitative approach, a cable trailing in a walkway used by infirm older patients may be evaluated as being a 'high' risk, because it is felt that somebody is bound to trip over it and the consequences would be severe. This is a useful, straightforward approach; though it is a relatively subjective method it can easily lead management to take action to control the risk, which is the purpose of risk assessments.

A qualitative approach to risk assessment is provided in the guidance to the Manual Handling Operations Regulations (MHOR) 1992. It requires the consideration of factors that lead to a manual handling hazard and the subsequent evaluation of the risk in terms of low, medium and high.

Section B: Lifting and carrying - More detailed assessment, where necessary

Questions to consider:	If yes, tick appropriate level of risk			Problems occurring from the task (Make rough notes in this column in preparation for the possible remedial action to be taken)	Possible remedial action, eg changes that need to be made to the task, load, working environment etc. Who needs to be involved in implementing the changes?
	Low	Med	High		
Do **the tasks** involve:					
• holding loads away from trunk?					
• twisting?			✓	*1. Twisting when picking up the box.*	*Remind operator of need to move feet.*
• stooping?		✓			
• reaching upwards?			✓	*2. Stooping when placing box on pallet and stooping when picking box up from the conveyor.*	*Adjust pallet height – Review availability of rotating, height adjusting equipment and raise height of conveyor.*
• large vertical movement?	✓				
• long carrying distances?	✓				
• strenuous pushing or pulling?	✓				
• unpredictable movement of loads?	✓			*3. Sometimes extended reaching when placing boxes on pallet.*	*Provide better information and instruction.*
• repetitive handling?	✓				
• insufficient rest or recovery?	✓				*Review mechanical handling equipment to eliminate manual lifting.*
• a work rate imposed by a process?	✓				
Are **the loads:**					
• heavy?			✓	*4. Load too heavy. Is the weight of the load a problem for customers too?*	*Review product and customer needs with a view to improving product design.*
• bulky unwieldy?	✓				
• difficult to grasp?		✓			
• unstable/unpredictable?	✓			*5. Smooth cardboard boxes are difficult to grasp.*	*Provide boxes with hand grips.*
• intrinsically harmful (eg sharp/hot)?	✓				

Figure A7-8: Qualitative approach, in manual handling risk assessment. *Source: HSE, L23.*

Semi-quantitative

The semi-quantitative method is half way between the simplest approach, which uses qualitative methods, and the quantitative approach, which uses probability statistics.

Different people may have a different perception of risk, as a result of training, life experiences and background. Semi-quantitative methods try to establish parameters that ensure a more consistent approach to risk rating and attempt to overcome individual differences. Risk can be rated according to the probability (likelihood) and severity (consequence) of harm resulting from a hazard.

Thus, ***risk rating = Probability (likelihood) x Severity (consequence)***.

Where the:

Probability (likelihood): How likely that this harm will occur.

Severity (consequence): The degree or amount of any resultant harm.

Risk rating: The level of the remaining risk after current controls have been taken into account.

Probability (likelihood) categories

5.	Almost Certain	Absence of any management controls. If conditions remain unchanged there is almost a 100% certainty that an accident will happen (for example, broken rung on a ladder, live exposed electrical conductor, and untrained personnel).
4.	High	Serious failures in management controls. The effects of human behaviour or other factors could cause an accident but is unlikely without this additional factor (for example, ladder not secured properly, oil spilled on floor, poorly trained personnel).
3.	Medium	Insufficient or substandard controls in place. Loss is unlikely during normal operation; however, it may occur in emergencies or non-routine conditions (for example, keys left in fork lift trucks; obstructed gangways; refresher training required).
2.	Low	The situation is generally well managed - however occasional lapses could occur. This also applies to situations where people are required to behave safely in order to protect themselves but are well trained.
1.	Improbable	Loss, accident or illness could only occur under freak conditions. The situation is well managed and all reasonable precautions have been taken. Ideally, this should be the normal state of the workplace.

Severity (consequence) categories

The consequence can be assessed on a scale of 1 to 5.

5.	Major	Causing death to one or more people. Loss or damage is such that it could cause serious business disruption (for example, major fire, explosion or structural damage). Loss/damage in excess of (£___________).
4.	High	Causing permanent disability (for example, loss of limb, sight or hearing). Loss/damage in excess of (£___________).
3.	Medium	Causing temporary disability (for example, fractures). Loss/damage in excess of (£___________).
2.	Low	Causing significant injuries (for example, sprains, bruises, lacerations). Loss/damage in excess of (£___________) for example, damage to fixtures and fittings.
1.	Minor	Causing minor injuries (for example, cuts, scratches). No lost time likely other than for first aid treatment. Loss/damage in excess of (£___________) for example, superficial damage to interior decorations.

The amounts set in each of the above categories will depend on the size and type of organisation. Senior management should decide the loss figures for each category. Using the formula stated above (Risk Rating = Probability (likelihood) x Severity (consequence)) the risk rating can be calculated. It will fall into the range of 1 - 25. This risk rating is used to prioritise the observed risks.

The risk rating is then classified as follows:

Risk Rating	1-9	Low
Risk Rating	10-15	Medium
Risk Rating	16-25	High

There are many variations on this method and other variables can be used in the matrix, which may move the approach closer to the quantitative method.

Quantitative

These methods are mainly used for complex risk and are based on equations related to probabilities. In its simplest form it may be used to establish probability of failure of a component that has safety critical implications, such as a pressure relief valve. Data is sourced from the manufacturer on how often a failure can be expected. If they provide information that they had experience of 10 failures out of 10,000 valves in the last year this would provide a quantitative probability of failure of 0.001 for the valve. Two of the techniques that can be used as a quantitative method are Event Tree Analysis (ETA) and Fault Tree Analysis (FTA). Estimated failure rates are used in a sequence of events to evaluate the probability of a resultant failure.

Hazard and operability studies (HAZOP) can be used to identify the undesired events and the probability of failure rates may be worked out from the reliability rates of the components given by the manufacturer. They may also come from maintenance records of failure, be based on historical data or informed opinion.

ETA is often used for systems that have components that operate in sequence. For example, the risk (including probability) of a fire burning out of control, taking into consideration the existing control measures, i.e. a detector, alarm and sprinkler system, can be evaluated using ETA. The reliability data for each component part may be obtained from the manufacturer; therefore, the failure rate of each component can be calculated.

FTA can be used to show the probability of a top event (risk/loss) occurring by evaluating the risk at each level of the tree. For example, an investigation of an explosion following the release of a flammable gas, would consider all the root causes which lead to the release and similarly all the root causes which lead to the presence of a source of ignition. ***See also - Failure tracing methods - later in this section.***

Organisational arrangements for implementing and maintaining an effective risk assessment programme

It is important to establish clear organisational arrangements for implementing and maintaining an effective risk assessment programme. This will form a major part of the health and safety policy. Written ***procedures*** should be established for the risk assessment process which the organisation is going to follow. The procedures should clearly state who is responsible for carrying out risk assessments, identify the ***training*** and ***competencies*** they should have, what method should be used for collecting data, for example, template forms to be completed, method of evaluation, qualitative, semi-quantitative, quantitative and how the information is to be analysed and ***recorded***. The procedure should identify those who are ***responsible for prioritising and authorising follow-up actions***, and the method for ***monitoring*** progress. The procedure should specify the frequency and circumstances when the risk assessments should be ***reviewed***.

Acceptability/tolerability of risk - principles in HSE's 'Reducing risks, protecting people' (R2P2)

ACCEPTABILITY OF RISK

There is risk in all parts of life and everyday risks are generally acceptable. The risk of tripping over your feet as you walk along an obstacle-free surface would be acceptable. Acceptability of risk relates to societal, individual, organisational and regulatory acceptability. Essentially risks that are acceptable are ones where no additional controls to reduce the risk may be necessary. It is necessary for organisations and regulators to determine the acceptability of risk when determining the organisation's compliance with legislation.

Where the duty in law is qualified by 'so far as is reasonably practicable (SFAIRP)', the acceptability of risk will depend on the level of risk. The higher the risk the less acceptable it is. Of particular importance in the interpretation of SFAIRP is Edwards v The National Coal Board (1949). This case established that a computation must be made in which the quantum of risk is placed on one scale and the sacrifice, whether in money, time or trouble, involved in the measures necessary to avert the risk, is placed in the other. If it can be shown that there is a ***gross disproportion*** between them, the risk being insignificant in relation to the sacrifice, the person upon whom the duty is laid discharges the burden of proving that compliance was not reasonably practicable.

> *"In seeking to apply this case law, when regulating or producing guidance on compliance with duties qualified by all injunctions embodying the concept of 'reasonable practicability' such as SFAIRP, ALARP (as low as reasonably practicable), ALARA (as low as reasonably achievable), HSE believes that such duties have not been complied with if the regime introduced by duty holders to control risks fails the above 'gross disproportion' test."*

Figure A7-9: Application of 'reasonably practicable'. *Source: HSE, Reducing risks, protecting people (R2P2).*

Organisations may set their own levels of acceptable risk using, for example, a risk rating method. If on a scale of 1 to 25, the risk rating up to category 8 may be deemed by the organisation to be acceptable. However, this may be seen as an oversimplification, as the concept of SFAIRP includes the fact that if additional action can easily be taken by the organisation, and it would reduce the risk rating further, then this must be done.

The HSE consider acceptable, tolerable and unacceptable (intolerable) risk in their framework approach, 'Tolerability of Risk' (TOR). This is depicted by the following diagram, which shows that there is a graduation between these three conditions and that where the risk lies on the graduation depends on the perspective of the person considering it.

> *"Getting it right depends to a large extent on the criteria adopted for deciding whether a risk is unacceptable, tolerable or broadly acceptable. It is, therefore, not surprising that a lot of effort has been spent in developing such criteria."*

Figure A7-10: Deciding tolerability. *Source: HSE, Reducing risks, protecting people (R2P2).*

The HSE, in the document 'Reducing risks, protecting people' explain the three main criteria it uses when considering tolerability. An equity-based criterion starts with the premise that all individuals have unconditional rights to certain levels of protection.

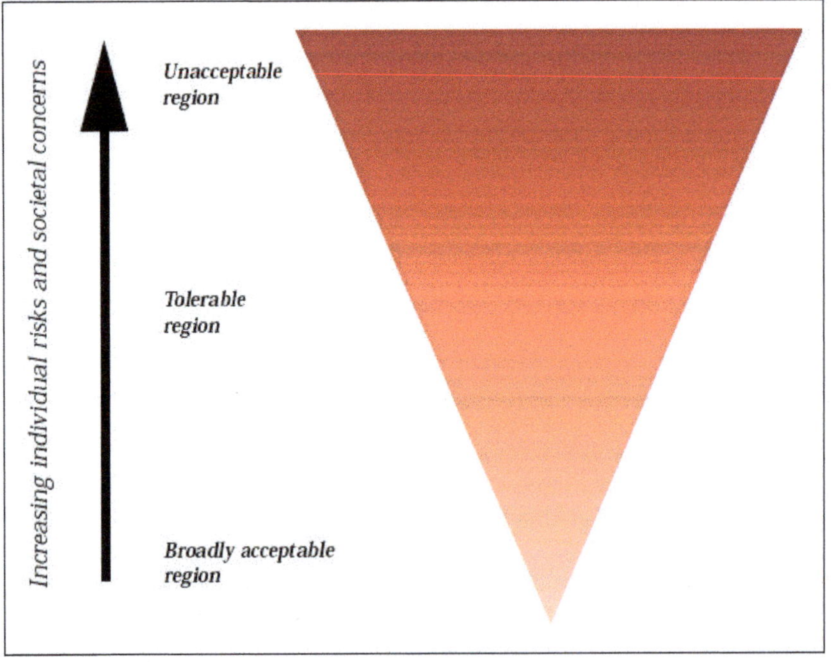

Figure A7-11: Framework for tolerability of risk. *Source: HSE, Reducing risks, protecting people (R2P2).*

This leads to standards, applicable to all, held to be usually acceptable in normal life. A utility-based criterion considers the benefits of the measures to prevent the risk of harm, and the cost of the measures. The technology-based criterion 'essentially reflects the idea that a satisfactory level of risk prevention is attained when 'state of the art' control measures (technological, managerial, and organisational) are employed to control risks whatever the circumstances.'

TOLERABILITY OF RISK

To say that a risk is 'tolerable' does not mean that it is acceptable. Tolerability relates to a person's willingness to accept a risk so that they might gain certain benefits. Thus saying that risks are tolerable does not mean that they are negligible, or not in need of proper control, or are something to be ignored. Even tolerable risks need constant review in order to reduce them further if possible.

Individual risk is concerned with the consequences relating to an individual. For example, what is the chance that a person and his/her family might be injured or suffer some other loss through a large-scale accident involving a power station? This is considered by calculating the risk to any individual who lives within a certain radius of the station.

The risks that are regulated by society as a whole must also be considered. Societal risks are based on the concept of securing general benefits for the community. The increased risks posed to people living near a power station, for example, are offset by the benefit that society enjoys through the use of electricity. These risks are still considered or minimised through appropriate controls and the geographical site of individual power stations. An estimation of societal risks can be measured by considering, for example, the chance of a large accident causing a defined number of deaths or injuries.

The HSE document, ***'The Tolerability of Risks from Nuclear Power Stations'***, broadly defines a risk of death of 1 in 1,000 per annum as the most that is ordinarily accepted by substantial groups of workers within any industry in the UK. This level is exceeded only by fishermen and small sub-groups such as divers, demolition workers and helicopter pilots.

To put this into context, the following table gives some indication of levels of fatal risk:

1 in 100	Risk of death from 5 hours of solo rock climbing every weekend.
1 in 1000	Risk of death due to work in high-risk groups within relatively risky industries such as mining.
1 in 10,000	General risk of death in a traffic accident.
1 in 100,000	Risk of death in an accident at work in the very safest parts of industry.
1 in 1 million	General risk of death in a fire or explosion from gas at home.
1 in 10 million	Risk of death by lightning.

Figure A7-12: Levels of fatal risk (average figures approximated). *Source: HSE - "The Tolerability of Risks from Nuclear Power Stations".*

Thus, it would seem reasonable to say that a risk of death from accidents at work greater than around 1 in 1,000 is unacceptable to any other than fairly exceptional groups of workers. Similarly, a risk of death arising from work activities to a member of the public of 1 in 10,000 would be the maximum that should be tolerated. The individual acceptance of these risks depends on a number of factors (for example, who is in control, who benefits and how much benefit accrues).

"HSE believes that an individual risk of death of one in a million per annum for both workers and the public corresponds to a very low level of risk and should be used as a guideline for the boundary between the broadly acceptable and tolerable regions. Indeed many activities which people are prepared to accept in their daily lives for the benefits they bring, for example, using gas and electricity, or engaging in air travel, entail or exceed such levels of residual risk."

Figure A7-13: Tolerability of risk of death. *Source: HSE, Reducing Risks, protecting people (R2P2).*

The border between the graduation of tolerable and intolerable risk is not a clear one. Society can become focused by hazards that pose relatively quite low levels of risk to most individuals, but could impact particularly on vulnerable groups, such as the young or the elderly or particularly susceptible individuals. Furthermore, exposure to an activity may result in a low level of average risk to any one individual, but the totality of such risks across the affected population would not be socially or politically acceptable.

INTOLERABLE RISKS

Most examples of where risks have been found to be intolerable and action taken to prohibit them relate to substances. The prohibition of 2-naphthylamine in 1967 eliminated a potential risk of cancer. Similarly, the Control of Lead at Work Regulations (CLWR) 2002 prohibit certain glazes. In the same way, the Control of Asbestos at Work Regulations (CAW) 2006 established prohibitions for the importation, supply and use of certain asbestos products. The regulations did not prohibit the use of asbestos entirely as products that had an established history of being in use before the regulations were introduced were exempt. However, through the introduction of the regulations, the process of risk reduction had begun and the risk was reduced to a tolerable level as perceived at the time.

Subsequently European legislation in the form of a European Union Regulation called Registration, Evaluation, Authorisation and restriction of Chemicals (REACH) has been made. REACH is supported by the European Regulation (EC) No 1272/2008 on classification, labelling and packaging of substances and mixtures (CLP) which came into force on 20 January 2009 and became effective in the UK in 2015. The method of classifying and labelling chemicals it introduced is based on the United Nations' Globally Harmonised System (GHS).

The CLP Regulation ensures that the hazards presented by chemicals are clearly communicated to workers and consumers in the European Union through classification and labelling of chemicals.

REACH prohibits the supply and use of asbestos fibres and articles containing asbestos fibres that have been added intentionally. Extending prohibitions further than previous legislation. This shows the continuing and increasing intolerability of asbestos related risks. However, some minerals may contain asbestos as impurities and are not covered by this prohibition, for example olivine, vermiculite, talc, dolomite, hermatite, iron ore, basalt, marble, chrome ore, magnetite and wollastonite. Also, the use of articles containing asbestos fibres that were already installed and/or in service before 1 January 2005 continue to be permitted by REACH until they are disposed of or reach the end of their service life. Furthermore, REACH contains a derogation in Article 67 allowing supply and use for scientific research and development. This depicts European society's current level of intolerability to asbestos risks.

A7.4 - Systems failures and system reliability

Meaning of the term 'system'

A system is a set of inter-related elements that starts with an input that undergoes some process and results in an output, which has a monitoring (feedback) loop that evaluates the input, process and output in order to make adjustments that ensure the intended resultant output is provided.

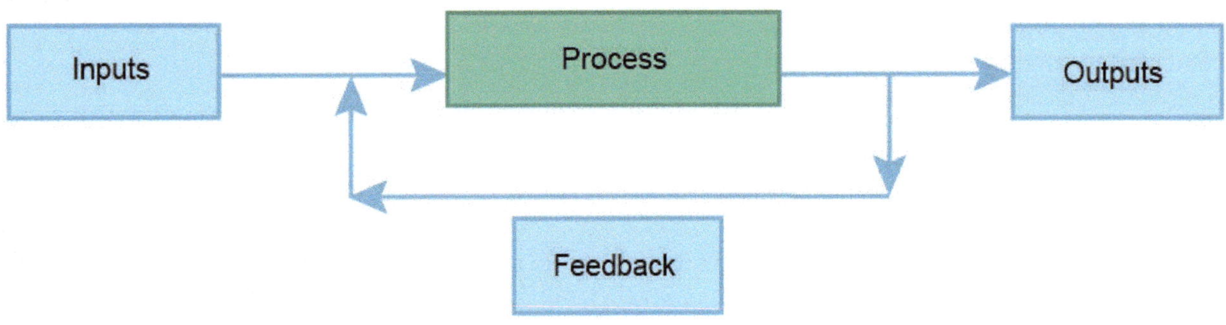

Figure A7-14: System. *Source: RMS.*

Principles of system failure analysis

HOLISTIC AND REDUCTIONIST APPROACHES

The ***holistic*** approach examines the system as a whole. The rationale behind this is the argument that the behaviour of a system cannot be satisfactorily explained merely by the study of its component parts. Thus, the philosophy behind this approach is that the system is synergistic - that is, the whole is more than the sum of its parts. An example of the holistic approach is a hazard and operability (HAZOP) study, which most effectively examines the impact of failure on the whole system, although it can be used at the sub-system level. Thus, the impact malfunction of components is considered in terms of the effect on the system as a whole.

The advantage of holistic analysis is that it draws attention to the relationship between components and the qualitative aspects of the system that may not be appreciated by the reductionist approach.

The ***reductionist*** approach to analysis of any system involves dividing the system into its individual component parts. An individual component is then isolated and the factors that influence it are examined. Thus, the reductionist approach to failure analysis begins at the component level, by identifying basic modes of failure for each component within the system. Each component failure is then evaluated for its effects on the system. An example of this technique is failure mode and effects analysis (FMEA).

DIFFERENCES BETWEEN SYSTEMIC AND SYSTEMATIC ANALYSIS

Systemic:	Of the body as a whole.
Systematic:	Methodical; according to plan, not casually or at random.

This is similar to the differences between the holistic and reductionist approach to complex failure analysis considered above.

A ***systemic*** analysis considers the whole system whereas a ***systematic*** analysis considers the component parts of the system in a logical, methodical way that considers each stage of the system in turn. Systemic analysis allows for an intuitive approach that may perceive relationships in an apparently unconnected array of activities.

APPLICATION TO ACTUAL EXAMPLES

Systems may be 'hard' or 'soft'. A 'hard' system is, for example, a piece of equipment, a vehicle, a plant that processes a chemical, a smoke or heat detector, an automatic fire alarm system, etc. A 'soft' system is a management strategy, for example, to manage contractors, recruit new staff, deal with emergencies, etc. Systems are made up of sub-systems; parts that work together to make the system work.

A forklift truck (FLT) may be considered a system that operates within an environment. It can be analysed as to its failure within that environment by consideration of the task it performs and its movements: hitting overhead pipes, overturning, a wheel going down a pothole, etc.

The FLT (the system) can be broken up into a number of sub-systems. The lifting mechanism and the braking mechanism are two of the sub-systems, i.e. parts of the greater whole. Analysis of the lifting mechanism, for example, using FMEA, will show which parts may fail and how. The failure of component parts in the lifting mechanism, the chain for example, may cause the lifting mechanism to fail while carrying a load or not be able to operate at all. The effect on the system, the FLT, may just be that it cannot perform its duties as a lifting machine and there may be some product damage. Analysis of the braking mechanism could show that failure of the component parts, the brake pads for example, may make the brakes inoperable. The effect on the system (the FLT) could be catastrophic and cause loss by way of personal injury, plant and building damage, loss of product and the inoperability of the FLT.

An analysis of the system and sub-system would have shown the probability of a failure of a weak temporary link (solvent pipeline) between vessels at ***Flixborough,*** where a major flammable chemical explosion occurred. A HAZOP and/or an FMEA would have shown how the temporary change introduced could have adversely affected the system.

A system to manage contractors will be made up of sub-systems for example, choosing the contractor, site induction, issuing work orders, safe procedures and permits-to-work, supervising them, etc. On analysis, each of these sub-systems can be further broken down so, for example, issuing permits to work becomes a system that can be broken down into sub-systems such as support documentation, permit authorisers and so on until a further breakdown is no longer possible. Controls can be considered at each stage.

The ***Piper Alpha*** explosion was due, in part, to management failure within a permit-to-work system. The permit-to-work sub-system had not been analysed to find out what effect its failure in operation could be on the whole system.

If a car and passenger ferry such as the ***Herald of Free Enterprise*** leaving port is considered as a system, then for the numerous problems that developed, for example, management attitude 'to turn the ship around quickly', failure to follow standard procedures in the sub-systems inevitably led to disaster.

Use of calculation in the assessment of system reliability

GENERAL POINTS

Reliability is not confined to single components only; the evaluation of systems, simple or complex, is needed. Evaluation techniques are used for designing reliable systems or for gaining reasonable assurance in advance that a design will meet certain safety and reliability requirements.

System reliability calculations are based on two important operations:

1) As precise as possible a measurement of the reliability of the components used in the system environment.
2) A calculation of the reliability of some complex combinations of these components.

The calculation of the reliability of components will take account of the probability of failure, for example where a component had a probability of failure of 3 in every 100 this could be represented by a probability of failure of 0.03. This can then be translated to derive a reliability for the component of 0.97.

SERIES SYSTEMS

Generally, the reliability of the series system is computed by multiplying the probability of each component part that it will survive its operating life. Multiplying the probability of survival of A and B in the diagram will give the probability of the system surviving.

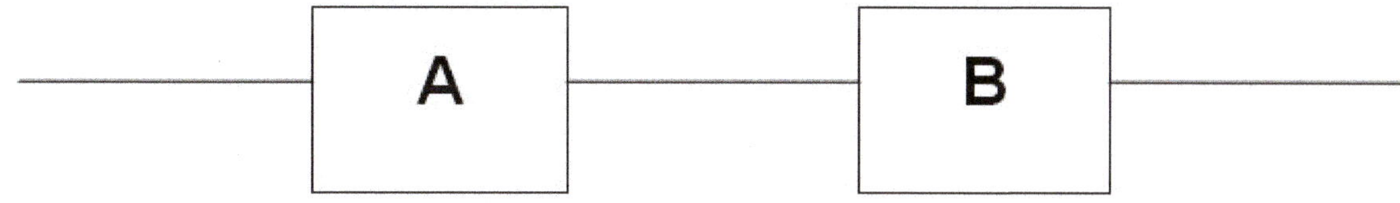

Figure A7-15: Series system. *Source: RMS.*

If there are number of components in the series, the system reliability is given by:

Rs = R1 X R2 X R3 X......X Rn

For example, if A had a reliability of 0.98 and B had a reliability of 0.97 the reliability of the whole system, a series system would be:

Rs = 0.98 x 0.97 = 0.95

A series system has drawbacks, in that one failed component causes failure to the whole system. The accident at Markham Colliery is an example of this, where the one failed component (central pin in the winding mechanism to the lift cage) caused the whole system to fail and hence led to the disaster.

PARALLEL SYSTEMS

If very high system reliabilities are required, for example, on lifts or aircraft, the designer must duplicate components and sometimes whole circuits to fulfil such requirements. Use must be made of parallel reliabilities, called parallel redundancy. In series systems, all components must function for the system to operate. In parallel systems, it is necessary for only one component to operate for the system to operate. The systems can be evaluated in a similar, but more complex, way to the series equation in order to work out the probability of success or failure.

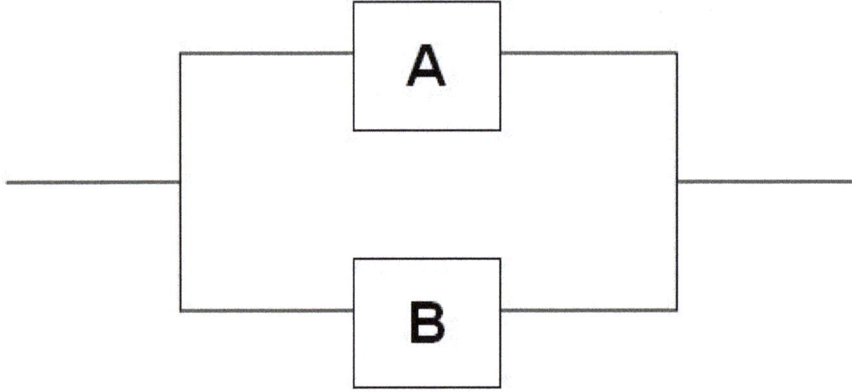

Figure A7-16: Parallel system. *Source: RMS.*

If there are number of components in parallel, the system reliability is given by:

Rs = 1 - (1 - R1) x (1 - R2) x (1 - R3) ..... x (1 - Rn)

For example, if A had a reliability of 0.95 and B, being an identical component, had a reliability of 0.95 the reliability of the whole system, a parallel system would be:

Rs = 1 - (1 - 0.95) x (1 - 0.95)

Rs = 1 - (0.05) x (0.05)

Rs = 1 - 0.0025

Rs = 0.99 (2 decimal places)

Three components in parallel are represented by:

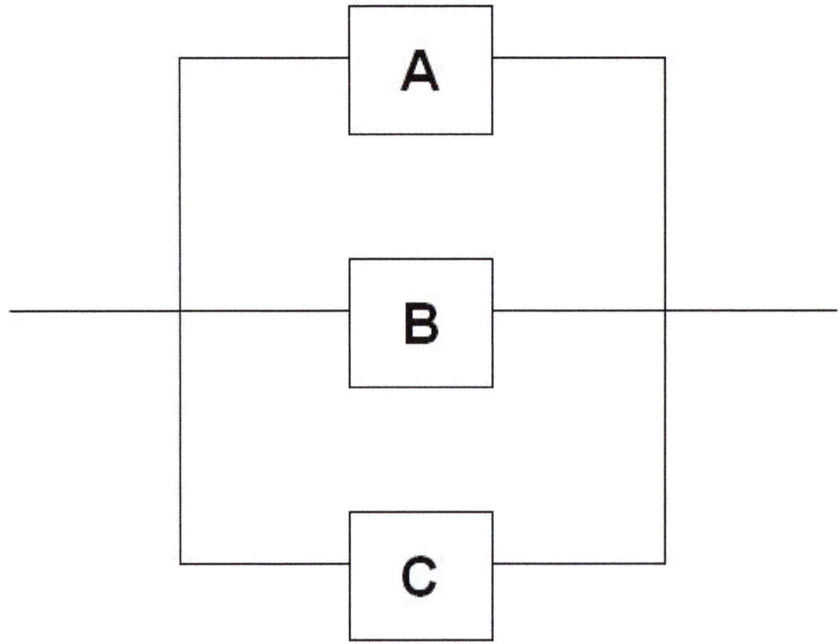

Figure A7-17: Three components in parallel. *Source: RMS.*

MIXED SYSTEMS

The reliability of a mixed system involving series and parallel components can be calculated as follows:

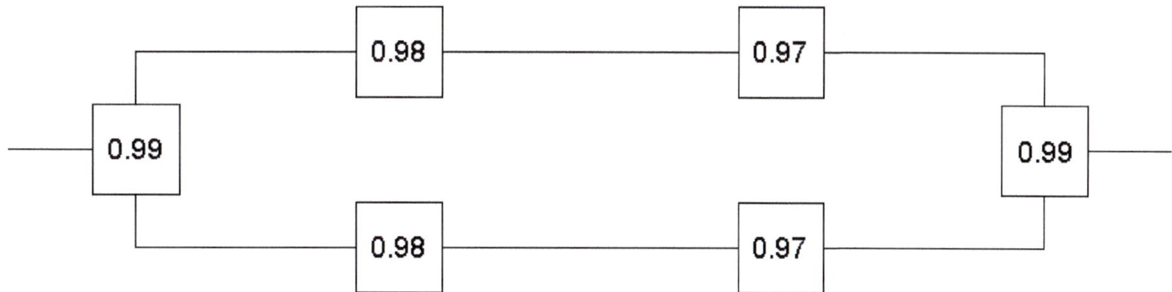

Figure A7-18: The reliability of a mixed system (series and parallel). *Source: RMS.*

Where there are systems with components in parallel, with a switch component in and a switch component out, the reliability of the whole system can be calculated as follows:

Reliability of each of the parts of the parallel system, comprising of two components in series, is 0.98 x 0.97 = 0.95

Reliability of the two systems in parallel is:

= 1 - (1 - 0.95) x (1 - 0.95)

= 1 - (0.05) x (0.05)

= 1 - 0.0025

= 0.99 (2 decimal places)

Reliability of the whole system, including the two switches, is then calculated as a series system, with the parallel part of the system represented by a reliability of 0.99:

= 0.99 x 0.99 x 0.99

= 0.97

Reliability of the whole system = 0.97, which is less reliable with the inclusion of the two switches, but is more reliable than the single series system (which would have been 0.99 x 0.98 x 0.97 x 0.99 = 0.93).

Probability of failure of whole system = 1 - 0.97 = 0.03

There are some cases when not all reliability models can be reduced to the series, parallel and stand-by. There are combinations of components that are neither series, parallel, nor stand-by. Again, there are formulae available to evaluate the reliability of these systems.

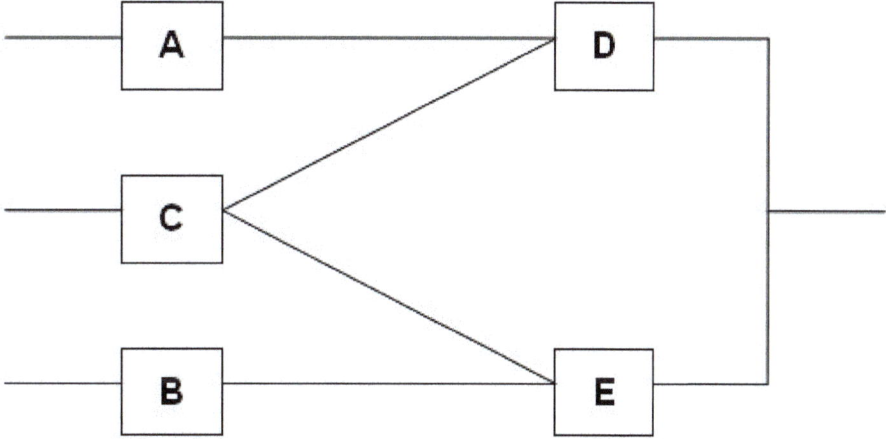

Figure A7-19: Mixed system. *Source: RMS.*

In the above system, the two equal paths, A-D and B-E, operate in parallel so that if at least one of them is good, the output is assured. But, because units A and B are not reliable enough, a third equal unit, C, is inserted into the system so that units D and E are supplied with the necessary input. Therefore, the following operations are possible:

A ⇨ D or **B ⇨ E** or **C ⇨ D** and **C ⇨ E**

COMMON MODE FAILURES

Failure can occur when an external factor affects the systems. For example, in 1980 a leg sheared off an oil platform (rig), the platform tilted and the generator was knocked off. The generator supplied all the electrical power. The rig was plunged into darkness, the resulting fire could not be dealt with and there was no means of escape. The systems for dealing with the various emergency situations should have been so completely apart from each other that the risk of them all being affected by the same external factor could have been avoided.

PRINCIPLES OF HUMAN RELIABILITY ANALYSIS

Human reliability analysis (HRA) was developed as a means of quantifying the interaction between human and engineering systems. It aims to improve the understanding of the contribution made by all people to engineering systems.

This is significantly greater than the previous man-machine interface analysis. HRA is a systematic evaluation of the factors that influence the performance of humans in the workplace and is used to identify potential human errors, their causes and their subsequent effects.

Human behaviour

Human behaviour is controlled by the requirements of the task, the working environment in which the task is performed and the skills, experience, abilities and training of the human performing the task.

Human knowledge can be assigned to one of three categories:

1) Skill based knowledge and action.
2) Rule based knowledge and action.
3) Formal reason based knowledge and action.

The skills required from and exercised by an operator of equipment include sensing, perceiving, prediction, familiarity with controls, and decision making. The table below summarises some of the errors made by people contributing to the failure of engineering systems over their lifetime.

Examples of errors

Designer	*Procurer*	*Manufacturer*	*Distributor*	*End user*
Failure to select most appropriate control-display component.	Incorrect specification of system type.	Incorrect production of design.	Incorrect handling and storage.	Use of incorrect operating and maintenance.
Human is assigned a task that they cannot do or that is best done by machine.	Incorrect specification of parameters.	Use of wrong machining and fabrication techniques.	Incorrect monitoring of environment during transit.	Use of incorrect parts in maintenance.

Figure A7-20: Examples of errors. *Source: RMS.*

Applications of human reliability analysis

For example, in the computer industry, the requirement for reliability safety critical software has led to wide studies for reducing errors by software developers. Based on the findings of HRA, computers are used to check the design decisions of humans and hence improve reliability of software.

In the transport industry, analysis of incidents involving the transport of dangerous substances showed that over 40% of these incidents were directly caused by human error, rather than by mechanical failure.

Implementation of human reliability analysis

The following steps are taken when implementing HRA.

1) List the system functions to be performed.
2) For each function, list all the actions that must be performed by operators to implement that function.
3) Determine whether any operator tasks have already been specified by a customer procurement document.
4) Determine sub-tasks that must be performed to implement higher order tasks.
5) Describe each task and sub-task in terms of a verb that indicates the nature of the action being performed by the task, for example, monitor, check, read, etc.
6) For each task and sub-task identify error-likely situations.
7) Rank the errors in the order of severity.
8) Document the results.

Type and nature of results

The results of HRA are both qualitative and quantitative. They are qualitative in the respect that they give a listing of the errors and the contributing factors that are likely to be encountered during normal and emergency operation.

They also identify systems interfaces that are affected by specific errors. Human reliability analysis gives quantitative results in the respect that it provides relative ranking of errors based on probability of occurrence and severity of consequences.

Methods for improving system reliability

GENERAL POINTS

Regular testing is important to identify fail to danger situations before a demand is placed on the system, but there may be a danger in testing systems too frequently.

Three risk factors need to be considered if the testing of systems is carried out more frequently:

1) Continuance of plant operation whilst protective devices are removed for testing.
2) The likelihood of error/damage in removal/replacement of equipment by test engineers.
3) Potential exists to replace a good component with a component that is defective.

1984 Bhopal, India

A major disaster occurred to the local population following the uncontrolled release of highly toxic vapour when a storage tank containing methyl isocyanate was contaminated with water (major respiratory injury results at exposure to levels at parts per billion). 2,500 people were killed and it is estimated that approaching 500,000 people were injured from the release. The most significant aspect that contributed to the loss was that the plant was operating whilst the protective trip systems and devices were disconnected for testing and maintenance work. It is essential to include testing/maintenance considerations at the design stage, when conducting HAZOPs.

Typically, the trip system/devices should be tested to a frequency that gives at least the same reliability as the protective device.

The human element is a very important consideration when calculating the frequency of the test of protective systems and devices, and as a minimum, task analysis should be carried out. For high potential risk considerations other techniques such as psychological profiling will often be required. Formulae are available to calculate these probabilities, but these are outside the scope of this book.

USE OF RELIABLE COMPONENTS

By considering the data available, the most reliable components can be chosen. The most expensive may not always be the most suitable. For example, the reliability of a system may be lessened because a valve has a probability of failure of 0.05. Using the same level of quality valve, a second valve can be placed in parallel so that if one fails the other will come into use. This means, for a dangerous situation to arise, both valves will have to fail together, the probability of which is 0.05 X 0.05 = 0.0025. The probability has gone from 1 in 20 to 1 in 400.

Another method of improving reliability is derating. This means having a safety margin. For example, a resistor rated for a current of 2 amps is run at 1.4 amps. The derating is 1.4/2 = 0.7. This is a typical figure for electronic parts.

QUALITY ASSURANCE

Components that are made to a specification within a quality assurance system, where rigorous testing is carried out are more likely to be reliable. It is important that reliability is built in from the design stage, through manufacture, through the building of the system, its use, making changes, and maintaining the system and its component parts.

PARALLEL REDUNDANCY

System reliability can be improved by duplication of critical components in parallel, so that when one component fails (becomes redundant) the other operates to maintain control for example dual braking fitted to a road vehicle. When designing parallel systems consideration should also be given to diversity, which is where alternative mechanisms provide the desired action, for example a pneumatic system may be provided in parallel to support an electronic system.

STANDBY SYSTEMS

A stand-by system is where a component or unit is operating and one or more units are standing by to take over the operation, should the primary one fail. The supporting components or units are normally idle and begin to operate only when the primary unit fails. An example of this is a second water supply pump at a water treatment plant provided to be available by valve switching when required following a failure or maintenance of the primary pump.

These systems therefore require failure sensing and switchover devices to put the stand-by component or unit into operation. The reliability of the sensing/switching device must be taken into consideration, as well as the failure rates of the stand-by units. Standby is therefore only appropriate when the system has the capacity to deal with the time for switchover, which may be either manual or automatic.

MINIMISING FAILURES TO DANGER

Minimising failure to danger is a design consideration at the HAZOP stage. This requires consideration of the operation of safety critical equipment and devices at the time of failure of the device, detector communication link or responder.

Consider a remotely operated steam control valve to a reactor. Such valves are usually controlled using compressed air operating the valve against a spring. A typical arrangement would be that the valve closes at zero supply pressure and opens when compressed air at a known pressure is applied.

At the design stage it would need to be determined which is the more critical to the process, failure to open or failure to close. The reliability of the equipment and air supply would be assessed and the design would ensure the correct sequence of events to ensure a failure to a safe condition results.

Certain high risk facilities, for example, the nuclear industry, may need complex duplication of protective devices to meet the societal level of assurance. Such systems are designed to include 'majority voting systems' (MVS). Majority voting systems use voting logic. For example, consider the temperature of a reactor; it may be decided that this needs to be controlled critically. Any message from the thermocouple that shows overheating will set a sequence in play to shut down the reactor. However, some of the messages from the sensor are spurious. The solution is to install three thermocouples to monitor a single process variable. A voting logic analyser will receive readings from each of the sensors, but any response action (for example reactor shut down) is only initiated by any two of the three sensors. The logic behind this is that there is a probability that one sensor could send a spurious message, but the probability that two will send spurious messages at the same time is remote. Conversely, there is a probability that one sensor might not pick up a change in temperature, but the probability that two will not at the same time is remote.

DIVERSITY

Diversity is concerned with the identification of common mode failure. Common mode failure may cause a parallel system failure, for example if the parallel system requires compressed air and both components are supplied from the same air supply, then failure of the air supply will result in the failure of both components in the parallel system. A single failure of electrical supply may affect many different supply voltages to a plant, including low voltage instrumentation, if not provided with separate back up. Care needs to be taken when considering the need for either redundancy or diversity. The best design systems strive to reduce the need for protective devices as the more protective devices that are included the more likely you are to have spurious trips.

For example, a power press utilises three protective devices to safeguard the operator: electrical, mechanical and pneumatic. If an electrical failure occurred the electric interlock would fail, but in such a diverse system, the mechanical interlock would protect the operator from injury. If both the electrical and mechanical protective devices failed, then the pneumatic interlock would become the protective device. A common fault in the system would be unlikely to affect all three types of safeguarding system. The machine would be guarded by at least one of the systems and the probability of all failing would be very low.

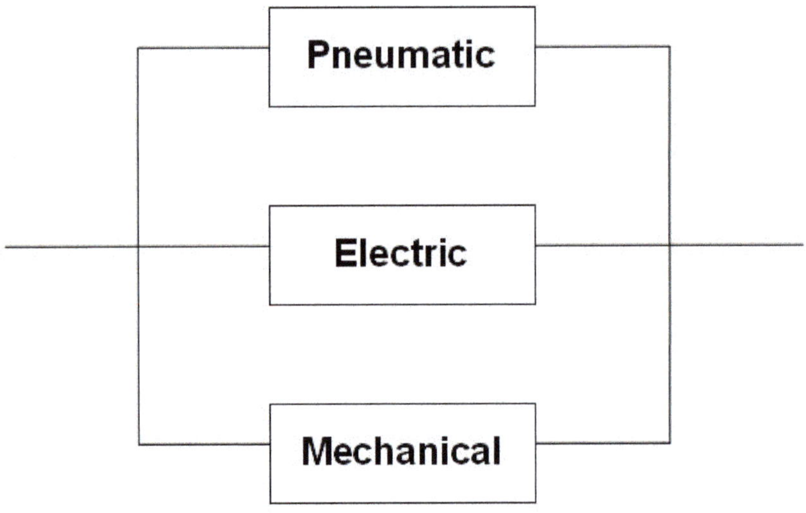

Figure A7-21: Diverse system. *Source: RMS.*

PLANNED PREVENTIVE MAINTENANCE

Plant maintenance systems

The risk management approach to maintenance is to minimise costs (of all types, including accidents) and to maximise productivity and equipment life. There are various strategies that can be adopted including emergency maintenance, breakdown maintenance, shutdown maintenance and planned preventive maintenance, including routine condition monitoring.

Although there may be some short-term gains in adopting a total breakdown maintenance strategy (no disruption of production, or use of spare parts, no or low maintenance staff costs) this will become increasingly less cost-efficient as time passes. Breakdown maintenance would be appropriate for non-safety or production critical items where the cost of maintenance outweighs the need for increased reliability.

No matter what maintenance strategy is adopted, some equipment failures will occur. Although the precise timing or location of the failure cannot always be accurately forecast, the fact that a breakdown will occur at some time is mostly predictable - either by examination of historical records or via the risk assessment process. It is important therefore to have a suitable system for emergency maintenance in place.

This may be achieved through the use of:

- Contingency plans.
- Model (or generic) risk assessments.
- Safe systems of work (for example permit to work system).
- The provision of appropriate skills and training.

Shutdown maintenance, where the renewal work is carried out when the process is stopped, is still common to the chemical and engineering sectors of industry. Traditionally in the engineering sector this occurred during the Easter break. Chemical plants usually have a cycle whereby they shutdown, for example, every two years. The frequency and duration of the maintenance period is dependent on the nature of the plant and processes being operated. This has the advantage of allowing maintenance to take place without the pressures of production. Many organisations combine the shutdown period with a general process employee holiday. Unless adequate management arrangements are in place, this can lead to a lowering of normal safety standards, particularly when third party maintenance staff is involved.

Planned preventative maintenance seeks to take action which tries to keep the condition of the component at its best by carrying out frequent care of the component for example lubrication, adjustment, cleaning. Monitored maintenance seeks to monitor the conditions of components, for example, noise or frequency levels from a gearbox to estimate bearing wear to determining the point just before failure, at which point it would be replaced. The main benefits of planned preventative maintenance are:

- Extended useable life of components.
- Assurance of reliability.
- Confirmation of condition of components.
- Reduced risk/loss producing failure events.
- Ability to carry out work at a suitable time.
- Better utilisation of maintenance staff.
- Less peaks and troughs in maintenance activity.
- Less standby facilities required.
- Less expensive (last minute) contracted facility required, for example temporary hire of equipment such as fork lift truck.

Routine condition monitoring is the comparison of the state parameters of the plant. Traditionally in engineering systems important parameters have been temperature, coolant and vibration. For example, when a local exhaust ventilation system (LEV) is installed the state parameters (for example capture velocity and static pressures) will be measured during the commissioning process. These are then used as a benchmark against which the efficiency of the system and thus the need for maintenance is measured. This has the advantages of ensuring that the plant is working efficiently, allows early detection of any deterioration and the carrying out of maintenance tasks at the optimum time. In turn, condition monitoring may be extended to the components that comprise the system. This is particularly relevant to the monitoring of critical components such as the motor/fan within an LEV system. Here the motor and its associated bearings can be monitored to identify any deterioration in performance that might threaten the efficiency of the system as a whole.

The role of statutory examinations of plant and equipment

The role of statutory examinations is to provide a means to identify the condition of critical equipment that, if not maintained to an acceptable standard, would create serious risk. The examinations take the form of a physical examination of the equipment by a competent person to identify patent defects, such as surface corrosion or physical damage, in the condition of equipment and its safety devices.

Statutory examinations have been prescribed for a considerable period. Traditionally the period of 14 months (for example LEV under COSHH 2002) was considered an appropriate period. This had the twin advantages of allowing for the movement of Easter, the traditional maintenance period, and gives a variation in the time of year that the examination is carried out. However, having fixed intervals between examinations with universal application takes no account of the actual conditions of use. The trend, however, appears to be moving away from prescriptive statutory examinations towards risk based assessment by a competent person (for example written systems of examination for pressure systems and lifting equipment). The latter example illustrates the prescribed versus risk bases approach. The Lifting Operations and Lifting Equipment Regulations (LOLER) 1998 provide that operational lifting equipment is to be thoroughly examined:

- At least every 6 months for lifting equipment for lifting persons or lifting accessories.
- At least every 12 months for other lifting equipment.
- In either case, in accordance with an examination scheme.
- On each occurrence of exceptional circumstances liable to jeopardise the safety of the lifting equipment.

Note that the law setting out the requirements for statutory examinations does not precisely define a 'competent person' nor does it stipulate that they should be independent of the employer. Although statutory examinations can be delegated to an outside organisation, this does not absolve the employer from the responsibility to ensure that they are carried out.

MINIMISING HUMAN ERROR

Human reliability analysis (HRA) can be used to develop training schemes for skill based behaviour and associated physiological factors, the design of controls, workplace, buildings, environmental conditions, transportation, and communications. A good example is the design of aircraft cockpit control layouts in which human performance at the operator level reaches the highest levels of criticality.

The concept of HRA could be used by management to devise management controls. The collapse of Baring's Bank and the huge losses suffered by a Japanese investment bank of the New York bonds market and Black Monday were attributed to human errors in monitoring and controlling the operations of the traders. HRA is increasingly used in the development of expert and machine intelligent systems to improve the performance of humans and equipment. At the design stage, HRA can be used to identify the features of hardware and job design that are likely to produce a high rate of human error.

Application of HRA requires skilled analysts who understand the plant operation and the tasks performed by operators. The lack of reliable data on operator's behaviour in error-likely situations is a major drawback for the application of the technique.

Analysts must consider that any analysis of operator behaviour is likely to bring about a modification of the behaviour. Operators will modify their behaviour and this can have significant effects on the HRA. One technique to reduce effect on behaviour is to train someone close to the operators in the technique, such as their line manager. The presence of someone known and often in the workplace has less effect on behavioural change, than for example an 'outsider' such as a consultant. Many of the same problems exist in other types of analysis, for example job safety analysis, and can be alleviated where the operators have confidence in the approach being taken and the analysts involved.

Failures of components and hardware can be expressed as probabilities. These may be derived from various studies such as in-service or test failures. Predicting the probability of human failure is far more complex. However, if risk assessments are to be done then human reliability needs to be considered and quantified.

When attempting to quantify human behaviour, it is necessary to be clear about what task or step is to be performed, when this task/step is to be performed and how much time there is available to perform it. The next stage is the consideration of how the person can deviate from the step/task. The most common deviations are when the step/task is not carried out correctly, when a step/task is forgotten in a sequence or is performed out of sequence, or when the step/task is not completed in the time required.

Error rates can be affected by a variety of conditions. As previously mentioned, observations of task performance can alter error rates, as can environmental conditions, training and skill levels, and various other stressors. Error rates are likely to be high when the person is under very little stress (bored), or under great stress (emergency situations). Human beings also have the ability to recover from errors, often demonstrating what appears to be an innate ability to control errors of great consequence compared to those of small consequence. Error rates for a control room situation may typically be as follows:

Error rate	***Situation***
1 in 1	Impending disaster, rapid action needed, panic
1 in 10	No impending disaster apparent, busy, signals, alarms
1 in 100	Quiet but busy, relaxed
1 in 1000	Familiar, routine tasks

Figure A7-22: Error rates for a control room. *Source: RMS.*

For example, consider a situation where studies have been carried out, which may be a number of task analyses, and found out that under no stress conditions there is an error in reading a gauge of 1 in 100 operations (1% of readings). Under pressure, this increases to 1 in 10 operations (10% of readings). If there are 400 readings taken per year, then the probability of an error under no stress and stress conditions within the next 1,000 hours is as follows:

Error rate under ***no stress*** conditions - based on 400 readings per year and 1 error in 100 readings (1%).

Error rate per year $= \frac{\text{Number of readings per year x 1}}{100} = \text{error readings}$

= 400 x 1/100 per year = 4 errors per year

= 4 errors in 24 (hours) x 365 (days) = 4 errors per 8760 hrs

Error rate per 1000 hrs $= \frac{4 \times 1000}{8760}$ errors per 1000hrs

Therefore error rate per 1000 hrs = 0.46 errors per 1000 hrs

Probability of error under ***no stress*** P_{f1000} = $1 - e^{-0.46}$

= 0.37

Error rate under ***stress*** conditions - based on 400 readings and 1 error in 10 readings (10%)

Therefore error rate 1000 hrs = 4.6 errors per 1000 hrs

3) Probability of error under ***stress*** P_{f1000} = $1 - e^{-4.6}$

= 0.99

This data can be used in risk assessments to quantify risk once the effect of the error is known. It can help reduce the amount of subjectivity that is usually present when dealing with human error. Once the probability of an accident happening is known, then it is relatively easy to balance the cost of the accident against the cost of controls.

A7.5 - Failure tracing methodologies

Principles and techniques of failure tracing methods in the assessment of risk

Not every situation can be covered by a general or generic risk assessment. General assessments will not identify failures that may lead to a downstream catastrophe, for example, failure of a signalling circuit that leads to a train derailment. There is always a need for complex analysis, which may be qualitative, quantitative, or a mixture of both.

Many specialist risk assessment techniques have been developed, predominantly within those industries generally regarded as 'high risk' (for example, nuclear, chemical and petrochemical industries). The effectiveness of these techniques has been demonstrated over the past years and they have moved into the mainstream of safety management techniques. These techniques adopt a structured approach to identify high-risk areas and tasks and those elements of a process that contribute to overall risk.

The main 'advanced' risk techniques, which adopt a failure tracing methodology, are:

- Hazard and operability studies (HAZOP).
- Fault tree analysis (FTA).
- Event tree analysis (ETA).

These techniques offer significant benefits in all areas of risk management not just health and safety.

HAZARD AND OPERABILITY STUDIES (HAZOP)

HAZOP Studies aim to identify hazards and operability problems that could reduce a manufacturing plant's ability to achieve its target production in safety. ICI Limited initially developed it for use in their chemical plants and HAZOP is now used as a standard assessment technique in other types of systems and complex operations such as nuclear power plant operation.

HAZOP can be used on new plant, part of the plant or existing facilities, but is always best done at the design stage. Any change or modification to such plant, temporary or permanent must also undergo a HAZOP to ensure plant safety is not jeopardised by the change.

Features of HAZOP

HAZOP is a team-based assessment and a major feature of a HAZOP is the use of a multidisciplinary team of specialists. An experienced facilitator is required to guide the team and ensure that each discipline can make its contribution. A typical team would include a safety specialist, engineering specialists and operational staff; other specialists could be utilised depending on the operation under assessment, for example, building and services engineers, the process relies on significant in depth discussion and reflection by the team.

The initial efforts of the team are directed towards identifying the deviations from the design intent and establishing solutions and remediation. This allows everyone to become familiar with the operation and encourages a free flow of ideas within the team. Checklists are used to maintain the relevance of the process and reduce the opportunity for ideas to be lost or forgotten. The team will identify a range of hazards that vary from little consequence to life or multi-life threatening.

Definitions

The key words used in HAZOP are, 'guide words', 'study nodes', 'intention', 'deviations', 'causes' and 'consequences'.

- Guide words - are used to qualify the intention in order to guide and stimulate discussion.
- Study nodes - are defined as the locations on the plant or services at which the process parameters are investigated for deviation from the design intent.
- Intention - describes how the plant is designed to operate.
- Deviation - are the departures from the intention.
- Causes - are the reasons why deviations may occur.

- Consequences - are the results of the deviations.

Example - HAZOP guide words, meanings and applications

Guide word (Deviation)	*Meaning*	*Parameter*	*Deviation*
No	Negation of the design intent	Flow	No flow
High	Greater than the design intent	Pressure	High pressure
Less	Quantitative decrease	Temperature	Low temperature
Reverse	Logical opposite of the intent	Open	Close

Figure A7-23: HAZOP guide words, meanings and applications. *Source: RMS.*

The steps of a HAZOP

There are ten steps to carrying out a HAZOP:

1. Define objectives and scope

Management should establish these and typical examples would include checking the safety of a proposed plant or equipment changes.

2. Select the team leader

The team leader should be independent and experienced and should be capable of identifying study nodes, problems, and also encourage and support team members.

3. Select the team

The team should be skilled in the disciplines relevant to the operation.

4. Define the physical boundaries

The team should define the precise physical area that the HAZOP will cover.

5. Collect data

Drawings, flow charts, plant layouts, technical drawings, equipment manuals should be collected and identified.

6. Process the data

For some processes the amount of data processed can be quite small. However, for alterations to existing operations the amount can be substantial.

7. Design review

Checklists are applied to the study nodes where the operation has defined design intent. The area between the nodes is considered, as there may be equipment there which changes the conditions between nodes. At each node the guide-words are applied which identify deviations. Guide-words ensure that the design is thoroughly examined. At this stage the study may uncover gaps in the available information or lack of particular knowledge on the part of the team members.

8. Record the results

All ideas must be recorded to ensure a complete record of the discussions of the team resulting in a decision or modification. This is often of immense use in the future in any further modifications.

9. Implement design modifications

The team leader delegates particular modifications to the appropriate team specialist.

10. Reporting

The report should be accurate and identify major deviations from the design intent, recommended modifications and where possible estimates of necessary expenditure.

General comments on HAZOP

HAZOP is a powerful assessment tool detecting deviations through a methodical approach using specialists guided and aided by a formal system. Although formal the approach encourages freethinking among the team members and the freedom to develop new guide words means that the approach can be used in all situations.

The approach requires a team to be gathered and will fail if specialists with the appropriate skills and expertise are not available.

FAULT TREE ANALYSIS (FTA)

FTA is a logic based assessment process used to identify and analyse the events, which can lead or contribute to accidents. The procedure was developed by Bell Telephone Laboratories for the US Air Force for use with the Minuteman ICBM system to determine the sequence or combinations of events that could lead to an unauthorised launch of the weapon. FTA describes sequences of events, either human failure or equipment malfunctions, by working backwards from a Top Event or undesired occurrence. The FTA is constructed by using this reverse logic to establish the relationship between cause and event. It is easy to follow the flow of events and sub events in each particular failure and it also indicates where the placement of sensors or controls will have maximum impact on the progression from cause to effect. FTA has three key aspects that differentiate it from other assessment systems:

1) Problem formulation.
2) Key definitions.
3) Graphical representation.

Problem definition

As with many systems of analysis failure to identify the top event accurately allows the analysis to drift from the necessary path of investigation and consequently fails to identify the combinations of causes leading to the undesired event. The top event should be described in terms of:

'What' ***'When'*** ***'Where'***

A fire or explosion is too general a top event; a specific and clear description is required, for example, an explosion occurring during the off-loading of fuel oil from a bulk storage tank.

Key definitions

Design faults and failures. These are equipment failures that occur in the environment for which the equipment was designed and intended. For example; plant such as a pump used for flammable liquids being installed in such a way that there is insufficient access for maintenance or repair; electronic systems such as those utilised for the control of the movement of robotic paint sprayers. If the designer did not take into consideration the effect, for example, of the loss of electrical power, then the robotic arm might not come to rest in a safe position and may malfunction on start up.

Operation faults and failures. An environment for which the equipment was not designed and intended, for example, a conventional smoke alarm may be an unsuitable fire detection device in an area where a corrosive atmosphere is present, whereas a heat sensor might be more suitable because it is less likely to be affected.

Signal faults and failures. Signal faults and failure of protective devices are commonly associated with spurious trips of plant; they may result from the choice at design stage of unsuitable equipment to measure, for example, over pressurisation or over temperature. Often this can result from the use of a single detection device which provides an incorrect reading, leading to unnecessary plant shut down. If two or more detectors arranged in a voting system were used it might reduce the effect of single failure of one detector. All types of faults and failures appear in a fault tree, since the main aim of a fault tree analysis is to identify the basic contributing failures.

Graphical representation

Symbols fall into two categories, logic symbols and event symbols. These represent the relationships that exist between equipment failures, human errors and a specific accident. There are also ***OR*** gates, ***AND*** gates, ***DELAY*** gates, and ***MATRIX*** gates. The two most basic elements of FTA are the ***AND*** gate and the ***OR*** gate. The ***AND*** gate means that all identified inputs are required simultaneously for the end event to occur. The ***OR*** gate means that any one of a number of possible inputs can cause an output event. The most used event symbols are the rectangle, circle and diamond. Rectangles represent a fault event resulting from the combination of more basic faults acting through logic gates. The circle designates a basic system component failure or fault input that is independent of all other events designated by circles and diamonds. The diamond symbol describes fault inputs that are considered basic in a given fault tree; however, this fault tree is not developed further either because current information must be augmented or the consequences are insufficient to warrant further work.

Logic symbols used in fault tree construction

Logic symbol	***Graphical representation***	***Description***
OR Gate	Outputs Inputs	The ***OR*** Gate indicates that the output event occurs if any of the input events occur. When calculating probabilities input components to an ***OR*** Gate are ***added***.
AND Gate	Outputs Inputs	The ***AND*** Gate indicates that the output event occurs only when all the input events occur. When calculating probabilities input components of an ***AND*** Gate are ***multiplied***.

Figure A7-24: Logic symbols used in fault tree construction. *Source: RMS.*

Event symbols used in fault tree construction

Logic symbol	*Graphical representation*	*Description*
BASIC Event		The ***BASIC*** Event represents a basic equipment fault or failure that requires no further development into more basic faults or failures.
INTERMEDIATE Event		The ***INTERMEDIATE*** Event represents a fault event resulting from the combination of more basic faults acting through logic gates.
UNDEVELOPED Event		The ***UNDEVELOPED*** Event represents a fault tree event that is not examined further because information is unavailable or because its consequence is insignificant.

Figure A7-25: Event symbols used in fault tree construction. *Source: RMS.*

Example - fault tree analysis

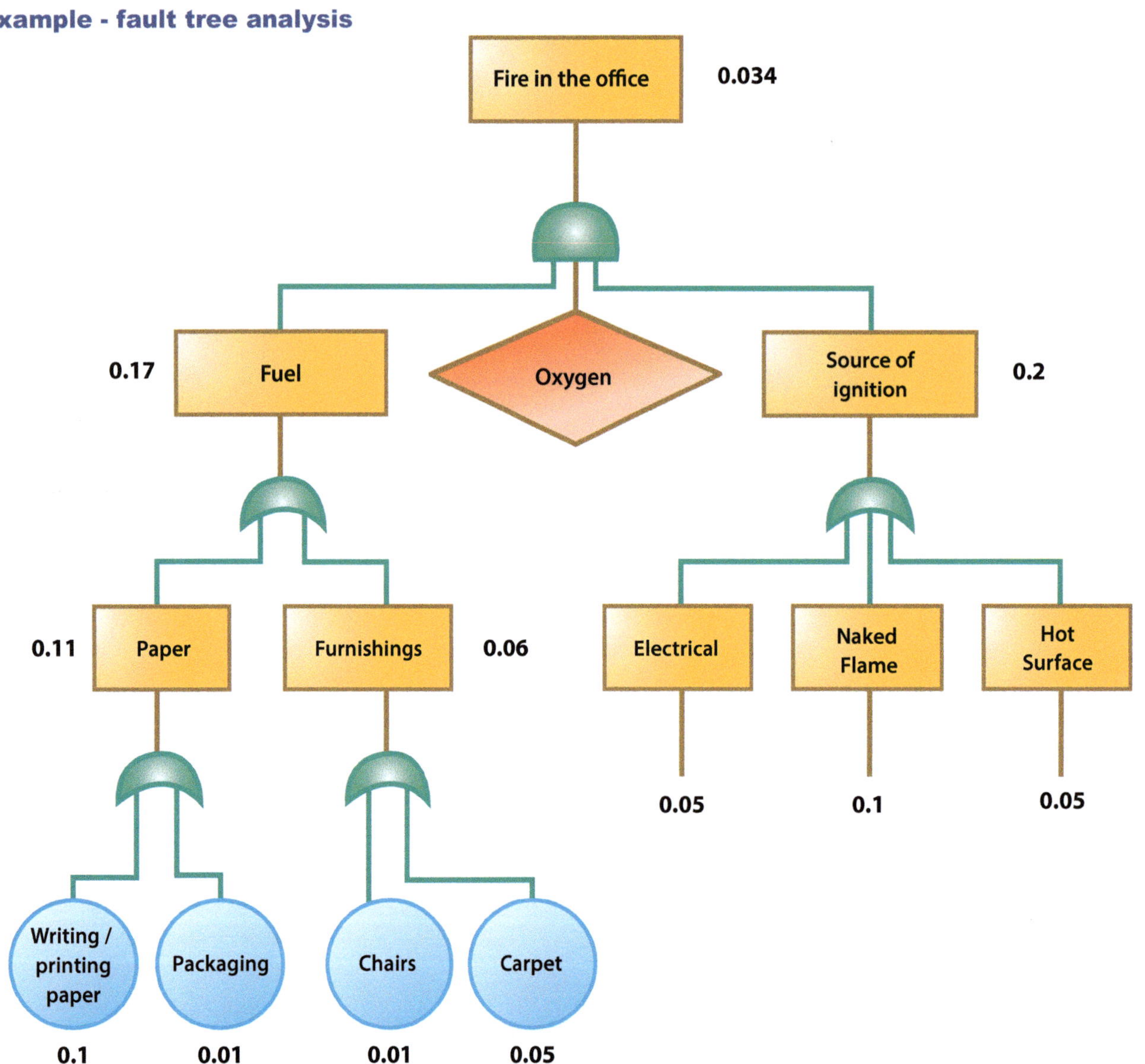

Figure A7-26: FTA example - showing probability figures. *Source: RMS.*

The steps of fault tree analysis

1) Identify the top event (undesirable event)

The top event should be the major failure of the system under consideration (for example, fires or explosions). The analyst determines the immediate and necessary causes that result in the top event. Normally, these are not basic causes but are intermediate faults that require additional development. If the analyst can immediately determine the basic causes of the top event, then the analyst is spending time on an assessment that is too powerful for the problem.

2) Decide on level of resolution

The analyst decides the amount of detail to be included in the fault tree by specifying the level of resolution for the fault tree events. For example, a motor-operated centrifugal pump can be included as a single piece of equipment, or it can be described as several hardware items (for example, casing, impeller, volute, shaft, packing, etc.).

3) Define the analysis boundary conditions

The analysis boundary conditions describe the system in its normal, successful state. Time should be spent identifying unacceptable events and existing events.

4) Define system physical boundaries

The boundaries of the physical system include the equipment that will be considered in the fault tree and the interaction of the equipment with other processes, utilities and support systems.

5) Define the system initial conditions

The initial equipment configuration and conditions describe the configuration of the system and the equipment that is assumed for the fault tree analysis. The analyst specifies which valves are open and closed, which pumps are on or off, etc. for all equipment within the physical system itself.

6) Construct the fault tree

The starting point in fault tree analysis is to determine whether the fault (or input) has the capacity to act as an ***AND*** gate or as an ***OR*** gate.

Fault tree construction begins at the top event and proceeds, in a logical approach, level by level, until all fault events have been analysed to their basic contributing causes. The immediate causes of the top event are shown in the fault tree with their relationship to the top event. This has significant similarities to the Domino model of accident causation.

7) Determine the minimal fault tree

The completed fault tree provides useful information by showing the interfaces of equipment failures that could cause an accident. However, except for a simple fault tree, it is necessary to solve the fault tree in order to identify all the combinations of equipment failures that can lead to an accident. To solve a fault tree means to obtain the minimal development lines for the fault tree.

8) Rank the elements

Ranking the development lines in the FTA is important for setting priorities for designing modifications.

9) Quantify the fault tree

Quantification of the fault tree is carried out by assigning failure data and probability to each event on the tree.

General comments on FTA

FTA is a very useful tool for studying the routes by which a hazard can occur, although its implementation requires skilled analysts and increasingly the use of computers to deduce the development lines. These may also be called ‘cut sets’. Quantification of the fault tree depends on the accuracy of the failure data and its availability.

EVENT TREE ANALYSIS (ETA)

Unlike fault tree analysis, event tree analysis is a ‘forward thinking’ process, that is, the analyst begins with an initiating event and develops the resulting sequences of events that describe potential accidents, accounting for both the successes and the failures of the safety functions as the accident sequence progresses.

Event trees provide a methodical way of recording the accident sequences and defining the relationships between the initiating events and the subsequent events that combine to result in an accident. They are an element of safety management much favoured by the Health and Safety Executive.

The steps of an event tree analysis

Five steps are necessary to perform ETA:

1) Identify an initiating event of interest.

2) Identify the safety functions or controls designed to deal with the initiating event.

The safety functions can be thought of as the plant's defence against the occurrence of the initiating event, and they usually include:

- Safety systems that automatically respond to the initiating event.
- Alarms that alert the operator when the initiating event occurs.
- Operator actions.

The safety systems automatically respond to the initiating event and can shut down the systems. The alarms alert the operator when the initiating event occurs. The operator responds with corrective actions according to laid down procedures or trained for responses.

3) Construct the event tree

The event tree displays the logical progression of an accident. The event tree begins with the initiating event and proceeds through the successes and/or failures of the safety functions that react to the initiating event. Only two possibilities are considered when evaluating the response of the safety functions, quite simply success or a failure.

4) Describe the resulting accident event sequences

The accident event sequences represent a multitude of incidents that can result from the initiating event. One or more of the sequences may represent a back to normal safe operation, some may represent a controlled shut-down and others may result in an accident.

5) The analyst then ranks the accidents based on the severity of their outcomes

Example - event tree analysis

A success of a safety function is defined as its ability to prevent the initiating event from progressing further, thus preventing an accident. The failure of a safety function is defined as its inability to stop the progression of an initiating event or later its course so that other safety functions can respond to it.

Consider the case of a toxic gas escape in a section of a process plant. The relating safety functions are:

- Gas detector.
- Audio alarm.
- An automatic shutdown system.

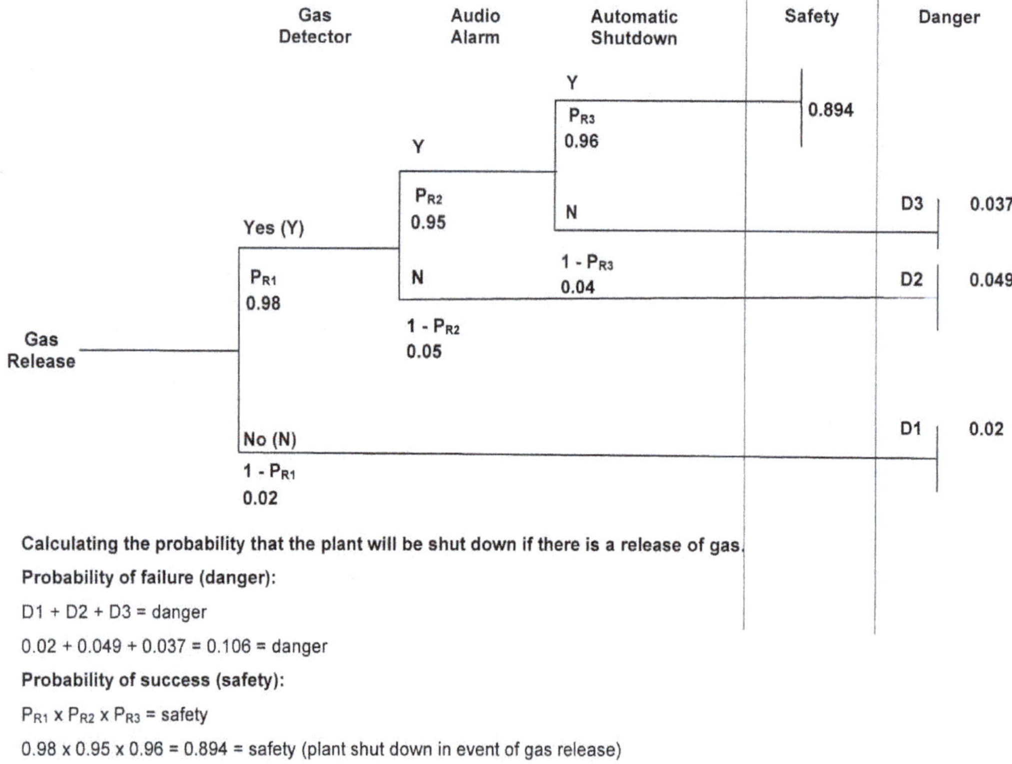

Figure A7-27: ETA example - with probability figures. *Source: RMS.*

The initiating event is listed on the left-hand side of the page and the safety functions are listed in a chronological order across the top of the page. The line underneath the initiating event description represents the progression of the accident path from the occurrence of the initiating event up to the first safety function or control (gas detector).

The next step is to evaluate the first safety function. Only two possibilities are considered - success and failure of the safety function. The analyst must decide whether the success or failure of the safety function affects the course of the accident. If the accident is affected, a branch point is inserted in the event tree to distinguish between the success and the failure of the safety function. Usually an upward path denotes the success of the safety function and a downward path denotes the failure of the safety function.

Every branch point developed in the event tree creates additional accident paths that must be evaluated individually for each of the subsequent safety systems. When evaluating a safety function on an accident path, the analyst must assume the conditions dictated by the path of the accident up to the safety function.

General comments on ETA

Event trees are valuable in examining the consequences of failure, because these are greatly influenced by factors such as operator intervention and weather conditions, which can be expressed more readily in event tree form.

ETA is not usually used for the analysis of system failure because of the limited available data, making quantification of failure rates difficult.

BOW TIE RISK ANALYSIS (BTRA)

The bow tie method of risk analysis provides a visualisation of the relationships between the causes of adverse events, the escalation of such events, the controls preventing the event occurring and the preparedness measures in place to limit the consequences of the event.

In addition, the process links the preventive and mitigating measures to tasks, procedures, responsible individuals and competencies. This highlights the connection between risk controls (technical, procedural or behavioural) and the management system necessary for ensuring their effectiveness.

The exact starting point of the Bow Tie Methodology is unknown but it is believed that they were originally called 'Butterfly diagrams' and evolved from the Cause Consequence Diagram of the 1970s. It is then thought that David Gill of Imperial Chemical Industries (ICI) developed the methodology and the earliest mention of the bow tie methodology appears in the course notes of the ICI Hazan training in 1979.

The Royal Dutch/Shell Group developed the technique in the mid-1990s as a result of the Piper Alpha disaster. Shell is acknowledged as the first major company to integrate the whole bow tie methodology fully into its business practices but as the 1990s came to an end, the approach became a standard method within many other companies.

The success of the diagram derived from the bow tie methodology is that it is easy for the non-specialist to understand. The idea is a simple one of combining the cause (fault tree) and the consequence (event tree).

When the fault tree is drawn on the left hand side and the event tree is drawn on the right hand side with the hazard drawn as a 'knot' in the middle, the diagram looks a bit like a bow tie.

The structured approach of the bow tie forces an assessment of how well all initial causes (threats) are being controlled and how well prepared the organisation is to recover should things start to go wrong. It also provides an opportunity to analyse how well controls deal with factors that may escalate the risk. It highlights the direct link between the controls and elements of the management system.

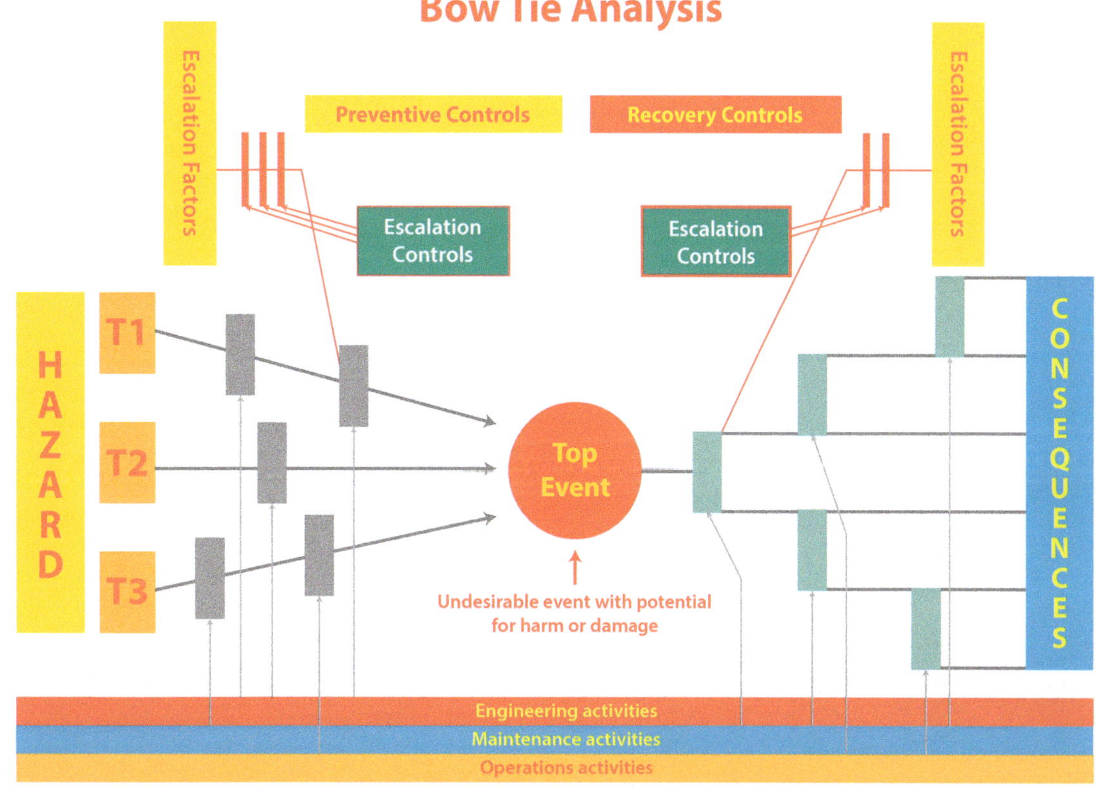

Figure A7-28: Bow tie diagram. *Source: Lee and Burdekin, CAAS*

This page is intentionally blank

Risk control

Learning outcomes

On completion of this element, candidates should be able to demonstrate understanding of the content through the application of knowledge to familiar and unfamiliar situations. In particular, they should be able to:

A8.1 Explain the use of common risk management strategies.

A8.2 Outline factors to be taken into account when selecting risk controls.

A8.3 Explain the development, main features and operation of safe systems of work and permit-to-work systems.

Contents

Relevant statutory provisions

Health and Safety at Work etc. Act (HASAWA) 1974

Sources of reference

Reference information provided, in particular web links, was correct at time of publication, but may have changed.

Guidance on permit-to-work systems. A guide for the petroleum, chemical and allied industries, HSG250, HSE Books, ISBN: 978-0-7176-2943-5, http://www.hse.gov.uk/pubns/books/hsg250.htm

Occupational health and safety management systems - requirements (OHSAS 18000 series),

OHSAS 18001, ISBN: 978-0-5805-9404-5

OHSAS 18002, ISBN: 978-0-5806-2686-9

The above web links along with additional sources of reference, which are additional to the NEBOSH syllabus, are provided on the RMS Publishing website for ease of use - www.rmspublishing.co.uk.

A8.1 - Common risk management strategies

Concepts within a health and safety management programme

Risk control may be considered in four broad categories. These are avoidance, reduction, retention or the transfer of risk. These risk control approaches are sometimes known as the 4Ts - Terminate (avoid), Treat (reduce), Tolerate (retain) and Transfer.

AVOIDANCE

Even though it may involve giving up a commercially beneficial activity the organisation decides that the risks outweigh benefits. The risk may involve unacceptable harm to customers, employees or local residents. The risk may be avoided completely by, for example, changing to a radically different and possibly more expensive production method. Another example of an avoidance strategy is to pay employees directly into their bank accounts in order to avoid the possibility of a wages robbery.

It should be appreciated that the avoidance of one risk may introduce others. For example, the decision to avoid siting a factory on a flood plain may mean an increased risk of wind damage due to building on an elevated site. The avoidance strategy is usually considered at the planning stage of the operation in order to avoid potentially substantial costs.

It could be decided to avoid the risks associated with the use of trichloroethylene to de-grease metal by purchasing components that have been treated and are ready for use.

REDUCTION

The risk assessment process is concerned with the probability (likelihood) and severity (consequence) of potential losses. The reduction strategy is therefore concerned with the reduction of these two variables. This can take a number of forms:

Reduction through organisational planning

Here the three main areas to be considered are the organisation, the task and the individual.

Reduction of risk through organisational planning can be achieved by focusing redesign of the task, education and training, planning, salvaging, and contingency/emergency planning.

- ***Redesigning the task*** can reduce risk, for example, placing a component on a bench at waist height to be welded instead of the welder stretching up to weld above head height. In some cases, the welding process may be automated, reducing the risk still further, although introducing new, but different risks.
- Education and training concentrate on the promotion of a positive health and safety culture and the minimisation of the negative consequences of human behaviour.
- Planning deals with minimising the organisation's exposure to key elements such as suppliers or customers.
- Salvaging deals with minimising loss in the same way that a first aider minimises injury. This requires the provision of trained personnel and equipment, for example, to deal with a flood.
- The last part of this reduction process deals with contingency and emergency planning.

See also - 'NEBOSH National Diploma Unit C Study Book' - 'Element C4.4 - Emergency planning'.

Reduction through health and safety control measures

Reduction of risk through health and safety control measures can take the form of physical safety devices to prevent loss (locks, alarms, machine guards) or devices to minimise the consequences of loss (sprinklers, fire resistant construction, electrical trip devices). ***Automation of processes*** can reduce the exposure of workers to hazards by removing them from the higher risk parts of the process. In addition, procedural controls such as permits to work, good housekeeping and storage can reduce risk.

TRANSFER

Transfer can take two forms:

1) The transfer of activities or supplies to another organisation.
    - For example, contracting higher risk work activities to ***specialist contractors*** or contracting supply of parts/materials to specialist manufacturers. For example, a component which requires a finishing process which involves hazardous substances, such as chromium plating may be supplied by a specialist manufacturer, thus transferring people and environmental risks.
2) The assignment of financial or legal responsibility to another organisation.

The most common form of the latter method is via ***insurance***. Risks may be transferred in full or partially. The options available to a driver buying motor insurance are a good example of this strategy. These range from fully comprehensive cover through to third party cover.

Similarly an organisation can opt for something between zero and full insurance cover. Transfer of risk can also be achieved by assigning legal liability, for example, transferring liability for damage by contract or conditions of sale where this is permitted under civil law (note the Unfair Contract Terms Act 1977).

RETENTION

With knowledge

An alternative name for active risk retention may be 'self-insurance' where potential losses are funded by the organisation. This may arise through a deliberate decision to stand some or all of the potential losses or via the formation of a captive insurance company. Many retail chain outlets do not insure their stores for fire; the cost of insuring all their stores is far greater than the likely cost of losing one store.

Another retention strategy is the area of 'claims management'. Here an organisation decides to settle claims up to a certain value directly, for example, claims made by an injured employee. This allows the organisation to control costs and insurance premiums and to get actively involved in the claims management process.

Without knowledge

Risk may be retained by the organisation without their knowledge, either through oversight or through inertia. This may not only arise from failing to identify or evaluate risk adequately or at all, but also failure to keep insurance cover in line with current market values.

Organisations may believe that they have successfully transferred liability to a contractor, but find that when this is tested in law they are held legally liable. This was the case with Associated Octel, who contracted specialist contract cleaners for cleaning storage vessels that had held chemicals. ***See 'Element A9 - Organisational Factors' for more information on the case 'R v Associated Octel Co Ltd [1996] 4 All ER 846'.***

Failure to identify critical employees or unique equipment that are not easy or impossible to replace to the same standard may be an example of where risk is retained without knowledge. It has been identified across Europe that there are large numbers of engineers approaching retirement which will result in a significant skill shortage in the power generation and supply sectors within the next ten years.

Circumstances when each of the above strategies would be appropriate

A combination of engineering, technical and operational control measures may be required if the task the person undertakes involves dangerous fumes (for example, rubber manufacturing). In this case, automation of processes to remove employees from the most hazardous areas may be preferential to measuring fume levels providing local exhaust ventilation systems and personal protective equipment (respirators) for employees.

Mitigation of risk can also be managed through insurance policies. The management of risk seldom achieves a risk free solution - a residual measure of risk remains often despite extensive engineering controls. Natural disasters, such as hurricanes, tsunamis, typhoons and earthquakes can defeat even the safest system of work.

Case Study - Engulfment in a 'free-flowing solid' example

Following a major earthquake, a 15-metre tsunami disabled the power supply and cooling of three Fukushima Daiichi reactors, causing a nuclear accident on 11 March 2011. All three cores largely melted in the first three days. Eleven reactors at four nuclear power plants in the region were operating at the time and all shut down automatically when the quake hit. Subsequent inspection showed no significant damage to any from the earthquake. The reactors proved robust seismically, but vulnerable to the tsunami.

The original design basis tsunami height was 3.1m for the Daiichi plant based on assessment of the 1960 Chile tsunami and so the plant had been built about 10 metres above sea level with the seawater pumps 4m above sea level. The government and 12 utilities are contributing funds into a new insurance institution to pay compensation to individuals and businesses claiming damages caused by the accident. The institution received JPY 7 billion ($91 million) in public funds as well as a total of JPY 7 billion from 12 nuclear plant operators, the Tepco (plant operators) share of JPY 2379 million ($30 million) being largest.

Before any maintenance of plant starts, the employer needs to decide if the work needs to be undertaken by specialist contractors. The employer should not undertake work where they are not competent to do so. As in the Associated Octel case prosecution case study the cleaning out of vessels containing flammable solids, liquids, gases or dusts, will need specialist help and advice to do this safely.

Selection of an optimum solution based on relevant risk data

The selection of the optimum solution (i.e. the avoidance, reduction, transfer or retention of risk) will depend upon the available risk data, the perspective of tolerability of the risk and the ability of the organisation to withstand loss. The ability for a multi-national conglomerate to withstand and absorb losses is likely to be vastly greater than a small engineering company.

It should be noted that, as with all decisions and policies that managers take, the proposed solutions for handling risks should be consistent with corporate objectives. Thus, the health and safety practitioner must be attuned to the culture, vision and goals of the organisation.

The starting point therefore in the decision making process is the setting of corporate objectives. These objectives can be ascertained through the identification of hazards, the assessment of risk, and then deciding on appropriate ways of handling them.

The factors to be considered in the selection of the best solution include:

- Stakeholders' and the organisation's health and safety aspirations.
- Legal and other constraints.
- The organisation's financial objectives.
- Cash budgeting and liquidity.
- Workforce relations and related issues.
- Ethical and social considerations.
- Effectiveness of proposed solution.

ASPIRATIONS OF STAKEHOLDERS AND THE ORGANISATION

The health and safety aspirations of the organisation will greatly influence the corporate perspective of tolerance of risk and therefore the risk management strategy. Those organisations with high corporate social responsibility perspectives will expect the organisation to reduce risks well below minimum legal requirements, avoid activities that remain at a significant risk and maximise transfers by contracting and insurance. This type of organisation would take considerable effort to identify risks so that they did not retain risks without knowledge.

LEGAL AND OTHER CONSTRAINTS

The selection of optimum risk management solutions can be influenced by both civil and criminal law. For example:

- The need for machine guards dictated by the Provision and Use of Work Equipment Regulations (PUWER) 1998.
- Contractual conditions and obligations.
- Strict liability for escape of things brought onto premises under the Rylands v Fletcher rule. ***(See 'Element A9 - Organisational Factors' for information on relevant example prosecutions).***
- The need to have compulsory insurance (for example, employer's liability).

The need for legal compliance should be regarded as the minimum standard which the organisation must achieve in terms of risk control. The selection of controls may also be dictated by external factors such as the agencies upon which the organisation depends, for example, banks (capital loans), customers and the insurance companies.

THE ORGANISATION'S FINANCIAL OBJECTIVES

Most private sector organisations would place the need to make a profit high on their list of corporate objectives. This can vary from the wish to make enough money to survive (i.e. cover costs) through to the objective of maximising profits. The selection of solutions will therefore need to be consistent with the organisation's goals.

The health and safety practitioner must aim to reduce the potential losses from risks, but at the same time balance that need by keeping the cost of controls in proportion. For example, an organisation could transfer their risks either by the purchase of insurance or by contracting out a process; however, the costs of doing so are likely to outweigh the expected losses. It should be remembered that insurance companies and specialist contractors are also in business to make a profit.

CASH BUDGETING AND LIQUIDITY

Business financial philosophy dictates that cash is treated as a resource that must be controlled like any other. The aim must be to have enough cash available at the time it is needed. Even in organisations that have large cash reserves, this has a cost.

Money held in reserve could be used for investment outside the business and thus earn interest. The unexpected need to meet the costs of a risky event may disturb this balance. For example, a fire can result in:

- Replacement cost of property and equipment (capital assets).
- Disruption of production and sales - possibly leading to lack of customer confidence and future orders.

This may put an intolerable strain on the business and even threaten the organisation's survival. It has been estimated that some 70% of businesses that have a major fire fail within two years of the event. The transfer of risk to an insurance company (property and consequential loss), with known premium costs, therefore allows a greater degree of certainty in the event of the unexpected costs. The balance must therefore lie in the amount of potential loss involved and the cost of transfer.

WORKER RELATIONS AND RELATED ISSUES

Although the province of the human resources manager, there are several aspects of worker relations that concern the health and safety practitioner and the selection of the appropriate risk controls. These are safety and security including:

- The cost of accidents.
- The need for employer's liability insurance premiums.
- Security (for example, pilfering).

Frequent accidents not only have associated costs but can also lead to a lowering of morale and consequent reduction in efficiency and higher unit costs.

Alternatively, the costs of control can include a resistance by employees to use them (for example, wearing personal protective equipment, using guards).

Although not all the costs associated with accidents are necessarily borne by the organisation (for example, hospital treatment), accident prevention should always dominate these conflicting needs. The need to implement controls to reduce accidents is dictated by legal obligations as well as discharging the organisation's social responsibilities.

Similar considerations apply to security issues. The cost of pilfering can be quite high (in some cases it is thought to exceed 5% of turnover) so the cost of stringent controls can be justified. The imposition of stringent security checks, which affect all employees, to prevent theft by the few who might steal can cause great resentment and antagonise many employees.

ETHICAL CONSIDERATIONS

Finally, the moral argument states that people must matter and that risk control decisions cannot be made purely on financial or legal grounds. The price that an individual pays for lack of adequate control can range from personal injury to death.

On the other hand, excessive costs can lead to job losses. The manager deciding on the optimum solution must have a clear idea how far the organisation is prepared to go beyond the bare legal minimum and the need for survival. This stems from the organisation's culture, vision and goals.

EFFECTIVENESS OF PROPOSED SOLUTION

Avoidance of risk is clearly effective in removing the risk, but may carry significant costs to extricate the organisation from the risk and may have commercial impacts or effects on the organisation's service obligations.

Tolerating the risk can only be effective where the risk is low and therefore the probability and severity of losses from the risk are low. The option to transfer risk to another party can have some variability in its effectiveness, for example, insuring against a risk does not necessarily remove the effects on the organisation's image related to the loss experience.

In the same way, a contracted relationship may leave residual risks to the organisation's image, as the organisation often carries ultimate responsibility for the management of the risk. Reduction, and therefore self-management of the risk, may be most effective for those risks that relate to the main operation of the organisation and have most experience of managing successfully.

The principles and benefits of risk management in a global context

Risk management may be defined as: 'The minimisation of the adverse effects of pure and speculative risks within an organisation'.

Pure risks can only result in a loss to the organisation, whereas speculative risks may result in either gain or loss. The objective of risk management is to minimise the risk of loss by safeguarding the organisation's assets: people, equipment, materials, environment, methods, services/products, reputation and money.

The role of risk management in the workplace is to:

- Consider the impact of certain events that can give rise to loss of performance of the organisation.
- Devise alternative strategies for minimising these risks.
- Relate these alternative strategies to the general framework used to manage the organisation.

This role is fulfilled by establishing a risk management process to coordinate effort to meet the objective/role. The benefits of risk management are that it provides a systematic and therefore more assured approach to the identification of risks - making the acceptance of risk without knowledge less likely.

It will provide a structured approach to the evaluation of risk and an opportunity to make a reasoned decision related to the risk, allowing risks to be addressed in an efficient and effective manner. Risk management will enable cost effective and timely management of resources to enable risks to be controlled to a satisfactory level and that controls have been implemented and are/remain effective.

Internationally it is of great benefit for organisations to use a risk management approach because they may not be able to rely on national legislation or guidance to direct them to establish necessary risk controls. Similarly, any organisation operating within a specific country that has a limited health and safety structure may develop an effective health and safety approach by working from the primary principles that a risk management process will deliver.

Even in countries with well-developed health and safety structures it has proved difficult to provide legislation and guidance on every risk an organisation may encounter. Organisations are expected to take the lead and identify the risks they create, or that they are affected by, and manage them. The risk management process can be summarised in ***figure ref A8-1.***

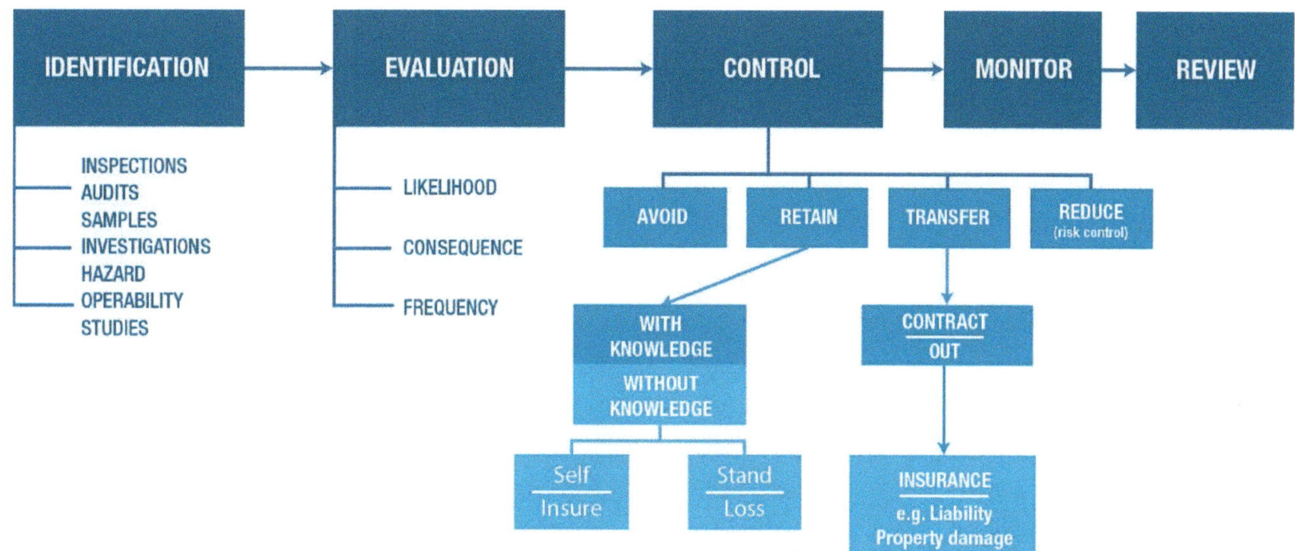

Figure A8-1: Risk management process. *Source: RMS.*

The IOSH Publication 'Business Risk Management' introduces the subject: 'Increasingly, organisations are taking a holistic (i.e. global) approach to managing risk and this represents both a challenge and an opportunity for health and safety professionals ... well-informed health and safety professionals are more able to make the case for considering health and safety risks within the wider context of business risk'.

A global approach is highlighted by the Financial Reporting Council (FRC) guidance, which requires major organisations to manage their significant risks in a suitable manner. Systems for regular review of risks must be in place, together with statements in company annual reports confirming the effectiveness of these systems.

Health and safety hazards must be included where they represent significant risks to the security and well-being of the organisation. These may affect wider financial losses and damaged reputation, such as those sustained by BP after the Gulf of Mexico disaster.

The development of global business risk management therefore combines loss control (risk reduction), insurance (risk transfer), good corporate governance - balancing the interests of the many stakeholders in the organisation, and formal integration of the risk management systems (for example, health and safety, quality, environmental effects).

The link between the outcomes of risk assessments and the development of risk controls

A suitable and sufficient risk assessment should consider a range of control measures and provide a basis for the selection of control measures. Risk assessment can be a useful tool, which can save or optimise the use of resources, by determining the effectiveness and costs of different control options, improving the decision-making process and providing a basis for allocating resources in the most effective manner.

The risk assessment process should provide the following in relation to control measures:

- Identification or clarification of existing and potential control measure options:

1) Evaluation of effects of control measures on risk levels.
2) Basis for selection or rejection of control measures and the associated justification of adequacy.
3) Basis for defining performance indicators for selected control measures.

The risk assessment should evaluate the range of control measures in terms of viability and effectiveness to provide a basis for selection or rejection of each control measure:

- Viability relates to the practicability of implementing the control measure within the facility.
- Effectiveness relates to the effect of the control measure on the level of risk. For example, the reliability and availability of control measures influence the likelihood of an incident occurring, while the functionality and survivability of the control measures during the incident influence the consequences.

By evaluating options for control measures within the risk assessment the employer should be able to decide what additional benefit is gained from introducing additional or alternative control measures.

A8.2 - Factors to be taken into account when selecting risk controls

The general principles of prevention

MANAGEMENT OF HEALTH AND SAFETY AT WORK REGULATIONS (MHSWR) 1999

In order to control the risks identified by the risk assessment, employers and the self-employed need to introduce preventive and protective measures.

Regulation 4 of the MHSWR 1999 requires that:

> *"Where an employer implements any preventive and protective measures he shall do so on the basis of the principles specified in Schedule 1 to these Regulations."*

Figure A8-2: Regulation 4 of MHSWR 1999. *Source: Management of Health and Safety at Work Regulations (MHSWR) 1999.*

The text of the schedule is reproduced below:

Schedule 1 - General principles of prevention

This Schedule specifies the general principles of prevention set out in Article 6(2) of EU Council Directive 89/391/EEC.

- Avoiding risks.
- Evaluating the risks which cannot be avoided.
- Combating the risks at source.
- Adapting the work to the individual, especially as regards the design of workplaces, the choice of work equipment and the choice of working and production methods, with a view, in particular, to alleviating monotonous work and work at a predetermined work-rate and to reducing their effect on health.
- Adapting to technical progress.
- Replacing the dangerous by the non-dangerous or the less dangerous.
- Developing a coherent overall prevention policy which covers.
- Technology, organisation of work, working conditions, social relationships and the influence of factors relating to the working environment.
- Giving collective protective measures priority over individual protective measures.
- Giving appropriate instructions to employees.

The principles of prevention outlined above are designed to direct the approach of duty holders in identifying and implementing the measures needed. The measures are not intended to be prescriptive; however, they should be applied when it is reasonable to do so. The intention is to ensure that risks are adequately controlled to an acceptable level.

It should be noted that risks assessments and the application of the general principles of prevention should not be used as a justification for reducing controls in a situation where high standards already exist. The following table lists the principles of prevention and comments on their application.

Measure	***Comment***
Avoiding risks.	If risks are avoided completely then they do not have to be either controlled or monitored. For example, not using pesticides or not working at height.
Evaluating the risks which cannot be avoided.	Carry out a suitable and sufficient assessment of risks.
Combating the risks at source.	Repairing a hole in the floor is much better than displaying a warning sign.
Adapting the work to the individual, especially as regards the design of workplaces, the choice of work equipment and the choice of working and production methods, with a view, in particular, to alleviating monotonous work and work at a predetermined work-rate and to reducing their effect on health.	This emphasises the importance of human factors in modern control methods. Factors to consider include natural lighting and ventilation, positioning of controls and reducing physical stressors such as manual handling and noise. If the well-being of the person is dealt with, there is less chance of the job causing ill-health and less chance of the person making mistakes which lead to accidents. Alleviating monotony by rotating staff work can help the individual to remain alert and pay attention to the task.
Adapting to technical progress.	Can lead to improved and safer working conditions, for example, new, non-slip floor surfaces.
Replacing the dangerous by the non-dangerous or the less dangerous.	For example, using a battery operated drill rather than a mains powered tool.
Developing a coherent overall prevention policy which covers technology, organisation of work, working conditions, social relationships and the influence of factors relating to the working environment.	Taking a holistic stance to the control of risk, which includes consideration of the organisation (system), the job (task analysis) and the individual (human factors).
Giving collective protective measures priority over individual protective measures.	Giving priority to creating a safe place of work before a safe person strategy. For example, fire arrangements for a large shopping mall include automatic detection, fire spread, fire suppression, direction and assembly.

Measure	Comment
Giving appropriate instructions to employees.	Ensure all workers, employees and self-employed, understand what they have to do, in normal and non-normal conditions.

Figure A8-3: General hierarchy of controls and application. *Source: RMS.*

Determine the technical/procedural/behavioural control measures using the general hierarchy of control

Risk control measures can be categorised by their type:

- Technical (place) (job).
- Procedural (system) (organisation).
- Behavioural (person) (person).

ELIMINATION (TECHNICAL)

Eliminate the hazard at source, for example:

- Change a work at height process such as cleaning windows so that it is done from the ground.
- Stop using a piece of noisy equipment.
- Using a non-hazardous substance instead of a hazardous one.

Technical control measures include:

- Equipment - design (for example, guarding) and maintenance.
- Access/egress - provision of wide aisles, access kept clear of storage items.
- Materials (substances and articles) - choice of packaging to make handling easier.
- Environment (temperature, light, dusts, noise) - local exhaust ventilation (LEV).

SUBSTITUTION (TECHNICAL/PROCEDURAL)

Substitution occurs when one hazard, for example, a toxic substance is replaced by something less hazardous. An example is using acrylic paint instead of VOC (volatile organic compound) paint. Sometimes this is classified as elimination since, effectively, the first hazard has been removed from the workplace.

Another approach to substitution is to change the form of a substance so that it is less likely to be inhaled or absorbed into the body. Immersing powders in a liquid to form a slurry or paste rather is an example of this idea.

SIGNS/WARNINGS AND/OR ADMINISTRATIVE CONTROL (PROCEDURAL/BEHAVIOURAL)

Administrative controls are those that involve changes to procedure, training installing signs and giving warnings through instruction. Administrative controls are not as effective as engineering controls because there is a heavy reliance on workers to follow the rules; putting up a sign saying 'No Entry' will not prevent someone from gaining access to a danger area.

Procedural control measures include:

- Policy and standards.
- Rules and procedures.
- Permit-to-work and safe systems of work.
- Maintenance.
- Authorisation and coordination of actions.
- Purchasing controls.
- Accident investigation and analysis.
- Emergency preparedness.

ENGINEERING CONTROLS (TECHNICAL/BEHAVIOURAL)

Engineering control measures include for example:

- Limiting the exposure of workers to dust in the atmosphere by the use of local exhaust ventilation.
- Limiting the exposure of workers to noise, vibration or chemical by task rotation, so that they each get exposed to a small amount instead of one worker to a lot.

Remove the person from the hazard, for example:

- Use robots for tasks that have risk, such as spraying of isocyanate paints.
- Separate people from hazards, like noisy equipment or dust.

Contain the hazard by enclosure, for example:

- Enclose dangerous parts of equipment.
- Do chemical mixing in an enclosed glove box.
- Enclose noisy equipment in sound absorbing material.

Behavioural control measures include:

- Information and training to improve awareness, knowledge, skill, competence.
- Attitude, perception, motivation, communication.
- Supervision.
- Health surveillance.
- Personal protective equipment.

Reliance on only behavioural measures (safe/healthy person strategy) is the weakest of controls. The preferred strategy involves the use of technical measures to provide a safe/healthy place of work; by ensuring a safe/healthy place of work all people that find themselves in it will gain protection.

In practice, the most successful organisations use a combination of the three measures (strategies), with the emphasis on using technical measures to make the workplace safe/healthy and supporting this with procedural measures and behavioural measures.

Many organisations that feel they have invested effort in getting the technical and procedural measures right have subsequently focused on the behavioural measures - studies have shown that behavioural measures are a critical aspect of effective management of health and safety.

Provide personal protective equipment (PPE), for example:

- Respiratory protective equipment, gloves or goggles for exposure to substances.
- Personal hearing protection for noise.

In some cases, for example when providing protection from biological hazards, hygiene regimes and inoculation are also important control measures.

All of the control measures can be supplemented by the provision of information, instruction, training and supervision; including in the form of safe systems of work and permits to work.

This hierarchical approach reflects the accepted view that eliminating and controlling risk by using physical engineering controls and safeguards is more reliable than relying solely on people. In practice, what makes one control more effective than another may depend on a number of factors, including:

- The number of people protected by the risk control measure. This reflects the risk control principles set out in the Management of Health and Safety at Work Regulations (MHSWR) 1999. It is better to use a risk control measure which will protect everyone who could be exposed to the hazard, for example, it is better to put a sound absorbing enclosure around a noisy source than to rely on everyone exposed to wear personal hearing protection.
- The level of risk reduction obtained by the use of that particular control.
- The cost to introduce and maintain the control measure, in time trouble and money. Some controls may have high initial costs and low on-going costs, or vice versa.
- The level of reliance the risk control has on human behaviour. Controls that rely on human behaviour may fail because it may not be properly used or used at all. Controls that remain in place at all times, other than at times of maintenance, would be preferable.
- The amount that the control relies on adjustment, maintenance, cleaning testing or replacement in order to remain effective.

Factors that affect choice of control measure are covered in more detail later in this section.

Factors affecting choice of control measures

LONG-TERM/SHORT-TERM

Short-term measures may be applied until further controls are arranged. For example, if there is a noise problem, hearing protection in the form of ear defenders or plugs may be issued until engineering methods of reducing the noise levels are devised, which could be a long-term project.

Painting yellow lines on the floor may be the short-term control for protecting pedestrians from forklift trucks, while the long-term controls involve re-routing the traffic routes.

The decisions are generally based on ease of implementation, legal requirements and cost. In the first example, the risk of noise-induced hearing loss has to be minimised, but the Control of Noise at Work Regulations say it should be done wherever possible by engineering means.

In the second example, the better solution may be very costly, so the company may factor it into its budget over a two-year plan.

APPLICABILITY

It is important not to take a blanket approach to the use of controls. The Safety Signs and Safety Signals Regulation require that only valid signs should be displayed. This means that signs must only be used where they are applicable.

For example, the requirement would mean the removal from display of an electrical hazard sign fitted to the door of an electrical distribution panel which has been isolated and discontinued in use.

PRACTICABILITY

If the legal requirement is 'practicable', then cost is not an issue when deciding on control measures - the measure must be taken if it is technically feasible.

This requires the employer or the competent person on behalf of the employer keeping up to date with modern technology. For example, this will require knowing what the latest and possibly better machinery guarding techniques are.

COST

If the legal requirement is qualified by 'so far as is reasonably practicable', then the cost is balanced against the risk. The choice of control measures will then depend on the level of risk. For example, if a floor occasionally got wet and two people had to walk over it, then the risk of slipping might be evaluated as low.

The control measures could be a warning sign, a mop or some sawdust to sprinkle on the wet part, with the cost as minimal. However, if the floor had different levels, often got wet and fifty people were moving around the area, the risk might be evaluated as medium or high. The control measures could then be installing a non-slip surface, looking at long-term measures to prevent the floor getting wet, issuing shoes with non-slip soles, etc. The cost would be substantially more than the low risk measures.

Even if there is a medium or high risk that needs to be controlled, there is nothing wrong with considering cheaper versions of control measures if they do the same job. It may also be cheaper and better if a task is changed rather than putting in expensive control measures. For example, reducing the time operators spend at a noisy machine will reduce their personal exposure to a lower level than the use of expensive engineering techniques.

When considering costs it is necessary to consider the whole cost, for example, noise reduction equipment for machinery may involve significant capital costs and use of personal protective equipment (PPE) may seem a cheaper option, but the cost of replacement PPE, training and supervision over forthcoming years has to be taken in to account.

EFFECTIVENESS OF CONTROL

The general hierarchy of control measures indicates a hierarchy of effectiveness. The most effective control is one that eliminates the hazard or substitutes it for one of acceptable risk.

Other controls will have a more limited effectiveness; provision of local exhaust ventilation (LEV) is a more effective control than the provision of personal protective equipment. However, the LEV's effectiveness can decline over time, so will need resources to inspect and maintain it. Personal protective equipment has limited effectiveness because it is particularly dependent on the user wearing it, how well it fits the user and correct replacement when it is contaminated or worn out.

LEGAL REQUIREMENTS AND ASSOCIATED STANDARDS

The choice of controls will need to take account of any specific legal requirements, for example, the requirement to guard dangerous parts of machinery. Reference should always be made to HSE Codes of Practice/Guidance, relevant British Standards and Trade Associations for such issues as guarding options and maintenance requirements.

COMPETENCE OF PERSONNEL AND TRAINING NEEDS RELEVANT TO PREFERRED CONTROLS

When choosing control measures it is important to consider the current competence of workers and the additional training that will be required on the introduction of the new measures. Measures that require the introduction of complex equipment can create the need for workers to learn new competencies.

For example, where the risks from manual handling tasks are reduced by the introduction of fork lift trucks, workers will need significant training in order to be able to operate them and not all workers may be capable of learning the new skills involved.

Similarly, where control measures require new systems of work, for example, a permit-to-work system, those required to use the system of work will need significant training and supervision. Training may have to be repeated periodically to ensure competence is retained.

A8.3 - Safe systems of work and permit-to-work systems

Safe systems of work

MEANING

A safe system of work is concerned with the integration of people, equipment and materials in the correct working environment (PEME) to produce the safest possible conditions in a specific work area.

LEGAL AND PRACTICAL REQUIREMENTS

Section 2(2)(a) of the Health and Safety at Work etc. Act (HASAWA) 1974 clearly requires the provision and maintenance of plant and systems of work, that are, so far as is reasonably practicable, safe and without risks to health. Employers also have a general duty of care at common law.

A safe system of work (SSOW) results from a formal consideration of all the hazards and controls involved in a particular task. Where the hazard potential is low and the controls are simple or automatic (for example, interlocking guard on a photocopy machine) the work instructions may be informal, i.e. verbal. Where the hazard potential is high and controls are complex (for example, entry to a confined space) a SSOW must be formal, i.e. written.

COMPONENTS (PEME)

People

Selection, mental and physical capability, willingness to conform with the systems of work, training, induction, job orientation, change in work practice, adequate knowledge, and skill, safe behaviour; supervision, adequately trained, positive motivation, resistance to pressures to behave in an unsafe manner.

Equipment

Good design; safety through specification of plant, machinery and equipment; ergonomic factors and planned maintenance.

Materials

Safe in both the raw state and as the finished product; appropriate quality standards; system to manage purchase, receipt, storage and use through to safe disposal of waste products and by-products of manufacture.

Environment

Effective control of heating, lighting and ventilation; safe levels of noise and vibration; effective control of dust, fumes, radiation, chemicals and biological hazards; effective means of access and egress; good standards of welfare amenity provision, sanitation, hand washing, showers, clothing storage, catering, drinking water, and first aid.

DEVELOPMENT AND IMPLEMENTATION

Development of a safe system of work

The development of a safe system of work involves the following steps:

1) Assess the task (task assessment/analysis).
2) Identify the hazards and assess the risks.
3) Define safe methods.
4) Implement the safe system.
5) Monitor and review.

Assess the task

Assessment of the task must consider not only the work to be done but also the environment where it is to be done, to allow a full consideration of the hazards to be made. This may involve a detailed review known as Task Analysis. Task analysis consists of a formal step-by-step review of the work to be carried out. All aspects of the task should be considered and recorded in writing to ensure that nothing is overlooked.

Typical considerations include:

- How is the task carried out?
- Where the task is carried out?
- What is used?
- What are the current controls?
- Are the controls adequate?
- Do operators use the controls correctly?
- Factors that may affect behaviour of workers/supervisors (error considerations).

How the progress of work will be monitored should be considered, in particular the health and safety arrangements. Any special requirements for monitoring should be specified during the planning stage, for example, gas contamination testing, temperature or pressure levels monitoring, measurement of emissions, foreseeable emergencies that may arise.

Hazard identification and risk assessment

The identification of hazards and the assessment of risks are key factors arising from task analysis. The Management of Health and Safety at Work Regulations (MHSWR) 1999, Regulation 3, requires a suitable and sufficient risk assessment to be made of all risks to which employees, and others who may be affected by them, are exposed. Where a significant risk is identified through the general risk assessments a more formal detailed analysis is often required to develop a safe system of work.

Task analysis may help with this. Each step in the task should be analysed to determine associated hazards, the controls in place to limit the probability of contact with the hazard and the severity of any effects of contact. The resultant risk should be evaluated and any additional controls implemented prior to defining the safe methods. Where possible, hazards should be eliminated at source.

Define the safe methods

- The person in control of work should be clearly identified.
- Specify responsibilities at various stages.
- Any competency requirements for those conducting the work should be identified.
- The need for protective or special equipment should also be identified, as should the need for the provision of temporary protection, guards or barriers.

- Adequate emergency procedures should be in place, or developed, to control likely incidents, for example, fire, spillage.
- If there is a possibility that injury could result during the task, emergency aid and rescue methods should be considered.
- The safe system of work should define who does what, where, when and how.
- The system should be checked against three main criteria:

1) It should adequately control hazards associated with the task.
2) It should comply with company standards.
3) It should comply with relevant legal standards.

Implementation

Once the system has been developed and agreed, preparation for implementation can proceed.

The process of implementation will involve:

- The provision for the communication of relevant information to all involved or affected is fundamental to any system of work.
- The person in charge of work ensuring that all stages of the system of work are clearly understood and implemented.
- If problems arise which necessitate modification to the system, formal approval and documentation should be made.
- Any permanent record of any monitoring must be kept and regularly checked by a member of the management team.
- Control of work to ensure conformity with the safe system of work. Various methods of control are available. Selection and application of the work control method should reflect the level of risk in the activity, for example, permit-to-work systems, method statements and supervision.

Safe system of work	***Bench drilling machine***
Task	Operation and use.
Special equipment	Swarf (metal shavings produced through machining) rake, clamping bolts and/or work vice; 'wear protection' signs; chuck/drill guard and/or trip device.
Personal protective equipment	Eye protection (BS EN 166).
Method	***General precautions*** 1) Do not operate machine unless you have been trained in its use and are authorised to do so. You must know the functions of all controls. In particular how the machine is stopped and isolated. 2) Ensure all pulleys, belts and shafts are fully guarded. 3) The drill bit guard must be used at all times whilst the machine is in motion. 4) If the machine is fitted with a trip device, ensure that it is set correctly (i.e. extended to the correct length and a max. of 75 mm from the bit). ***Before use*** 5) Check that the chuck key (used to tighten or loosen the drill bit) has been removed before starting. 6) Ensure all safe guards are in position and all protective clothing is worn. 7) Ensure safe access/egress to machine at all times. 8) Ensure all clothing is tightly fastened. Long hair fastened back or contained in a cap. Jewellery (for example, necklaces etc.) should be removed.

Figure A8-4: Safe system of work example. *Source: RMS.*

Importance of training

The Provision and Use of Work Equipment Regulations (PUWER) 1998, require that supervisors be adequately trained to enable them to identify the hazards and control strategies associated with work under their control. Where hazards cannot be eliminated or the risks reduced procedures to ensure a more formal safe method of work known as a permit-to-work system must be devised.

Task analysis is not only necessary to identify the hazards and controls, but is essential for preparing written procedures and specifying, by means of comprehensive training plans, the skill and knowledge content of the work to be carried out. The training plans will not only identify the sequence of work, but often the performance outcomes. Sometimes timeliness is an important consideration in performance outcomes, particularly in relation to certain chemical manufacturing processes.

The result of the analysis is incorporated into training programmes. Where training requirements have been identified by the analysis, it should be conducted or confirmed. For certain high risk tasks this will often involve a course to develop knowledge and understanding.

This is then followed by practical application of what was learnt, either utilising a work simulator (for example, train driver/air craft pilot) or close one to one supervision (for example, fork truck driver). The provision and availability of any required equipment or materials necessary for training should be confirmed.

Permit-to-work systems

ESSENTIAL FEATURES

A permit-to-work system is a formal written system used to control certain types of work that has high hazard potential.

A permit-to-work (PTW) system is more than a permit document and includes the people, equipment, materials and the environment involved in the system. The people aspect of the permit-to-work system includes the role, responsibilities and competence of those involved.

The equipment will mean arrangements regarding hazards from equipment being used or worked on, such as isolation of moving parts and other sources of energy.

The permit-to-work system will also include arrangements to minimise the effects of materials and substances that may be encountered or created by the work. The environment will include such things as limiting access to the area and provision of safe lighting.

In addition, the permit-to-work system includes mechanisms to control the work to ensure it is done in the correct way, for example, supervision and confirming signatures on a permit-to-work document.

Requirements of the system

- Must be formal and documented.
- Simple to operate.
- Have commitment of those who operate and are affected by it.
- Provide concise and accurate information.
- Liaison with controllers of other plant or work areas whose activities may be affected by the permit-to-work.
- Boundary or limits of work must be clearly defined and/or marked.
- Confirmation that the safe system of work is being conformed to.
- Contractors undertaking specific tasks must be included in the permit-to-work system, including any briefing prior to commencement.
- Provides for training in the system for all working under it and affected by it.

The permit-to-work document

A permit-to-work document:

- Specifies the work to be done.
- Pre-determines a safe procedure for the work.
- Is a clear record that foreseeable hazards have been considered in advance.
- Defines the appropriate precautions and the sequence in which they are to be carried out.
- Set out responsibilities.
- Confirms critical steps have been taken and by whom.

A permit-to-work document should only be as complicated as the work requires. One permit-to-work document could not cover all situations without being needlessly complicated for some.

GENERAL APPLICATION

A permit-to-work system should be used whenever the method by which a task is to be done is likely to be critical to the health and safety of those involved, other nearby workers, the public or the plant itself.

A permit-to-work system is often required when the safeguards available in normal working are no longer available. Before maintenance work is begun, consideration should be given as to whether a permit-to-work is required.

The type of work may involve:

- Hot work of any type or the use of tools or equipment that may create sparks, in an area where there is the possibility of fire or explosion.
- Entry into a confined space.
- The disconnection or opening of pipelines or vessels that have contained flammable, toxic or harmful substances.
- Work on machinery or electrical equipment.
- Working at height.
- Working in excavations.
- Work on pressure vessels.

The Health and Safety Executive suggest in their document ***'HSG250, Guidance on permit-to-work systems'***, that permit-to-work systems are normally considered most appropriate to:

- Non-production work (for example, maintenance, repair, inspection, testing, alteration, construction, dismantling, adaptation, modification, cleaning etc).
- Non-routine operations.
- Jobs where two or more individuals or groups need to co-ordinate activities to complete the job safely.
- Jobs where there is a transfer of work and responsibilities from one group to another.

OPERATION

The role of a permit-to-work system is to ensure that full and proper consideration is given to the risks of the particular work that it relates to and to control the way of working in order to ensure a safe system of work is followed.

The operation of a permit-to-work system will involve:

- Ensuring the proper authorisation of specified work.
- Confirmation of the identity, nature, timing, extent and limitations of the work.
- Establishing criteria to be considered when identifying hazards and what they are.
- Confirmation that hazards have been removed, where possible.
- Confirmation that control measures are in place to deal with residual hazards.
- Confirmation that work has been started, suspended, conducted, and finished safely.
- Confirmation of who has control of the location and equipment relating to the work, particularly when it passes between parties.
- Controlling change and considering other work activities that might interact with specified work.
- Providing a record that the steps in the safe system of work have been followed.

EXAMPLE			***ENTRY INTO CONFINED SPACES***
Possible Lay-Out For A Permit-To-Work Certificate **Permit-To-Work Certificate**			
PLANT DETAILS (Location, identifying number, etc)		ACCEPTANCE OF CERTIFICATE Accepts all conditions of certificate	Signed Date Time
WORK TO BE DONE			
WITHDRAWAL FROM SERVICE	Signed Date Time	COMPLETION OF WORK All work completed - equipment returned for use	Signed Date Time
ISOLATION Dangerous fumes Electrical supply Sources of heat	Signed Date Time		
CLEANING AND PURGING Of all dangerous materials	Signed Date Time	EXTENSION	Signed Date Time
TESTING For contamination	Contaminations tested Results Signed Date Time		
I CERTIFY THAT I HAVE PERSONALLY EXAMINED THE PLANT DETAILED ABOVE AND SATISFIED MYSELF THAT THE ABOVE PARTICULARS ARE CORRECT (1) THE PLANT IS SAFE FOR ENTRY WITHOUT BREATHING APPARATUS (2) BREATHING APPARATUS MUST BE WORN Other precautions necessary: Time of expiry of certificate: Delete (1) or (2)	Signed Date Time	THIS PERMIT-TO-WORK IS NOW CANCELLED. A NEW PERMIT WILL BE REQUIRED IF WORK IS TO CONTINUE Signed Date Time	
		RETURN TO SERVICE	I accept the above plant back into service Signed Date Time

Figure A8-5: Example of a permit-to-work for entry into confined spaces. *Source: HSE guidance note on permits-to-work.*

The permit-to-work system is typically formalised by pre-printed forms (permit-to-work document). These are used to specify the hazards likely to be present and the controls that are in use (for example, electrical isolation, to prevent electric shock) or the controls that need to be used by the individuals at risk (for example, safety harness to prevent falls from a height).

The permit-to-work document will also specify the location and boundaries of the work to be done, the time period during which work may be carried out and any arrangements to deal with foreseeable emergencies. The permit-to-work document may only be issued or extended by an appointed competent person(s).

To be fully effective those who receive the permit-to-work document, upon which conditions of work are obligated, should be similarly appointed and competent. On completion of the work, the persons responsible should formally sign off the permit-to-work document and state that the work has been completed satisfactorily or, if the work is incomplete, what is necessary to ensure a safe re-instatement of the work process. The appointed issuer of the permit-to-work document should formally receive the work back and sign to indicate that all isolations etc. have been reinstated. The permit-to-work system should be formally monitored and records kept of compliance and effectiveness.

Use of risk assessment in the development of safe systems of work and safe operating procedures

Risk assessment will identify the hazards and controls related to a given task, enabling them to be considered in order to define a safe system of work or safe operating procedure. If the risk assessment process is conducted in a step-by-step manner this can help identify the hazards and controls for each step, making it easier to construct the safe system of work or safe operating procedure. The task analysis methods already described lend themselves to this approach.

It is essential when drawing up safe systems of work and procedures, that a control is identified, and specified, for each hazard. This will enable a logical link between the risk assessment and safe systems of work or safe operating procedures.

Where generic safe systems of work and procedures are used they can be compared with the risk assessment appropriate to a specific location or circumstances to enable the identification of hazards not covered by the generic approach. This will enable the generic systems or procedures to be made more specific and the definition of a safe system of work or safe operating procedure for the work undertaken at that specific location.

Element

A9

Organisational factors

Learning outcomes

On completion of this element, candidates should be able to demonstrate understanding of the content through the application of knowledge to familiar and unfamiliar situations. In particular, they should be able to:

A9.1 Explain the types of health and safety leadership, their advantages, disadvantages and likely impact on safety performance.

A9.2 Explain the organisational benefits of effective health and safety leadership.

A9.3 Explain the internal and external influences on health and safety in an organisation.

A9.4 Outline the different types of organisation, their structure, function and the concept of the organisation as a system.

A9.5 Explain the requirements for managing third parties in the workplace.

A9.6 Explain the role, influences on and procedures for formal and informal consultation with employees in the workplace.

A9.7 Explain health and safety culture and climate.

A9.8 Outline the factors which can both positively and negatively affect health and safety culture.

Contents

Relevant statutory provisions

Health and Safety (Consultation with Employees) Regulations (HSCER) 1996

Health and Safety Information for Employees Regulations (IER) 1989

Safety Representatives and Safety Committees Regulations (SRSC) 1977

Trade Union and Labour Relations (Consolidation) Act (TULRCA) 1992 (as amended)

Health and Safety (Consultation with Employees) Regulations (Northern Ireland) (HSCERNI) 1996

Sources of reference

Reference information provided, in particular web links, was correct at time of publication, but may have changed.

Consulting workers on health and safety, Safety Representatives and Safety Committee Regulations 1977 (as amended) and Health and Safety (Consultation with Employees) Regulations 1996 (as amended), L146, HSE Books, ISBN: 978-0-7176-6461-0, http://www.hse.gov.uk/pUbns/priced/l146.pdf

Financial Reporting Council - Guidance on Risk Management, Internal Control and Related Financial and Business Reporting (September 2014), https://www.frc.org.uk/Our-Work/Publications/Corporate-Governance/Guidance-on-Risk-Management,-Internal-Control-and.pdf

Hersey-Blanchard Situational Leadership Theory, http://www.leadership-central.com/situational-leadership-theory.html#axzz3p0SLB4U1

Leading health and safety at work, INDG417, IOD and HSE publication, HSE Books, http://www.hse.gov.uk/pubns/indg417.pdf

Occupational health and safety management systems - requirements (OHSAS 18000 series),

OHSAS 18001, ISBN: 978-0-5805-9404-5

OHSAS 18002, ISBN: 978-0-5806-2686-9

The above web links along with additional sources of reference, which are additional to the NEBOSH syllabus, are provided on the RMS Publishing website for ease of use - www.rmspublishing.co.uk.

A9.1 - Types of safety leadership and their likely impact on health and safety performance

The meaning of safety leadership

A safety culture starts with leadership; leadership drives culture, which in turn drives behaviour. Although leadership has been defined from various perspectives, all of them agree on the following characteristics:

- Leadership is a process that involves influence.
- Leadership occurs in groups and involves common goals and purpose.

As an organisation matures, it usually passes through stages in its attitude to safety management. It is the basic duty of senior management to provide resources, dedicated safety personnel, safety training and incident investigations. The organisation should then develop safety management systems, set safety objectives and introduce reactive and proactive monitoring. The aim of the arrangement is to ensure continual improvement in safety.

The next stage should concentrate on what is sometimes called 'behavioural based safety' - ensuring individual ownership of safety. An additional ingredient is necessary for this to happen: 'The Safety Leader. Managers at all levels need to act as safety leaders. This is summed up (particularly in the oil and gas industry) as 'Walk the Walk - Talk the Talk'.

According to T.R. Krause (Leading with Safety: A Behaviour-Based Approach 2005), there are seven ways in which the characteristics and associated behaviours of leaders can influence safety culture:

- Credibility - what leaders say is consistent with what they do.
- Action orientation - leaders act to address unsafe conditions.
- Vision - leaders paint a picture for safety excellence within the organisation.
- Accountability - leaders ensure employees take accountability for safety-critical activities.
- Communication - the way leaders communicate about safety creates and maintains the safety culture of the organisation.
- Collaboration - leaders who encourage active employee participation in resolving safety issues promote employee ownership of those issues.
- Feedback and recognition - recognition that is soon, certain and positive encourages safe behaviour.

Types of safety leadership, their advantages, disadvantages and likely impact on safety performance

TRANSFORMATIONAL TYPE

Transformational leadership is a type of leadership style that can inspire positive changes in those who follow. Transformational leaders are generally energetic, enthusiastic, and passionate. Not only are these leaders concerned and involved in the process; they are additionally focused on helping every member of the group succeed as individuals related to group success.

Both leaders and followers enable each other to advance to a higher level of morale and motivation. Through the strength of their vision and personality, transformational leaders are able to inspire followers to change expectations, perceptions, and motivations to work towards common goals.

Acting as a role model, inspiring and motivating employees to work safely and showing concern for employees' welfare enhances a number of safety outcomes including fostering perceptions of a positive safety climate, promoting higher levels of employee participation in safety activities, compliance with safety rules and procedures and safety citizenship behaviours - participation in safety committees, looking out for colleagues' safety.

TRANSACTIONAL TYPE

Transactional leadership is also known as managerial leadership, it focuses on the role of supervision, organisation, and group performance. Transactional leadership is a style of leadership in which the leader promotes the compliance of followers through both rewards and punishments..

This type of leadership is effective in crisis and emergency situations, as well as for projects that need to be carried out in a specific way. There are times during 'normal' operations when it can be seen as overbearing or repressive.

Transactional leaders have a desire for a positive safety climate, positive safety behaviours and reduced accident rates.

TRANSFORMATION VERSUS TRANSACTIONAL

In terms of human characteristics, a purely transformational personality may not be practical enough to deal with workers who are 'difficult' or 'resistant'. A blend of transformational and transactional characteristics is more likely to produce an effective safety leader.

SERVANT

Servant leadership was developed by Robert K. Greenleaf in 1970. Servant leaders serve the people being lead, which implies that employees are an end in themselves rather than a means to an organisational purpose or 'bottom line'. Servant leadership is meant to replace command and control models of leadership, to be more focused on the needs of others.

Servants devote themselves to serving the needs of organisation members, focusing on meeting the needs of those they lead, developing employees to bring out the best in them, coaching others and encouraging their self-expression; they facilitate personal growth in all who work with them and listen well to build a sense of community and joint ownership.

Servant leaders are felt to be effective because the needs of followers are so looked after that they reach their full potential, hence perform at their best. This style of leadership considers that the first responsibility of a leader is to define the requirements and the last task is to say thank you to the followers.

In between, the leader is a servant of the health and safety requirements in the same way as the followers. The style is non-intrusive but it must be accompanied by a close monitoring process to ensure that the goals are being effectively approached since the followers may not be the best placed to recognise the leader's goals.

SITUATIONAL AND CONTEXTUAL

The situational leadership model is a leadership theory developed by Paul Hersey, professor and author of the book 'The Situational Leader' and Ken Blanchard, leadership trainer and author of 'The One Minute Manager'.

The theory was first introduced as 'Life Cycle Theory of Leadership' during the mid-1970s and was renamed 'Situational Leadership Theory'.

The basis of the situational leadership theory is that there is no single 'best' style of leadership. Effective leadership is task-relevant and the most successful leaders are those who adapt their leadership style to the maturity of the individual or group they are attempting to lead or influence.

Maturity is considered to be the capacity to set high but attainable goals, the willingness and ability to take responsibility for the task and relevant education and/or experience of an individual or a group for the required task.

Effective leadership varies, not only with the person or group that is being influenced, but it also depends on the task, job or function that needs to be accomplished. The Hersey-Blanchard Situational Leadership Model rests on two fundamental concepts; leadership style and the individual or group's maturity level.

Behavioural attributes of an effective leader

Safety leaders exhibit a variety of behaviours that, often unintentionally, influence fellow workers to improve their safety behaviour, such as:

- Being an example, by knowing and following the rules.
- Avoiding complacency for the dangers of the job.
- Reporting safety hazards, violations, and incidents.
- Maintaining open communication with colleagues and management about safety concerns.
- Implementing change to improve safety and working conditions.
- Encouraging co-workers to exhibit safe behaviours.
- Showing colleagues small tokens of appreciation for a safe job well done.
- Making colleagues aware that unsafe behaviour is unacceptable.
- Being involved in safety initiatives and committees.

An excellent report from the Health and Safety Executive has these conclusions.

Transformational and transactional theories of leadership have received consistent support in the literature suggesting that they are effective approaches for the management of safety. Specifically, combining transformational and transactional leadership styles have been shown to have several safety benefits and are crucial for the development of a positive safety.

Both leadership styles involve active forms of leadership and emphasize the importance of not turning a blind eye (this 'idiom' means to deliberately ignore undesirable information) to safety through demonstrating an active involvement and commitment to safety.

A transformational leadership style requires managers to take an active role in safety and demonstrate behaviours, such as articulating a safety vision for the organisation and explaining how it can be achieved, acting as a safety role model and exemplifying the importance of safety in both words and deeds, showing concern for employees' welfare and respond and act upon their safety concerns (for example, Yukl, 2010; Flin & Yule, 2004).

A transactional leadership style requires managers to explain performance expectations and set high safety performance standards; also to recognise and reward positive safety behaviours and practices.

Training interventions may be an effective way of helping managers to develop these leadership skills. For instance, safety-specific transformational leadership training has been shown to be an effective approach to improving occupational safety (Mullen & Kelloway, 2009).

Analysis of the causes of major incidents suggest that leaders should pay attention to the importance of open and trusting safety communications with the workforce.

Leader behaviours as well as the quality of relationships that they develop with employees are crucial for developing trust and promoting open safety communications.

Leaders can influence the extent to which employees will communicate safety issues and concerns by developing good working relationships characterised by openness, support and mutual respect (for example, Mullen, 2005; Kath et al., 2010b).

In summary - the most effective safety leadership may require a combination of management styles intelligently applied according to the needs of the organisation.

A9.2 - Benefits of effective health and safety leadership

Purpose of HSE/Institute of Directors (IOD) guidelines 'Leading health and safety at work'

As part of the Government's 'Revitalising Health and Safety' initiative the HSC and Institute of Directors (IOD) produced guidance for directors relating to their health and safety responsibilities. This was called 'Leading Health and Safety at Work'.

It was produced in order to reflect the important role that senior managers play in successful health and safety management.

"Health and safety is a fundamental part of business. Boards need someone with passion and energy to ensure it stays at the core of the organisation."

Figure A9-1: Quote from health and safety leaders. *Source: HSE/IOD, INDG417.*

The guidance reflects a common management approach, plan, deliver, monitor and review. . The main points in their leadership guidance are set out in the following section.

PLAN

Core actions

- Board aware of risks.
- Set policy as integral part of culture, values and standards.
- Policy sets out board and individual responsibilities for leadership.
- Communicate, promote and champion health and safety.

Good practice

- Health and safety on the agenda for board meetings.
- Chief Executive gives clear visibility of leadership.
- Consider appointing one board member as health and safety director (champion).
- Setting targets.
- Consider appointing non-executive director as scrutineer to ensure robustness.
- Integrate within corporate governance structures.

Figure A9-2: Guidance for Directors. *Source: HSE/IOD.*

DELIVER

Core actions

- Establish management systems to protect employees, customers and the public.
- Take responsibility and ownership.
- Ensure health and safety is resourced.
- Obtain competent health and safety advice.
- Ensure risk assessments are carried out.
- Ensure risks are dealt with sensibly, responsibly and proportionately.
- Involve employees or their representatives in decisions.
- Consider new processes, practices, and personnel.
- Make board decisions in the context of the health and safety policy.
- Design in health and safety when implementing change.

Good practice

- Board should be visible where work is done and seen to be following health and safety rules.
- Consider health and safety when making management appointments.

- Introduce health and safety procurement standards.
- Ensure health and safety performance of partners, suppliers and contractors contributes positively.
- Establish a risk management/health and safety committee chaired by a senior executive.
- Provide health and safety training for the Board.
- Support worker involvement above the legal duty - improve participation and commitment.

MONITOR

Core actions

- Ensure reporting of preventive (training, maintenance) and incident data.
- Ensure periodic audits of effectiveness of management structures and risk controls.
- Impact of changes is reported timely.
- Introduce procedures for changed legislation and other external developments and events.

Good practice

- Report health, safety and well-being in annual report.
- Board extra workplace visits to gain information for review.
- Celebrate good performance at central and local level.
- Ensure audits are seen as a positive Board and management tool.

REVIEW

Core actions

- Review performance at least annually.
- Examine whether policy reflects current priorities, plans and targets.
- Examine effectiveness of reporting to Board.
- Consider shortcomings and effect of Board decisions.
- Decide actions to address weaknesses.
- Conduct immediate reviews when major shortcomings or incidents occur.

Good practice

- Monitor sickness absence to identify symptoms of problems.
- Use data collected to benchmark with others.
- Appraise senior managers on their contribution to health and safety.
- Consider reports on performance and actions of contractors.
- Consider involving workers in monitoring.

HEALTH AND SAFETY LEADERSHIP CHECKLIST

- How does the board demonstrate commitment to health and safety?
- Is there a periodic board-level review of health and safety?
- Is competent health and safety advice available at all levels, including the board?
- Are all staff - including the board - sufficiently trained and competent to carry out their health and safety responsibilities?
- Is the workforce, particularly safety representatives, consulted properly on health and safety matters?
- Are health and safety concerns communicated to the appropriate level including, as necessary, the board?
- Are systems in place to ensure the organisation's risks are assessed, and that sensible control measures are established and maintained?
- Are audits or assessments undertaken to inform the board about what the organisation and contractors actually do?
- Does the board receive information regularly about health and safety, for example, performance data and reports on injuries and work-related ill health?
- Are targets set to improve health and safety and is the organisation's performance benchmarked against others in its sector or beyond?
- How are changes in working arrangements, which have significant implications for health and safety, brought to the attention of the board?

Leadership as a core element of effective health and safety management

The HSE have published guidance note 'HSG65 - Successful Health and Safety Management' in which they introduce the subject by emphasising that effective leadership is essential in organising and controlling the management process.

In ***figure ref A9-3*** the strong link between leadership and management is illustrated. Protecting the health and safety of employees or members of the public who may be affected by the organisations activities is an essential part of risk management and must be led by the board of directors of the organisation.

Failure to comply with these requirements can have serious organisational and individual consequences. Sanctions include fines, imprisonment and disqualification from directorships that can have an immediate impact on the organisation and its employees as well as the individual director.

By not managing health and safety as a key business risk in board decisions, catastrophic results can follow. Many high-profile safety cases over the years have been rooted in failures of leadership.

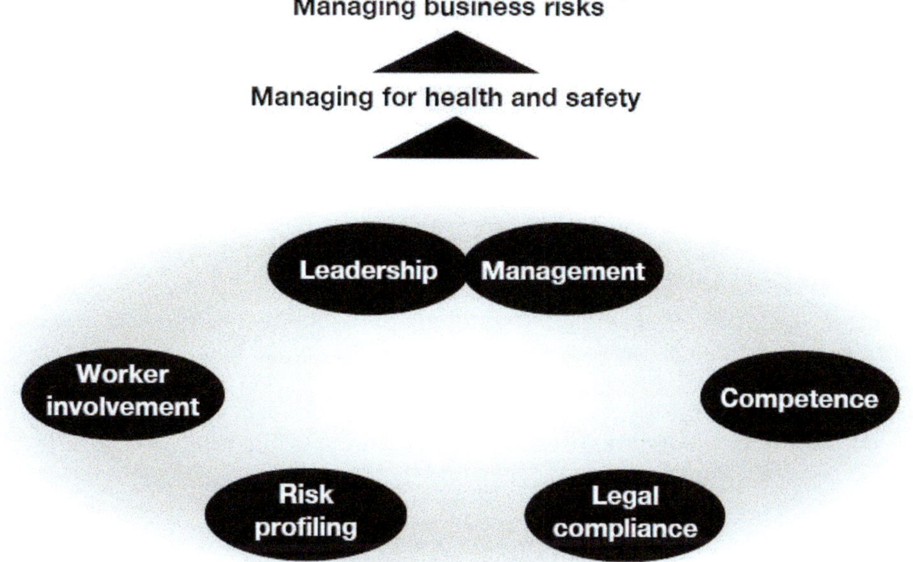

Figure A9-3: Core elements of safety management. *Source: HSG65 HSE UK.*

Benefits of effective safety leadership

The Chartered Management Institute in the UK recognises the benefits of effective leadership in any area of management including health and safety:

- Leading people by providing purpose and inspiring trust.
- Managing change.
- Communicating information and knowledge to aid decision making.
- Managing activities and resources on time, within budget and meeting any standards that are required.
- Encouraging 'self-management' using influence and persuasion.
- Enabling workers to achieve personal goals.
- Creating high performance teams.
- Capturing the ingenuity of people to ensure innovation.

For the benefits of good health and safety as detailed in the HSE's guidance 'Leading health and safety at work', ***see figure ref A9-4***:

Addressing health and safety should not be seen as a regulatory burden alone, it offers significant business benefits including:

- *Reduced costs and reduced risks - employee absence and turnover rates are lower, accidents are fewer, the threat of legal action is lessened.*
- *Improved standing among suppliers and partners.*
- *A better reputation for corporate responsibility among investors, customers and communities.*
- *Increased productivity - employees are healthier, happier and better motivated.*

Figure A9-4: Benefits of good health and safety. *Source: HSE, INDG 417.*

The link between effective leadership and employee engagement

Achieving the right combination of skills, talent and experience is essential to the success of all organisations. However, when the employee attends work they need to be engaged with the job role and use their skills, talent and experience to positive effect.

Imagine your organisation with 30% of the lights not working or 30% of the computers not working! Employees not being engaged with the organisations activities can have a significant influence on organisational performance.

A range of studies indicate that having an engaged workforce can lead to significant improvements in productivity, reduced absenteeism, increased workforce retention, improved income growth, increased innovation and an improved health and safety record.

An effective leader must positively inspire confidence in the organisations future and that of the employees and empower employee engagement.

An extract from an article by Jack W Riley further clarifies this relationship between engaged employees and effective leadership.

> *Employee engagement - the extent to which employees are motivated to contribute to organizational success, and are willing to demonstrate commitment, loyalty, and a willingness to go beyond basic requirements to accomplish tasks and organizational goals is critical for business success. However, nurturing a motivated workforce in the face of challenging economic times that include downsizing, furloughs, and salary reductions creates new challenges. With reduced resources and grim economic pressures, leaders are still expected to motivate high levels of performance. In prosperous times, it is equally essential to maintain high levels of engagement to retain the company's best talent. In periods of turbulence and change as well as in times of economic growth, leaders who understand what drives employee engagement can build a workforce that is motivated to perform. Conversely, how employees view company leadership can have a significant impact on their engagement level and overall views of the company.*
>
> *To drive higher levels of engagement, it is important to first understand the behaviours that support employee engagement: Leaders must inspire confidence in the future of the organization. The Bureau of Labor Statistics projects a shortfall of 10 million workers in the United States by this year and a slowdown in the Labor force growth rate through 2018. As organizations require more work from fewer people, it's imperative to have leaders who can motivate remaining employees to perform their best. Leaders who minimize uncertainty by securing employees' perceptions of the organization's future, and support the belief that there is a future for the employees, will be able to drive higher levels of engagement.*
>
> *Managers must demonstrate respect and recognition for their employees. Employee recognition is a critical component of achieving higher levels of engagement. Having reinforcement from managers and others whom employees hold in high regard is essential. People like to be appreciated, acknowledged, and respected and know that their efforts are making a difference for their organization. Managers need to understand that recognizing employees for their contributions is more than just nice to do, it is an imperative.*
>
> *In addition to receiving praise for a job well done, employees want to know that their opinions are valued, that they have input into decisions, and that even when they make mistakes, they can count on the support and respect of their managers. Individuals must have a job that is interesting and exciting to them. Management and leadership play a significant role in employee engagement, but it's equally important for employees to have enthusiasm for the work that they do. Beyond mastering the job's tasks, employees need to feel pride in what is being accomplished and share in the excitement when goals are achieved. The more employees feel they are building their own skill sets and feel competent and in control of their work product, the more likely they are to demonstrate high levels of engagement.*

Figure A9-5: Quote from Jack W. Riley © 2010 Wiley Periodicals, Inc. *Source: Published online in Wiley InterScience (www.interscience.wiley.com).*

Encouraging positive leadership and supporting managers at all levels

ROLE OF THE ORGANISATION

Psychologist Fredrick Herzberg discovered that employee satisfaction and dissatisfaction are not opposites. For managers and leaders to be satisfied in their jobs, you must first remove the causes of dissatisfaction, and then add factors that contribute to satisfaction. For example, are there policies in the organisation that could be causing dissatisfaction for leaders? Is each person's salary competitive? Are there traditions that are resistant to change? Is there an owner of the organisation who opposes change or will not relinquish authority and control? Are the shareholders excessively powerful?

The culture and values of the organisation will produce a corresponding response from the management team - either positive or negative.

ROLE OF THE SAFETY PRACTITIONER

The safety practitioner has the task of 'enabling' and 'supporting' line management in their management of health and safety.

> *"Every employer shall, subject to paragraphs (6) and (7), appoint one or more competent persons to assist him in undertaking the measures he needs to take to comply with the requirements and prohibitions imposed upon him ..."*

Figure A9-6: Regulation 7 of MHSWR 1999. *Source: Management of Health and Safety at Work Regulations (MHSWR) 1999, UK.*

Often the role of the safety practitioner is reduced to a list of responsibilities. The safety practitioner has the responsibility for:

- Ensuring the company is aware of statutory obligations and recommended Codes of Practice.
- Advising management of their responsibilities for accident prevention and avoidance of health hazards.
- Interpreting and keeping management and employees informed of new and developing legislation and other standards.
- Through fine management and supervision, advising where improvements in health and safety standards or practices are appropriate.

- Regular health, safety and housekeeping inspections which cover buildings, plant, equipment, services, and fire arrangements, to ensure conformity with regulations and company policies.
- Maintaining statutory safety records and making statutory safety returns, in addition to maintaining health and safety records required by the company.
- Advising on possible hazards when considering the introduction of new machinery, new materials, new processes, or changes in existing ones.
- Overseeing and reviewing all accident investigations and preparing statistics to assist in monitoring health and safety performance.

Although these details are important, the role is similar to that of the leaders of the safety culture. The safety practitioner must understand the existing culture and support the leaders in their objectives in achieving continual improvement.

This may involve being a 'communication bridge' with the workforce. Visible and verbal presence in the workplace is essential. The safety practitioner should have qualities of character that combine (among others) approachability, communication skills, diplomacy, humour, patience and the realistic assessment of human nature. The safety practitioner must have the enthusiastic support of the leadership team.

Influence of corporate social responsibility and business ethics on health and safety management

Business ethics are moral principles that guide the way a business behaves. The same principles that determine an individual's moral duty also apply to business. For a business to act ethically means distinguishing between 'right and 'wrong' and then making the 'right' decision. Unethical business practices include 'insider trading' the purchase of stocks and shares when the organisation has confidential information relating to the trade, bribery or corporate gifts to an individual to gain orders and the use of copyrighted materials etc. are examples of unethical business practices. However, good ethical practices are harder to define since the organisation must make a competitive return for its shareholders and keep its employees in employment. The organisation also has wider responsibilities for the environment - responsibly disposing of hazardous waste may be expensive but it is ethically and legally the right choice for the organisation to make.

PRINCIPLE OF 'CORPORATE SOCIAL RESPONSIBILITY'

Corporate social responsibility (CSR) is about understanding the organisation's impact on the wider world and considering how it can be used in a positive way. CSR can also be good for improving the on-going profitability and effectiveness of the organisation. Organisations cannot exist in isolation - their employees are dependent upon them for their livelihood and customers need their products and services. The local community is affected by their operation and their demands will impact on suppliers of raw materials and others such as suppliers of water/sewage, electricity and gas. They may also have an impact on the environment. The principle of corporate social responsibility requires organisations to focus on a number of important aspects.

EMPLOYEES

For the responsible business, consideration of employees means doing more than simply complying with legal requirements, for example, investing in health and well-being programmes for workers.

SUPPLIERS

Responsible organisations must consider the choice of supplier and what is expected of them, for example, trading with suppliers who pollute the environment is as irresponsible as the organisation doing it itself. Where possible, the use of local suppliers should be considered. This helps support the community and also reduces carbon emissions from deliveries. Customers openness and honesty about products and services. It is important that customers are told what they want to know and what they need to know, including what steps the organisation takes to be socially responsible. If something goes wrong, it should be acknowledged and the problem dealt with efficiently, timely and gracefully.

LOCAL COMMUNITY

Many socially responsible organisations involve their employees in working with the local community, for example, some encourage employees to volunteer for community activities and provide paid time off for this. Working in sympathy will improve community relations and can help employees to improve motivation and their interpersonal and team participation skills.

IMPACT ON THE ENVIRONMENT

The current responsible ethos is to reduce the impact on the environment, for example, creating recyclable products and sourcing responsibly. It also includes:

- Using recycled materials and sustainable renewable sources.
- Minimising packaging.
- Buying locally to save fuel costs.
- Creating an efficient (and fuel-efficient) distribution network.
- Working with suppliers and distributors who take steps to minimise their environmental impact.

The Influence of the Financial Reporting Council guidance on internal control

The Code published by the Financial Reporting Council (FRC) applies to accounting periods beginning on or after 17 June 2016 and applies to all companies with a Premium listing of equity shares regardless of whether they are incorporated in the UK or elsewhere.

Figure A9-7: UK Corporate Governance Code. *Source: FRC.*

THE MAIN PRINCIPLES OF THE CODE

Leadership

Every company should be headed by an effective board which is collectively responsible for the long-term success of the company. There should be a clear division of responsibilities at the head of the company between the running of the board and the executive responsibility for the running of the company's business. No one individual should have unfettered powers of decision.

The chairman is responsible for leadership of the board and ensuring its effectiveness on all aspects of its role. As part of their role as members of a unitary board, non-executive directors should constructively challenge and help develop proposals on strategy.

Effectiveness

The board and its committees should have the appropriate balance of skills, experience, independence and knowledge of the company to enable them to discharge their respective duties and responsibilities effectively. There should be a formal, rigorous and transparent procedure for the appointment of new directors to the board. The board should undertake a formal and rigorous annual evaluation of its own performance and that of its committees and individual directors. The board is responsible for determining the nature and extent of the principal risks it is willing to take in achieving its strategic objectives. The board should maintain sound risk management and internal control systems.

Accountability

The board should present a fair, balanced and understandable assessment of the company's position and prospects. The board is responsible for determining the nature and extent of the principal risks it is willing to take in achieving its strategic objectives. The board should maintain sound risk management and internal control systems.

The board should establish formal and transparent arrangements for considering how they should apply the corporate reporting, risk management and internal control principles and for maintaining an appropriate relationship with the company's auditors.

Remuneration

Executive directors' remuneration should be designed to promote the long-term success of the company. Performance-related elements should be transparent, stretching and rigorously applied.

There should be a formal and transparent procedure for developing policy on executive remuneration and for fixing the remuneration packages of individual directors.

Relations with shareholders

There should be a dialogue with shareholders based on the mutual understanding of objectives. The board should use general meetings to communicate with investors and to encourage their participation.

A9.3 - Internal and external influences

Internal influences on health and safety within an organisation

OVERVIEW

Internal influences on health and safety will once again depend on the maturity of the organisation with regard to health and safety. This maturity and how it is displayed may be described as the culture of the organisation. The culture may be positive or negative, with many shades of grey in between.

FINANCE

Finance is a primary resource of an organisation and all factors come back to this. Even a small health and safety initiative requires a person's time and time has to be paid for. The provision of adequate finance for the phase of development of the organisation is important. Perhaps more important is to use the resource of finance wisely. It is necessary to take a risk-focused approach to health and safety to ensure priorities are resourced first. The absence of this could be a negative influence on health and safety in the long run. Finance provision tends to be planned and set on an annual cycle, which means that health and safety needs not planned for may have to wait until the next year's budget allocation. It is therefore essential to plan health and safety measures well in advance.

Finance for the cost of accidents is best allocated against budget holders that were the cause of the accident. It defuses the budget holder's focus on health and safety matters if the cost is absorbed into a central point.

PRODUCTION TARGETS

Production targets are an integral part of running an organisation. The setting of targets (objectives) is a positive way of influencing an organisation, provided they are realistic and achievable. They are only considered realistic and achievable if they can be met in a safe and healthy manner. It is not uncommon for employees (including managers) to relegate health and safety by taking risks to achieve a production target. Analysis of the situation often reveals such things as ill communicated health and safety objectives and well defined production targets. This, together with more senior management attention (monitoring of) to whether production targets are being met, leads to focus on production to the exclusion of health and safety.

Individuals may choose to take risks to meet their personal production targets, for example, when involved in 'piece' work on a production line.

TRADE UNIONS

Internal influences of trade unions, where they exist in organisations, depends mainly on the views of the individuals that comprise the members, but in particular their representatives. In situations where appointed representatives deal with all union matters their efforts on health and safety may be diluted, particularly at times when pay is being reviewed. If the representative is appointed to deal with health and safety alone they may be able to make more contribution and be of a less reactive nature. Many union representatives have received significant training on health and safety matters and therefore have an opportunity to influence health and safety through knowledge. In some organisations union representatives have received more training than managers, which can lead to misunderstanding and conflict.

ORGANISATIONAL GOALS

Organisations are required to set out their health and safety goals in their health and safety policy statement. The purpose of the policy is to inform staff and others who may be affected by the organisation's activities how health, safety and welfare issues will be addressed by the organisation.

Goals should be established to meet both legal and the moral obligations, to reduce the organisation's impact on the health, safety and welfare of all those concerned. Without clear health and safety goals other clearer goals will tend to take priority over the organisation's resources and effort.

It is important to set out how the goals will be achieved, by whom and in what period. The process used to achieve the goals is usually through the setting of ***'smart'*** (specific, measurable, achievable, relevant and timely) objectives. The process should recognise the need for ongoing continuous improvement, since technological improvements and societal pressures will always place the challenge on organisations to do better.

CULTURE

Culture relates to the values and beliefs of an organisation and may have a positive or negative influence on health and safety. If the culture of the organisation is one that has a natural, unconscious bias to productivity over health and safety, or a tendency for short-term thinking and a reactive approach, this can create a strong negative influence on health and safety. An organisation whose culture is to consider issues that affect it in the short, medium and long term and has a strong planning and proactive approach that includes and understands the importance of health and safety will have a strong positive influence.

External influences on health and safety within an organisation

OVERVIEW

There are many external influences on an organisation. They will have a varying influence depending on the current status of the management of health and safety in the organisation. The status of health and safety management may be observed to be in one of three broad stages, 'young', 'immature' and 'mature'.

The young organisation will tend to be driven by events that are occurring and the pressures put on it by external organisations such as the enforcing authorities. The young organisation tends to see remedies as technical in nature and can be said to be operating at level 1.

The immature organisation will tend to be driven by unplanned events, but is beginning to establish systems and practices in anticipation of events. The choice of preventive systems tends to be those required to comply with the law, for example, conducting risk assessments and to be procedural in nature and is therefore seen to be operating at level 2.

The mature organisation has spent considerable resource establishing active systems and practices. Unplanned events that result in actual loss are in-frequent. Enhanced systems and practices are being established as the organisation observes opportunities for improvement. Attention is tending to be focused on preventive systems and practices that are behavioural in nature and is therefore seen to be operating at level 3.

LEGISLATION

In the past, legislation has tended to be seen as a punitive influence. More recently, with the introduction of legislation that encourages preventive action and self-development, this perspective has softened. Balancing this is the move from specific prescriptive action for the organisation to comply with to management actions that must be carried out as duties by individuals (and the organisation).

For legislation to influence organisations they must either want to comply or there has to be a real prospect of punishment for non-compliance.

PARLIAMENT/HSE

Parliament and the Health and Safety Executive (HSE) can influence an organisation on a strategic basis. Parliament and the HSE have an influence from the point of view that they decide what laws are appropriate and when they are to be introduced. As such they can control the quantity of law and the scope/extent of a given law. If organisations successfully lobby these bodies, they may be able to gain modification or delay. In addition, legislation is supported by government and HSE policy and initiatives.

These are designed to promote awareness and positive action by organisations. There have been initiatives focussing on the management of health and safety and on specific risks like asbestos. These influence organisations to focus on these issues and set targets for improvement.

The initiatives communicate to organisations that these issues are important and the government/HSE feel strongly enough that they are doing something about it. The inference being, that organisations should also be doing something about the issues if they are acting responsibly and enforcement agencies will look out for good practices.

ENFORCEMENT AGENCIES

They have a significant role in influencing an organisation's level of performance with regard to health and safety. If the focus of the enforcement agency is on technical specifics, then organisations will tend to follow this lead and deal with these issues. It is therefore important that the enforcing agencies demonstrate the value of and interest in seeing organisations use technical, procedural and behavioural preventive measures.

The relative influence of the enforcement agencies is highly dependent on them being sufficiently field active to contact a significant number of organisations. It would not be a balanced influence if the only time they were seen was when there had been an accident; they therefore have to be seen by organisations before accidents occur, in order to encourage planned preventive actions.

COURTS/TRIBUNALS

Courts and tribunals have an influence on health and safety after the event, in that they consider health and safety based matters after an issue has arisen, for example, appeal against a prohibition notice. Because they represent a neutral perspective they provide a source of arbitration and moderation of extreme views/practices concerning health and safety.

CONTRACTS

Contracts establish an opportunity for organisations to set standards for health and safety performance of suppliers/contractors. It represents the equivalent of a contract of service for employees. The system of awarding contracts, of itself, does not have a positive or negative influence on health and safety.

Influence rather depends on the standards expected of the person that establishes the contract and then the level of monitoring/enforcement of the contract. In many organisations it has been used successfully to establish good health and safety standards in work activities that the client organisation has had trouble achieving with a history of resistant employees.

CLIENTS/CONTRACTORS

Work undertaken by contractors is normally covered by a civil contract. It is good practice for health and safety requirements to be written into the contract. Because health and safety responsibilities are defined in criminal law, they cannot be passed to another party through a contract.

However, clients must satisfy themselves that contractors are competent to carry out work without risks to health and safety. Clients should also ensure that contractors are aware of the health and safety standards that are expected of them.

TRADE UNIONS

Trade unions have had a significant influence on health and safety in previous years. In recent times their influence has subsided nationally as the number of members has reduced though they have remained influential in some industries/organisations. Union activities have, in recent times, tended to be distracted by other employment matters.

There is now a renewed union interest in health and safety matters as it is beginning to be seen as a worthwhile way of servicing their members' needs. Unions have always maintained a profile of member support in making claims from employers with regard to injury at work. As claims consciousness increases their role and influence will increase. Trade Unions, through the Trade Union Congress (TUC), are represented on the Health and Safety Commission (HSC). Through this forum they have an influence on strategic aspects of health and safety, contributing to the development of Regulations and Approved Codes of Practice.

INSURANCE COMPANIES

Insurance companies have become increasingly aware that they may have under-valued the risks related to some companies they are insuring. This has caused them to look again at the risks and the factors that lead to claims. In conjunction with this they influence organisations to minimise the risks in order to control their level of premium. The new look includes a focus on the status of management of health and safety in the organisation insured, which has been found to be a useful predictor of future loss.

PUBLIC OPINION

Public opinion tends to fall into two parts: strategic influence by the general mass of public with regard to its tolerance of specific workplace hazards or situations (for example, Display Screen Equipment or major disasters). Local influences tend to surround acceptability or unacceptability of the practices of a specific organisation. This is most acute following an accident and has had the effect of causing closure of some smaller organisations.

A9.4 - Types of organisations

The concept of the organisation as a system

Consider in conjunction with HSG65 and BS OHSAS 18001.

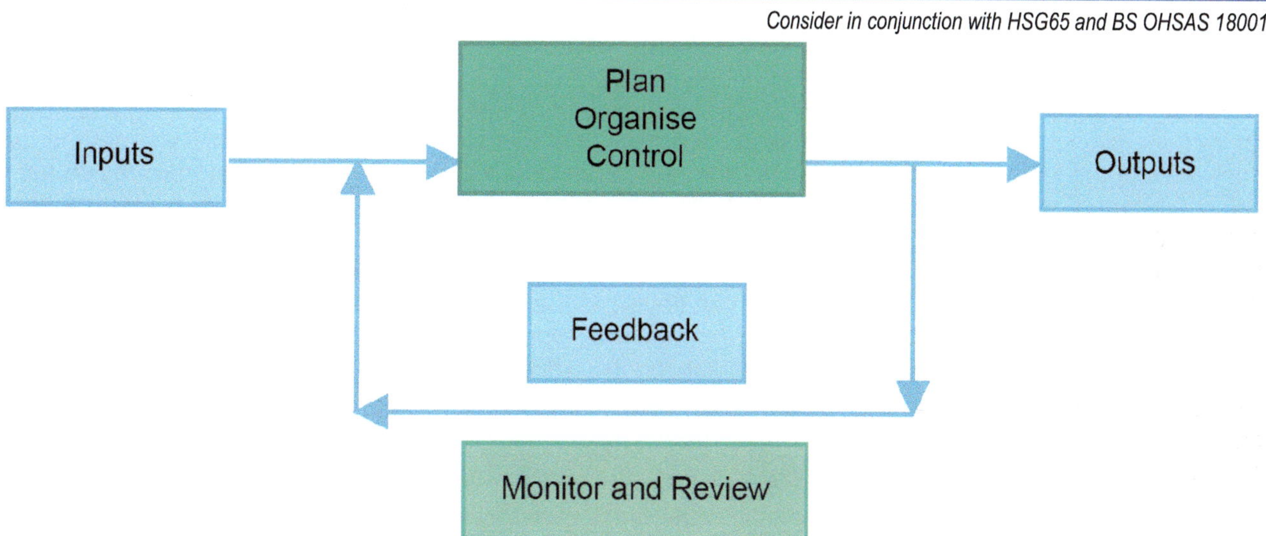

Figure A9-8: The organisation as a system. *Source: RMS.*

As may be seen by the above diagram an organisation may be represented as a system comprising inputs that are processed to provide outputs, with a feedback loop to adjust the process to ensure the correct output is consistently produced.

If we overlay the health and safety components on this, we find the following:

Inputs	Includes resources in the form of raw materials, time, money, people in conjunction with goals (derived from moral, legal and economic sources) and information. Routes into the organisation include design and development of structures, equipment and materials, acquisitions, purchase of products and services, and recruitment.
Process	Includes actions related to planning for health and safety, organising, controlling. The process part of the system can be broken down further into planning, organising, controlling of the following three main strategies: 1) Technical measures. 2) Procedural measures. 3) Behavioural measures.

Outputs	Includes planned outputs such as provision of products and services, many resultant by-products (for example, waste) and information for external use. This must be seen in conjunction with the health and safety outputs; negative outputs (if the process is not successful) such as injury accidents, ill-health, fines and positive outputs such as good health, reduced insurance premiums and public respect.
Feedback	Includes monitoring on a short period basis through meetings and inspections, longer term such as accident investigation or audits and review of general/specific actions of the process.

Organisational structures and functions

The effective functioning of an organisation is dependent on a number of factors:

- Type of organisation - formal or informal.
- Members of the organisation understand the goals.
- Members of the organisation identify positively with the goals.
- Members know what their role and function is in the organisation.
- Members are competent (and confident) to fulfil their role and function.
- Members communicate effectively.
- Members co-operate to reach the goal.
- The organisation controls the actions of members in order to maintain direction towards the goal.

FORMAL

The most common formal structure is hierarchical, with authority being dispersed through the lines of the organisation that separate into different roles and then 'down' through the structure to the members of the organisation that carry out specific tasks.

Formal structure to an organisation does not establish a style of organisational working; this is established by the culture and may be autocratic through to democratic.

INFORMAL

Within a formal structure of an organisation informal organisations are created. Their purpose (goal) is varied and may include personal relationships, habitual work-related contact, influence/lobby groups, getting round the structure to avoid bureaucracy, and to achieve objectives and goals not identified by the formal organisation.

Health and safety might be a good example of the last purpose, in that it may be achieved despite the formal organisation rather than because of it. Necessary improvements that could not get formal backing and budget might be achieved by an informal organisation that uses subtle means to spread/hide costs and get the task done without the knowledge of the formal organisation. Similarly, the informal organisation may decide not to follow the formal requirement for a particular health and safety measure and cohesively resist, and on some occasions agree to hide this fact.

Informal organisations tend to rely on verbal communications, unless it is unlikely that the written word will be discovered by the formal organisation. They can have positive or negative influences on success. They can undermine established systems and elevate the effective importance of people that have the right informal connections and/or a strong personality, but little authority. This can mask persistent bad practice, such as not planning for health and safety, allowing a person to get something done ahead of those that plan and use the formal system to achieve things.

LARGE OR SMALL ORGANISATIONS

Small organisations tend to have less formal organisational structure, which allows roles to be re-assigned more easily, on a temporary or longer-term basis. Individuals adopt a role and tend to stay in it for a long time. Whereas larger organisations tend to be more rigid, people are appointed to a post with a role, which is fixed and does not tend to change. People often have more opportunity to change their role in larger organisations as they may have more posts that are vacated and require to be filled.

Smaller organisations may tend to be less specific with authority; responsibility and accountability for health and safety, in larger organisations they will tend to have a more defined approach. However, authority, responsibility and accountability may be well defined in larger organisations, but it may not include health and safety.

Large organisations tend to have a number of layers of management, which can make communication and implementation of health and safety goals more difficult, whereas in smaller organisations this may be more easily achieved.

In larger organisations it is easier for the senior management to become more remote from where the work is carried out, with the consequence that their commitment to health and safety may become less visible and their understanding of the risks affecting workers become diluted. In smaller organisations, senior management tend to be more visible and are likely to be more closely involved in the work activities and therefore aware of health and safety issues affecting workers.

In larger organisations, communication upwards to senior management about risks and the need for resource to resolve them can be more difficult, resulting in inertia. Also, it is easier for a manager in a large organisation, particularly one with multiple locations, to remain unaffected by the health and safety goals and aspirations of the organisation, and their lack of performance can go unnoticed. This is more difficult in a smaller organisation.

ORGANISATION CHARTS

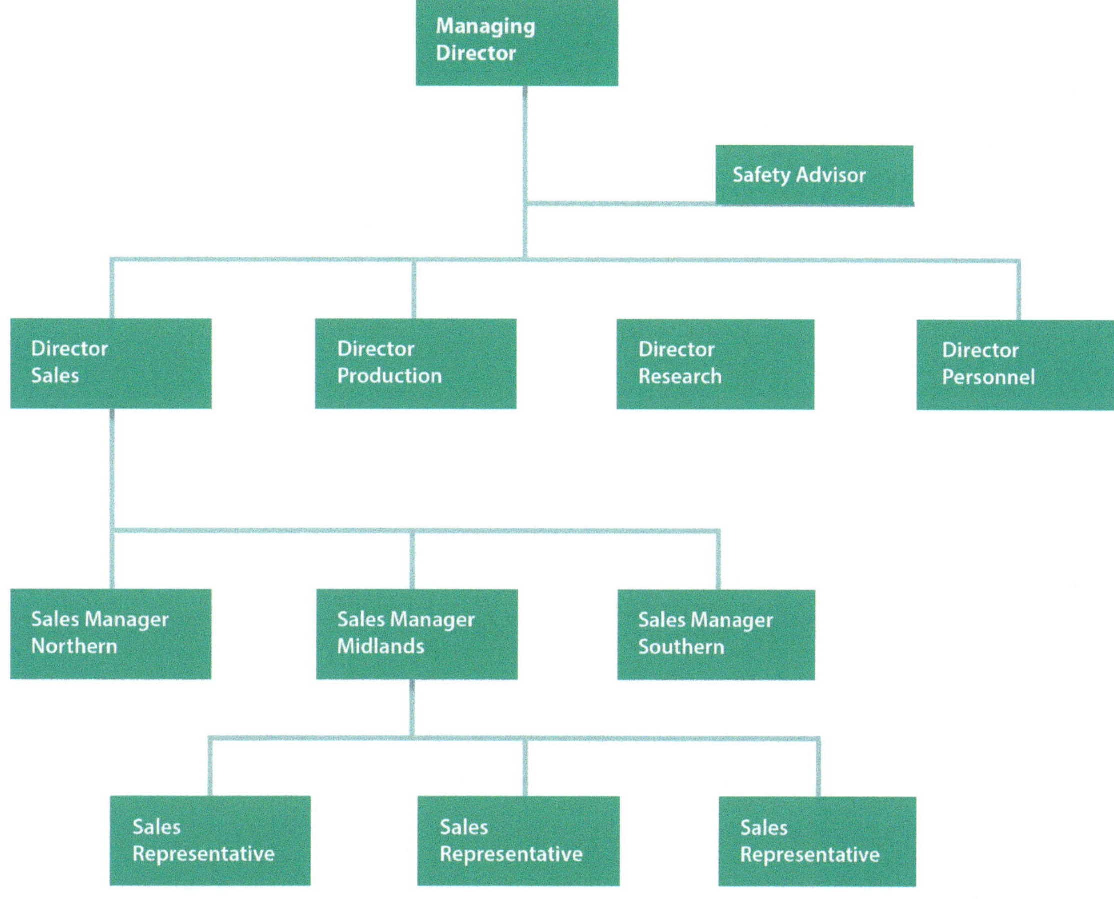

Figure A9-9: Organisational chart - ORG Ltd. *Source: RMS.*

ROLE OF MANAGEMENT

The management's role should be to complement the health and safety policy of the organisation. The specific responsibilities of management should be:

- To recommend to the senior management team improvements or changes in health and safety practices and their priority.
- To plan health and safety into their work activities.
- To interpret the policy to subordinates/employees and ensure that they are understood, action is taken to organise for it and it is implemented effectively.
- To work with technical staff and health and safety practitioners in monitoring, reviewing and improving work systems, with the object of improving health and safety performance.
- To monitor the work of subordinates/employees and stimulate their interest and involvement in health and safety.
- To control the work activities of their subordinates/employees so that health and safety is assured.
- To review reports and statistics and investigate adverse results prior to initiating corrective action.
- To take part in particular investigations of hazards and accidents as required.

The relationship of line management to health and safety practitioners should be clearly established. There is much debate on the appropriate management reporting position in the organisation for the health and safety practitioner.

This will depend largely on the existing organisational structure, but the following factors should be borne in mind:

- The safety practitioner should report to an effective senior, who is sufficiently senior to call for action right across the organisation.
- They should not report through to a 'Production' Manager as it may produce an intolerable health and safety versus production conflict. Many organisations, therefore, favour administrative or industrial relations based reporting lines.

- In large organisations a number of health and safety practitioners may be employed and their inter-relationships are important. Each large organisation has its own culture, which will affect the type of structure adopted. Some are highly centralised, others have complete autonomy.

HIERARCHICAL VS FLAT MANAGEMENT STRUCTURES

Hierarchical structure

A hierarchical organisation follows the shape of a pyramid. Every employee in the organisation, except one, usually the CEO, is controlled by someone else within the organisation.

Advantages of hierarchical structure

- Employees understand the levels of leadership within the organisation.
- There is a clear route for promotion to motivate employees to perform well.
- Employees feel loyal to their departments and work for its success.

Disadvantages of hierarchical structure

- Communication across different departments is sometimes difficult.
- Rivalry between departments may be damaging to the overall success of the organisation.
- Increased bureaucracy often damages an organisation's flexibility.
- Salaries for multiple layers of management increase an organisation's costs.

Flat structure

A flat organisation refers to an organisation structure with few or no levels of management between management and staff level employees. The flat organisation supervises employees less while promoting their increased involvement in the decision-making process.

Advantages of a flat structure

- It promotes the employees' sense of responsibility in the organisation.
- It speeds communication and coordination throughout the organisation by removing hierarchical routes and decisions can be made more quickly and easily.
- Eliminating the layers of middle management reduces an organisation's wage bill.

Disadvantages of a flat structure

- Employees have no immediate 'boss' to discuss personal needs.
- The exact job function of employees may not be clear.
- Larger organisations may have to divide into smaller units if they wish to change into a 'flat' structure.

Potential conflict between organisational goals and those of the individual

There can be a wide difference between the goals of the individual and those of the organisation. The differences can result in conflict, for example, where the level of productivity is seen by the individual to be more important to the organisation than health and safety issues. This is particularly the case where this leads the individual to be exposed to risks that they find unacceptable.

Where the goals of the organisation include a desire for good health and safety performance this can have a potential for conflict in situations where individuals do not see this as in their short-term interests. For example, a manager may identify that in meeting these goals it is going to involve considerable personal effort, and that of the manager's team. The manager may not feel the personal effort involved is worthy and may prefer to put effort into other things than health and safety.

Where the organisation's goals represent a change in which health and safety may not have been considered before, managers may identify that to meet these goals they will have to bring about significant change in the perspective and behaviour of their workforce, which they see as very difficult. They may prefer to ignore the organisational goals. The manager may be encouraged to take such action if the organisation has had many initiatives in the past that have faded away; they are likely to see health and safety in the same way.

Where organisational health and safety goals have been established workers may find themselves in conflict with the organisation's goals because they are not adopting the appropriate behaviour towards health and safety. Workers may prefer to take risks because they see that doing so makes their life easier and gets work done; this may lead to them receiving a reward for productivity.

A particular organisational goal may be to reduce the number of reported accidents, but this may conflict with the individual's goal to ensure accidents affecting them are reported in order that they gain compensation and the accident is prevented. The organisational goal can become confused and lead to conflict because the individual is torn between reporting an accident and compromising the goal. This can be exacerbated when bonuses are paid to a group of individuals if reported accidents are below the target level.

Conflict between the organisation and the individual can also be said to have a positive function. It can result in rules and norms being identified and boundaries established. Each understands the role of the other better and can assess the authority the other possesses. This 'knowing where you stand' can aid stability, which can be beneficial to both parties. However, in general, the aim is to reconcile the differences which cause conflict.

The goals of the individual and organisational may be summarised as follows:

Individual goals	***Organisational goals***
Physical and mental well-being.	High productivity.
Reward for their contribution, particularly financial.	Low error conditions - health and safety, environment and quality.
Job satisfaction.	Low absenteeism.
Personal development.	Co-operation.
Achievement.	Harmony.
Respect from work-group.	Constructive disagreements.
	Low labour turnover.
If goals are met, a contented, productive workforce.	If goals are met, a successful, efficient organisation.

The human relations approach tends to see human needs as relatively fixed, and so it is up to the organisation to create an environment in which both its needs and those of its employees can be met. This can be achieved by analysing the human needs and promoting ways which allow people to achieve them whilst, at the same time, meeting the needs of the organisation.

Integration of the goals of the organisation with the needs of the individual

INTEGRATION OF GOALS

In order to integrate the goals of the organisation with those of the individual it is important that a sense of involvement or ownership is achieved. This must be part of the wider issue of creating a positive health and safety culture, in which all play a part. Where individuals and the management form a collective commitment to health and safety, the organisation develops the positive synergy of a positive health and safety culture.

In a report published by the ***CBI, 'Developing a safety culture'***, a number of dominant themes emerged:

- The crucial importance of leadership and the commitment of the chief executive.
- The executive health and safety role of line management.
- Involvement of all employees.
- Openness of communication.
- Demonstration of care and concern for all those affected by the business.

Worker health and safety representatives can also make a substantial contribution to the development of a positive health and safety culture. They are important in the communication between workers and management. In fact, effective communication is probably the most important health and safety performance indicator.

Creating the sense of involvement required to bring the goals of the individual and the organisation closer together is not an easy task. J. Neumann (1989) found in her research that resistance to participation by the workers is usual where the participation effort by the organisation is seen as a 'bolt on' adjunct to the basic operation of the organisation.

The most important reason for not participating, she found, was when the real decisions of the organisation were made elsewhere and the individual jobs of the workers did not require participative decision making. Early attempts failed because the workers' representatives were insulted by the agenda which consisted of 'lifts and lavatories', and the managers felt despair at the lack of interest. It is important that participation is not restricted to trivial issues if a sense of worth and ownership is to be created and nurtured. Ways should be found of encouraging participation and ownership if the situation of shared goals is to be achieved.

A method of achieving shared goals, which in turn can develop a positive safety culture, is the Japanese approach to individual management known as 'Japanisation'. Whatever its merits and demerits it is worthy of consideration. Japanisation involves what Friedman calls 'responsible autonomy', which encourages the worker to adapt to changing situations rather than the system of 'direct control' which limits by close supervision and keeps worker responsibility to a minimum.

It is a system of 'paternalism' or 'familyism' where the organisation is treated as a family: a social organisation that operates in an economic environment. It is also based on lifetime employment from school/college/university to retirement.

Quality is central to the approach and workers are highly trained and have the responsibility of getting it right first time. Workers participate in discussions and problem solving. Management are expected to be team members and there are no outward shows of status differences. The constant aim of all is for an improvement in performance. The objective is that it becomes second nature to behave in a manner that conforms to the shared goal of quality. Health and safety is incorporated into the system and an accident is seen as a failure in quality assurance. The health and safety of those using the product is also included in the quest for quality.

There would be several problems if this method were adapted in its entirety, but there is a move in the UK to a flatter management structure and workforce consultation. What is interesting about the Japanese approach is that the workers at all levels are strongly committed to common goals.

AUTHORITY, RESPONSIBILITY, ACCOUNTABILITY

In organisational terms, managers are delegated authority by those appointed more senior than themselves. In the case of the most senior manager of an organisation, authority may be delegated by shareholders/stakeholders to someone like the chief executive officer (CEO). In turn, the CEO may delegate authority to others in the organisational structure. Authority provides rights to act in a particular way if they so wish, for example, a manager may be given authority to run a department of an organisation or to fulfil expenditure against a budget.

Authority also carries a responsibility, for example, a manager given authority to run a department will be responsible for running it in a manner that fits expectations of the person who gave the authority. Ultimately, this will mean meeting the most senior manager's expectations, such as those for health and safety. Therefore, the manager will be responsible for running the department so that it meets organisational goals and health and safety is assured. Similarly, a manager may be given authority to hold health and safety committee meetings, but will also be given a responsibility to carry them out.

When a manager's execution of authority is evaluated, to see if they have met their responsibilities, they will be accountable for their success or failure, the consequences of which could reflect the full range of disciplinary or commendation action that the organisation makes available to the manager holding the subordinate to account.

'Accountability' means ultimate 'responsibility', and relates to the person who is held to account if something is not done, does not work, or fails to achieve its objective.

Figure A9-10: The term accountability. *Source: BS OHSAS 18002.*

It is important that organisations establish clear health and safety responsibilities for managers and workers in order that they meet the organisations' goals. Responsibilities that are written down are clearer to managers and enable them to be held to account more easily. Responsibilities should be specific enough to enable the manager to understand the actions required to fulfil them and to enable their measurement.

A9.5 - Requirements for managing third parties

Identifying third parties

Managers and worker employees work for, and are employed by, the organisation directly, whereas third parties are those that do not work for the organisation directly. The term third parties will include the following categories of people.

Contractors are that class of third party who do work for the organisation to the extent that is set out in a contract. The workers are employed by the contracted organisation. This can include the self-employed who may be working under a contract for service. It will also include those that work as a sub-contractor to contractors. Contractors may have a long-term contract for service, for example, a contract caterer providing lunches at a school, or short term, such as an electrician called in to the school to replace a light fitting.

Visitors are those classes of persons that are allowed on an organisation's premises because they are vested with authority to do so or are invited to access them by the organisation. Those allowed to visit premises by vested authority would include enforcing authorities like the HSE and Fire Authority. Visitors invited to access premises will include people from other organisations who visit in order to sell goods or services, those that have a technical interest in what the organisation does and work experience students. In some situations, volunteers may be considered to be visitors.

Trespassers are not invited to visit an organisation's premises, but they do so of their own volition. Trespassers may be opportunist individuals that take a short route through the premises, children attracted by what they see as an interesting play area, sales people or others who are not invited or individuals with illegal intent.

Members of the public include neighbours of the organisation, people that are affected by a work activity conducted outside the organisation's premises, people who come onto an organisation's premises to purchase goods and those affected by the goods or services provide by the organisation.

Agency workers

Agency workers are temporarily hired by their employer (the agent) to work in a client organisation. The agency and the client both have responsibilities under law.

MHSWR 1999 Regulation 12

Every employer shall ensure that the employer of any employees from an outside undertaking who are working in his undertaking is provided with comprehensible information on the risks to those employees health and safety arising out of or in connection with the conduct by that first-mentioned employer of his undertaking; and the measures taken by that first-mentioned employer in compliance with the requirements and prohibitions imposed upon him by or under the relevant statutory provisions.

Every employer shall ensure that any person working in his undertaking is provided with appropriate instructions and comprehensible information regarding any risks to that person's health and safety which arise out of the conduct by that employer of his undertaking.

Every employer shall ensure that the employer of any employees from an outside undertaking who are working in his undertaking is provided with sufficient information to enable that second-mentioned employer to identify any person nominated by that first mentioned employer to implement evacuation procedures as far as those employees are concerned; and take all reasonable steps to ensure that any employees from an outside undertaking who are working in his undertaking receive sufficient information to enable them to identify any person nominated by him to implement evacuation procedures as far as they are concerned.

Other employers in shared premises

Where two employers share premises but carry on separate businesses, they have duties to each other in law.

MHSWR 1999 Regulation 11

Where two or more employers share a workplace (whether on a temporary or a permanent basis) each such employer shall co-operate with the other employers concerned so far as is necessary to enable them to comply with the requirements and prohibitions imposed upon them by or under the relevant statutory provisions; (taking into account the nature of his activities) take all reasonable steps to co-ordinate the measures he takes to comply with the requirements and prohibitions imposed upon him by or under the relevant statutory provisions; and take all reasonable steps to inform the other employers concerned of the risks to their employees' health and safety arising out of or in connection with the conduct by him of his undertaking.

Internal rules and procedures concerned with the selection, appointment and control of contractors

SELECTION AND APPOINTMENT OF CONTRACTORS

Assessment of contractor

Acquiring the services of a competent contractor with relevant professional qualifications and a proven safety record is fundamental to any risk control programme. This is also necessary in order to comply with the OLA 1957 and OLA 1984 (or the Occupiers' Liability (Scotland) Act (OSLA) 1960), the general duty of care under section 2 of HASAWA 1974 and the requirement to undertake a risk assessment under Regulation 3 of the MHSWR 1999. It is fundamental to the MHSWR 1999 in particular.

This should be a proactive exercise. As far as is possible the types of works and activities that are likely to involve contractors working in the premises should be identified, for example, electrical, mechanical, gas, maintenance, etc.

It will be possible from this to draw up a list of professionally suitably qualified contractors and carry out an investigation into their safety policies and their safety performance at previous locations. Employers should additionally check out their qualification to do the required works. For example, for work involving gas appliances, the contractor should be registered with the HSE Gas Safe scheme.

The employer should produce a 'check list' to carry out this vetting process. In order to obtain any further information which may be required, the check list should include the following:

- A copy of contractor's health and safety policy document.
- Details of the Contractors' 'Competent person(s)', for example, internal health and safety practitioner(s) or consultant(s)).
- Details of assessments which have been carried out and the risk control measures that have been introduced, for example, MHSWR 1999, COSHH 2002, manual handling, etc.
- References from previous employers of the contractor.
- Details of any accidents or incidents that occurred in any other works contracts (check with the relevant HSE area office if these are reportable under the Reporting of Injuries, Diseases and Dangerous Occurrences Regulations (RIDDOR) 2013).
- Details of the contractor's emergency procedures for their employees, for example, fire, accident, injury and first aid.
- Information on any prosecutions, prohibitions and improvement notices issued, a time limit may be specified, for example, the last three years.
- Details of health and safety training provided.
- Brief details of safe systems of work (for example, permit-to-work).
- Brief details of how the contractor controls their sub-contractors.
- Copies of the contractors' Public Liability and Employers' Liability insurance certificates.

The Construction (Design and Management) Regulations (CDM) 2015 placed many of these duties upon 'clients'. Following the above vetting, a 'short-list' of competent contractors can be produced.

This procedure could prove to be invaluable, as often the services of contractors are needed in emergency situations, when sufficient time is not available to carry out these vetting procedures.

Consultation prior to commencement of contract

Before any contract work is commenced, a responsible person representing the Principal Contractor must discuss with the occupier the health and safety precautions necessary as far as the occupier's workforce is concerned and any other third parties on site.

This requirement was highlighted by the decision in R v Swan Hunter Shipbuilders Ltd [The Times, July 1981] where both the main contractor (Swan Hunter Shipbuilders Ltd) and a sub-contractor (Telemeter Installations Ltd) were fined £3000 and £1500 respectively for failing to advise and inform the on-coming sub-contract labour of the dangers produced by oxygen-enriched atmospheres in confined spaces. As a result 8 men were burnt to death whilst working below decks on HMS Glasgow. This incident took place in 1976.

Hence, except on clearly defined 'green field' sites, where the Principal Contractor is likely to have primary responsibility for health and safety, both occupier and Principal Contractor should ensure that:

- The site of operations is clearly defined, if necessary on a location plan, including those areas where contractors' staff are not permitted to enter.
- Agreement is reached as to whether any amenities for example, catering, washing, sanitation, first aid and clothing storage, are to be made available for contractors' employees.
- Where appropriate, the contractor has full information concerning the occupier's processes or activities which may affect or involve contract work for example, specific hazards, health and safety precautions to deal with the hazards, hygiene requirements, parking restrictions.

Use of owner's equipment

Contractors should generally be expected (by agreement between parties) to provide all their own tools, plant, equipment and materials necessary for the satisfactory performance of work.

On no account should use be made of the owner's electricity, gas or air mains without appropriate authority. Where such permission is granted, the method of connection should be approved by the appropriate manager.

Unexpected situations requiring a contractor

It is foreseeable that an unexpected situation can occur where the services of a contractor who is not on the 'approved list' are required. In these circumstances it may be possible to establish a proposed contractor's suitability by contacting organisations that have previously used the contractor.

Whatever arrangements the employer has for appointing a contractor, reasonably practicable documentary evidence should be produced as proof that appropriate steps were taken to ensure the contractor's competency to carry out the prescribed works. Additionally, information stating that appropriate risk assessments were carried out and risk control measures were introduced should also be able to be reproduced if necessary.

CONTROL OF CONTRACTORS

Liaison

A suitable member of the client's staff must be made responsible for liaising with the contractors on a daily basis. Similarly, a member of the contractor's staff must be responsible for supervising their work.

Concerns about the competence or effectiveness of the supervision must be raised immediately.

Access

Contractor's employees will need to be accounted for in case of fire or other emergency. Several options are available to control access: signing-in systems, pass-card systems, electronic (swipe) systems, etc.

Permits to work

The principal advantage of a permit to work system is that it forces clients and contractors into contact with each other and ensures that all risks and appropriate control measures are identified before work commences. However, the permit to work system must be properly managed if it is not to become a purely bureaucratic exercise.

The permit to work system may cover any or all of the following aspects.

Contractors' general permit

This will be used for all work which does not involve hot work, electrical equipment, confined space entry or working alone. This permit is suitable for control over access to the occupier's premises, including a booking 'in and out' system, thereby ensuring that the occupier is fully aware of the time when the contractors are at work.

The occupier can also use the contractor's general permit to work to advise contractors of emergency procedures, the requirement to maintain escape routes free from obstruction and accident reporting procedures.

Hot work permit

This should be used for work that at some stage will require welding, cutting or use of flame or spark producing tools.

Electrical permit

A dead equipment electrical permit to work should be used for work on dead electrical equipment. A live equipment electrical permit to work should only be used when it is necessary to work on live electrical equipment, when it has been established that it is reasonable to do so.

Confined space entry permit

This permit should be used for work which at some stage requires entry into a vessel or other confined space.

Lone workers permit

This should be used for work which will be carried out by a person working alone in a remote part of the premises, for example, in a lift motor room, plant room, or electrical switch room.

Issuing a permit to work

Permits to work may be issued by the occupier and/or the contractor. The contractor's general permit to work is most likely to be issued by the occupier to exercise control over access to the premises.

However, for works of a specialised nature, which the occupier cannot reasonably be expected to understand, for example, electrical work, the contractor would issue the permit to work.

Accepting a permit to work

It is the responsibility of the person(s) carrying out the work to comply with the conditions laid down on the work permit. Only work stated on the work permit is permitted to be carried out.

Monitoring

The activities of contractors should be included in all routine inspections and audits carried out by the host employer.

Conformity with personal protective equipment (PPE) standards, suitability of equipment and compliance with permits and/or method statements are aspects which should be given special consideration.

In the case of major failings, repetitive problems or failure by the contractor to take remedial action the host will need to take strong action.

Accident and incident reporting and investigation

The host employer needs to be made aware of any significant accidents or incidents since all such occurrences have the potential to affect his own employees.

The responsibility for investigation is a matter to be negotiated by the host and contractor at the earliest opportunity but, if the contractor accepts the responsibility then the host must be sure that the contractor has sufficient skills and/or knowledge to carry out the task. Accidents or incidents that are reportable to the enforcing agency will be of particular interest to the host.

The contractor should be aware of the requirements of the Reporting of Injuries, Diseases and Dangerous Occurrences Regulations (RIDDOR) 2013, coupled with the company's internal system for reporting, recording and investigating all such incidents.

All incidents, including traffic accidents, involving contractors and sub-contractors should be reported to the occupier.

Review meetings

Regular meetings should be held throughout the duration of long contracts. The health and safety performance of contractors is just one of the items that should be on the agenda. During such reviews the host must decide whether or not to continue to use the services of the contractor.

Completion reports

Work completion reports help the client to build up a picture of the quality of the contractor's work with regard to health and safety and other relevant performance. They would take account of such features as: the methods of work, the quality of health and safety management and any significant incidents or problems that occurred.

Responsibilities for control of risks

Responsibilities for control of risks associated with a contractor on site rests principally with the organisation that has primary control of the site, which would usually be the client.

In some cases the Principal Contractor will have the main role in providing effective control of contractors on site on behalf of the client. This is emphasised by the Construction (Design and Management) Regulations (CDM) 2015.

As can be seen by the diagram in ***figure ref A9-11,*** both the client and contractor owe each other a duty to control risks arising from their work activities. They both have similar duties to consider the risks that affect visitors and members of the public.

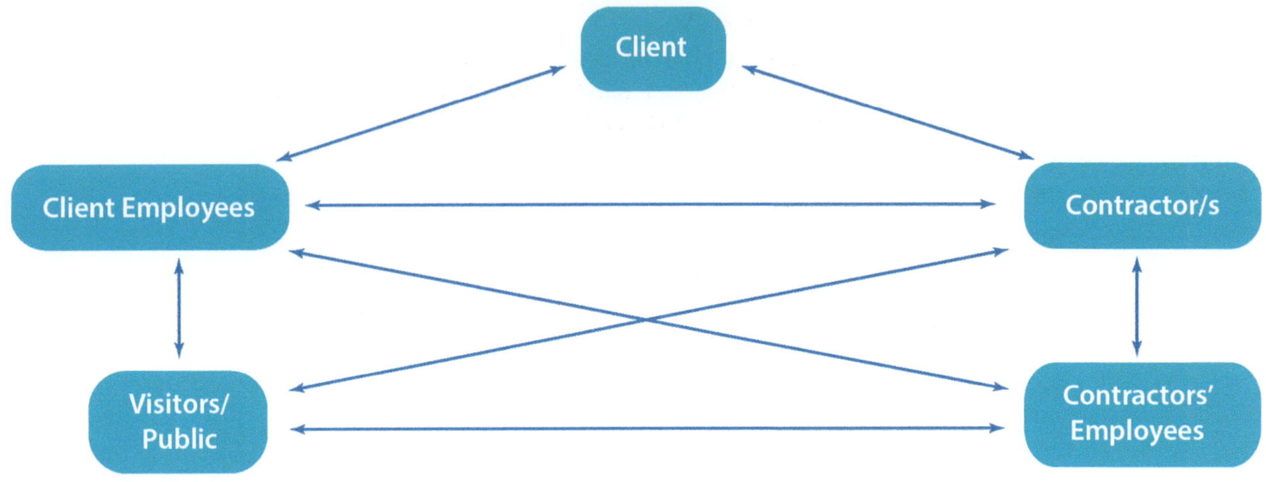

Figure A9-11: Flow chart and summary of duties to other party duties under Health and Safety at Work etc. Act 1974. *Source: RMS.*

RISK ASSOCIATED WITH CONTRACTORS ON-SITE

Plant and machinery

Only in specifically controlled circumstances should a contractor remove the guard, fencing or other safety equipment/devices to machinery and plant, and any guards and/or safety devices should be reinstated and operational before the plant or machinery is handed back for use.

Plant and machinery belonging to a contractor should be adequately guarded before being operated on the premises.

In the case of electrically-operated hand tools, such tools should be of the low-voltage type and connected to a 110 volt mains-isolated circuit where applicable. Alternatively, contractors should provide their own step-down transformer (250/110 volts, 50 cycles) with the mid-point to the secondary winding efficiently earthed.

In all cases, the metalwork on portable equipment and any flexible metallic covering of conductors should be efficiently earthed and, in all other respects, constructed and maintained in accordance with the Electricity at Work Regulations (EWR) 1989 and the accompanying Memorandum of Guidance.

Cables supplying portable apparatus should be of correct size and properly connected to approved type plugs and sockets. Makeshift and/or unsafe connections are dangerous and should not be permitted.

In the case of welding equipment, contractors should limit direct current (DC) to 40 volts and ensure the correct use of earth leads for the article being welded.

Noise

Where noisy equipment is used in close proximity to a working area on site, operators should be provided with suitable hearing protection (ear muffs) and required to wear same where sound pressure levels are above the second action level (85 dB(A) Lep,d) as outlined in the Control of Noise at Work Regulations (CNWR) 2005.

Noise from equipment should be minimised in working areas, for example, by the use of bag mufflers on pneumatic drills.

Contractors should also take all reasonably practicable steps to prevent noise nuisance to the inhabitants of the neighbourhood.

Fire protection

Every year fires of varying degrees of severity cause loss of life and devastation on building sites. Reference to the 'fire triangle' will show the three requisites for fire to take place i.e. an ignition source, fuel and air. Ignition sources on construction sites are many and varied - welding and cutting activities, people smoking whilst on site or in site huts, cooking on open flames in site huts, the use of blow lamps and gas or liquid-fuel fired heating appliances.

The fuel can take numerous forms - timber, bitumen impregnated paper, plastics, flammable liquids, such as paraffin, and combustible refuse produced during erection of buildings and in other site work. Oxygen, naturally present in the air, will, on a windy day cause fire to spread at a rapid rate. The need, therefore, for effective fire protection procedures cannot be overemphasised.

Fire protection procedures will vary to some extent according to the type of site for example, a green field site, where no work has taken place previously, and an existing site, where modification and extension to existing buildings may be taking place.

In the case of green field sites, planning of fire protection procedures should take place well before site work commences.

Here consideration should be given to the following:

- Provision of access for fire brigade appliances.
- The location of buildings and the separation of high risk buildings from other buildings.
- The provision of adequate space between buildings and site huts.
- The actual construction of buildings for example, site huts, canteens, offices.
- Flammable materials stores, equipment stores, etc.
- The establishment of prohibited areas where smoking and the use of naked lights are forbidden.
- The system for the storage and disposal of combustible and flammable refuse.
- The provision of a separate flammable materials store suitably constructed and identified as such.
- The availability of water supply for fire brigade appliances and on-site fire-fighting.
- The provision of adequate fire appliances, located according to the fire risk involved, and an effective fire alarm system.
- Evacuation procedures in the event of fire, including the training of site operators in such procedures.
- The appointment of fire wardens and the operation of fire patrols, particularly at night and weekends.
- The system for liaison with the local fire and rescue authority.

Many of the above requirements apply to existing sites. All contractors' employees should understand the fire warning system currently in operation. Instructions on action to be taken in the event of fire should be made available to contractors and work force and be clearly displayed on site. Fire-fighting equipment installed by the occupier should also be made known to contractors or, alternatively, be provided by the contractors.

Where any work involves interference with, or the removal of, fire appliances, alarms or systems, prior notification should be given to the contractor.

The use of petroleum, petroleum mixtures, liquefied petroleum gas (LPG), cellulose and other highly flammable and/or explosive substances should comply with the requirements of the Dangerous Substances and Explosive Atmosphere Regulations (DSEAR) 2002.

Where any form of protective material or covering is used, whether against dust or climatic conditions, it should be fire-resistant or treated with fire-resistant solution.

'Hot work' activities for example, operations involving the use of oxy-acetylene welding and cutting equipment, electric arc welding, blow lamps and other flame-producing equipment, should not be commenced until authority to do so has been given by the occupier. In certain cases this may require the issue of a permit-to-work (hot work permit). Gas cylinders, particularly acetylene, should be stored in a manner required by DSEAR 2002.

The contractor should provide and ensure the use of welding screens to give protection to all persons in respect of any arc flash caused by his operators. Combustible materials, such as paper, timber, rags and plastics, should be stored in suitable refuse containers. Non-combustible blankets i.e., fibreglass, should be used to afford protection against 'welding spatter'. The burning of refuse on site should be prohibited.

Dangerous substances and wastes

Two aspects must be considered here:

- The requirements of the Control of Substances Hazardous to Health (COSHH) Regulations 2002 (as amended) with regard to protecting all persons on site from exposure to hazardous substances.
- Requirements of the Environmental Protection Act (EPA) 1990, and associated Regulations, covering pollution of air, water and ground, and waste disposal procedures.

A wide range of toxic and other dangerous substances can be encountered in construction activities, including asbestos, lead and many flammable substances. Site management must be aware of the hazards associated with such substances, ensuring correct storage, use and disposal.

Personal protective equipment

Personal protective equipment should be provided and worn wherever hazards to the contractors' workforce exist. This includes safety helmets, eye and face protection, respiratory protection and safety boots. Contractors should be aware of the requirements of the Personal Protective Equipment at Work Regulations 1992 relating to employers and workers.

Contractors' vehicles

The speed of vehicles within premises on which construction work is taking place should be consistent with safety and drivers should not exceed any speed limits for example, 10 mph, displayed. Drivers should also comply with any traffic direction systems and signs in use on site. Vehicles used by contractors or their employees should be parked only in specifically designated locations.

All vehicles should be prohibited from moving during times when access roads are crowded with people arriving at or departing from the premises, as the risk of accidents is greater at these times. This prohibition includes cranes, dumpers, concrete vehicles and other lorries.

All vehicles and trailers used on site must be in an efficient state, in sound working order and in a good state of repair, and must not be used in an unsafe manner.

Site clearance

On completion of work, contractors should be required to remove all unused materials and leave the site clean and tidy. This may include reinstatement of perimeter fencing, removal of mud and debris from roads, removal of waste building materials and site refuse, and the levelling of disturbed ground.

On no account should items such as empty gas cylinders, oil drums or paint cans be buried. Excavations and trenches should be filled in and the ground levelled.

Window cleaning and external painting

Cleaning the outer surfaces of windows and painting buildings can present a number of hazards. Contractors should be aware of the Work at Height Regulations (WAH) 2005. In order to ensure safe working in either of these situations:

- Where it is practicable contractors should use other means of access than ladders, or work from the ground.
- Contractors should ensure that any ladder, safety harness or other appliance used, or intended for use, by his employees, is of sound construction, adequate strength, and sufficient length and properly maintained.
- Contractor should satisfy themselves that any structural handhold and/or foothold likely to be used by his employees is secure, warning employees, in relevant cases, that such a handhold or foothold is not to be used.

RISKS ASSOCIATED WITH VISITORS

The responsibility for control of risks associated with visitors lies with whoever is in control of the premises. If this is an employer, then they will have responsibility. This will include conducting a risk assessment related to visitors and may have to consider various types of visitors. There is a significant difference in risk between a visitor attending an office for a meeting and a visiting driver unloading materials on a construction site.

It is important that those in control of premises foresee risks and implement controls. One of the most important controls is that which manages access to and egress from the site. This enables access to be restricted and supervision of the visitor to the level relating to the degree of risk.

Requirements to provide information

HAZARDS AND RISKS TO THIRD PARTIES

A lot of modern legislation requires the provision of information to third parties, notably the HASAWA 1974 (Section 3); the Management of Health and Safety at Work Regulations (MHSWR) 1999 (Regulations 11 and 12); and the CDM Regulations 2015 which require both the designer and the client to make information available to the planning supervisor.

The R vs. Swan Hunter Shipbuilders Ltd case emphasised the duty of an employer (client) under Section 2 of the HASAWA to take reasonably practicable steps to ensure the health and safety of their employees.

In the case it was identified that this could extend to the provision of information the employer had on safe systems relating to the work contractors did, where this was reasonably practicable. The employer may therefore need to share information with contractors to ensure the health and safety of their own employees and third parties.

Review of contractor performance

Reviewing is about learning from mistakes and identifying good practices that can be used next time. The review should take account of the entire contracting process and not just the quality of the finished work. Evaluation of the process will allow improvements in the health and safety system to be made. It may be necessary to review at weekly, monthly or quarterly intervals depending on the nature of the work and the level of risk. Regular review will allow management to decide whether the work should continue or if the contractor should be used again.

A9.6 - Consultation with employees

The role and benefits of consultation within the workplace

The role of consultation with employees is to provide them with an opportunity to comment on health and safety issues that might affect them in the workplace. It also provides the employer with an insight into the impact of health and safety issues on employees.

The main benefits of consultation with employees are to provide a mechanism of involvement in influencing the health and safety performance of the organisation. This can lead to shared values, beliefs and goals. This in turn can lead to a more co-operative approach to health and safety and reduced conflict between organisational and individual goals.

The duty of employers to consult all employees via safety representatives was enshrined in sections 2(4) to 2(7) of the HASAWA 1974. However, section 2(5) which gave all employees the right to elect safety representatives was repealed by the Employment Protection Act (EPA) 1975. Thus, only those workplaces where workers are represented by officially recognised trades unions are covered by the HASAWA 1974. This requirement was translated into the Safety Representatives and Safety Committees Regulations (SRSC) 1977.

The MHSWR 1992 amended the SRSC Regulations 1977 by stipulating the matters on which the employer must consult; these are:

- Measures introduced into the workplace which may affect employees' health and safety.
- The appointment of competent persons.
- Any health and safety information that the employer is obliged to provide by law.
- The planning and organisation of any health and safety training that the employer is obliged to provide by law.
- The health and safety consequences of the planning and introduction of new technology into the workplace.

The SRSC Regulations 1977 only apply to safety representatives from recognised trades unions leaving many, non-unionised, workers without any consultation rights. This situation was rectified to an extent by the introduction of the Health and Safety (Consultation with Employees) Regulations (HSCER) 1996. It should be noted that the roles and functions of trade union representatives remains unchanged and that the two Regulations are designed to sit side-by-side with each other, their application depending on union status.

Formal consultation

FUNCTIONS AND RIGHTS OF EMPLOYEE REPRESENTATIVES

Trade union appointed representative (Safety Representatives)

The Safety Representatives and Safety Committees Regulations (SRSC) 1977 are concerned with the appointment by recognised trade unions of safety representatives, the functions of the representatives and the establishment of safety committees.

Representatives are appointed when a recognised trade union notifies the employer in writing. Representatives must have been employed throughout the preceding 2 years or, where this is not reasonably practicable, have had at least 2 years' experience in similar employment.

Similarly, employees cease to be representatives when:

- The employer has been notified in writing by the trade union.
- The representative ceases to be employed.
- They resign.

Functions

The SRSC Regulations 1977 grant safety representatives (SR) certain functions as outlined below.

Functions are activities that safety representatives are permitted to carry out by legislation, but do not have a 'duty' to perform and therefore are treated as advisory actions. As a consequence, the representatives cannot be held accountable for failing to carry out these activities or for the standard of the advice given, when performing their functions. They are, however, still employees and have the same consequent duties as any other employee (for example, their duties under HASAWA 1974 Sections 7 and 8). Their functions as safety representatives are:

1) Represent the employees in consultation with the employer.
2) a) Investigate potential hazards and dangerous occurrences at the workplace and to examine the causes of accidents at the workplace.
   b) To investigate complaints by any employee they represent relating to that employee's health, safety or welfare at work.
   c) To make representations to the employer on matters arising out of the above.
   d) To make representations to the employer on matters of general health, safety and welfare affecting the employees at the workplace.
   e) To carry out inspections of the workplace.
   f) To represent the employees, they were appointed to represent in consultation at the workplace with inspectors of the Health and Safety Executive and of any other enforcing authority within the Act.
   g) To receive information from inspectors in accordance with section 28(8) of the HASAWA 1974.
   h) To attend meetings of safety committees during which he/she attends in his capacity as a safety representative in connection with any of the above conditions.

The ACOP to the regulations establishes that in order to fulfil their functions they should:

1) Take all reasonably practical steps to keep themselves informed of:

- The legal requirements relating to the health and safety of persons at work, particularly the group or groups of persons they directly represent.
- The particular hazards of the workplace and the measures deemed necessary to eliminate or minimise the risk deriving from these hazards.

- The health and safety policy of their employer and the organisation and arrangements for fulfilling that policy.

2) Encourage co-operation between their employer and his employees in promoting and developing essential measures to ensure the health and safety of employees, and in checking the effectiveness of these measures.
3) Bring to the employer's notice, normally in writing, any unsafe or unhealthy conditions, or unsafe working practices, or unsatisfactory arrangements for welfare at work, which comes to their attention whether during an inspection/investigation or day to day observation.
4) The report does not imply that all other conditions and working practices are safe and healthy or that the welfare arrangements are satisfactory in all other respects. Making a written report, does not preclude the bringing of such matters to the attention of the employer or his representative by a direct oral approach in the first instance, particularly in situations where speedy remedial action is necessary. It will also be appropriate for minor matters to be the subject of direct discussion, without the need for a formal written approach.

Rights

a) To time off with pay during working hours to:

- Perform functions.
- Undergo training in aspects of the functions.

b) To the use of facilities and assistance, as may reasonably be required for the purpose of carrying out their functions.
c) To inspect and take copies of relevant documents which relate to health, safety and welfare.
d) Information in the employer's knowledge necessary to enable them to fulfil their functions, except exempted information.

This should include:

- Information about the plans and performances of the undertaking and any changes proposed, in so far as they affect the health and safety at work of their employees.
- Information of a technical nature about hazards to health and safety and precautions deemed necessary to eliminate or minimise them, in respect of machinery, plant, equipment, processes, systems of work and substances in use at work. This should include any relevant information provided by consultants or designers or by the manufacturer, importer or supplier of any article or substance used, or proposed to be used, at work by their employees.
- Information which the employer keeps relating to the occurrence of any accidents, dangerous occurrences or notifiable industrial disease and any statistical records relating to such accidents, dangerous occurrences or cases of notifiable industrial disease.
- Any other information specifically related to matters affecting the Health and Safety at work of his employees, including the result of any measurements taken by persons acting on his behalf in the course of checking the effectiveness of his health and safety arrangements.
- Information on articles or substances which an employer issues to home workers.
- Any other suitable and relevant reasonable facility to enable the representatives to carry out their functions.

Exempted information is:

- Any information the disclosure of which would be against the interests of national security.
- Any information which he could not disclose without contravening a prohibition imposed by or under an enactment.
- Any information relating specifically to an individual, unless he has consented to its being disclosed.
- Any information the disclosure of which would, for reasons other than its effect on health, safety or welfare at work, cause substantial injury to the employer's undertaking or, where the information was supplied to him by some other person, to the undertaking of that other person.
- Any information obtained by the employer for the purpose of bringing, prosecuting or defending any legal proceedings.

e) If two or more safety representatives request, to establish a health and safety committee.

Elected representatives (Representatives of Employee Safety)

Where there are employees not represented by the SRSC Regulations 1977, the employer must consult those employees in good time on matters relating to their health & safety at work. In particular they must be consulted on:

- The introduction of any new measures which may affect their safety and health.
- Arrangements made by the employer for appointing or nominating competent persons in accordance with Regulations 6(1) and 7(1) of the MHSWR 1999.
- Any health and safety information the employer is legally obliged to provide to workers.
- The planning and organisation of any health and safety training required under particular health and safety laws.

- The health and safety consequences for employees of the introduction of new technologies into the workplace.

Employers can consult either directly with employees or, in respect of any group of employees, one or more elected representatives of that group. These are referred to as 'representatives of employee safety' (RES).

If the latter option is chosen, then employers must tell the employees the name of the representative and the group he/she represents. An employer that has been consulting a representative may choose to consult the whole workforce. However, the employer must inform the employees and the representatives of that fact.

Functions

Representatives of employee safety have the following functions:

- To make representations to the employer on potential hazards and dangerous occurrences at the workplace which affect, or could affect the represented employees.
- Make representations to the employer on general matters of health and safety affecting the employees they represent, in particular on matters they are consulted about by the employer.
- To represent the employees, they represent in workplace consultations with enforcing authority inspectors.

Rights

a) To reasonable training in order to carry out their duties. Employers must meet the costs of the training and any travel and subsistence.

b) To take time off, with pay, during working hours in order for them to carry out their functions.

c) To time off, with pay, during working hours where this is required for any person standing as a candidate for election as a representative. Employers must also provide suitable facilities for the representatives to carry out their duties.

d) To facilities and assistance as may reasonably be required to carry out their function. The guidance to the regulations suggests that facilities could include:

- Lists showing the names and workplaces of employees in a representative's constituency.
- Communications, distribution and photocopying facilities for the representative to communicate with represented employees.
- Access to senior management to discuss health and safety issues whenever the representative(s) might reasonably wish to do so.

e) To information within the employer's knowledge as is necessary for them to carry out their functions and any RIDDOR 2013 record relating to the workgroup they represent. Exempted information is:

- Any information the disclosure of which would be against the interests of national security.
- Any information which he could not disclose without contravening a prohibition imposed by or under an enactment.
- Any information relating specifically to an individual, unless he has consented to its being disclosed.
- Any information the disclosure of which would, for reasons other than its effect on health, safety or welfare at work, cause substantial injury to the employer's undertaking or, where the information was supplied to him by some other person, to the undertaking of that other person.
- Any information obtained by the employer for the purpose of bringing, prosecuting or defending any legal proceedings.
- Information that does not relate to health and safety.

The Trade Union and Labour Relations (Consolidation) Act 1992 is an Act of Parliament which regulates British Labour law. The Act applies in full in England and Wales and Scotland, and partially in Northern Ireland.

The law contained in the Act (TULRCA 1992) contains four main principles:

- Defines trade unions and states they are the subject of legal rights and duties.
- Protect the right of workers to organise into, or leave, a union without suffering discrimination or detriment Provide a framework for a union to engage in collective bargaining for a better workplace or business standards with employers.
- Protect the right of workers in a union to take action, including strike action and industrial action short of a strike, to support and defend their interests, when reasonable notice is given, and when that action is 'in contemplation or furtherance of a trade dispute'.

The Act has undergone significant amendments to comply with international labour rights standards.

The SRSC Regulations has largely incorporated several sections of the TULRCA. Section 168 of the TULRCA 1992 specifically covers time off for carrying out trade union activities. An employer shall permit an employee of his, who is an official of an independent trade union recognised by the employer, to take time off during his working hours for the purpose of carrying out any duties. Section 168A extends the time off allowance to cover 'learning' or training to perform his duties.

Section 169 of the Act covers payment for time spent not at work that a representative may take carrying out his trade union duties. Section 170 of the act allows the trade union representative to be paid for the time spent conducting his trade union duties.

FUNCTIONS OF HEALTH AND SAFETY COMMITTEES

If two or more appointed safety representatives request in writing the formation of a safety committee, the employer must implement this request within three months. Consultation must take place with the representatives making the request and the appointing trade union.

A basic requirement for a successful safety committee is the desire of both employee and management, to show honest commitment and a positive approach to a programme of accident prevention and the establishment of a safe and healthy environment and systems of work. For any committee to operate effectively, it is necessary to determine its objectives and functions.

Objectives

- The promotion of safety, health and welfare at work by providing a forum for discussion and perhaps a pressure group.
- To promote and support normal employee/employer systems for the reporting and control of workplace problems.

Functions

- To review accident and occupational health trends.
- To review recurring problems revealed by health and safety audits.
- To consider enforcing authority reports and information releases.
- To consider reports on matters arising from previous health and safety committee meetings.
- To assist in the development of health and safety rules and systems of work and procedures.
- To review health and safety aspects of future development and changes in procedure.
- To review health and safety aspects of purchasing specifications of equipment and materials.
- To review renewal/maintenance programmes.
- To monitor health and safety training programmes and standards achieved.
- To monitor the effectiveness of health and safety communications within the workplace.
- To monitor the effectiveness of the health and safety policy.

This may be summarised as review and recommend on the overall direction of the health and safety programme, on specific aspects of the programme, on difficulties encountered in its implementation, and to monitor the programme in both a specific and overall manner.

Composition

The membership and structure of the health and safety committee should be settled in consultation between management and the trade union representatives concerned. This should be aimed at keeping the total size as compact as possible, compatible with the adequate representation of the interests of management and employees. Management representatives will naturally be appointed by the management. Employee representatives will either be appointed by a recognised Trade Union (HASAWA 1974 Section 2(4)) or, where non-union employees are present, elected by their colleagues.

The committee suggested in HASAWA 1974 Section 2 (7) will probably be the 'Company Health and Safety Committee'. There is nothing to prevent the formation of smaller committees for parts of an organisation or different sites in order to maintain the company health and safety committee at a reasonable size.

FORMAL CONSULTATION DIRECTLY WITH EMPLOYEES

Where there are employees who are not represented by safety representatives under SRSCR 1977, the Health and Safety (Consultation with Employees) Regulations (HSCER) 1996 require the employer to consult with them in good time on matters relating to health and safety, in particular with regard to:

a) The introduction of any measure at the workplace which may substantially affect the health and safety of those employees.

b) His arrangements for appointing or, as the case may be, nominating persons in accordance with regulations 6(1) and 7(1)(b) of the Management of Health and Safety at Work Regulations 1992 (a), (b).

c) Any health and safety information he is required to provide to those employees by or under the relevant statutory provisions.

d) The planning and organisation of any health and safety training he is required to provide to those employees by or under the relevant statutory provisions.

e) The health and safety consequences for those employees of the introduction (including the planning thereof) of new technologies into the workplace.

It is the employer's choice as to whether they consult using a Representative of Employee Safety or directly with employees. In a situation where the employer is consulting with a RES they must inform the employees of the names of the RES and the groups they represent. If consultation starts with the RES but ceases and the employer decides to consult with employees directly they must inform the RES.

The choice of direct or representative consultation is the responsibility of the employer and which one is chosen may depend on the subject being consulted on, the size of the organisation and resources available. Some health and safety matters may be common to all employees and therefore consultation with the RES appropriate. Other matters may be very much a personal view, such as an employee's perspective on the culture of the organisation, and therefore consultation directly is more appropriate. Matters that only affect one or two employees might also lend themselves to direct consultation.

Informal consultation

Experience within many organisations has shown that employee health and safety representatives can either be very positive or, conversely, seriously undermine the health and safety effort. The outcome depends on the attitudes of those involved and is a good indication of the culture of the organisation.

As a minimum, any organisation must comply with the legal obligations outlined above. Implementation should also play an integral part in the drive to improve health and safety performance. This can be achieved by going beyond the legal minimum in order to positively involve staff at all levels.

Treating representatives differently depending on his or her trade union status, for example, could promote both confusion and division amongst staff. This would not demonstrate the commitment to employee involvement in all aspects of health and safety required as part of any management system.

The advantages and benefits of consulting employees has long been recognised by good organisations who have recognised that employees who have had their thoughts and opinions taken into account will often be better motivated (and therefore more productive) than those who do not.

This, less formal consultation, can be achieved by encouraging communication between individual managers and their employees and by using a mixture of mechanisms, such as discussion groups, health and safety circles, departmental meetings and employee discussions.

INFORMAL CONSULTATION MECHANISMS

Discussion groups

Meetings between individuals at the same level within the organisation to discuss their mutual safety concerns. This encourages free expression and promotes the efficient use of resources.

Safety circles

Safety circles usually involve a group of people in a work area, who consider a specific problem that has been observed in order to establish a prompt solution to it. The safety circle can also be used to develop improvement ideas. The circle usually involves those that are most affected by the issue, a team leader and someone that can add technical health and safety input. The use of safety circles engenders a high degree of local ownership of health and safety issues.

Departmental meetings

These are additional to the organisation's health and safety committee which has been formed under, for example, the SRSC Regulations 1977. They can be used to deal with local matters and can be a good way for managers to demonstrate a personal commitment to the health, safety and welfare of their staff.

Employee discussions

Employees will often have a view regarding the way they are asked to work, the risks involved or the suitability of equipment for the tasks they have to carry out. These views are often expressed to front line managers, but sometimes do not get the consideration they may deserve.

A proactive organisation will use the technique of health and safety tours, where senior managers talk to employees in the workplace about issues which affect their daily lives to identify these breaks in communication. The employee discussions that result may reveal potentially high risk health and safety issues that might not have been brought to the organisation's attention until an accident occurred.

Email and Web-based Forums

In larger organisations, it may be possible to set up discussion forums using this kind of technology. The advantages of empowering employees in the decision making processes of the organisation must be matched with the resources to respond to the employees' contributions.

It is also possible that some employees who do not have easy access to the technology will feel excluded.

Behavioural aspects associated with consultation

The consultative process depends on both personal relationships and the culture of the organisation. Conflict will exist if managers view consultation on health and safety matters as a burden which interferes with production.

It is management's responsibility to take executive action and to have adequate arrangements for regular and effective checking for health and safety precautions and for ensuring that the declared health and safety policy is being fulfilled.

The work of consultative mechanisms such as safety committees should supplement these arrangements; it cannot be a substitute for them. Alternatively, employees can have a negative effect on the process if they view consultation as a way of getting back at managers without having to go through any grievance procedures.

In essence the attitudes towards safety representatives and committees on both sides of the management 'divide' say a great deal about the culture of the organisation.

The general attitudes of groups of employees can play either a positive or negative role in the consultative process. The social need of most individuals to both belong to, and be accepted by, a group is a powerful motivating factor. The need to belong to a group provides our need for friendship and affiliation and supports both self-esteem and identity.

The attitudes of the peer group can play a powerful role in either helping or hindering the consultation and thus the safety effort. Those who manage organisations are well advised to influence both organisational and individual performance by working with and through groups.

PEER GROUP PRESSURES

For example, as previously mentioned, there is a normal process of individual development as a child progresses towards adulthood. This can become skewed due to a lack of "suitable" role models or a breakdown in societal controls. Here the dominant influence during later childhood and early adolescence comes from a peer group which itself is not fully socially developed.

This is illustrated by the gang culture which dominates certain areas in the USA. Thus, certain attitudes towards risk taking may be shared by members of one adolescent culture, which are markedly different from those of other age groups.

Individuals tend to seek out social groups which suit their personality and attitudes, but when there is a mismatch the individual will be changed by the group to a greater or lesser extent depending upon the importance that the individual attaches to acceptance by the group.

In a multi-cultural society, it is important for the health and safety practitioner to understand the influences that are shared by a particular group within the workforce. The individual believes that these culturally-shared values are correct and perhaps the only 'right' way of behaving.

Thus, these cultural values and needs can often over-ride the 'common' perception of health and safety needs. For example, the need for a Sikh to wear appropriate head-dress over-rides the need to wear a safety helmet.

Another example is the wearing of neck-ties by male workers even though they are working with rotating machinery with a real risk of entanglement. Again, the need to conform to the cultural idea of what is respectable over-rides health and safety needs.

DANGER OF TOKENISM

Tokenism is the convenient inclusion of a representative from a marginalised group into a mainstream body. Tokenistic appointments are frequently made to fall in with anti-discriminatory legislation. Token inclusion of health and safety representatives, or other pressure groups, on advisory boards and in other situations is often not a genuine attempt to listen and learn from 'other people'. They are, rather, ways of avoiding such genuine attempts.

If tokenism is allowed to prevail there is a danger of arranging the selection of an employee to be involved in consultation that management already identify as either not very opinionated or follows management perspective on the matter.

Care should be taken to involve people that will give an honest opinion, even if it is not the one management want to hear. The process of consultation does not mean that the employer has to agree with the opinion or follow the suggestions of those consulted.

POTENTIAL AREAS OF CONFLICT

In the workplace, the majority of people will conform to the norms of behaviour of that particular occupational group/society. Traditional managers encouraged a norm of behaviour based around the work ethic, which in some cases led to a conflict with safety values. The aim of the safety practitioner, therefore, is to encourage a balance of production versus safety.

Often workplace groups are influenced by internal rather than external factors. These are based on such concepts as 'peer pressure' and 'group leaders'.

These can have both a positive and negative influence. The good manager will recognise and use these influences to encourage safe behaviour.

The role of the health and safety practitioner in the consultative process

Although the prime responsibility for health and safety at work rests with management and the other categories of worker, they often need the help and services of specialist and qualified advisers.

The health and safety practitioner is predominantly an advisory role and is primarily concerned with the detailed implementation of the health and safety management system.

The health and safety practitioner should act as an advisor to both managers and workers. Ideally, the health and safety practitioner should act as an 'ex-officio' member to any established health and safety committee, providing advice and guidance as required.

A9.7 - Health and safety culture and climate

Definition of health and safety culture and climate

Definition of culture:

"A word used to describe typical patterns of social, emotional and intellectual behaviour deriving from a shared set of beliefs and values, which are adaptive to the physical environment."

Figure A9-12: Definition of culture. *Source: Unknown.*

"The safety culture of an organisation is the product of individual and group values, attitudes, perceptions, competencies, and patterns of behaviour that determine the commitment to, and the style and proficiency of, an organisation's health and safety management."

Figure A9-13: Safety culture. *Source: ACSNI.*

A health and safety culture can be said to be the collection of attitudes, values and beliefs that members of an organisation hold in relation to health and safety. When taken together they produce an organisational culture that may be positive, negative or neutral.

The style of management is crucial to the type of safety culture within an organisation. Strong leadership and a commitment to health and safety are essential for a positive safety culture. If senior management is only interested in productivity, then a negative health and safety culture is sure to follow.

Climate is usually regarded as being more superficial than culture in that it involves the current position of a company (Glendon & Stanton, 2000). Guldenmund (2000) suggests that safety climate refers to the attitudes towards safety within an organisation while safety culture concerns the underlying beliefs and convictions of those attitudes, in other words the prevailing values of the social group.

Figure A9-14: Definition of climate. *Source: Health and Safety Laboratory, Safety Culture: A Review of Literature.*

The health and safety climate of society may be considered to be the social, political, business and other factors that the organisation is influenced by that lead to a given culture existing within the organisation.

An organisation's health and safety climate is usually taken as its current position on health and safety; some organisations measure the culture at intervals to determine how positive towards health and safety it is.

Risk management should be viewed from the moral, legal and economic standpoints and the senior manager who is committed to this will encourage a positive health and safety culture.

As has been shown in the previous pages, the 'organisation' aspect of health and safety management has the most influence on human behaviour. The individual will fit into the culture, positive or negative, and act accordingly.

A positive health and safety culture once achieved should be self-perpetuating, if commitment from the top is continued.

Influence of health and safety culture on behaviour and the effect of peer group pressure and norms

The term 'peer' is usually used to refer to those people that work with an individual. Peers are usually at the same level in an organisation as the individual, but they may be members of the same group, for example, members of a health and safety committee or management team.

In psychology, the Asch conformity experiments were a series of laboratory experiments, directed by Solomon Asch in the 1950s, that demonstrated the degree to which an individual's own opinions are influenced by those of a majority group (peer pressure).

Students were shown a card with three lines of different lengths and asked to compare them to a card with a single line on it. The students had to identify which of the three line on the card was the same length as the single line on the other card.

All except one of the students had been briefed by Asch to nominate an obviously short line. The one student, who answered last and had not been briefed, agreed with all the other students - an example of succumbing to peer pressure.

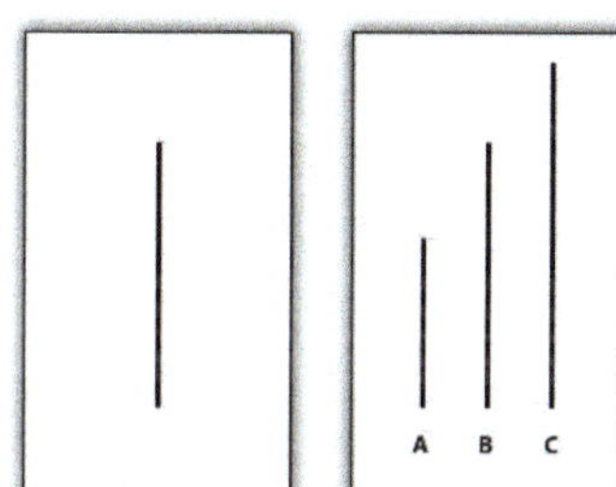

Figure A9-15: Cards of the type used by Asch in his conformity experiment. *Source: RMS.*

Many individuals are influenced by what people think of them and may act in a particular way to ensure that the people around them feel positive towards them. In group situations this can include the individual acting in the same way as the group to maintain a feeling of belonging in the group. This desire for belonging and to be liked can lead to an individual consciously or subconsciously behaving in a similar way to those they work with, their peers. An individual's peers will also try to influence the individual to behave in a similar way to the way they behave. This influence of peers may lead to peer pressure to behave in a particular way and can be a very strong influence when peers act as a group. This can have a significant effect on an organisation's health and safety culture as the behaviour of people in an organisation is an important part of what influences the culture.

Wherever individuals interact or communicate, their behaviour is governed by ***norms*** (standards or expectations) and they strive for a common purpose. If they ignore a norm they are classified as ***deviants*** and pressure is put on them to conform. Such pressure can take the form of verbal abuse, physical assault, silence and physical exclusion.

Most individuals, when in a group, want to be accepted by the group. They especially want to be accepted by their peer group, and will act according to the group norms. These peer group norms are extremely influential and can work for or against health and safety.

For example, if a young person's peer group norm is that wearing hearing protection is not 'cool', then that young person will not wear it. Peer pressure is so influential that even enforcing the wearing of hearing protection will probably mean that the person will wear it only when they are closely supervised and remove it as soon as the supervision is gone.

In an organisation where it has a culture that is negative to health and safety those individuals that have values and beliefs that are positive to health and safety will be influenced by the effect of the group norm created by the negative culture. They are likely to be pressured by the peer groups they are involved with to conform to this negative norm.

For example, a manager expressing concerns about health and safety in an operational meeting where all other manager's beliefs and values of health and safety are that it gets in the way of productivity is likely to experience pressures that challenge the manager's concerns and seek to trivialise them.

This can take many forms including derision, ridicule or the concerns may simply be ignored to indicate their insignificance. This pressure from group members of the meeting can have a strong influence on the manager's behaviour, even if it may not change the manager's beliefs and values.

The pressure may continue outside the meeting where the manager may be isolated by the group from discussion, involvement and contribution to decisions. This pressure would usually continue until the behaviour of the manager was that of the norm and health and safety concerns were relegated so they were not raised in the operational meetings and potentially not at all.

Peer group pressure to conform to norms can also have a positive effect in favour of health and safety where this is the norm. A group may be cohesive and resist the unsafe or unhealthy behaviour of a new manager or worker of a group. The peer group might exert pressure on the non-conforming individual to adhere to the positive norm.

It is therefore important to anticipate the influence of peers and plan to use this influence to promote health and safety, helping to establish a positive health and safety culture. This will usually involve obtaining the commitment of the senior manager and using it to influence other managers, who in turn influence those they work with. In this way the influence of peers can have a significant positive effect on an organisation's health and safety culture.

Impact of organisational cultural factors and associated values on individual behaviour

BLAKE & MOUTON

Country club management

Comfortable and friendly work atmosphere. Togetherness is the key word. Likes to be liked.

Production is incidental to lack of conflict and good fellowship. Does not push people for production - you can lead a horse to water, but you cannot make him drink.

Tries to avoid direct disagreements or conflict. Production problems are glossed over. New ideas avoided since they might cause stress.

This style can easily grow up in a quasi-monopolistic situation where profit is not important, or work is done on a cost-plus basis.

Task management

Sees people as a commodity just like machines. The manager's job is to plan, direct and control the work of those subordinate to him. Subordinates do what is required but no more. Any creative energy of subordinates goes into how to defeat the system rather than how to improve it. Disagreements are ruled out and suppressed rather than settled. This win/lose thinking leads to the development of militant trade unionism and union/management conflict.

Swinger

Management by crisis, staff never know just how the manager will react. It is possible for such a manager to 'get it right', but rarely.

Dampened pendulum

Marginal shifts round a happy medium.

Middle of the road style - pushes the staff enough to get acceptable production, but yields enough to maintain acceptable morale. To aim fully for both is too idealistic.

A compromiser - always aims for a balanced solution rather than the most appropriate one.

Impoverished management

Sees effective production as unobtainable because people are lazy, apathetic and indifferent. Sound and mature relationships are difficult to achieve because, human nature being what it is, and conflict is inevitable.

- Little concern for either production or people.
- Avoids responsibility or personal commitment.
- Often frustrated, having been passed over for promotion after years in a routine job.
- Waiting to retire. (Has retired but is waiting to leave).

Team management

- High concern for people and the task, seeks to integrate his staff round the task.
- Morale is task related, high sense of job satisfaction.
- For most people this is an ideal to aim for.

ORGANISATIONAL STRUCTURES AND ORGANISATIONAL ROLE CULTURES

Charles Handy (1985) popularised a method of looking at culture which some scholars have used to link organisational structure to organisational culture. He describes:

- ***Power culture*** which concentrates power in a few pairs of hands. Control radiates from the centre like a web. Power Cultures have few rules and little bureaucracy; swift decisions can ensue.
- In a ***Role culture***, people have clearly delegated authorities within a highly defined structure. Typically, these organisations form hierarchical bureaucracies. Power derives from a person's position and little scope exists for expert power.
- By contrast, in a ***Task culture***, teams form to solve particular problems. Power derives from expertise so long as a team requires expertise. These cultures often feature the multiple reporting lines of a matrix structure.
- A ***Person culture*** exists where all individuals believe themselves superior to the organisation. Survival can become difficult for such organisations, since the concept of an organisation suggests that a group of like-minded individuals pursue the organisational goals. Some professional partnerships can operate as person cultures, because each partner brings a peculiar expertise and clientele to the firm.

Power culture

Power cultures are most often found within small to medium sized organisations. Within a power culture, control is the important element; decisions in a power culture are centralised around one key individual. The key individual will need to be involved in all aspects of decision making.

The results of group consultation can quickly be considered and a decision made. Power cultures can be a positive force for health and safety if the key individual can be convinced of its value; all decisions will then reflect the importance of health and safety. Conversely, if that individual does not see the value of health and safety, decisions that productivity are more important may compromise health and safety, leading to risk taking.

Role culture

Common in many large organisations today is a role culture, where the various functions and each individual within the function are assigned a particular role. Employees focus on their particular role as assigned to them and if the tasks are matched carefully to the skills of the individual this will increase greatly the productivity of the organisation as a whole.

Role cultures rely on the authority derived from someone's role, whether it is positive to health and safety or not. The authority of the individual is not necessarily dependent on their competence or alignment with organisational goals, but is derived from the role they take and position in the hierarchy.

Where roles are ill-defined, or do not emphasise health and safety, the individual in the role may not take any specific action to manage risks. Where roles are well defined, the individual is held to account and this includes health and safety; this can have a positive impact on a manager's behaviour and the risks in the workplace. Role cultures, because of their hierarchical bureaucracies, can resist positive change to a health and safety culture.

Task culture

A task culture is concerned with a team based approach to complete a particular task. This type of organisational culture may be present in research, academic or project based organisations. Because of their reliance on the significant value of each individual in the group all people in the group have to be persuaded of the importance of health and safety. If they are not persuaded, individuals will tend to do what they see is in the best interest of the task, and this may or may not include health and safety.

Person culture

Person cultures are commonly found in charities or non-profit organisations. Here the focus of the organisation is on a specific aim. Successful person cultures attract like-minded individuals and their combined will often enables them to achieve their goals, where a more commercial structure might fail. Individuals are not usually motivated by individual gain, but are driven to achieve the collective goal of the organisation.

Sometimes this may result in the conscious acceptance of risks to themselves or others, such as less able volunteers carrying out tasks for the organisation that they may not be fit enough to do, for example, collection of bric-a-brac items for sale in charity shops which might be difficult or heavy to carry. If individuals in a person culture are convinced of the value of health and safety they become strong advocates, actively promoting the value and creating a strong positive influence.

Indicators of culture

EFFECTIVE COMMUNICATION

Effective communication is an indicator of a positive health and safety culture; therefore, effectiveness of communication can be used as a measurement.

Indicators of positive health and safety culture related to effective communication include where the organisation:

- Provides good (clear, concise, relevant) written materials (health and safety bulletins, posters, guidance).
- Provides good briefings on current issues day to day and in formal health and safety meetings.
- Provides mechanisms for listening and feedback.

LEADERSHIP AND COMMITMENT

Evidence of commitment by personnel at all levels of the organisation can be measured. The evidence can be shown by ***the clear identification and acceptance of responsibility*** for health and safety from the top. Du Pont has been quoted as saying: *"The chairman takes the role of Chief Health and Safety Officer."* Indicators that can be considered include management membership of the health and safety committee, and their participation in the committee.

Another indicator is where there is evidence of managers' responsibilities being accepted and taken seriously, for example, the following of health and safety rules like wearing personal protective equipment where necessary. Measurement of what the organisation is achieving compared to the standards that have been set in their health and safety policy can be done by conducting an ***audit.*** Another indicator of a positive health and safety culture is where non-compliance situations are dealt with promptly.

Indicators of positive health and safety culture related to effective leadership and commitment include where the organisation:

- Makes health and safety always a priority; work can be stopped if workers do not feel safe or healthy.
- Have senior managers that make regular useful visits to the workplace.
- Has senior managers that discuss health and safety matters with frontline workers.
- Has management that will stop production for health and safety reasons regardless of cost.
- Spends time and money on health and safety, for example, to provide protective equipment, safety training, and conduct health and safety culture workshops or audits.
- Will not tolerate violations of procedures and actively try to improve systems so as to discourage violations, for example, plan work so that short cuts are not necessary to do the work in time.

EQUAL PRIORITY

This includes evidence that health and safety is treated as an equal partner alongside other important business issues such as quality, finance, production, etc. The health and safety policy should be integrated with other corporate policies for example, purchasing, training, etc.

Indicators of positive health and safety culture related to equal priority include where the organisation:

- Keeps up to date with new ideas on health and safety.
- Ensures employees can get health and safety equipment and training if needed.
- Ensures everyone is included in decisions affecting health and safety and are regularly asked for input.
- Ensures it is rare for anyone to take shortcuts or unnecessary risks.

ACCIDENT INVESTIGATION

An indicator of a positive health and safety culture is where an organisation investigates all accidents and ill-health, including near misses, does something about it and gives feedback. The scope, findings and who the investigations involve can also be an indicator.

CONSULTATION

Proactive involvement of employees and/or their representatives in decision making for example, when selecting access equipment for work at height. Discussion with employees about work methods and conditions of work will result in fewer or no complaints.

Indicators of positive health and safety culture related to effective consultation include where the organisation:

- Consults widely about health and safety matters.
- Does more than the minimum to comply with the law on consultation.
- Seeks workforce participation in:
  - Setting policies and objectives.
  - Risks and their controls.
  - Accident/near miss investigations.

TRUST

Indicators of positive health and safety culture related to trust include where the organisation:

- Ensures employees can raise a health and safety concern, knowing the organisation will take it seriously and they will tell them what they are doing about it.
- Has management and employees that can be open and honest about safety.
- Does not simply find someone to blame when things go wrong.
- Has morale that is generally high.
- Encourages all employees and contractors to challenge anyone working on site about health and safety, without fear of reprisals.
- Keeps their promises.
- Treats the workforce with respect.

SPECIFIC TANGIBLE OUTPUTS INDICATORS

- Accidents/incidents/ill-health.
- Absenteeism.
- Level of compliance with health and safety rules and procedures.
- Staff turnover.
- Sickness rates.
- Complaints about working conditions.

Identification of these specific indicators may be easily made with 'direct labour' employees, but this can present a greater challenge where workers are mainly contractors.

Correlation between health and safety culture and climate and health and safety performance

'Organisations with a positive safety culture/climate are characterised by communications founded on mutual trust, by shared perceptions of the importance of safety and by confidence in the efficacy of preventive measures'. There are various performance indicators of where on the safety culture/climate continuum an organisation is; the continuum ranging from negative, through to neutral and on to positive. Indicators should be of the type that can be measured and compared in order to be meaningful.

For example, knowing the number of lost-time accidents a department has in any one year is only meaningful if it is then compared with the number from a previous year, or compared with a similar department in the same organisation or in the same industry. The departments with which the comparison will be made should have the same type of reporting system. It is important to only compare like with like.

The correlation between health and safety culture/climate and health and safety performance can be illustrated by research that has been conducted. There are two categories in this field of research: correlation studies and intervention studies.

Correlation studies may use the method of comparing low accident rate with high accident rate plants or departments in order to identify the factors that could explain the difference. Much work has been done in the mining industry using this method.

For example, Peters (1989) carried out a review of the organisational and behavioural factors associated with mining safety. His methods of collecting information were: direct observation of work methods and physical environment, interviewing miners and issuing questionnaires to a wide range of managers and officials who had a good working knowledge of how things were done.

Intervention studies look at the effects of interventions by studying before/after comparisons. For example, after the introduction of a safety programme in the range of forestry and logging organisations in Columbia, it was found by Painter and Smith (1986) that there were dramatic improvements in performance.

The accident frequency rate was reduced by 75% and the workers' compensation costs were reduced by 62%. In further research, Lauriski and Guyman (1989) found that after a safety management programme had been introduced at the Utah Power & Light Company, lost time injury rates were reduced by 60% over a period of five years. From 1980 to 1988, the accident frequency rate was reduced from 40 to 8 per annum, while production more than doubled.

Research has shown that improvements in health and safety management are influential in achieving a positive health and safety culture/climate. This leads to reduced accident rates, which is seen as a positive step forward, which is a further influence on the health and safety culture/climate.

Measurement of the culture and climate

Developing and promoting a positive safety culture/climate is an important aspect of health and safety management. A health and safety culture is an intangible thing, which has tangible manifestations. It is these manifestations that can be measured.

An organisation that desires a culture where achievement is paramount would most likely favour a system based on results, which is the most favoured system by many organisations. This can be manifested in what is actually being appraised, for example, personal traits, behaviours, results or methods of appraisal used for example, objective or subjective methods. The 'objective' measures output of workers while the 'subjective' evaluates their performance; the 'subjective', however, is more commonly in use by many organisations.

There should be a clearly defined system for measuring the health and safety culture/climate. Targets should be set, which are both measurable and achievable, and performance measured against them. There are a number of ways of measuring the culture and climate; they include health and safety climate assessment tools, perception surveys, findings of accident investigations, effectiveness of communication, and evidence of commitment at all levels.

HEALTH AND SAFETY CLIMATE ASSESSMENT TOOLS

Health and safety climate assessment tools can be used when seeking to establish the adequacy of the steps that the organisation is taking to ensure that everyone in the organisation is genuinely committed to the successful implementation of the health and safety programme. By using a climate assessment tool periodically the climate of the organisation can be measured and the current position compared with the climate at an earlier reference point. Assessment tools can be designed to probe at a greater depth than other indicators of health and safety culture.

The following table is part of a scheme devised by the ACSNI Study Group, which is intended to be an in-depth probe into a health and safety culture.

1. Review of organisational culture

Has the organisation evidence to demonstrate that:

1.1. Communications at all levels are founded on mutual trust?

1.2. All personnel understand, and agree with, corporate goals and the subordinate goals of their work group?

1.3. All personnel understand, and agree with, the means adopted to achieve corporate and work group goals?

1.4. The work practices of the organisation are under continuous review to ensure timely responses to changes in the internal or external environment?

1.5. Managers and supervisors demonstrate care and concern for everyone affected by the business?

Figure A9-16: An abstract from a health and safety culture prompt-list. *Source: ACSNI Human Factors Study Group.*

At ground level, lists can be devised of what is expected in the workplace and used in inspections. The degree of compliance can be used to indicate the interest and attitude to health and safety. Generally, the more safeguards in place and the cleaner the workplace, the more likely there will be a positive health and safety culture.

The HSE have developed a computer based 'Climate Survey Tool' (CST) to provide organisations with practical help to promote employee involvement in health and safety and to improve aspects of their existing health and safety culture and climate.

The questionnaire was designed by the HSE to record views on some important aspects of the management of health and safety within their organisation, and on some issues which are recognised as being important in occupational accident and ill-health prevention.

The questionnaire explores three discrete groups (managers, supervisors and the workforce) so that results can be compared, as there are often wide differences in the views of the three groups on the same issue. Managers, supervisors and the workforce are asked to express the extent to which they agree or disagree with statements, on a five-point scale.

The questionnaire covers the following important factors:

- Organisational commitment.
- Health and safety orientated behaviours.
- Health and safety trust.
- Usability of procedures.
- Engagement in health and safety.
- Peer group attitude.
- Resources for health and safety.
- Accidents and near misses reporting.

The questionnaire can also be tailored to enable respondents to write their own suggestions, which they believe would give the biggest improvement to health and safety in the organisation.

PERCEPTION SURVEYS

Perception surveys can show how employees view all aspects of the organisation and their role in it. They may perceive that what controls their lives at work is what has been labelled 'external or internal loci of control'.

There are those who believe that what happens is down to luck, chance or fate, which is external control, and those who believe that events are either wholly or partially down to themselves, which is internal control.

Research has shown that those who were 'internal' had a greater sense of responsibility for their own and others' health and safety.

The extent to which employees believed their health and safety depended on their own skills and care compared with the conditions and control set by the employer is another factor that can be considered when measuring perception. A number of different attitudes have been shown to relate to low accident rates.

FINDINGS OF ACCIDENT/INCIDENT INVESTIGATIONS

Findings of accident/incident investigations can be used as an indicator of health and safety culture. Root cause accident/incident investigation can show where things are going wrong, for example, by considering:

- Technical, procedural and behavioural aspects and controls.
- Validity of management controls.
- Level of commitment to working safely.

The procedure for dealing with the findings of the accident/incident investigation, that is, the recommendations for improvement and prevention, is another indicator of the health and safety culture.

For example:

- People responsible for action should be named.
- Completion dates for each action.
- Follow up for actions not completed.
- Findings not communicated to the workers.

EFFECTIVENESS OF COMMUNICATION

Effective communication is a factor in achieving a positive safety culture; therefore, effectiveness of communication can be used as a measurement.

Considering a practical example: as an organisation's health and safety policy must be communicated to its employees, asking them about it will give an indication as to how well it has been communicated. Communication can be measured at all levels of the organisation with each person knowing and understanding their area of responsibility.

Various means of communicating can be brought into the equation to measure and determine which is the most effective medium for the information to be conveyed.

EVIDENCE OF COMMITMENT BY PERSONNEL AT ALL LEVELS IN THE ORGANISATION

Evidence of commitment by personnel at all levels of the organisation can be measured. The evidence can be shown by the acceptance of responsibility for health and safety from the top, as in the statement of intent in the health and safety policy; as Du Pont say: the chairman takes the role of 'Chief Health and Safety Officer'.

Areas that can be considered are: membership of the health and safety committee and attendance; responsibilities accepted and taken seriously - whatever they should do they do; rules apply to everyone, for example, the wearing of necessary PPE; etc.

Measurement of what the organisation is achieving compared to the standards that have been set in the health and safety policy can be done by safety audit. Promptly dealing with shortfalls is another indicator of a positive health and safety culture.

THE SUBJECTIVE AND OBJECTIVE NATURE OF CULTURE AND CLIMATE

Organisations with a poorly developed safety culture and a poor climate will have a polarisation between the views of the management and the workforce.

The management will regard the workforce as obstructive and reactionary to their attempts to improve safety standards. The workforce will regard the management's attitude as detached and dictatorial.

However, in an analysis of culture and climate following the Ladbroke Rail Disaster (HSE Publication RR367), the authors identified the following steps of improvement that can be taken:

- Top-down communication: the general consensus is that effective communication from management to staff is the key to successful health and safety leadership. This can be achieved by a visible safety policy statement, emphasis on safety issues and procedures via newsletters, videos, warning sheets, and the communication of major accident risks.
- Safety reporting: communicating a problem or concern is only one step in the route towards a good safety culture. Feedback mechanisms should be in place to respond to the reporter (if required) concerning any actions taken.
- Horizontal communication: the organisation should provide a system for the effective transfer of information between individuals, departments and teams. In practice, appropriate safety information is available when required.

- The effect of the improvements on the subjective attitude of the workforce was summarised as follows:
  - If the workforce feel that they are responsible for their own safety, and that of others, they are likely to feel a sense of involvement within the company.
  - Ownership for safety can be increased by providing effective training, and providing opportunities for employees to be responsible personally for areas of safety.
  - Safety specialists should play an advisory or consultancy role.
  - It should be easy for staff at all levels to report concerns about decisions that are likely to affect them, particularly during an organisational change.
  - Feedback mechanisms should be in place to inform staff about any decisions that are likely to affect them.

A9.8 - Factors affecting health and safety culture

Factors that may promote a positive health and safety culture

Organisations need to produce a culture that promotes management and employee commitment to health and safety, and emphasises that deviation from corporate safety goals, at whatever level, is not acceptable.

Producing such a culture requires clear, visible, ***management leadership and commitment*** to health and safety from the most senior level in the organisation. The commitment should be not just a formal statement, but be evident in the day-to-day activities of the organisation. This commitment must be known and understood by managers and employees. Individuals may be reluctant to err on the side of caution in matters that have health and safety implications if their decisions to do so are likely to be subject to unwarranted criticism from their superiors or their peers.

The attitude of a strong personality at a senior level within the organisation may have either a beneficial or an adverse effect on a health and safety culture. Inevitably, junior employees will be influenced by that person's example.

Health and safety procedures soon fall into disuse if there is no system of ensuring that they are followed. Too often procedures lapse because of management neglect, or employees are discouraged from working to them by peer groups or other pressures, such as production targets. Where managers become aware of deficiencies in health and safety procedures but do not act to remedy them, the workforce readily perceive that such actions are condoned. It is essential that organisations create a ***high business profile for health and safety.***

Individuals may not understand the relevance of procedures or appreciate their significance in controlling risk. ***Involvement*** and consultation in the making of rules and decisions will ***promote ownership*** of health and safety.

Sometimes procedures are faulty, irrelevant, or lacking in credibility. When accidents happen managers cannot blame individuals for taking short cuts which seemed safe and were allowed to become routine, if they have not explained the importance of, or ***provided sufficient information*** on the procedures they originally laid down.

To promote a proper working culture, it is essential to have an effective system for monitoring health and safety that identifies, investigates and corrects deviations. The introduction and operation of such systems requires considerable effort by managers and only by allocating adequate resources can they be confident that failures will be prevented or controlled.

In short, the organisation needs to provide:

- Clear and evident commitment, from the most senior management downwards, which promotes a positive health and safety culture, in which management's objectives and the need for appropriate standards are communicated and in which constructive exchange of information at all levels is positively encouraged.
- An analytical and imaginative approach identifying possible routes to human factor failure. This may well require access to specialist advice.
- Procedures and standards for all aspects of critical work and mechanisms for reviewing them.
- Effective monitoring systems to check the implementation of the procedures and standards.
- Accident/incident investigation and the effective use of information drawn from such investigations.
- Adequate and effective supervision with the power to remedy deficiencies when found.

The organisation should identify and set standards through key performance indicators (KPI's) for health and safety. The standards should be achievable and designed not to compete with other organisational performance standards such as those set for production/service or quality. The standards must be agreed at the highest level within the organisation and standards for establishing management control must be established. The management controls must be designed to send a clear signal that health and safety is an equal partner to the other organisational objectives.

Management controls may take many forms, but should include system checks, such as random examination of completed permits to work, observations of high risk work activities or periodic tours.

The organisation should identify the safety critical tasks and establish suitable controls. The task analysis should not only take account of engineering controls, such as guard design or fume extraction requirements, but the requirements of the personnel involved in the task. This will include ensuring that the relevant

knowledge, skill and work experience is established, before an individual is put to work. Any deficiencies must be clearly identified and any necessary ***training*** must be provided.

Factors such as individual aptitude, dexterity and physical ability/endurance may also be important. High-risk tasks may utilise simulation equipment to allow skill to be developed, at no risk to the individual or others, for example, the use of aircraft flight simulators. Similarly, it may be necessary for the trainee to be under close supervision (an instructor flies with a new pilot of an aircraft) until their skill can be demonstrated as appropriate through their displayed actions and ability.

Factors that may promote a negative health and safety culture

Just as a positive health and safety culture starts with commitment from the top, a negative health and safety culture will develop from lack of it. There are, however, other factors involved which may lead to a negative health and safety culture.

When a company is going through a period of ***organisational change***, it is a time of upheaval, personal as well as corporate. Individuals and groups tend to be resistant to change, especially when they are unsure of the need for it. Lack of proper communication can lead to rumours of closure, redundancy or changes in the company's structure. The resulting fall in morale and period of ***uncertainty*** may lead to a lack of belief in what the company is committed to.

Reorganisation can also lead to people changing their position in the company structure with more, less or different responsibilities than previously. Without proper communication and necessary training this can lead to a mistrust of the company and its aims and objectives.

The company may state the aims and objectives, but the actions do not seem to be a clear way of achieving them. The aims and objectives may state a commitment to health and safety, yet the changes in work patterns do not allow for safe working. These conflicting messages will lead to a ***lack of confidence in the organisation's objectives and methods.*** This could be from the point of view that production is seen as all important and safety must be secondary, or an add-on only if time allows for it.

A practical example would be: if a company decided on a speed limit of five miles an hour for the fork lift trucks as a control to prevent accidents, but then increased the amount of product a driver had to move on a shift, it would be seen as mixed signals.

On one hand, the company is showing commitment to health and safety by restricting the speed of the vehicles, but on the other, no one seems to care that the increased workload means that the drivers must break the speed limit to get their job done. ***These management decisions prejudice mutual trust and lead to mixed signals regarding commitment to health and safety.*** This promotes a negative health and safety culture. The management and the employees then have differing aims and objectives and energy is exerted each fighting the other.

Effecting cultural change

Most industries must recognise the importance of a positive health and safety culture and the need to review and, when necessary, take steps to improve the culture. For progress to be made, careful ***planning and communication*** is essential. A piecemeal approach is time consuming, can be costly and is usually ineffective. As the ACSNI Study Group said: *"While the outcome of well-conceived plans to improve the safety culture of an organisation may be revolutionary, the plans themselves should be evolutionary. A gradualist (step-by-step) approach is essential."*

The major steps of the plan should be:

- To review the existing health and safety culture.
- To decide the aspects that has the highest priority for change.
- To decide on actions that may change those aspects, and to launch those actions.
- To repeat the previous three steps indefinitely.

The fourth step implies that the effects of each step are checked and ***performance is measured regularly***. The results may mean that new actions or new priorities are chosen. This means that the process is continuous and not just a momentary enthusiasm.

Achieving the long-term goal of developing a positive health and safety culture can be daunting, but a step-by-step approach can make it more manageable. The most urgent issues are addressed first and this process itself will generate a wider effect. At each of the steps, the methods chosen should be interactive and encourage open communication.

A wide range of people must be involved, from different levels of the organisation and having different skills and backgrounds. All members of the organisation must feel a strong sense of ownership of the need and drive to improve safety standards.

This ***direct action*** to promote change is likely to succeed in developing a positive health and safety culture, as the continuing process will show management's interest in the activities and their willingness to support them in terms of time, trouble and finance. Each problem that is successfully dealt with will add to the positive development of the safety culture.

As the safety management plan unfolds, there will be increased awareness of hazards and risks, and the methods of controlling risk, which will come from, for example, ***risk assessments and training.*** This increased awareness, along with the knowledge that the company is carrying out its plans for health and safety, can reap ***cultural dividends***.

This will require strong leadership and ***strong employee engagement*** and ***ownership at all levels***. If it is also seen that everyone's ***performance is measured*** to ensure health and safety remains a priority at all levels, then it will give credence to the general plan. The spin-off from this is all levels of staff are working for the same aims and objectives, resulting in mutual trust and a visible commitment.

As part of the communication and consultation processes, ***feedback is essential***. It should be part of the overall plan and involve everyone. 'How are we doing?' and 'How am I doing?' are necessary question to ask and feedback is needed to answer them. Development and improvement will only come from the feedback of the findings of inspections, surveys, audits and prompt lists. It must be part of the continuing process in the plan to effect change in the health and safety culture.

Problems and pitfalls

There are a number of problems and pitfalls to be aware of when trying to change the health and safety culture.

The two main barriers are:

1) ***Attempting to change culture too rapidly*** by adopting every conceivable measure that might lead to improvement. This becomes too demanding a task and may lead to a loss of interest because not enough seems to be happening. From the employees' point of view, it may be seen as just another short-lived scheme where a lot of things are happening at once, but then it fizzles out. As discussed earlier, producing a positive climate requires clear, visible, management commitment to health and safety from the most senior level in the organisation. The commitment should be not just a formal statement but be evident in the day-to-day activities of the company. Failure to display long-term commitment will lead to an ***absence of trust in communications.***
2) ***Adopting too broad an approach.*** The breadth of the concept of changing the health and safety culture may make the task of managing improvement appear to be both abstract and daunting.

Further barriers include problems with communication and people being ***resistant to change.*** Absence of trust in communication will only be overcome as the methods of communication improve and when the topic of that communication is seen to be carried out and the results fed back. In simple terms, if what is said is to happen, happens and everyone knows about it, barriers to communication will start to fall.

People are naturally resistant to change and this barrier can be broken down by helping them see that the new ways will be beneficial to them. It requires giving full information, explaining the benefits, including them in making the changes thereby enabling them to take ownership, and take a gradual approach.

As previously mentioned, the step-by-step approach is essential to avoid the barriers which might impede progress.

Regulating the workplace by encouraging companies to develop a health and safety culture transcends the setting and imposing of external criteria. Instead, it concentrates on the organisation of the system and the internal climate. Rather than feeling that actions for improvement are imposed from the outside, individuals are encouraged to own those actions.

Developing and promoting a positive health and safety culture is an essential part of health and safety management. There must be commitment and clear leadership from the top. Aims, objectives and standards must be set, which are measurable and achievable.

Element

A10

Human factors

Learning outcomes

On completion of this element, candidates should be able to demonstrate understanding of the content through the application of knowledge to familiar and unfamiliar situations. In particular, they should be able to:

A10.1 Outline psychological and sociological factors which may give rise to specific patterns of safe and unsafe behaviour in the working environment.

A10.2 Explain the nature of the perception of risk and its relationship to performance in the workplace.

A10.3 Explain the classification of human failure.

A10.4 Explain appropriate methods of improving individual human reliability in the workplace.

A10.5 Explain how organisational factors could contribute to improving human reliability.

A10.6 Explain how job factors could contribute to improving human reliability.

A10.7 Outline the principles, conditions and typical content of behavioural change programmes designed to improve safe behaviour in the workplace.

Contents

Relevant statutory provisions

N/A

Sources of reference

Reference information provided, in particular web links, was correct at time of publication, but may have changed.

Reducing error and influencing behaviour HSG48, HSE Books, ISBN: 978-0-717624-52-2, http://www.hse.gov.uk/pubns/books/hsg48.htm

HEALTH AND SAFETY INCIDENTS

Kegworth Air Disaster, https://www.gov.uk/aaib-reports/4-1990-boeing-737-400-g-obme-8-january-1989

Herald of Free Enterprise, https://www.gov.uk/maib-reports/flooding-and-subsequent-capsize-of-ro-ro-passenger-ferry-herald-of-free-enterprise-off-the-port-of-zeebrugge-belgium-with-loss-of-193-lives

Learning from the Piper Alpha Accident: A Postmortem Analysis of Technical and Organizational Factors, http://engineeringfailures.org/files/Learning%20from%20the%20Piper%20Alpha%20Accident.pdf

Train accident at Ladbroke Grove Junction, http://www.railwaysarchive.co.uk/documents/HSE_Lad_IntRep003.pdf

Three Mile Island Accident, http://www.world-nuclear.org/information-library/safety-and-security/safety-of-plants/three-mile-island-accident.aspx

The above web links along with additional sources of reference, which are additional to the NEBOSH syllabus, are provided on the RMS Publishing website for ease of use - www.rmspublishing.co.uk.

A10.1 - Human psychology, sociology and behaviour

Meaning of the terms: psychology and sociology

OCCUPATIONAL PSYCHOLOGY

Psychology is the study of the human mind. The individual receives information through the senses: sight, hearing, touch, smell and taste. The information is processed in the brain/mind and there will be resultant behaviour. Sometimes the behaviour is not what is expected, which suggests that something has gone wrong with either the information received or the processing in the brain/mind. Reasons why the communication stage has gone wrong are relatively straightforward: sensory impairment, for example, deafness or blindness, or due to wearing hearing protection or goggles or being in a noisy environment.

What is not so straightforward is what has happened at the processing stage. We cannot see into the human mind to know how it works nor see what has caused the behaviour we did not expect. Occupational psychology attempts to explain the processing in order to understand why people behave the way they do and so influence and predict human behaviour in the workplace.

Controlling behaviour and human error in the workplace is an important consideration in the management of risk. If we consider the human mind as a 'black box' then we can study the inputs and the outputs - but not the processes in the mind that produce them. These processes are governed by personality (the 'nature' versus 'nurture' factors).

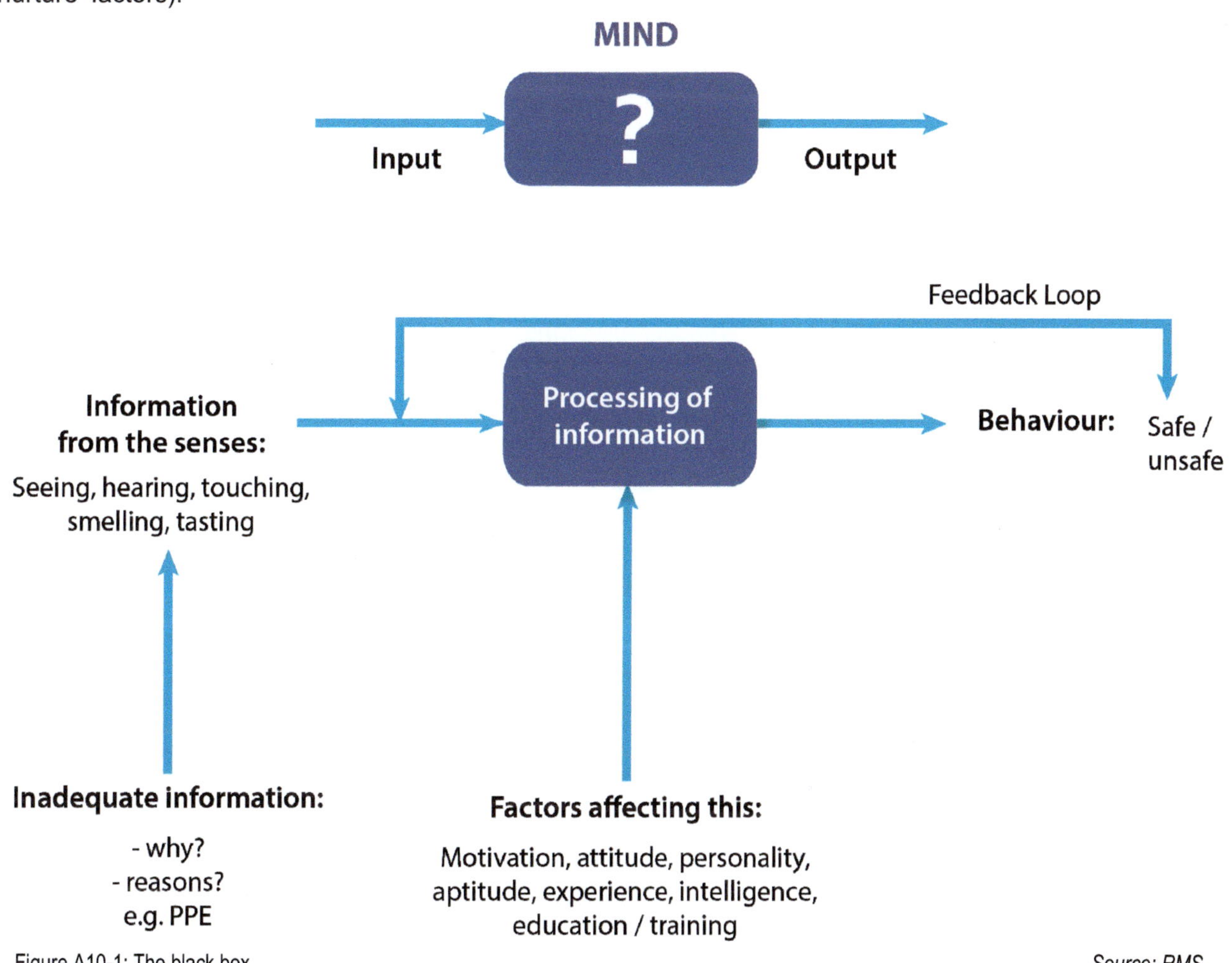

Figure A10-1: The black box. *Source: RMS.*

SOCIOLOGY

Sociology is the study of the development, structure and functioning of society. The question of whether each individual helps to form society or society forms each individual who is born into it, is central to the study of societies.

The workplace/organisation can be considered as a society, with each employee as a member. That society will function in a different way from others. It is the task of occupational sociology to describe the differences and the reasons for the differences in various workplaces, how each individual fits in to the society and how groups function.

Individuals form into groups which will affect the society more than each individual will, on most occasions. It takes a very strong individual to change society and in the workplace this may have positive or negative effects.

The influence on human behaviour of personality, attitude, aptitude and motivation

PERSONALITY

There are as many personalities as there are people. The individual develops with inborn tendencies, nature, and outside influences, nurture. According to Cattell's Trait Theory, there are more unique character personality combinations than the number of human beings that have ever lived. People behave according to their personality traits, which may be inherited as well as learnt. They may be: cheerful, pessimistic, slapdash, perfectionists, and/or the many variables Cattell suggests. Personality is extremely difficult to change, and probably it is unethical to even consider the prospect.

Research has been done on the link between personality and accident proneness, but with no clear-cut conclusions. The link between certain individuals and accidents is more likely to be because of sensory impairment, for example, colour blindness or short-sightedness, or because of the lack of compatibility between the person's build and their particular task.

Using accident proneness as a reason for accidents reinforces a blame culture and does not encourage looking for root causes.

It is part of the management of risk to choose people for a task according to their physical and mental capabilities. This must include a consideration of the person's personality. For example, a person who tends to be a lively, hyperactive type should not be given a job sitting watching a gauge for a crucial rise in temperature. Accidents resulting from the mismatch of individual and task are usually not the fault of the individual.

ATTITUDE

"A pre-set response to conditions that are presented to the person"

Attitudes are a set of factors that constitute ways in which individuals differ one from another. Attitudes are not directly observable and can only be assessed by observing behavioural expression (physical or verbal behaviour). Clearly, a person's attitudes will govern the way in which an object or situation is viewed and it will dictate the resultant response or pattern of behaviour. This is obviously very important when considering an individual's working patterns and any safety aspects associated with them.

Attitudes, like other aspects of individual differences, are formed (not necessarily consciously) as a result of a lifetime of experiences and although not totally immutable are not easily changed. A person's attitudes are not simply an aid to coping with the world; they are the means by which that person copes with the world. Any attempts to change such a fundamental part of an individual's personality will be resisted. His very being is under threat. Attitudes can be changed, although it is a difficult process. They have been changed within our society by pressure groups and government campaigns. Examples of where this has happened are drink driving campaigns, anti-smoking and the wearing of seat belts.

Smoking cigarettes was once considered the norm in our society, but a long campaign against it has changed society's collective mind. This has been achieved through information and education, price rises, the banning of smoking in most workplaces and public places, and peer pressure. It is now acceptable behaviour to complain to someone if they smoke near you. Passive smoking is considered a hazard to health.

In some cases, changes in other attitudes can affect the attitude to cigarette smoking and cause it to increase. It may be used by young women to keep their weight down. One attitude over-rides another.

"People's attitudes and opinions that have been formed over decades of life, cannot be changed by holding a few meetings or giving a few lectures."

Figure A10-2 Observation made by Chairman Mao Tse Tung. *Source: 'Little Red Book'.*

Examples of attitudes affecting safe working:

- It will never happen to me.
- It's only the price of a plaster.
- We have never had an accident.
- I know my limits.

Remedial action

- Train, and retrain when need for reinforcement is evident.
- Change by experience (involvement), for example, selection of personal protective equipment (PPE).
- Peer pressure.
- Role model.

Everyone at work should attempt to change their own and their colleague's attitudes to health and safety from - work safely because:

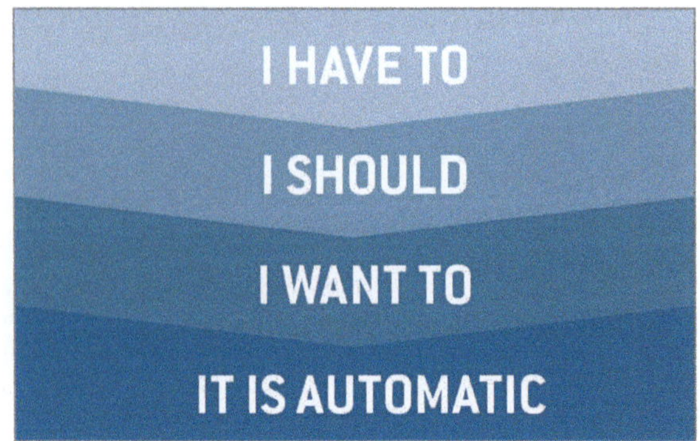

APTITUDE

Aptitude is closely linked to personality. Some people are particularly good at certain things, for example, an individual may be good at working with his hands, while another may say they 'could not change a light bulb'.

Aptitude can be developed over time as a skill, but it is more likely to be part of that person's characteristics. This again can be a factor when placing people in particular jobs. A person with no aptitude for precision work, but superb at felling trees is an accident waiting to happen if given the job of soldering electronic components.

MOTIVATION

"What makes people strive to satisfy their needs?"

In the context of the working situation there have been, over the years, many attempts to identify why people work.

The earliest approach (by F.W.Taylor) was that people worked for money and fear of losing their livelihood. Financial reward was seen as the prime motivator. The more they were paid the harder they worked. This led to a new management philosophy:

- Payment by results.
- Incentive schemes.
- Piece work.

From a safety management viewpoint, this theory is unsound since most bonus schemes encourage people to work unsafely by cutting corners, taking short cuts etc. with safe working practices the inevitable victim.

Money ***is*** important but other factors are more important, for example, social belonging, acceptance by one's peers.

Elton Mayo, as a result of a lengthy series of experiments carried out at the Hawthorne Works of the General Electric Company of America, came to the conclusion that it was social interaction and a sense of belonging to an important working group that provided the job satisfaction. This was more important to the workforce than financial reward. Work was seen as a 'social activity' not just as a means of earning wages. He found workers were interested in *"a fair day's work for a fair day's pay."*

Herzberg said that there was no one factor that determined the presence or absence of job satisfaction. He found what he called: 'motivators', for example, status recognition, and 'maintenance factors', for example, salary and working environment.

Key theories of human motivation

MAYO (HAWTHORNE EXPERIMENTS)

Elton Mayo, an Australian Industrial psychologist, conducted a lengthy series of experiments between 1927 and 1932 at the Hawthorne Works of the General Electric Company of America. They were originally intended to be a short investigation into the 'relation of quantity and quality of illumination to efficiency in industry'.

During the first experiments an experimental group was moved into a special area where illumination was increased and then decreased. A reference or control group was left where it was and was not subjected to any changes in illumination.

The observers predicted that output would vary with the illumination level. This did not prove to be the case, as output appeared to vary independently of the lighting level. Indeed in some cases, as the observed group's level of activity went up, sometimes so did that of the reference group (despite the fact they had not suffered from or benefited from any obvious physical changes in their environment).

These first observations drove the observers to devising a fresh series of experiments, which would control or eliminate the 'various additional factors, which affect production output in either the same or opposing directions to that which can be ascribed to illumination'.

The observers started to ask themselves what factors motivated individuals who worked in groups. They then extended the variables to include those factors which might be influencing the level of activity of the group. For example, in one experiment, an individual was allowed to select the other members of his group; as a result productivity was improved and the group appeared to work more harmoniously. Other variables the experimenters introduced included variations in:

- Rest pauses.
- Working periods.
- Systems of payment.

Eventually, the company called a halt to these investigations as they caused too much disruption. For example, the workers who did not benefit from the changes felt that they should have been involved. Finally, there was a downturn in production due to lack of demand for the product as a result of the Great Depression of the 1930s.

The conclusion drawn from these experiments was that social interaction and a sense of belonging to an important working group contributed a great deal to job satisfaction. This was more important to the workforce than financial reward. Work was seen as a social activity not just as a means of earning wages.

MASLOW (HIERARCHY OF NEEDS)

Each individual has needs to fulfil, from the basic need to eat to the need to attain his true potential. The drive to fulfil these needs is known as motivation.

Maslow defined motivation as a striving to satisfy a need. When we have a need (or a want) we are motivated to act in a way that will lead to a removal or diminishing of that need. We all have needs and strive to satisfy them. Maslow put forward the idea that needs can be categorised and that these categories can be placed in rank order.

Once one need has been satisfied, it ceases to act as a motivator and the individual then strives to satisfy the needs in the next category. Some needs are continuous, such as eating, while others can change at different times of one's life. Money may be of paramount importance to the parents of a young family, while recognition is more important to an older person. The model suggests that it would be difficult to fulfil the higher needs in the hierarchy if the lower needs were not fulfilled. For example, it would be difficult for a person to feel job satisfaction if they were worried about redundancy (security needs not fulfilled).

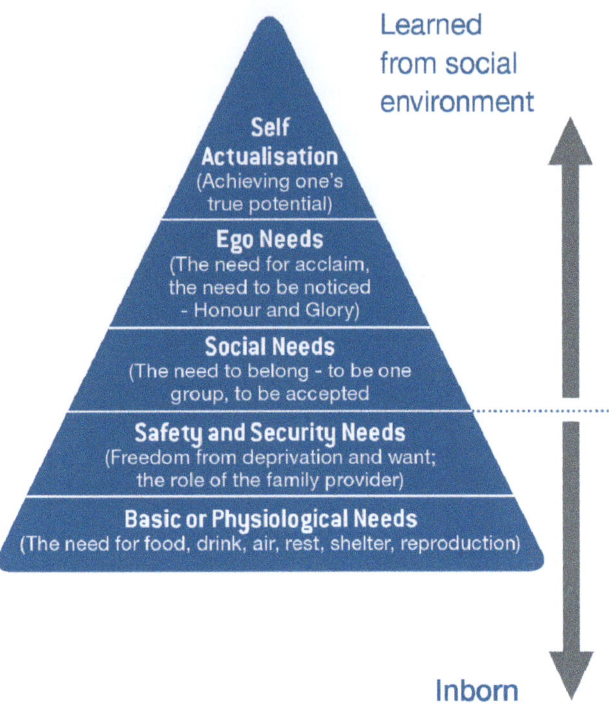

Figure A10-3: Maslow's hierarchy of needs. *Source: Maslow.*

VICTOR VROOM (EXPECTANCY THEORY)

Victor Vroom's expectancy theory was first proposed in 1964. Vroom's ***expectancy theory*** states that an employee's motivation is an outcome of how much an individual wants a reward (valence), the assessment that the likelihood that the effort will lead to expected performance (expectancy) and the belief that the performance will lead to reward (instrumentality).

Valence is the significance associated by an individual about an expected (not actual) outcome and the satisfaction that an employee expects to receive after achieving the outcome. Expectancy is in effect the faith the employee has that effort will result in better performance.

Expectancy is influenced by factors such possession of appropriate skills for performing the job, availability of right resources, availability of crucial information and getting the required support for completing the job. The individual evaluates the likelihood or probability that achieving the performance level will actually result in the attainment of the reward. Vroom terms this 'instrumentality'. Instrumentality is the perception the employee has that they will get their reward (even if a manager has promised them one).

The hypothesis proposed in Vroom's research is that if an employee perceives that high productivity leads to one of the employee's goals then they will be a high producer. Conversely, if the employee sees low productivity as a path to the achievement of his goals he will tend to be a low producer.

Following the expectancy theory, positive safety behaviour should be rewarded (instrumentality) and the implications of not following a safe system of work clearly outlines the risk of injury (valence) so that employees can relate clearly to the incident. The employees must be convinced that following a safe system of work will indeed make them safe; therefore, there expectancy of an outcome is achieved.

BLANCHARD (OPTIMAL MOTIVATION)

Ken Blanchard proposes that all employees are motivated but seeks to optimise that motivation. Optimal motivation comes from satisfying a need for autonomy, relatedness and competence in the pursuit of the achievement of meaningful goals. Based on self-determination theory, Blanchard's optimal motivation holds that humans have an innate tendency for growth and development, that humans have specific environmental and personal factors to thrive and flourish and that the universal needs of autonomy, relatedness and competence (ARC) needs to be satisfied.

Blanchard's interprets motivation as a spectrum, with suboptimal motivation being disinterested employees who focus on money and power and the individual participates because they are pressured into doing so. Optimal motivation would have the employee participating because they were able to assign other values (i.e. learning or having others learn from them).

An employee whose motivation is optimised would participate because they can add to what they believe is an important issue and they participating because the activity is enjoyable. Implications for safety training from this approach would be to involve employees in sharing experiences where they performed a safe act that prevented an injury (i.e. wearing PPE), convince attendees that safety was an important issue, make the activity of training enjoyable.

Effects on behaviour at work

EXPERIENCE

The experienced worker will know the hazards in the workplace and can make a decision on the risk he can take based on past experience. The problem with this is that the worker's accident-free past may well be down to luck and that may change at any time. They may no longer see the hazards because of familiarity with them. More than half the fatalities from electrocution involve so-called 'competent people'.

The experienced worker will, however, be more aware of what can cause harm, unlike the inexperienced young person. It would be unethical to cause someone to have an accident in order to heighten their awareness of risk or so they could gain experience. Awareness must be heightened in other ways, i.e. by training and education.

SOCIAL AND CULTURAL BACKGROUND

All individuals are different. These differences will influence patterns of work behaviour and may limit the effectiveness with which an individual carries out a job. They will also influence how safely the work tasks are carried out. These individual differences arise from an interaction between the 'inherited characteristics' (passed on from the parents) and the various 'life experiences' through which the individual passes from the moment of conception.

The factors that can influence behaviour include:

- Social background - individuals may be from social groups that work with and accept risks as a normal part of work.
- Cultural background - Eastern Europe migrant workers are much more likely to work long hours or take risks than endemic workers, for example, Chinese cockle gatherers in Morecombe Bay.
- Family influences.
- Geographical location.
- Pre-school influences.
- Education - opportunities, quality, support.
- Occupational factors - training and retraining.
- Hobbies and interests.
- Own family influences marriage, children.
- Ageing.

Any, many or all of the above will help to 'mould' the individual into a unique person different from all other individuals. The ways in which people differ are many and varied and it is important to bear this in mind from the point of view of work effectiveness and safety. It is vital to know what a particular job entails (the job description) and to specify the characteristics required to enable a person to perform that job effectively (the personnel specification).

EDUCATION

Levels of intelligence, training and education are closely linked. To a certain extent, a person's intelligence may limit their educational capabilities. Also, well educated people are usually more easily trained.

TRAINING

Training is important for the experienced and inexperienced. The experienced should receive refresher training to prevent complacency, and the inexperienced need training and education to enable them to work safely. Training is provided to show the person the safe and healthy method of working and education to show them why.

A10.2 - Perception of risk

Human sensory receptors and their reaction to stimuli, sensory defects and basic screening techniques

The main sensory receptors are the eyes, ears, taste, nose and skin (giving the sense of sight, hearing, taste, smell and touch).

All of the systems of sensory perception have a common structure, they consist of cells designed specifically to detect some aspect of the individual's surrounding environment (receptor cells), a series of neurons (nerve cells) that transmit the perceived information to the brain and a specific segment of the brain for receiving and analysing the information. Receptor cells can be classified according to the kind of stimuli to which they respond.

Chemoreceptors, for example, are cells that detect certain kinds of chemical substances, for example, receptor cells in the nose and mouth. Photoreceptors detect the presence of light and are present in the eye. Mechanoreceptors detect changes in mechanical energy, for example, those that occur during touch and hearing.

EYES

Reaction to stimuli of the eyes

The eyes are the most significant sensory receptor, vision enables us to:

- See a light at long distance in the dark - the light from a match up to 8 kilometres away (5 miles).
- In daylight to take account of movement in our extreme peripheral vision.
- To observe detail to 25 microns (one thousandth of an inch).
- To differentiate the seven colours of the rainbow (visible spectrum) in many shades and hues some 70% of all information is received through our vision.
- The binocular vision provided by two eyes enables humans to position items accurately.

Sensory defects of eyes

Common eye defects include short sightedness (myopia), where an individual is unable to differentiate detail at a distance because instead of light focusing on the retina it focuses in front of the retina.

Another defect related to the focusing of the eye is long sightedness, where the individual is unable to see items clearly close to the eye or see fine detail because the light received by the eye is focused behind the retina. This may be a hereditary abnormality of the eye or a natural part of the aging process (called presbyopia).

Another common eye defect is colour vision deficiency, which is the lack of ability to differentiate between colours, for example red and green. The retina contains light sensitive cells called cones; there are three types - red, green and blue.

Colour vision deficiency occurs when one or more of the cone types are faulty or missing, most people affected cannot distinguish some shades of red and green.

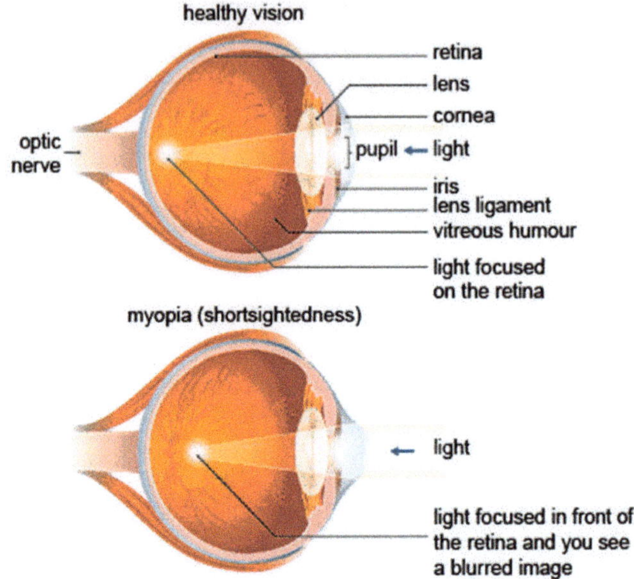

Figure A10-4: Short sightedness. *Source: BUPA.*

Colour vision deficiency is common among males and affects some 8% of the male population to varying degrees.

Colour vision deficiency is less common in females, affecting approximately 0.5% of females. Loss of peripheral vision may also occur, this is sometimes referred to as tunnel vision. In addition, the eyes often deteriorate over time or due to high exposure to non-ionising radiation, leading to such conditions as cataracts.

Cataracts cause the gradual loss of transparency of the cornea, the transparent membrane enclosing the lens of the eye.

Cataracts are a common defect caused by a variety of clinical disorders and exposure to ultraviolet radiation (sunlight or welding light) or extreme exposure to infra-red radiation (working with materials at high temperature such as molten metal or glass).

Basic screening techniques for eyes

Basic screening techniques for eyes include reading from a standard chart (called a Snellen chart) from a distance of six metres to identify focal vision defects like short sightedness. Colour vision deficiency can be screened by conducting specific colour vision tests, for example the Ishihara test. The Ishihara test uses plates made up of multi-coloured dots.

Some of the dots will be a different colour and depict a number. Where the individual has difficulty identifying correctly the numbers it can indicate colour deficiency. Peripheral vision may be tested by a practical vision test where an object is moved from the main field of vision towards the periphery until the individual can no longer see it.

Cataracts are identified by the direct observation of the eye by an optometrist (health professional) or ophthalmologist (specialist eye doctor).

EARS

Reaction to stimuli of ears

The ears are the next most important sensory receptor, sensing sounds of a varying intensity and frequency. This enables an individual to differentiate between the vowels and consonants of speech so that we are able to communicate and to appreciate music in fine detail. The position of two ears enables us to identify the direction of a particular source of sound.

Sensory defects of ears

Common sensory defects related to hearing include:

1) ***Conductive hearing loss*** - this is caused when sound cannot move freely from the outer ear to the inner ear. This may be caused by such things as a middle ear infection, a build-up of earwax (wax-like material produced by sebaceous glands inside the ear) or perforated ear drum.
2) ***Sensorineural hearing loss*** - this occurs when there is damage to the pathway between the inner ear and the brain. This may be caused by age related hearing loss (presbycusis) that progressively affects the function of the cochlea as part of the aging process; noise induced hearing loss or certain medicines like strong antibiotics.

A person may have hearing loss in one ear only (unilateral) or both ears (bilateral). Hearing loss will reduce the ability of the ear to recognise the consonants and vowels resulting in progress loss of understanding of the spoken word and individual isolation. It will also prevent the recognition of auditory signals designed to warn of danger, for example, fire alarms.

Basic screening techniques for ears

Basic screening techniques usually involve pure tone audiometry, where the individual wears a set of headphones attached to a machine called an audiometer. The audiometer produces sounds of different volumes and frequencies, and the individual is asked to indicate when they hear the sounds in the headphones. The level at which they can no longer hear the sound of a certain frequency is known as their hearing threshold.

Audiometric equipment is readily available and relatively easy to use for basic screening. Where it is not available less technical screening may be conducted with 'whisper tests', words spoken behind an individual to determine their ability to hear a low intensity, and tuning fork tests to determine their ability to hear certain frequencies.

TASTE

Reaction to taste stimuli

The tongue enables individuals to differentiate between acidity and levels of salinity and in conjunction with our nose a wide range of flavours.

Sensory defects affecting taste

The sense of taste is rarely lost, but can be affected by poisoning with strong acids or alkali.

Basic screening techniques for taste

Screening techniques related to taste are a specialist medical matter.

NOSE

Reaction to stimuli of the nose

The nose enables individuals to identify a wide range of odours and in conjunction with the tongue, a wide range of flavours.

Sensory defects of the nose

The nose sensory organ is easily damaged by exposure to alkaline, acidic, toxic dust or gas. Mild effects may recover with time; however, permanent loss will significantly reduce taste and will prevent the detection substances that may affect the individual. For example, harmful gas leaks, such as hydrogen sulphide associated with the oil extraction industry.

Basic screening techniques for the nose

Screening techniques related to the nose and its ability to smell are a specialist medical matter.

SKIN

Reaction to stimuli of the skin

The skin, the largest organ of the body, enables us to utilise the sensory process of touch. The skin also provides us with the ability to determine variations in ambient temperature, the direction and movement of the wind and the degree of humidity in the air.

Sensory defects of the skin

Damage to the skin may include abrasion, chemical destruction of cells or damage from radiation, heat or direct burning or mechanical damage from contact with sources of vibration. These effects will interfere with the correct operation of nerves located in the skin and reduce or prevent sensory functions.

Basic screening techniques for the skin

Screening techniques related to skin sensory defects are a specialist medical matter.

The process of perception

PERCEPTION OF DANGER

"The process by which people interpret information that they take in through their senses"

- A boring, repetitive job may result in 'day dreaming', which may result in a lowering of the impact of a stimulus.
- Warnings (or threats) may not be strong enough to get through the perceptual set.
- Patterns of behaviour and habits can be carried from one situation to another where they are no longer appropriate or safe, for example, we tend to drive too quickly after leaving a motorway.
- Individuals can get 'used to' a stimulus and, if it is not reinforced, it ceases to command the attention and is ignored.
- Intense concentration on one task may make attention to another stimulus difficult or impossible.

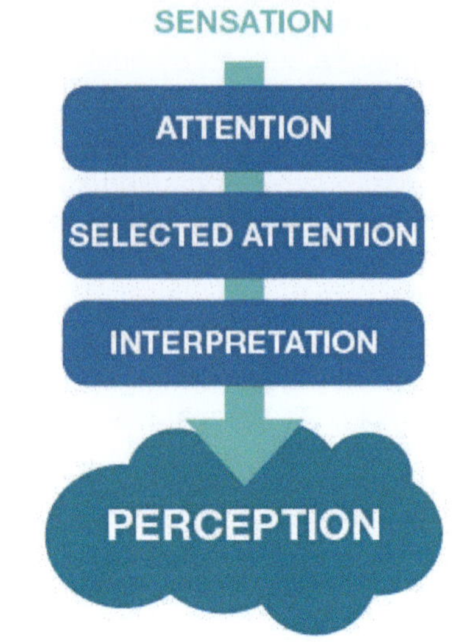

Figure A10-5: A model of perception. *Source: RMS.*

PERCEPTUAL SET

Within each individual these processes of attention and interpretation are closely interlinked. The factors which influence these processes and the way in which they operate are often referred to as the 'perceptual set' of the individual.

External determinants

(The nature of the stimulating conditions).

- Location.
- Senses and sense combinations.
- Intensity.
- Size.
- Colour.
- Tone.
- Motion.
- Novelty and change.
- Repetition.

These can, and do, act in combination.

Internal determinants

- Conscious or unconscious motives.
- Expectations.
- Capacity.
- Change in attention.
- Fatigue.
- Needs (deprivation).
- Culture.
- Sensitisation
- Habituation.
- Prejudice.

The ***perceptual set*** (which to a greater or lesser extent will govern the impacts of the external determinants and how they are interpreted).

PERCEPTUAL DISTORTION (PREJUDICE)

We cannot treat every sensory stimulus as new and unique. Every experience cannot be reacted upon as a new experience. We would go insane if we tried. In order to overcome the problem of dealing with the vast range of possible experiences, we adopt a system of categorisation. We learn to react to a particular category in a specific way and to make life bearable and workable; we categorise objects, experiences and situations based on our memory of similar things.

Sometimes we are forced to categorise a specific item that we have not previously experienced (in order to be able to deal with it) and we have to do this categorisation in the absence of adequate information. We have to pre-judge.

Without this process of pre-judging and categorisation we would not be able to cope with the world in which we live. There are two mechanisms that are of specific importance when considering practical applications of perception theory, perceptual sensitisation and perceptual defence.

Perceptual sensitisation

An individual can be or become 'sensitised' to certain stimuli if they are relevant, important or meaningful to that person. Sensitisation can be permanent or temporary and may or may not be accepted consciously by the individual. The stimuli to which the individual is sensitised have a much greater impact than would normally be expected.

It could be said that they are seen or 'perceived' as larger, brighter, more attractive or more valuable than other stimuli. For example, if our next-door neighbour is burgled, then the slightest sound in the middle of the night will cause instant wakefulness for several weeks afterwards, sounds through which we would normally have slept.

Perceptual defence

No one likes to feel threatened or anxious. The mechanism of 'perceptual defence' tends to protect the individual from such unpleasant situations by blotting out these threats and making them difficult to perceive at a conscious level.

The individual does not want to register the problem or difficulty so it is pushed into the unconscious. These two mechanisms can have an important influence on the perception of hazards and risks in the workplace. The safety advisor may well be 'sensitised' to observing safety hazards while those actually exposed to the danger are operating their 'perceptual defence' and may not accept that any danger exists.

Remember:

- We do not see what is there.
- We see what we expect to be there.
- We do not see what we do not expect to be there.
- We do not see what we do not want to be there.

Perceptual distortion can be influenced by:

- Information.
- Instruction.
- Training and drills.

Errors in perception caused by physical stressors

Human error can be defined as an inappropriate response to a signal. In the context of health and safety, the signal is any message from the environment (including colleagues) relating to an existing or potential danger, which requires a response in order to avoid or remove the danger.

Human error can arise from a variety of causes; examples are indicated in ***figure ref A10-6***.

Causes of error	*Action*
Signal not received Physical reasons (usually due to 'sender' problems rather than 'receiver') Background noise Too great a distance over which to communicate PPE (dust masks, ear defenders etc.)	Design of work environment
Psychological reasons (internal to the receiver) Non attention: Lack of concentration - boredom - fatigue	Staff selection
Over-concentration on one task to the exclusion of all other signals Desensitisation Systems overload - too much to attend to, unable to cope and signal missed	Job design
Signal misinterpreted Not the expected signal Signal perceived incorrectly (incorrect interpretation - sometimes called a 'false perception') Lack of understanding of the signal	Training
Inappropriate response to signal Lack of knowledge, understanding or training	Training and practice
Correct response - but inappropriate priority Lack of training (use of simulated situations as training exercises)	Training and practice
Signal ignored Deliberately ignoring a warning is rare but may arise from: ■ No appreciation of the consequences of inaction. ■ False priorities, for example, bonus earnings. ■ Attitude to safety (safety is for wimps - taking risks is macho).	Training attitude change (long term)

Figure A10-6: Causes of error. *Source: RMS.*

Note: Training regarding the appropriate course of action to be taken on receipt of a warning signal must also include the steps to be taken to pass on the warning to fellow workers.

Filtering and selectivity

The system receives inputs through the senses which may be in good working order, defective or defects may be imposed by PPE or clothing. There is always more information in the environment than the senses can deal with. The brain allows certain information through filters on the sense, which is partly conscious and partly unconscious. We shift our gaze over an area until it rests on an object of interest and we know in advance what characteristics to tune in to. This ability is known as 'perceptual set'. We can tune in to seeking out a defect in a machine or a work piece and blot out all other information. This other information may be important, but we may be unaware of it.

We can also tune out our hearing to certain sounds and be able to hear if a machine is running differently from usual. We can get used to certain sounds. Because they no longer signal anything to us, the filter does not allow them through. How we perceive the world is built up over years and we have a picture of what the world is like. This is called expectancy. We see what we expect to see which may be different from reality. We do not need to take in a lot of information about a particular scene to understand it because we have stored information about similar scenes. Problems occur where the real word differs from the expectation.

We have population stereotypes, which is expectancy that something will be a certain way. For example, we believe that pulling a lever down will turn something off, or red means danger and stop, or turning something clockwise will turn off a valve or turn the volume up. Stereotypes change from culture to culture. In Britain we put a light on by putting the switch down, but in the USA the switch is pushed up. If we came across designs that do not match up our stereotypes then we can easily be misled and make mistakes when carrying out a task automatically or making a choice in an emergency we tend to go for the stereotype. If this differs from reality then an error will occur. Standardisation is important to avoid these errors, but it must be total or it will make any expectations more likely to trap someone into an error.

A10.3 - Human failure classification

HSG48 classification of human failure

HSG48 'REDUCING ERROR AND INFLUENCING BEHAVIOUR' 1999

This is a model of types of human failure based on the work of Rasmussen and Reason. Human failures are divided into two types: errors and violations. A human error is an action that was not intended. Errors can be separated into two types: skill-based errors and mistakes. Skill-based errors can be further divided into slips of action and lapses of memory, and mistakes can be divided into rule-based mistakes and knowledge-based mistakes. Violations are deliberate deviations from the rules or procedures and can be separated out into three types: routine, situational and exceptional.

CAUSE OF HUMAN FAILURE

There are different types of human failures: errors and violations.

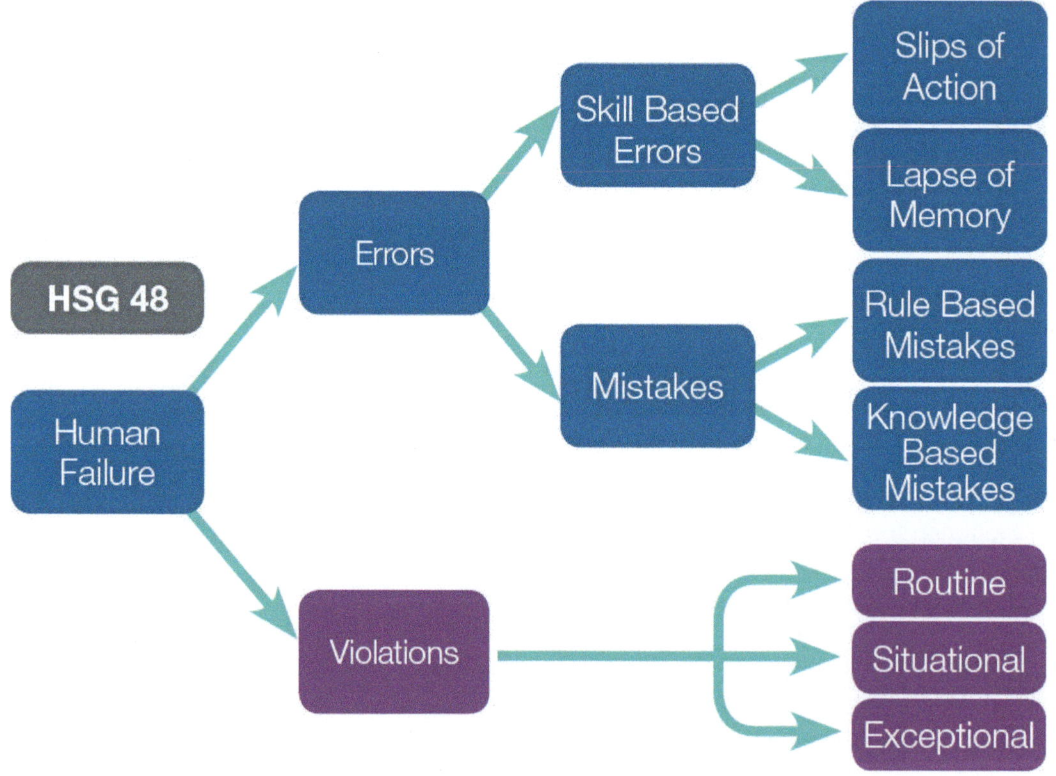

Figure A10-7: Causes of human failure. *Source: HSG48 'Reducing Error and Influencing Behaviour' 1999.*

Slips and lapses

Once we have learned a skill, there is little need for much conscious thought about what we are doing. We can carry out a task without having to think too much about the next step. We learn to ride a bike or drive a car in this way. We need to pay attention to the road and the traffic, but we manipulate the pedals and change gear without thinking about it. If our attention is diverted, we may fail to carry out the next action of the task, or we could forget the next action or lose our place resulting in an error.

Mistakes

Mistakes are a little more complex than slips and lapses. We may do the wrong thing believing it to be right. We have a tendency to use familiar rules or procedures, often when they don't apply. The wrong application of a rule to a situation can result in an error.

In unfamiliar situations we may have to apply knowledge-based reasoning. If this is miscalculated or the situation is misdiagnosed, then a mistake may occur. These errors typically occur with trained, experienced people, but also occur with untrained and inexperienced people. The untrained and inexperienced may base their decisions on misunderstandings and a lack of perception of risk.

Violations or breaking the rules

These violations are rarely acts of vandalism or sabotage, but are often carried out in order to get the job done. Many accidents, injuries and cases of ill health come about because of violations.

Routine violations are where breaking the rules or procedure has become the normal way of working. New workers come in and learn the incorrect ways, not realising they are wrong. The incorrect method may have come about because it is a quicker way to work or because the rules are seen as too restrictive. In one company it was felt that the work could not be finished on time if all of the rules were followed.

Situational violations may occur with pressures from the job: time pressure, extreme weather conditions and/or wrong equipment. Roof work may continue without edge protection because the correct equipment has not been provided. The person concerned may believe that he is doing a good job by making do with the wrong equipment and getting the job done quickly. Getting the job done without the proper controls may be condoned explicitly or implicitly by the supervisor, if nothing has gone wrong, which reinforces the violation. After a period of time, the situational violation may become a routine violation.

Exceptional violations occur when something has gone wrong. A decision has to be made to solve the problem and that might involve breaking a rule and taking a risk. It is erroneously believed that the benefits outweigh the risk. An employee, who has a meeting with a client and is running late, may decide to speed in his car to get there in time.

Cognitive processing; knowledge-based, rule-based and skilled-based behaviour

Input to the system or information to the person must be processed to be of use. This may be done on-line or off-line.

ON-LINE PROCESSING

On-line processing is the moment-to-moment decision making about what action to take when dealing with the environment the individual finds himself in. It involves the least thought in order to carry out the action, as the action will be a series of pre-learned steps. A human being could not function efficiently if everything he did had to be carefully thought through.

Some groups of actions work together to form a habit and through repetition they become a skill. Repetitive tasks become a skill that requires little thought. We seldom think about what we are doing when we change gear in a car or walk upstairs. Once set in motion it is difficult to halt the sequence of steps that make up the action until it has run its course. These learnt packages of actions rely on correct learning in the first place because it is difficult to insert new actions into them later.

OFF-LINE PROCESSING

Off-line processing requires a lot of 'thought energy' as the individual will consider the results of a number of actions before deciding on which course of action to take. Some courses of action will be rejected because the consequences will be predicted as being unpleasant. Many factors need to be manipulated in order to do this and it requires knowledge of a variety of facts. It is, therefore, a great skill that depends on intelligence and the amount of practice in using the skill.

Intelligence is difficult to define, but it is said to be based on the ability to learn, to manipulate concepts and to solve problems. Earlier this century it was common for child psychologists to carry out intelligence testing. This formed the idea of the 11+ exam, the results of which were used in deciding which secondary school a child went to. The child psychologist, Binet, who invented the term IQ (Intelligence Quotient), said that children with a high IQ, i.e. over 100, were able to do tasks quicker than the average child.

Intelligence can be subdivided into special aptitudes: numerical, verbal, musical, mechanical, etc. Tests can be devised for all these aptitudes.

Errors in off-line processing are said to be related to intelligence. There may be limitations on thinking through the courses of action to take and this can lead to unsafe decisions being made. Individuals may also make decisions based on illogical factors such as 'luck'. They also view things in a subjective way, the personality of the person having a great influence. The level of arousal in the brain can also affect decision making off-line. Low arousal in the brain is associated with drowsiness resulting in poor performance.

TYPES OF BEHAVIOUR LEADING TO ERRORS (RASMUSSEN)

There is a standard framework for classifying error - the skill-rule-knowledge based model, which comes from the work of Rasmussen. The three levels of behaviour show an increase in conscious control.

Skill-based behaviour

Skill-based errors involve slips or lapses. This is where highly practised action becomes routine or automatic action and a slip can occur in that routine.

Rule-based behaviour

Rule-based errors can occur where routines are selected from a large number of possible routines and the wrong ones are chosen.

Knowledge-based behaviour

Knowledge-based errors occur where people have to cope with what would happen if they tried particular sets of actions. This is the most complex situation and errors occur where the plan that is formed is mistaken. A shortcoming of the Rasmussen model is the fact that it does not cover violations. This is where a person deliberately carries out an action that is contrary to company rules. These types of error that have been proposed by Rasmussen and Reason have been developed in the HSE Publication HSG48, ***(see also 'HSG48 Reducing error and influencing behaviour 1999' earlier in this Element).***

Contribution of human failure to serious incidents

James Reason developed a model which proposed that most accidents/incidents can be traced to one or more of four levels of human failure: organisational influences, unsafe supervision, preconditions for unsafe acts and the unsafe acts themselves. In this model an organisation's defences against failure are represented as a series of barriers, with individual weaknesses in each of the barriers of the health and safety system (represented by holes) that are continually varying in size and position. The system as a whole produces failures when all individual barrier weaknesses align, permitting 'a trajectory of accident opportunity', so that a hazard passes through all of the holes (weaknesses) in all of the defences, leading to a failure. The model includes ***active failures*** and ***latent failures*** in the causal sequence of human failures that lead to an accident/incident or an error.

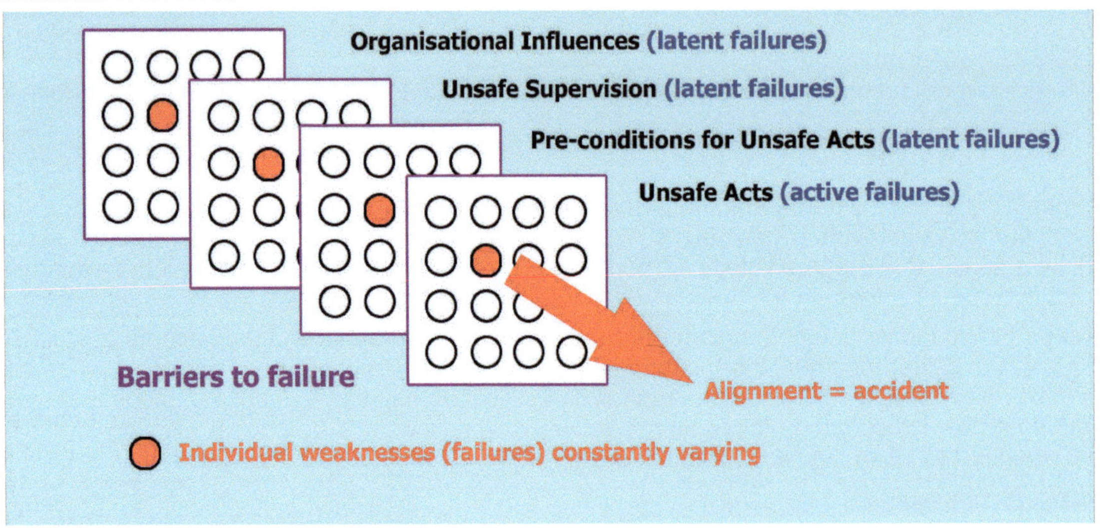

Figure A10-8: J. Reason's model of human failure. *Source: Reason/RMS.*

Active failures encompass the unsafe acts that can be directly linked to an accident/incident, such as in the case of rail accidents/incident due to driver errors. Active failures take a variety of forms: slips, lapses, mistakes and violations. Active failures have a direct and usually immediate effect on the integrity of the defences.

Latent failures are derived from latent conditions, sometimes called 'latent pathogens', within the health and safety system and may have lain dormant for a long time before they combine with active failures to create an accident/incident opportunity.

Latent failures comprise the first three levels of failure in Reason's model: organisational influences, unsafe supervision, preconditions for unsafe acts. Organisational influences include such things as reduction in expenditure on worker or management training in times of financial difficulty.

Unsafe supervision includes such things as an inexperienced worker being sent on their own to conduct a hazardous task in adverse conditions, such as poor weather and at night. Preconditions for unsafe acts include fatigue or inadequate communications. The latent conditions arise from poor decisions made by designers, builders, those that write procedures, those that plan and organise work and top level management.

Latent conditions can therefore, lead to error provoking conditions within the local workplace, for example, time pressures, inadequate workforce numbers and competence, fatigue, and inadequate equipment. Some of the latent conditions can also create long lasting weaknesses or holes in the defences of the health and safety system, for example, untrustworthy alarms and indicators, unworkable procedures, design and construction deficiencies.

When considering the contribution of human failure to serious accidents/incidents some have tended to take what may be called a person approach where the human failure focuses on the acts of the individuals immediately involved in the event, for example, in the case of a railway accident the driver of the train. However, focusing on the person approach often leads to looking no further for the causes of an adverse event once it has been identified that unsafe acts took place.

This would ignore the fact that most unsafe acts have a causal history that extends back in time and up through the levels/defences of the health and safety system and by focusing on the individual origins of error it tends to isolates unsafe acts from their system context. The system approach to human failure is based on the premise that humans are fallible and errors are to be expected. Errors are seen as consequences rather than causes, having their origins in systemic factors.

The same set of circumstances could have led to similar errors in another individual/person. The system approach therefore, fully considers the acts of those immediately involved in an adverse event, but also considers the human failures that lead to latent conditions that cause these actions.

"The Health and Safety Executive's (HSE) Accident Prevention Advisory Unit and others have shown that human error is a major contributory cause of 90% of accidents, 70% of which could have been prevented by management action."

Figure A10-9: Causes of human failure. *Source: UK, HSE, 'Improving compliance with safety procedures - reducing industrial violations'.*

KEGWORTH

8 January 1989

A British Midland B737 400 aircraft operated by British Midland Airways Ltd crashed at Kegworth, Leicestershire on 8th January 1989 at 2025 hrs.

The flight left Heathrow Airport for Belfast at 1952hrs with 8 crew and 118 Passengers (including 1 infant) on board. As the aircraft was climbing through 28,300 feet the outer panel of one blade in the fan of the No 1 (left) engine detached. This gave rise to a series of compressor stalls in the No 1 engine, which resulted in airframe shuddering, ingress of smoke and fumes to the flight deck and fluctuations of the No 1 engine parameters.

Believing that the No 2 engine had suffered damage, the crew throttled the engine back and subsequently shut it down. The shuddering caused by the surging of the No 1 engine ceased as soon as the No 2 engine was throttled back, which persuaded the crew that they had dealt correctly with the emergency. They then shut down the No 2 engine. The No 1 engine operated apparently normally after the initial period of severe vibration and during the subsequent descent.

The crew initiated a diversion to East Midlands Airport and received radar direction from air traffic control to position the aircraft for an instrument approach to land on runway 27. The approach continued normally, although with a high level of vibration from the No 1 engine, until an abrupt reduction of power, followed by a fire warning, occurred on this engine at a point 2.4 nautical miles from the runway. Efforts to restart the No 2 engine were not successful.

The aircraft initially struck a field adjacent to the eastern embankment of the M1 motorway and then suffered a severe impact on the sloping western embankment of the motorway. 39 passengers died in the accident and a further 8 passengers died later from their injuries. Of the other 79 occupants, 74 suffered serious injury.

The cause of the accident was that the operating crew shut down the No 2 engine after a fan blade had fractured in the No 1 engine. This engine subsequently suffered a major thrust loss due to secondary fan damage after power had been increased during the final approach to land. The following factors contributed to the incorrect response of the flight crew:

- The combination of heavy engine vibration, noise, shuddering and an associated smell of fire were outside their training and experience.
- They reacted to the initial engine problem prematurely and in a way that was contrary to their training.
- They did not assimilate the indications on the engine instrument display before they throttled back the No 2 engine.
- As the No 2 engine was throttled back, the noise and shuddering associated with the surging of the No 1 engine ceased, persuading them that they had correctly identified the defective engine.
- They were not informed of the flames which had emanated from the No 1 engine and which had been observed by many on board, including 3 cabin attendants in the aft cabin.

Contribution of human error

- When the captain informed the passengers that they had shut down the right hand engine (No. 2 engine), the passengers that had seen smoke coming out of the left hand engine assumed that the pilots had a different way of seeing left and right. It was as though they described left and right facing towards the cabin. No-one thought to challenge what they said they had done.
- Knowledge based error. The level of training they received to fly the new plane was brought into question. It was said that they were still using trial and error to find out what different instruments did.
- In the emergency situation, they fell back on well-learnt behaviour, which was not suitable for the new design plane. For example, one of the instruments that would have shown a problem was not considered, because in the old design that instrument was so unreliable they used to sometimes shut it down.
- Rule based error. The pilots applied a set of rules to finding the problem engine and it seemed to work. Every time they tried to reapply the rules, they were distracted by questions from the ground control operators.
- Perception of risk. Because that type of plane can fly on one engine, they may not have gone through every procedure available to them. The loss of one engine would not be viewed as a really serious situation.
- The new design control panel had smaller, computerised displays with a cursor instead of a large display with a pointer and a red area to show a dangerous situation, as the old design had. The pilots were not attuned to the new design and there was nothing to draw their eyes to a display showing the danger. The display that showed the problem was the size of a twenty pence piece and was positioned over to the side not in a prominent position.

HERALD OF FREE ENTERPRISE

6 March 1987

Less than half an hour after the ferry Herald of Free Enterprise sailed from Zeebrugge harbour for Dover, it capsized with 459 passengers on board. 188 lives were lost consisting of 150 passengers and 38 crew.

The subsequent inquiry revealed that the Herald left the harbour with both the inner and outer bow doors open. Water flooded into 'G' deck as the ferry increased speed thereby causing it to capsize and settle onto sandbanks with its starboard side above the waterline.

The Chief Officer, who had to be on the bridge 15 minutes prior to departure, was also responsible for ensuring that the bow doors were closed prior to sailing. The Assistant Bosun, Mr Stanley, had a duty to close the doors, but had fallen asleep after relief from maintenance and cleaning duties. The Bosun, Mr Ayling, noticed that the doors were open but did not close them, as it was not his duty. The Captain assumed that the bow doors were closed unless told to the contrary. Previous repeated requests for indicators to be fitted to the bridge had never been actioned - these would have cost about £500. There was pressure put on ferries to sail early. A company memo issued by the Operations Manager at Zeebrugge summed this up: *"There seems to be satisfaction if sailing is 2-3 minutes early. Where there is a full load every effort must be made to sail 15 minutes early - put pressure on the first officer if you do not think he is moving fast enough."*

There was therefore considerable pressure on the crew to meet departure times and the Herald sailed 5 minutes late on the 6 March.

The following factors also contributed to the actual incident:

- The emergency lighting was incapable of operating at large angles of heel and therefore did not work. This severely affected morale as well as hampering the physical rescue work.
- More than 1,000 lifejackets were locked away in order to prevent vandalism. The keys were in a small break-glass container adjacent to the locker door. The locker burst probably due to the buoyancy of the lifejackets. The resulting tangle of jackets and tapes along escape routes impeded people. The cold water also prevented people who had numb fingers from untangling the tapes in order to use the lifejackets.

A further cause of concern was the design of roll on - roll off ferries, which was considered to be inherently unsafe as they are top heavy. The Herald was a double deck ferry, but the berth-loading ramp had been designed for those of the single deck type. This meant that the ship's nose had to be trimmed by filling ballast tanks in order for the upper car decks to be loaded - a further cause of instability.

This 'head down' attitude caused increased stresses from bow waves when the vessel was underway which also affected the locking mechanism of the bow doors. It took up to 1½ hours to fill or empty ballast tanks and there was no indication to the bridge about what was going on during the process. A previous request by a Chief Engineer for quicker high capacity ballast pumps to be fitted to this class of ship had been denied on the grounds of cost (~£25,000).

The inquiry heard the following compounding factors:

- Ships routinely sailing with excessive numbers of passengers - sometimes way over the life saving capacity of the vessel.
- The Herald's complement of officers had been reduced by one-third (from 15 to 10) when it was transferred from the Dover-Calais run to the Zebrugge-Dover run on the basis that there was more time to 'relax' on the longer voyage.

- The crew's working schedule was 24 hours on duty and 48 hours off.
- There was no instrumentation to check the draught of the ship.
- There was no monitoring equipment (for example, CCTV) to observe problems in critical areas such as the car deck, engine room or superstructure doors.

The inquiry report found severe fault with the management system, which allowed the capsizing to happen:

"The Board of Directors did not appreciate their responsibility for the safe management of ships."

"...they did not apply their minds to clear orders for the safe operation of their ships..."

"...they did not have any proper consideration of what their duties were..."

"...there was a lack of thought about how the Dover-Zebrugge run should be organised..."

"...the entire body corporate was infected with sloppiness."

Contribution of human error

- The crew were working for 24-hour shifts, because they were able to rest on the journey. The assistant bosun had fallen asleep after being relieved of his normal duties. Having the responsibility for closing the bow doors seems to have been peripheral to his duties when he got really exhausted.
- A culture seems to have existed where they stuck to doing their own job. The bosun saw the doors open, but did not close them, as it was not his job.
- Perception of risk. Experienced crew get used to their workplace and fail to perceive any risk in it. Human beings cannot function if they are constantly aware of danger. We get round this by failing to see risk in something if it has never hurt us.
- Wrong priorities. Speed of leaving the harbour was given prominence in the company's communication. The priority was perceived as being speed rather than safety. This was further reinforced by the failure to spend money on safety features, for example, the indicators on the bridge to show the captain that the bow doors were closed.
- Routine violation. Setting sail with the bow doors open was breaking the rules, but became routine. This would be reinforced over time when it seemed to work and it appeared to be or was actually condoned by management.

PIPER ALPHA

6 July 1988

Piper Alpha, owned by Occidental Petroleum, was a key platform in the UK's North Sea oil production. It earned about £3.5 million per day and, at its peak, accounted for 10% of British North Sea oil production. Standing about 100ft above the surface of the water, it was designed to accommodate over 200 men. A series of explosions ripped through the platform during the night of 6th July 1988 killing 167 people. A pump had been shut down, using a permit to work, during the day shift in order to remove a safety pressure valve for recertification. A blind flange assembly was put in its place. At the end of the working day the suspended permit was returned to the control room but not displayed. During the night the working pump failed and the night shift started the pump that was fitted with the blind flange. Condensate entered the relief line and gas escaped from the flange assembly, which was not airtight. The gas exploded cutting out the main power supplies.

A public enquiry was held headed by Lord Cullen.

- The containment wall was fire, but not blast resistant.
- Two other rigs feeding into the same oil export line did not shut down until one hour after the initial mayday, which meant oil from the other rigs flowed back towards Piper and fuelled the fire.
- Gas pipelines ended in the area where the oil fire had started. They were eventually ruptured in the heat and the explosion engulfed the rig in thousands of tonnes of burning gas.
- People trapped in the accommodation block waited for helicopters to arrive but the first explosion made landing impossible.
- The routes to the lifeboats were blocked, and there was no message over the public address system telling them what to do.
- Most people stayed where they were until smoke and gas fumes overcame them. The survivors jumped 100 feet into the sea - contrary to the minimal training they had been given.

Permit to work system

- Valve locking off procedure not included in the permit.
- Work on the suspended pump had not been inspected by the designating authority (common occurrence).
- Although the permit system was monitored and frequently audited, failures had not been identified.
- Designating authorities (DAs) were not trained.

Common deviations:

- Multiple jobs on one permit.
- Failure to display permits in working area.
- Inaccurate description of work.

These problems had been highlighted to Occidental 12 months earlier in a report, so they knew about them. Lord Cullen's report concluded that Occidental had 'adopted a superficial attitude' to safety.

"The safety policy and procedures were in place: the practice was deficient."

Figure A10-10: Piper Alpha extract. *Source: Extract from Lord Cullen's report into the Piper Alpha disaster.*

Contribution of human error

- Knowledge based errors. The policy and procedures were there, but the personnel were not adequately trained. Their decisions were taken from an inadequate knowledge base.
- Misperception. The product was perceived to have priority over safety.
- Routine violations. Breaking the rules regarding the permit-to-work system had become common practice. This would be reinforced when management appeared to condone it either by overt means or by turning a blind eye.
- Most of the human error resulted from the influence of the organisation's negative health and safety culture. Managers of rigs had not been given authority to make decisions based on safety that could result in loss of revenue.
- Problems had been highlighted, but not remedied.
- In the emergency situation, there was no one person to take control and issue instructions. The personnel followed through their meagre training and went to the living accommodation. This was against the sort of action they would have taken if they had been in a situation where they could have thought it through. They travelled upward to the accommodation block even though heat and smoke rise. The ones who were saved were the ones who travelled downwards and then into the sea. Most of us do not think logically in an emergency situation.

LADBROKE GROVE

5 October 1999

At 0811 hrs on 5 October, two trains collided two miles outside London Paddington station at Ladbroke Grove Junction. One was a Great Western inter-city high speed train (HST) running from Cheltenham Spa to Paddington, the other a Thames Train 3-car diesel unit going from Paddington to Bedwyn, in Wiltshire.

The crash happened when the driver of the Thames Train went through a red light outside Paddington during the rush hour and collided head-on at 130 mph with the Great Western HST. The crash left 31 people dead, including the driver, and hundreds injured, some critically because of the rapid outbreak of fire in some of the HST carriages. The driver, Michael Hodder had only been qualified for 13 days when the accident happened. He had received no training from Thames about awareness of 'SPADS' (signals passed at danger), and had not been warned that the signal outside Paddington was a black spot. Eight drivers in the previous five years had failed to see the signal.

The company had not tested Mr Hodder's knowledge of the labyrinthine route that led into Paddington, nor was he given a map showing accident-prone signals. The layout at the junction is complex. To the west there are four running lines; to the east there are six bi-directional running lines; and at the junction there are connections between the various lines. An official report into the crash by Lord Cullen said Thames Trains' safety culture in relation to training was 'slack and less than adequate'. It criticised 'significant failures of communication within the organisation'.

Lord Cullen concluded in the report that it was likely that Mr Hodder was unable to read the signal accurately because of sunshine glinting on the track.

Contribution of human error

- This disaster shows how human error occurs within the organisational culture and the influences a negative culture will have on the individual.
- Knowledge-based error. The driver did not have the required knowledge base or experience to make an adjustment to his behaviour when approaching the Ladbroke Grove junction signals. Had he known it was a black spot, he would have been more aware that there could be problems and to watch out for them. He may also have slowed down in preparation for the possibility of a problem.
- Behaviour in the face of danger. According to the Hale and Glendon model, working at the rules level, which Mr Hodder must have been working at since he did not have enough time to develop all the skills or the knowledge of the job, would require an obvious signal for him to act on in order to bring the danger under control. If the red light could not be seen, then the warning of danger would not be obvious.
- The complexity of the connections between the various lines at the junction could have caused the driver to concentrate on where he was going and not on the signal. This is especially true when the driver is inexperienced, as Mr Hodder was.

THREE-MILE ISLAND

The Three Mile Island accident/incident was a partial nuclear meltdown which occurred at the Three Mile Island power plant in Dauphin County, Pennsylvania, United States of America on March 28, 1979. It resulted in the release of small amounts of radioactive gases and radioactive iodine into the environment.

The accident/incident began at on Wednesday 28th March 1979 with failures in the non-nuclear secondary system of the plant. The main feed water pumps stopped running caused by either a mechanical or electrical failure, which prevented the steam generators from removing heat. First the turbine, then the reactor automatically shut down.

Immediately, the pressure in the primary system, the nuclear portion of the plant, began to increase. In order to prevent that pressure from becoming excessive, the pilot-operated relief valve (PORV) located at the top of the pressuriser opened.

The valve should have closed when the pressure decreased by a certain amount, but it did not. Signals available to the operator failed to show that the valve was still open. As a result, cooling water poured out of the stuck-open valve, which allowed large amounts of nuclear reactor coolant to escape and caused the core of the reactor to overheat. As coolant escaped the instruments available to reactor operators provided confusing information. There was no instrument that showed the level of coolant in the core. Instead, the operators judged the level of water in the core by the level in the pressuriser, and since it was high, they assumed that the core was properly covered with coolant.

In addition, there was no clear signal that the pilot-operated relief valve was open. As a result, as alarms rang and warning lights flashed, the operators did not realise that the plant was experiencing a loss-of-coolant accident/incident. They took a series of actions that made conditions worse by simply reducing the flow of coolant through the core. Because adequate cooling was not available, the nuclear fuel overheated to the point at which the zirconium cladding (the long metal tubes which hold the nuclear fuel pellets) ruptured and the fuel pellets began to melt. It was later found that about one-half of the core melted during the early stages of the accident/incident.

Although the nuclear plant suffered a severe core meltdown it did not lead to a breach of the walls of the containment building and release the massive quantities of radiation available to the environment. The incident led to the authorised release, by the US Nuclear Regulatory Commission (NRC) of 150,000 litres of radioactive waste water directly in the Susquehanna River.

In the end the reactor was brought under control, although full details of the accident were not discovered until much later, following extensive investigations by both a Presidential Commission and the NRC.

Contribution of human error

The mechanical failures were compounded by the initial failure of plant operators to diagnose a stuck-open valve due to poor design of control panel and recognise the situation as a loss-of-coolant accident/incident.

This was due to inadequate training and human factors, such as poor design of the control panel, in particular the human-computer interaction design oversights relating to ambiguous control room indicators in the power plant's user interface. A significant factor was that a hidden indicator light led to an operator manually overriding the automatic emergency cooling system of the reactor because the operator mistakenly believed that there was too much coolant water present in the reactor. The confusion was compounded by the distraction of 100 alarms activating and a great number of flashing lights.

A10.4 - Improving individual human reliability in the workplace

Motivation and reinforcement

Maslow and Herzberg's theories of motivation deal with individual needs that apply to everyone. These are commonly known as content theories of motivation. The process theories of motivation deal with the process of work and consider that people are motivated by expectations that will vary from individual to individual.

This involves the identification of what performance is desired and the identification of the key motivating factors needed to achieve the goal. Two such process theories are expectancy theory and equity theory.

WORKPLACE INCENTIVE SCHEMES

Expectancy theory

Many behavioural scientists consider this theory as the most comprehensive and useful approach to understanding motivation in the workplace. Expectancy theory attempts to measure the strength of an individual's motivation to behave in particular ways. Managers should then ascertain what particular outcomes and rewards are most valued by the individual worker.

The job should then be designed so that performance is improved through increased motivation. A clear link should then be established between performance and outcomes/rewards.

Thus, in management terms, the key implications of using expectancy theory are that:

- The job, task or role should be designed to allow people the opportunity to meet their own needs.
- The pay/reward system should reward desirable performance (for example, long service payments should be avoided).
- The organisation should cater for the individual because the needs and values of each individual will be different (for example, promotion, job enrichment, salary).

Equity theory

This considers that the major input into job performance and satisfaction is the degree of equity (or inequity) that people perceive in their work situation. Inequity occurs when an individual's inputs and outputs are greater than or less than others in the workplace. Where this occurs, the individual will compensate accordingly. For example:

- Overpaid salaried workers will produce more in order to reduce inequity whereas underpaid salaried workers will achieve less in order to restore the balance.
- Overpaid piece-rate workers will produce higher quality and lower quantity than equitably paid piece-rate workers.
- Underpaid piece-rate workers will produce higher volume and lower quality than equitably paid piece-rate workers.

In practical terms, desirable safety performance can be rewarded in a number of ways:

- Bonus schemes based on desirable safety performance.
- Competitions with prizes.
- Award schemes - internal and external.

REWARD SCHEMES

The primary rewards associated with work are pay, status, promotion or intrinsic interest in the job. Perhaps the most fundamental of these is the need to earn enough money in order to live and, after that, other lifestyle benefits including recreation. Unlike other less tangible rewards, money can be counted and used as a measure of worth or appreciation. Earnings can be based on the quantity of goods produced (piece work) or be a guaranteed fixed salary.

Piecework has long been associated with the potential to cut corners in order to increase the reward with the classic example being the removal of a machine guard in order to increase production. This potential drawback in safety performance is balanced by the perceived benefit that the organisation will receive by obtaining the 'maximum' productivity from each worker.

The quality of the work however is not necessarily guaranteed by the piecework system that may allow the workers to produce high volume of low quality goods. Although the payment of a fixed salary reduces the pressure on an individual to cut corners in order to increase earnings, it does not follow that this will be the case. The organisational need to ensure that salaried workers produce at the required level may lead to a culture or appraisal system that places undue emphasis on production over safety. Whichever system of reward is in place it should be designed to ensure that safety performance is rewarded just as productivity is.

JOB SATISFACTION

Edwin Locke, in his process theory of motivation, advocates a goal setting approach to motivation. This establishes four main propositions:

- Stretch people by setting challenging goals, not too hard so they will fail, but attaining them will give a sense of achievement.
- Setting specific rather than vague goals. We adjust our behaviour when we know exactly what is expected of us.
- Participation in goal setting increases commitment to those goals. Managerially set goals can also add to job satisfaction, if they are fully explained and justified.
- Having knowledge of the results of what we do, i.e. feedback is also satisfying, whether it is seeing the finished product we have helped to build or being told we have done a good job.

Modern management theory places an increasing emphasis on the highlighting of employee concerns in order to improve decision-making and to maintain morale. Thus they have a key role in identifying causes of absenteeism, accident and illness incidence, high staff turnover, low productivity and poor industrial relations.

By its nature, the measurement of job satisfaction will involve consultation with individual employees. Many companies consider the use of tools such as attitude and opinion surveys a key component of organisational excellence. The job satisfaction survey is usually more than just a written questionnaire as it involves:

- Open exploratory interviews in order to gain understanding of the major issues involved.
- A closed written questionnaire which has been designed to obtain quantitative data.
- Further open interviews or workshops designed to explore or test issues identified in the written questionnaire.

Properly designed, this will enable the organisation to succeed in gaining valuable data, make better decisions and take more effective action.

APPRAISAL SCHEMES

Staff appraisal has long been a common practice by organisations although the setting of safety performance targets and the measurement of their achievement is a relatively recent innovation except in the most progressive of organisations. The common model is for managers to appraise subordinate staff although there is a trend towards employee appraisal of their managers and towards self-assessment.

An appraisal is a formal systematic method of staff assessment or development. Most appraisal systems are formal and usually involve set procedures, documentation, interview and feedback. In essence the prime purpose of the appraisal process is to establish controls on the behaviour of individuals and to bring about changes to behaviour if necessary.

There are six key elements to the achievement of this goal:

1) Auditing to discover the existing and future employment potential of the individual.
2) Manpower planning to ensure staff performance and future organisational management cohesion (for example, promotion potential for succession plans).
3) Training needs analysis by exposing inadequacies and deficiencies which could be remedied by coaching and training.
4) Staff motivation to achieve organisational goals and standards.
5) Individual development via praise or correction.
6) Fairness so that rewards such as promotion and money are distributed fairly.

Selection of individuals

IMPORTANCE OF INTERVIEW AND SELECTION

Recruitment and placement procedures should ensure that employees at all levels have the necessary mental and physical abilities to carry out their jobs. If the individual does not have the necessary skills, then he or she must have the ability to acquire them via on and off the job training (including experience). The procedure may require assessment of individual fitness by medical examination, physical fitness or aptitudes and abilities tests. The selection process should also include an assessment of training needs such as the extent and depth of induction training required. The most common personality assessment method used in the selection of staff is the interview. This is simply defined as a conversation that takes place under controlled conditions with the purpose of selecting a person for the job.

The purpose of the interview is to select the most suitable person for the job and, in doing so, the interviewer has three main functions:

1) To obtain information from the candidate.
2) Provide the candidate with information about the job.
3) Observe the interviewee's behaviour and reactions.

The flow of the interview itself can take many forms. One common format is:

1) ***Physical characteristics*** - key physical abilities and state of health (for example, eyesight, hearing).
2) ***Attainments*** - education, training, experience and achievements.

The selection process involves:

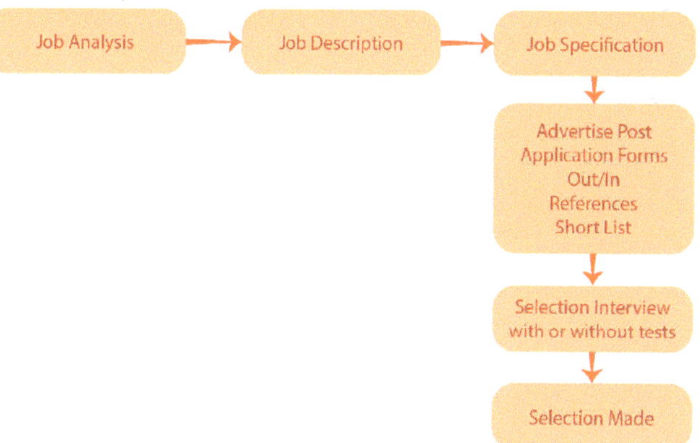

Figure A10-11: Job analysis flowchart. *Source: RMS.*

3) ***General ability*** - general intelligence and knowledge.
4) ***Special aptitudes*** - occupationally relevant talents (for example, scientific or mechanical).
5) ***Interests*** - may be of relevance as an indicator of intelligence or skills.
6) ***Personality*** - self-perception, social relationships etc.
7) ***Circumstances*** - family background, financial and current problems.

MATCHING SKILLS AND APTITUDES

It is important to determine the job profile requirements and the necessary skills before recruitment is started, often the job specification is vague because the departmental manager may also be considering an internal placement already familiar with the workplace. The manager may then not produce a concise job description and if ultimately an external candidate is chosen key health and safety factors may be overlooked. If the right person is to be selected it is important not to judge on personality or how quickly they work, but whether they are willing and able to do the work as described safely.

TRAINING AND COMPETENCE ASSESSMENT

Training is an essential element of ensuring that people have the skills, knowledge and attitudes to make them competent in the health and safety aspects of their work.

- Assess the skills needed to carry out tasks safely.
- Provide the means to ensure that all employees, including temporary employees, are adequately instructed and trained.
- Ensure that employees on especially dangerous work have the necessary training and experience to carry out the work safely.

Many employees are keen to undergo training as it demonstrates an investment by the organisation in them as an individual.

In addition to a natural desire by some, but not all, people to achieve new levels of skills or knowledge, employees may be motivated by the following factors:

- Pride in the job.
- Reward.
- Praise.
- Curiosity.
- Job satisfaction.
- Promotion prospects.
- Qualifications.

Whilst professional training will lead towards competent individuals it is important to assess how candidates have used their training in a practical way. If we are dealing with professional managers, engineers or health and safety practitioners then they will most likely be members of a professional body. Chartered membership usually requires the individual to maintain and update their core knowledge and skills on an ongoing basis, so these individuals may be employed with a degree of confidence; provided they have the practical experience as well.

Virtually all high risk trades, such as working with electricity, gas, cranes, will have their own competency arrangements and often licence arrangements, in a similar way to professionals they will also need to show practical application. Such trades will also require them to have continuous development, this will normally be achieved through refresher training schemes and re-licensing at regular intervals.

Competence assessment should focus on the needs of the job and be proportionate to the risks arising from the work. It is important that a systematic approach to assessment be taken which embraces knowledge, ability and experience. The frequency of competence assessment will depend on the likely amount of change and the risks related to the work of the individual.

FITNESS FOR WORK

Depending upon the nature of work, fitness and good health may be an essential requirement for a particular job; however when making a selection all applicants should be considered to avoid unnecessary discrimination. Consideration must be given to how the work might be carried out by someone who may be physically or mentally impaired, before they are rejected as unsuitable.

Health surveillance

Health surveillance may be a statutory requirement for those who work with lead alkyls and radiation, but for many work tasks it may be useful to monitor workers to ensure their work-related health is not impaired and addressed if found to be deficient, for example, for those who work in noisy environments a regular audiometric testing would be appropriate.

SUPPORT FOR ILL-HEALTH AND STRESS

Employees should be encouraged to report issues of ill-health or stress. Often this will enable an individual to be reassigned, either temporarily or permanently, to help with their particular problem. For this to be effective, individuals must feel that they are not disadvantaged by such admissions.

More successful arrangements will often include the appointment of an independent individual to report concerns to, such as an occupational nurse, trained person from human resources, independent senior manager or specialist health and well-being organisation.

A10.5 - Organisational factors

The effect of weaknesses in the safety management system on the probability of human failure

INADEQUACIES IN THE SETTING OF STANDARDS

Wherever possible common standards should be identified and implemented, this will reduce error when workers are transferred from one part of the business to another since they will be a common way of working. Good standards should avoid words such as 'should, as soon as possible' and replace them with definite command words such as 'always and never'.

Clear standards will instruct using words such as who (does), what (the task), where (the location), when, how (safe working procedure) and why (for motivation) and express what the required outcome will be.

INADEQUACIES IN POLICY

A policy statement which does not make a commitment to health and safety and seems to focus on the responsibilities of the employees can affect the attitude of the workforce and lead to a poor health and safety culture. Human error occurs more readily in a negative culture.

There could be inadequacies in the other sections of the policy following on from the statement. For example, failure to allocate responsibilities and failure to have proper arrangements can lead to human error.

INADEQUACIES IN PLANNING

Poor work planning will often results in inadequate time or equipment being allocated for the work to be done, often to meet unrealistic deadlines. Workers may then be compromised into taking risks and not follow all the necessary health and safety requirements that they should; for example, an urgent breakdown may require work at a height and in the absence of a scaffold tower, which had not been planned for, they may attempt to work from a ladder to get the job done, putting themselves at significant risk of a fall.

INADEQUACIES IN INFORMATION

Inadequate information may mean that operatives make uninformed decisions which can lead to error. For example, not having enough information on what would happen if two chemicals were mixed together caused an explosion that blinded Ken Woodward of Coca Cola Schweppes.

INADEQUACIES IN RESPONSIBILITIES

If health and safety responsibilities are not clearly defined, preferably in writing, then ambiguity in job roles will occur. Some individuals may take on health and safety responsibilities on their own volition, but this is not the way to allocate responsibilities as this is management by abdication. The keen volunteer may not have the skill or the experience to carry out the work safely and in their absence the work they do will often be neglected, for example, the testing of the fire alarms may not be done when the self-appointed person is absent.

INADEQUACIES IN MONITORING

There is an old maxim which states 'that which gets measured gets done'. It is not that all people are lazy, but over time they may become complacent and change the way they work; for example, in an office in the absence of a suitable step ladder they may use a chair. Each time this is done the violation is used it reinforces that it is the right thing to do. The seriousness of the deviation is only evident when the violation, using a chair with ***no*** castors to gain access at a height, is inadvertently transferred to the use of a chair ***with*** castors. In this instance, with the castors under load, the chair will slip when the individual reaches away causing them to fall (a rule based mistake). Employee and system monitoring will enable such infringements to be identified and corrected before harm is done.

INADEQUACIES IN IMPLEMENTATION

There may be systems of work devised, but they may not be implemented and the operators not properly supervised. This can lead to routine violations creeping in and errors more likely.

The influence of formal and informal groups within an organisation

Wherever people interact or communicate, their behaviour is governed by ***norms*** (standards or expectations) and they strive for a common purpose. If they ignore a norm they are classified as ***deviants*** and pressure is put on them to conform. Such pressure can take the form of verbal abuse, physical assault, silence, blacklisting and physical exclusion.

A group must have a minimum of two members and enjoy shared communication. They must have a collective sense of identity as well as shared goals that can only be achieved by the members working together. Groups can be classified as informal or formal. In formal groups the aims of the group are decided by the organisation whereas informal groups set their own agenda. Formal groups are created in order to achieve specific organisational objectives and are concerned with the co-ordination of work based activities. People are brought together (by management) on the basis of defined roles within the structure of the organisation.

The nature of the tasks to be undertaken and the skills involved are the main features of formal groups. Goals are defined by management, and certain rules, relationships and norms of behaviour established. Certain tasks can only be performed through the combined efforts of a number of individuals working together. The variety of experience and expertise among members of the group provides a synergetic effect, which can be applied to the increasingly complex problems of modern organisations.

MORENO'S PERSPECTIVE ON GROUPS

Team groups - these are fairly autonomous, terms of reference are usually broad and supervision is limited (for example, problem solving groups, maintenance crews, research teams).

Task groups - jobs are clearly defined and individuals assigned to specific positions. The group does have some limited discretion over methods or sequence of work as long as targets are met (for example, administrative or clerical workers).

Technological groups - members have very limited autonomy to determine or change operational activities. Content and methods of work are specified as well as the pace of work. Individuals will be assigned to specific jobs. There is little opportunity for individual discretion (for example, people working on an assembly line).

Decision-making groups and ***managerial groups*** are other possible categories.

Formal groups, such as the health and safety committee have an influence on the behaviour of individuals, which may be positive or negative, dependent on the status of the health and safety committee. A health and safety committee that is seen to get things done will have a positive effect on the health and safety culture.

The ***informal group***, a group of like-minded individuals who get together for support, may also have a positive or negative influence. The behaviour of the individuals will depend on the others in the group and what the group norms are. If the group are negative about the company, they influence each other to break rules and to act contrary to what they perceive management to want. This can create situations where violations cause errors to be made.

Within the formal structure of the organisation there will always be an informal structure. This informal organisation arises from the interaction of people working in the organisation, their psychological and social needs and the development of groups with their own relationships and norms of behaviour, irrespective of those defined within the formal structure.

FUNCTIONS OF THE INFORMAL GROUP SYSTEM

Some of the functions of the informal group system include:

- It provides satisfaction of members' social needs and a sense of personal identity and belonging.
- It provides for additional channels of communication; for example, information of importance to particular members is communicated through the 'grapevine' more quickly than through the formal system.
- It provides a system of motivation, for example, through status, social interaction, variety in routine or tedious jobs, and informal methods of working.
- It provides a means of highlighting deficiencies or weaknesses in the formal organisation, for example, areas of duties or responsibilities not covered in job descriptions or outdated procedures.
- Groups provide companionship and a source of mutual understanding and support from colleagues. This can help in solving work problems and also in mitigating stressful or demanding working conditions.
- Membership of the group provides the individual with a sense of belonging. The group provides a feeling of identity and a chance to acquire role recognition and status within the group.
- The group provides guidelines on generally acceptable behaviour. It helps to clarify ambiguous situations such as, for example the extent to which official rules and regulations are expected to be adhered to in practice, the rules of the game, and what is seen as the correct acceptable behaviour. It establishes its own set of norms, rules and values.
- The group may put pressures on members to resist demands from management on such matters as work output or changes in working methods.
- Group allegiance can serve as a means of control over individual behaviour. The group may discipline individuals who contravene the norms of the group.
- The group may provide protection for its membership. Group members collaborate to protect their interests from outside pressures or threats.

POSITIVE AND NEGATIVE ASPECTS OF GROUPS

Positive and negative aspects of groups include:

- The goals of the group may be at variance with those of the organisation.
- The group can work as well against the organisation as for it.
- Once a group has become fully developed and created cohesiveness, it may be difficult for management to successfully change attitudes and behaviour patterns of the group. It is important that the manager should attempt to influence the group in the early stages of its formation when members are establishing guidelines and standards and their own norms of acceptable behaviour.
- A strong, cohesive group may become critical or even hostile to outsiders or other groups. Lack of co-operation and inter-group conflict may arise to the detriment of the organisation as a whole.
- Group decision making can be costlier and time consuming than if taken by an individual, but:
    - Groups can provide a pooling of resources and can bring together a range of complementary knowledge and expertise.
    - Interaction among members can have a 'snowball' effect and provide future thoughts and ideas in the minds of others.
    - Group discussion leads to the evaluation and correction of possible decisions.
    - If full participation has taken place, decisions will have the acceptance of most members and they are more likely to be committed to decisions made and their subsequent implementation.

One might expect therefore, that 'better' decisions will result from group discussion. However there are two possible dangers: 'Groupthink' and 'Risky Shift'.

Groupthink

Described by Janis as:

"A deterioration of mental efficiency, reality testing, and moral judgement that results from in-group pressures. Groupthink results in the propensity of the group to just drift along."

Janis identifies a number of specific symptoms of groupthink:

- A sense of invulnerability with excessive optimism and risk taking. There is an unquestioning belief in the morality of the group.

- Pressures on individual members to conform and reach consensus means that minority or unpopular ideas may be suppressed. Members who oppose the group are stereotyped as evil, weak or stupid.
- The search for group consensus can result in rationalisation by members to discount warnings and there is an illusion of unanimity. There is self-censorship of any deviation from group norms or apparent group consensus.

(Groupthink is prevalent in large bureaucratic organisations, parliamentary departments, the civil service etc.).

The risky-shift phenomenon

There is a tendency for groups to make more risky decisions than would individual members of the group if acting alone. Presumably this is because individuals do not feel the same sense of personal responsibility for group decisions or their outcomes:

"A decision which is everyone's is the responsibility of no-one."

Possible explanations for the risky-shift phenomenon might include:

- People who are inclined to take risks might be more influential in group discussions than more reserved people.
- Risk taking is regarded as a desirable cultural characteristic that is more likely to be expressed in a social situation such as group working.

In spite of these possible problems, groups do seem to be more effective in problem solving than individuals when the problem requires a range of knowledge and expertise.

Informal groups	***Formal groups***
1. Qualifications for membership	
Personality, characteristics, temperament, attitudes, common (shared) interests, likes/dislikes Qualitative: immeasurable: subjective Complementary in social psychological terms Individuals 'earn' membership - cannot be 'placed' into group	Skills, experience, education, training capacity/capability Quantitative: measurable: objective Complementary in task requirement terms Individuals placed into groups to meet task needs
2. Leadership	
Leadership 'adopted' by group Leader must be 'accepted' by each individual in the group 'Un-named' leader 'Status' awarded internally - by group Not readily identifiable ('loudspeaker' may not be leader!) Flexible to meet varying situations Temporary - may change at any time Concerned for aspirations (needs) of the group	Leadership 'appointed' by external authority Leader need not be 'acceptable' to individuals Named leader Title - given by external authority Identifiable - badge, uniform, office, status symbols Inflexible - leader in all situations Permanent - may not change at any time Concerned for the needs (objectives) of the organisation
3. Communication	
2-way, from any member to all others Requests, consultation, advice, discussion Always verbal - face to face Slow - complicated network Understanding ensured Communication through each member Conviction of communicated message Decisions by consensus	1-way, always from the leader Directives, orders, rules, schedules Can be verbal - usually written Fast - straight from 'boss' to member Understanding expected and assumed Communication to the group Support expected from higher authority Decisions imposed from above
4. Size of group	
Depends on communication network	Depends on size and nature of task to be performed
5. Loyalty	
Always to group	Expected to be to the organisation
6. Identity	

Informal groups	*Formal groups*
Status, personal name, personal role, sense of belonging and being accepted	Title, job description, 'clock number'
7. Compliance by individual	
Unwritten rules, norms, values Pressure to comply imposed by group Invisible Variable sanctions	Written rules, standards, procedures Pressure to comply by organisation Prescribed sanctions Visible Organisation charts
8. Summary	
'Us'	'Them'
'Feelings'	'Facts'
'Social'	'Scientific'
'Democratic'	'Autocratic'

Figure A10-12: Summary of informal and formal groups. *Source: RMS.*

STAGES IN GROUP FORMATION

Tuckman has developed a model that represents the changes that take place in both the social and task aspects of group behaviour in the course of its formation and development.

There are four stages: Forming; Storming; Norming and Performing. Other writers have added their own interpretations of the processes that can be observed in these four stages. (Their contributions depend upon their individual standpoints and psychological perspectives.)

Stage 1: Forming

On the task dimension this is a period of orientation. Group members try to establish just what the parameters of the task are, how they should go about accomplishing it, what information they will need and so on. On the social dimension this is a period of testing and dependence. Members try to discover what kinds of interpersonal behaviour are appropriate. There is a tendency to look to the leader or some powerful (outspoken!) member of the group for guidance in this new situation.

Stage 2: Storming

Storming's characterised on the social dimension by internal conflict. There is often polarisation around key interpersonal issues. Members seem to be expressing their own individuality and attempting to resist group influence. With regard to the task, this stage is typified by emotional responses to the demands it seems to be making on the individual, particularly where the individual experiences a discrepancy between these demands and his own orientation.

Stage 3: Norming

The group begins to share ideas and develop cohesion. Members perceive themselves as part of a genuine 'group' which they wish to maintain and perpetuate. New standards and new roles emerge and are accepted. The emphasis is on harmony at all costs. Aspects of the task that are potentially conflict producing are avoided.

The task activities are typified by an open exchange of ideas and opinions. There is a willingness to listen and to accept the views of others.

Stage 4: Performing

The group has established a flexible and functioning structure of inter-related roles. The inter-personal aspects of the group's activity have been sorted out and now group energy can be channelled into the task. The task dimension sees the emergence of solutions to problems and constructive attempts at successful task completion. The task and social dimensions seem effectively to have come together at this stage.

At the performing stage any future development may not be easy. The group may need to splinter up and reform to meet the change. The established group may not readily accommodate new members (it may also not accept new managers with new ideas).

ROLES

Social roles are 'slots' in social systems which are occupied by individual persons. The most widely recognised roles are those that derive directly from the social structure:

- Kinship roles.
- Gender roles.
- Occupational roles.

A social role exists independently of the person who occupies it. It involves a recognised position or slot in the social structure and, additionally, a set of expectations about the behaviour of the person occupying it. Role positions do not exist in isolation from one another; they interlock and constitute social systems. For example, we could not have the role of 'mother' without the associated role of 'child'.

The expectations associated with a role vary:

- They differ in the degree to which there is general consensus of expectations covering all aspects of the role. For example, we may all agree that a mother should look after her child, but we do not agree whether she should teach her child to read and write or just play with him.
- Role expectations vary in their degree of permeation. Some roles pervade all areas of the occupant's behaviour, while others only affect the occupant's behaviour at particular times or in particular places (for example, we expect a nun to be pious and a paragon of virtue at all times. On the other hand, a bus driver can do what he likes in his spare time, as long as he performs his duties efficiently and safely during his hours of employment).
- Some role expectations may be backed by sanctions (for example, a mother can be sent to prison for neglecting her child).

Many aspects of role expectations are not backed up by sanctions but people still conform to them. Some role expectations may be formally laid down and a contract may have to be signed before the occupant takes up his position. The contract may specify precisely what the role expectations are (for example, a marriage ceremony). Some roles are covered by a code of conduct, for example, the medical profession. These codes do not specify particular expectations about behaviour but lay down certain guiding principles for covering a variety of contingencies.

Role expectations carry with them 'norms' of behaviour:

- What the majority of people do in this role.
- What would be ideal for people to do in this role?
- What all people must do in this role?

A norm can acquire a kind of super-human quality so that it appears more as a natural law than as a social contract (for example, gender roles).

In this way, a person in the position of a Bank Manager will:

- Usually be punctual and efficient.
- Ideally punctual and efficient.
- Neat and efficient.

These customs, ideals and rules will affect an individual strongly while they occupy a role position, because they will be expressed in many obvious and many subtle ways by the behaviours of the individual's role partners. The individual will feel compelled to behave in a way that other people, with whom they have contact, expect them to behave.

USE OF SOCIOGRAMS

A sociogram is a device that can be used to provide additional information regarding an individual and how they react with their peers. For example, it can show how a worker is viewed by their co-workers. It is a tool to assess social interaction and social perceptions. Questions, such as 'write down the name of the person with whom you would like to work on a project' are asked of each member of the group. The results are tabulated to show how many times each group member was chosen. This information is graphically plotted to identify popular workers, social isolates and workers who are disliked. They can also show interaction patterns over time. The diagram can be drawn as a series of concentric rings like an archery target, with one more ring than the greatest number of times any individual was chosen. The spaces are numbered from zero, starting from outside the last ring and working inwards. Each group individual's name is added to the diagram, inside the ring corresponding to the number of times they were chosen.

Arrows are drawn from each individual's name to the person selected by them.

Studying the diagram can show popularity and interaction preferences. This information can help to show natural leaders of groups who can be enlisted to help change the norms of a group who act negatively towards health and safety issues. It can also show which workers work better together, which is a good motivator. Unpopular workers can be helped with social skills and maybe, with the influence of the popular ones, become accepted in the group.

Workers lacking social skills could be helped by placing them in a job where some interaction is needed, such as team leader.

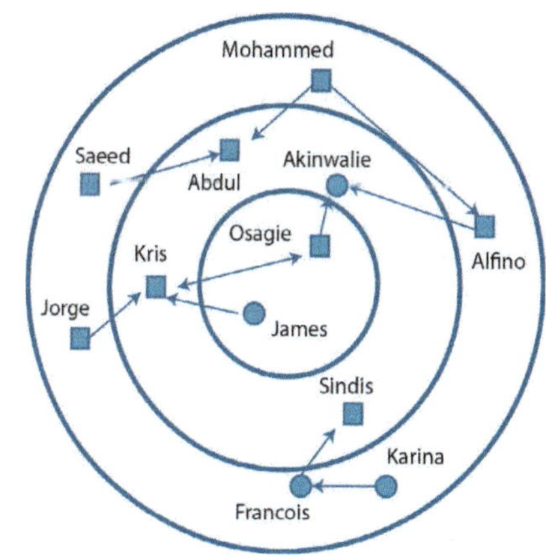

Figure A10-13: Sociogram. *Source: RMS.*

Organisational communication mechanisms and their impact on human failure probability

Organisational communication will be comprised of communication that will take place internally down from the top, horizontally amongst equals and from the bottom up; and with externals, from the organisation out and from the externals in.

Types of communication may be:

- Verbal: formal and informal meetings such as health and safety committee meetings or toolbox talks.
- Written: memos, work instructions, health and safe working procedures, rulebooks, health and safety committee minutes, the health and safety policy and the annual report.
- Electronic: intranet, e-mail.
- Posters: warnings, instructions, information.

SHIFT HANDOVER COMMUNICATION

Many organisations work 24 hours 7 days per week, and shift working has become a significant factor in highly developed societies. Many shift workers work in high risk occupations and their role is safety critical, not only to themselves but others. Communication at shift handover usually achieved by worker overlap allows critical work to be transferred without rush and safely.

ORGANISATIONAL COMMUNICATION ROUTES AND THEIR COMPLEXITY, RELIABILITY AND DEGREE OF FORMALITY

Communication can take many forms; informal communication is common, where a work instruction may be given by a supervisor, but rarely confirmed with the worker that it is understood. With repetitive tasks and skilled workers, familiar with their daily routines, such informal communication works efficiently with no adverse incidents occurring. More complex tasks, or work in different locations will require a more formal approach, involving method statement, risk assessments and the use of permits to work; such tasks may include, welding or hot work, excavation or demolition. Generally the greater the risk or the less frequency a task is carried out, the more formal the communication system used will be.

A10.6 - Job factors

Effect of job factors on the probability of human error

TASK COMPLEXITY

There is probably a normal distribution of task complexity associated with the probability of human error. Highly complex tasks have many opportunities for error; and error can sneak into simple tasks because of boredom and distractions.

EFFECTS OF PATTERNS OF EMPLOYMENT

Most people would agree that there is a clear link between the hours of work and degradation in human performance. The Working Time Regulations, with many exceptions and exclusions, place a general limit to the working week of 48 hours. Night workers should not exceed an average of 8 hours for each 24-hour period. Both of these 'limits' are averaged over a 17-week period. This would, therefore, seem to be the starting point for examining the effects that patterns of employment and shift work have on the individual.

Research has demonstrated a clear relationship between the length of the working day and performance, in terms of productivity, and accidents. There is, however, a great deal of individual variation interrelated with individual strength, health and nutrition. This makes a clear link difficult to establish but there appears to be a clear general deterioration if the working day exceeds 10 hours. For the traditional day-worker accident rates tend to increase at the start of the working period through to mid-morning. The rate then decreases as the morning progresses. There is a slight increase just after lunchtime towards mid-afternoon. The rate then levels off or slightly declines towards the end of the working day. Rest pauses perform both a physical and social function. Rest pauses should be of sufficient duration to allow the oxygen debt incurred by heavy work to be paid off. However, in light work frequent short breaks appear to improve performance and reduce human errors. The social function can often outweigh the physical benefits of rest pauses. The choice of working patterns and rest pauses made by the organisation is a balance between the needs of production and optimising human performance.

PAYMENT SYSTEMS

The most important factor for many individuals however is the payment considerations that can often over-ride those such as the disruption of family life produced by nightshifts. Thus many people who work 'anti-social' hours are compensated financially for doing so. This decreases in areas of high unemployment where competition for jobs is greater.

The bonus system of payment can create problems when employees are rushing to get the work done. Shortcuts will be taken and errors can occur.

SHIFT WORK

For night workers the accident rate tends to be high at the start of the working period and then drops towards the break periods. Again they tend to decline towards the end of the shift.

Shift working disturbs the individual's diurnal rhythm which produces impaired sleeping patterns. The most physically disruptive shift pattern appears to be a rapid three-way system. Ironically many workers prefer this to other systems because it disrupts their social life the least. The social effects of disruption to factors such as family life, however, appear to outweigh the physical effects.

Application of task analysis in predicting the probability and prevention of error

Task analysis can be used in the identification of hazards, risks and likely errors as well as development of a safe system of work. A type of task analysis which breaks the task into steps, the acronym SREDIM summarises the method.

It involves the following steps:

1) Select the task.
2) Reduce the task to steps.
3) Evaluate the risk at each step and put in appropriate controls.
4) Develop the safe system.
5) Implement the safe system.
6) Monitor and review.

Analysis of the task must consider not only the work to be done, but also the environment where it is to be done, to allow a full consideration of the hazards and risk of errors to be made. Task analysis consists of a formal step by step review of the work to be carried out. All aspects of the task should be considered and recorded in writing to ensure that nothing is overlooked.

Typical considerations would be:

- How is the task carried out?
- Where is the task carried out?
- What is used?
- What are the current controls?
- Are the controls adequate?
- Do operators use the controls correctly?
- Behavioural factors operators/supervisors (error considerations).

The objective is to establish the hazards and controls at each stage of the procedure to ensure a safe result. How the progress of work, in particular health and safety arrangements, will be monitored should be considered. Any special requirements for monitoring should be specified during the planning stage, for example, gas testing, temperature or pressure levels, measurement of emissions.

The task analysis will help provide an opportunity to put measures in place to reduce errors before the safe system of work is defined. This may include provision of specific information to help decisions, electronic prompts to take action or detailed skill training.

Task analysis is a general term which encompasses a wide range of techniques. A particular type of task analysis is a hierarchical one that breaks down the overall objective into a series of tasks by successively describing each step. At each of these steps a plan is developed to inform the operator how each task is performed.

A practical advantage of this approach is it allows whatever detail is appropriate to performing the task. Each level can be 'questioned' about the possibility of error and its consequences identified by developing performance influencing factors (PIFs). PIFs would typically include limited memory capacity, reliance on accepted diagnostic rules etc., which may give rise to human error. PIFs can be influenced organisational factors such as the presence of time stress, the quality of training and procedures and the amount of distractions.

The role of ergonomics in job design

INFLUENCE OF PROCESS AND EQUIPMENT DESIGN ON HUMAN RELIABILITY

The layout of controls and displays can influence the safety of the system.

Typical problems are:

- Switches which can be inadvertently knocked on.
- Control panel layouts which are difficult to understand.
- Displays which force the user to bend or stretch to read them properly.
- Badly identified controls which the operator could select by mistake.
- Emergency stop buttons which are difficult to reach.

Systematic analysis of how people actually use equipment can highlight problems that should be designed out, underlining the importance of manufacturers, designers and installers applying ergonomic principles. When designing control systems it is important to consider what people expect (their stereotypical expectations).

At the person-machine interface the operator will have to manipulate parts of the machine to pass a message to it; the machine will act according to the message received and pass a message back to the operator. The operator must be trained to communicate in a way the machine understands and the machine must be manufactured to be able to act according to instructions and communicate in an understandable way with the operator.

It would be impossible to carry out a task if every time you pressed the red button you got a different response. We all know the frustration when using a computer and you cannot get it to do what you want simply because you do not know the right key to press or the right icon to click on. There has been an attempt in recent years to make computer software 'user friendly', which means the operator and the computer can understand each other. The ***man-machine interface***, emphasising the performance of the human information processing system. Note that the broken lines underline that displays and controls are links between the operator and the machine.

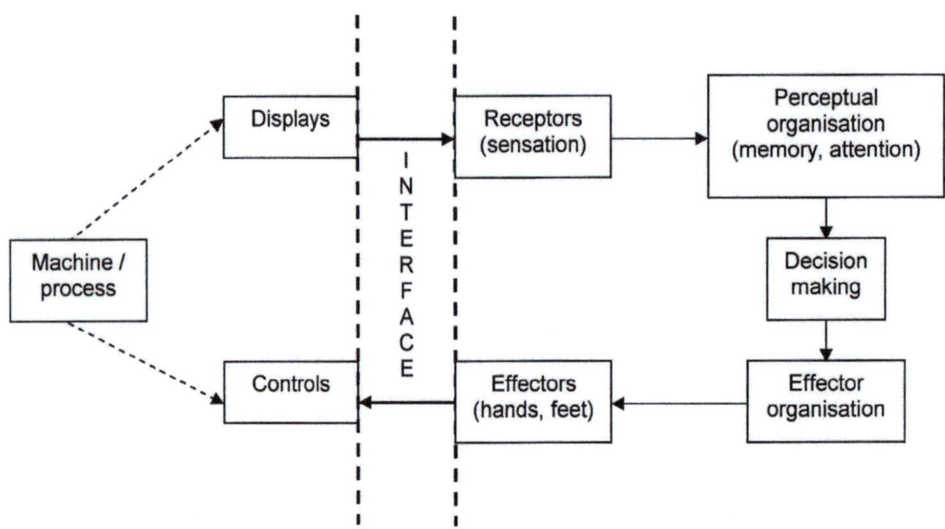

Figure A10-14: Man-machine interface. *Source: Ridley, J; Safety at Work; Fourth Edition; 1994.*

Figure A10-14 shows the system with the imaginary plane between the human and the machine, known as the interface. The message passed by the machine must be of a type compatible with the human senses. The human must process that information, make a decision and then give a response.

If too much information is given the short-term memory will not be able to handle it and error may occur. It is vital that the design of the machine takes into consideration the limitations of the human information processing system. Displays and controls must be designed with human expectations in mind. Which way do we expect to turn a tap or valve to close it, clockwise or anti-clockwise?

In an emergency situation in a chemical plant this is crucial. Emergency situations do not allow for lengthy thinking time and humans act instinctively, their actions based on past experiences and expectations.

Figure A10-15: Tap. *Source: RMS.*

The designer must consider questions such as:

- What colour do we expect the start button to be?
- Where do we expect the emergency stop to be?
- Do we expect the hot tap to be on the left and the cold on the right or vice-versa?

The decisions someone takes based on what they expect can have consequences ranging from a 'near miss situation' to a catastrophe.

Panel 1

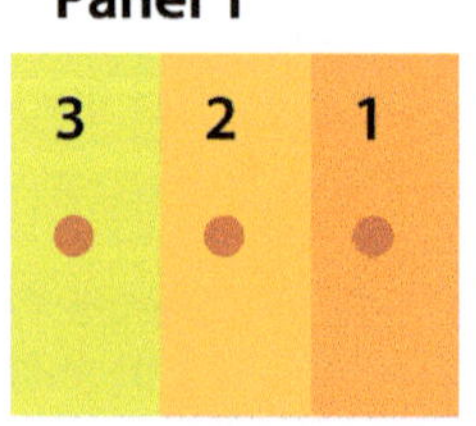

Panel 2

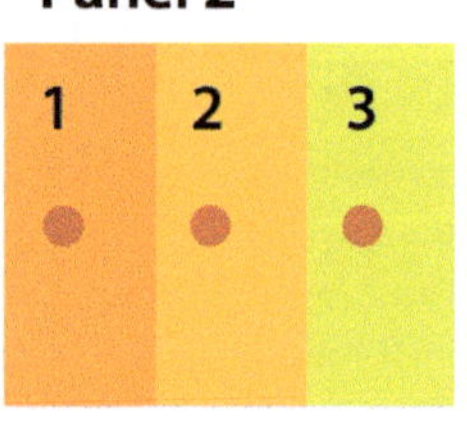

Figure A10-16: Panel/s diagram. *Source: RMS.*

The design of control boards should take into account the expectations of human beings, their information processing limitations and their physical capabilities.

Operators have taken decisions based on misplaced expectations with tragic results. Control buttons on a panel, which were a mirror image of each other, were the cause of an operator choosing the wrong one in an emergency situation. They were set up this way to make the wiring easier.

In the UK switches are pressed down for on, while in the USA they are pressed up for on.

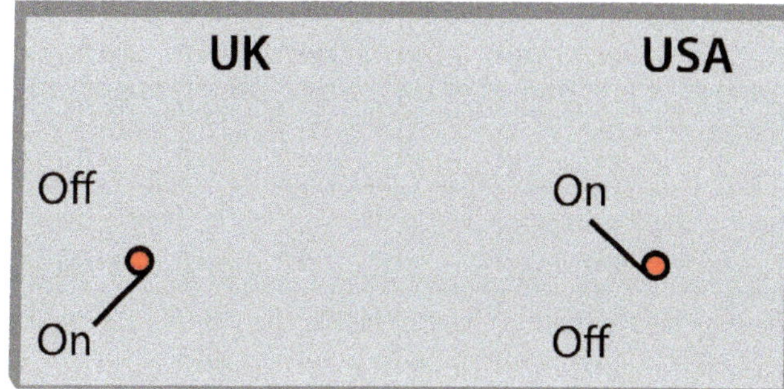

Figure A10-17: Switch diagram. *Source: RMS.*

If we consider 'fail safe' devices, it would be more logical to have the USA systems for switches. Gravity would cause a switch to drop if it was faulty, and with the UK system this would be a 'fail to danger' situation, that is, the switch would fall causing the power to be present when it may not be desired or safe. In most accidents people have been found to be the instigators or have not adequately intervened in the situation. For almost all cases this has not been due to wilfulness, but for a variety of reasons a wrong decision was made. On the other hand was it not for the quick thinking and skilful intervention of people, there would be many more accidents. Many very poor procedures are held together by skilful operatives. A further consideration in the design of control systems is the science of anthropometry. This is the science of the measurement of human bodily dimensions applied to the space in which the operator works. The human body is flexible and can adapt to operating in uncomfortable positions. Over time damage will result because of the poor posture caused by badly designed workspaces.

THE EMPLOYEE AND THE WORKSTATION AS A SYSTEM

General points - definition

Ergonomics can be defined as:

"The study of the relationship between the person, the equipment with which they work and the physical environment in which this person-machine system operates."

It is a broad area of study, which includes the disciplines of psychology, physiology, anatomy and design engineering. Ergonomics has a person at the centre of the study where their capabilities and fallibilities are considered in order to, ultimately, eliminate the potential for human error. It is also the study of ways to prevent the so-called 'ergonomic illnesses': work-related musculoskeletal disorders. These areas of ill health are usually insidious and may lead to disability. They stem from poorly designed machines, tools, task and workplace. Research has shown that there are about one million cases of work-related musculoskeletal disorders annually in England and Wales.

ELEMENTARY PHYSIOLOGY AND ANTHROPOMETRY

The general design of machines and workspace has not always taken account of the human bodily dimensions.

This was demonstrated by researchers at the Cranfield Institute of Technology who created 'Cranfield Man'.

They examined the positions of controls on a horizontal lathe and compared them to the bodily dimensions of an average human being.

The large discrepancies between the two sets of measurements led the researchers to create the man who could use the lathe with comfort.

'Cranfield Man' - 1.35m tall with a 2.44m arm span:

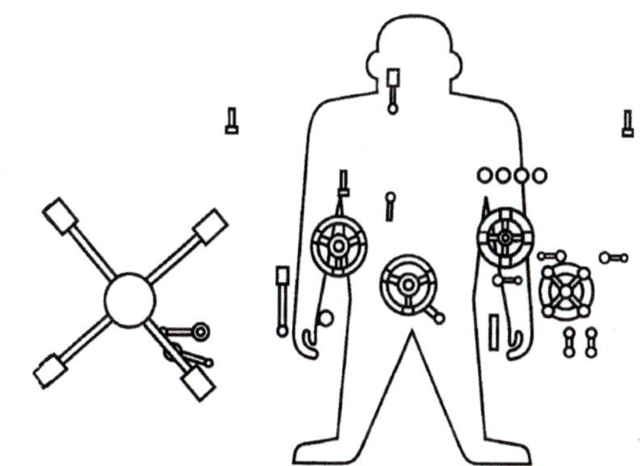

Figure A10-18: Cranfield man. *Source: Stranks, J; Human Factors and Safety; 1994.*

The physical dimensions of the average operator compared with those of 'Cranfield Man':

Average Operator	***Dimensions***	***Operator who would suit these controls ('Cranfield Man')***
1.75m	Height	1.35m
0.48m	Shoulder width	0.61m
1.83m	Arm span	2.44m
1.07m	Elbow height	0.76

Figure A10-19: Physical dimensions comparison. *Source: Stranks, J; Human Factors and Safety; 1994.*

Anthropometry can also find useful application when considering the need to prevent access to dangerous parts of machinery. Here, the measurements of bodily dimensions can be used in the safe placing of barriers from danger points and from necessary, permissible openings in fixed guards. Placed beyond a certain distance, a person should not be able to reach into the danger zone.

THE DEGRADATION OF HUMAN PERFORMANCE RESULTING FROM POORLY DESIGNED WORKSTATIONS

Man works within certain boundaries, which must be recognised for all situations. However, even when human beings work within their limitations, there will still be degradation in performance. Why, when and how degradation occurs needs to be understood in order to make allowances for or to remedy the situation.

The aims of ergonomics, therefore, are to design the equipment and the working environment to fit the needs and capabilities of the individual, i.e. fitting the task to the individual, and to ensure that the physical and mental well-being of the individual is being met. This involves a consideration of psychological and physical factors, including the work system, training, body dimensions, intelligence, noise, temperature and lighting.

The ergonomic concept of the design of a working system involving both man and machine with a view to, say, preventing possible accidents, is to observe both man and machine together as a combined working unit, observing the areas where correlation is important and the extent of the effectiveness of that relationship.

Thus at the design stage it is necessary to concentrate on what is required from the equipment and what from the human, when the machine and its user are working together.

Ergonomically designed control systems in relation to human reliability

CLASSICAL, ERROR AND SYSTEM ERGONOMICS - EXAMPLES OF APPLICATIONS

Some tasks are better performed by people than machines, and vice versa. For example, a computer can make mathematical calculations far quicker than a human being. A human, in cases of unplanned events, can assess a situation and make an informed judgement as to how to proceed.

A computer, however, would continue with the program until it could go no further and then grind to a halt or crash. The consideration of what is best done by what or whom is known as the allocation of function. The decision to use people, to automate a process or any combination in between these extremes does not rest solely on health and safety factors. The reasons may also rest on social and political issues. Human beings are creative creatures and may feel under-utilised when just tending machines.

A comparison between people and machines can be seen in ***figure ref A10-20***:

Advantages/people	***Disadvantages/machines***
Adaptable and flexible	Relatively inflexible
Can detect minute stimuli and assess small changes	Can detect if programmed, but not assess
Can interpolate and use judgement	Can do neither
Can synthesise and learn from experience	Can do neither
Advantages/machines	***Disadvantages/people***
Can operate in hostile environments	Lower capability
Fast response to emergency signals	Slow response
Can apply large forces smoothly	Can apply large forces coarsely
Information storage: large short-term memory	Easily distracted: limited short-term memory
Perform routine repetitive task reliably	Not reliable for this
Compute fast and accurately	Compute slowly and inaccurately
Can operate for long periods without maintenance	Suffer relatively soon from fatigue and monotony

Figure A10-20: Comparison between people and machines. *Source: Stranks, J; Human Factors and Safety; 1994.*

A schematic representation of the shift in the man-machine interface and change in allocation of function in different types of vehicle production is shown in ***figure ref A10-21***. This represents three types of manufacturing process in the production of motor vehicles. The first represents the manufacture of a car mostly by hand machining. The worker is skilled and creative and satisfaction is derived from the high quality finished product. The second is where the worker assembles pre-formed car body parts using power tools. This is repetitive and does not allow for creativity.

The third represents the automation of the car building process where the person has a monitoring and maintenance role. Each process requires a different input from the person and has different problems. Often the part played by the person is just doing what is left over after the machine has done its bit. This results in a fragmented and piecemeal job for the person, who is often left to hold together a badly designed system. This allocation of function is susceptible to safety problems. Whatever the allocation of function, ergonomics can be used to find solutions to the problems.

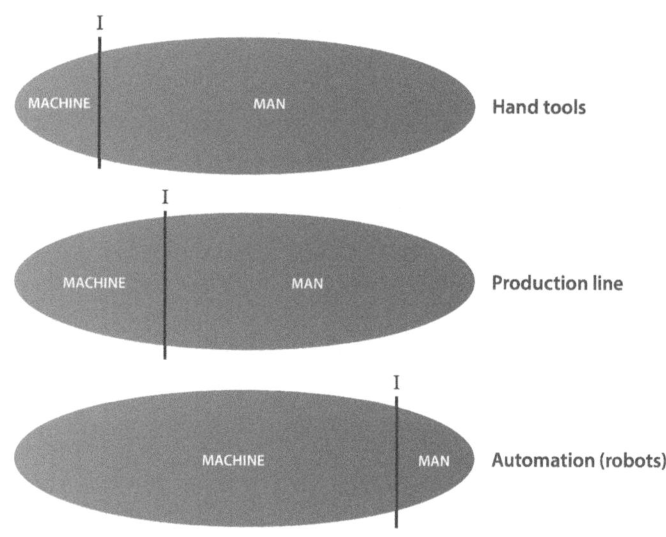

Figure A10-21: Man/machine interface. *Source: Ridley, J; Safety at Work; Fourth Edition; 1994.*

The allocation of function can be decided in a number of ways. Fitts drew up a list of the relative advantages of men and machines. The 'Fitts List' shown in ***figure ref A10-22*** has been modified by Singleton.

Property	***Machine performance***	***Human performance***
Speed	Much superior. Consistent at any level. Large constant standard forces and power available.	Lag one second. 2 horse-power for about ten seconds. 0.5 horse-power for a few minutes. 0.2 horse-power for continuous work over a day.
Consistency	Ideal for - routine, repetition and precision.	Not reliable - should be monitored. Subject to learning and fatigue.
Complex activities	Multi-channel.	Single-channel. Low information throughput.
Memory	Best for literal reproduction and short-term storage.	Large store multiple access. Better for principles and strategies.
Reasoning	Good deductive. Tedious to re-programme.	Good inductive. Easy to re-programme.
Computation	Fast, accurate. Poor at error correction.	Slow. Subject to error. Good at error correction.
Input	Can detect features outside range of human capabilities. Insensitive to extraneous stimuli. Poor pattern detection.	Wide range (1012) and variety of small stimuli dealt with by one unit, for example, eye deals with relative location, movement and colour. Affected by heat, cold, noise and vibration. Good pattern detection. Can detect very low signals. Can detect signal in high noise levels.
Overload reliability	Sudden breakdown.	Gradual breakdown.
Intelligence	None. Incapable of goal switching or strategy switching without direction.	Can deal with unpredicted and unpredictable. Can anticipate. Can adapt.
Manipulative abilities	Specific.	Great versatility and mobility.

Figure A10-22: An updated version of the Fitts List (from Singleton). *Source: Ridley, J; Safety at Work; Fourth Edition; 1994.*

EXAMPLES OF APPLICATIONS

Production process control panels

The need for good ergonomics is safety critical as can be illustrated by the incident at Three Mile Island nuclear accident. One of the factors leading to the reactor core damage and radioactive water loss at Three Mile Island was a poorly designed control panel that featured:

- Controls that went off scale.
- Key indicators that were sited out of view.
- Hundreds of alarms that were illogically arranged with no means of isolating the unimportant ones were provided.

This information overload did not allow control staff to react as the situation demanded. In a control room design, for example, the designer must consider the room layout in order to ensure:

- That all operators can see each other and all relevant instrumentation.
- That audible warnings are clearly understood and cannot be masked by other sounds.
- That workstations are designed to ensure that displays are clear and that controls are easily identified and operated.
- That lighting is provided which allows clear definition and does not compromise visual displays through reflections or glare.

Routemaster drivers' cabs

London Transport was alive to the ergonomic principles in the design of the driver's cab for the Routemaster Bus. The fact that drivers themselves were asked to take part in tests leading to the eventual design accepted as successful was in itself a good thing from the angle of psychological strategy.

The driver's cab and the layout of the controls received most thought and the opinions of physiologists and other medical authorities were sought in order to evolve the best relative positioning of controls, so that the driver could sit in the most suitable position for obtaining the best field of vision and enabling quick responses for movement of the various controls.

A mock-up of the cab was made in which various adjustments could take place, such as the angle between the seat and the squab, the position and angle of the footplate, the pedal travel and the steering column angle. A sufficient amount of adjustment was allowed to cover the sizes of most men of between 5'6" (1.68 metres) and 6'1" (1.83 metres).

A number of men, including drivers' representatives, were asked to adjust the various components to what seemed to them the most suitable positions, and such positions were recorded. From such information, various points of detail were decided and incorporated in the final design of the Routemaster.

Crane cabs

Similarly, the British Iron and Steel Association carried out work along ergonomic principles on the design and layout of crane cabs. In the traditional type of cabin, even though the driver stood, it was often difficult for him to obtain a satisfactory view of operations; the control handles moved in awkward semi-circles and he was unprotected from the dust, fumes and heat of the shop in a foundry (he is likely to be standing over the controls for a proportion of his working time).

In the redesigned cab, the driver sits on a seat which is adjustable to 90% accuracy of all possible sizes, a full view of the working area is possible, the controls are within easy reach and move in straight lines to permit ease and delicacy of control, and the provision of filtered and refrigerated air, where necessary, ensures cool and comfortable working conditions.

Figure A10-23: Traditional crane cab design. *Source: Engineering Lifting Systems LLC.*

Figure A10-24: Re-designed crane cab. *Source: Engineering Lifting Systems LLC.*

Aircraft cockpit

Poor design of controls and instruments in aircraft cockpits means that pilots have often made fatal mistakes, such as operating the flaps instead of the undercarriage and misreading altimeters.

A particular example of this occurred in the UK when a passenger aircraft crashed in 1989 killing 47 people.

The pilot reported engine trouble and was diverted to a nearby airport. A fan blade in the left engine had broken and was causing noise, vibration and smoke.

There was confusion about which engine had become faulty, which led the pilot and co-pilot to shut-down the right engine, leaving the aircraft to glide. For a period of time the smoke from the left engine was reduced and the symptoms appeared to have been corrected.

As more fuel was pumped into the damaged left engine it burst into flames. The aircraft crashed a small distance from the end of the runway.

Figure A10-25: Kegworth air crash. *Source: UK Press Association.*

There were many technical issues that contributed to the crash, one of which related to the design of the aircraft instruments in the cockpit. This version of the aircraft, a Boeing 737, had a number of modifications compared to previous versions of the aircraft, including changes to the instrument design and layout. In particular, the engine vibration indicators that would have enabled the correct identification of the faulty engine had been changed, making them less obvious and harder to read. At one point during the Kegworth flight, the dial indicating the vibration in the left engine rose to the maximum and stayed there for three minutes, but the pilots failed to see it. The earlier version of the aircraft had dials with a large, clearly visible mechanical pointer, making it easier to read, ***see figure ref A10-26***, the arrows show the engine vibration indicators.

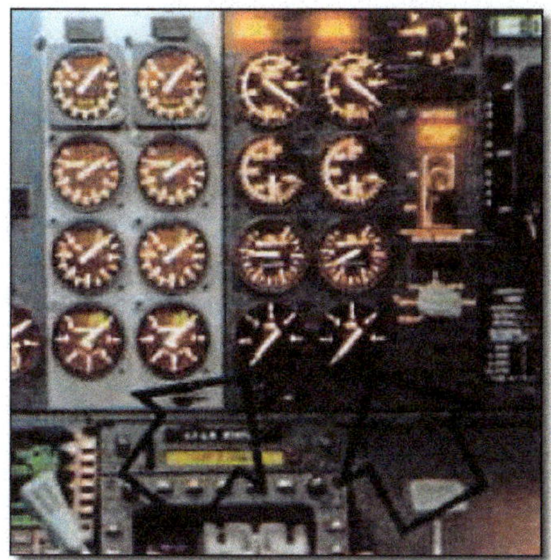

Figure A10-26: Earlier version of Boeing 737 dials. *Source: UK Air Accidents Investigation Branch.*

Figure A10-27: Dials of later version of Boeing 737. *Source: UK Air Accidents Investigation Branch.*

In the version of the Boeing 737 that was involved in the crash the engine vibration indicators had been made smaller, ***see figure ref A10-27***, and a less defined design of display was used, ***see figure ref A10-28***.

A 'stylish' design was used in the later version of the Boeing 737, it had a thin dotted line moving around the number indicators. Although the engine vibration indicators provided accurate information on the condition of the engines the design meant the information was less obvious to the pilot and co-pilot.

The aviation industry has been the focus of much ergonomics research, because of the ever-increasing mental demands on flight crew. To give some perspective, a 1940s aeroplane typically had 20 instruments, a 1950s jet about 50, a 1960s airliner about 180, and current airliners have around 450.

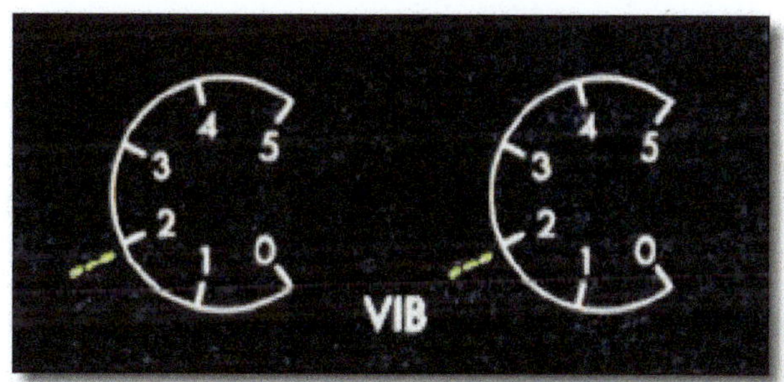

Figure A10-28: Vibration indicator dials of later version of Boeing 737. *Source: UK Air Accidents Investigation Branch.*

Ergonomics factors are important in the design of aircraft cockpits. The layout and function of cockpit displays and controls are designed to increase pilot situation awareness, without causing information overload. The pilot of an aircraft must be able to easily interface with all the controls.

The layout of instruments and controls must be in a logical pattern and emergency controls must be easily activated and, perhaps more critically, not activated in error. Pilots, male/female and differing nationalities, will have a range of anthropometric characteristics that the seat and layout of controls in the cockpit must be able to accommodate. In the past, many cockpits, especially in fighter aircraft, limited the size of the pilots that could fit into them. Now, cockpits are being designed to accommodate from the 1st percentile female physical size to the 99th percentile male size.

Modern aircraft design has adopted a fully digital 'glass cockpit'. A glass cockpit is an aircraft cockpit that features electronic (digital) instrument displays, typically large liquid crystal display (LCD) screens, instead of the traditional style of analogue dials and gauges.

Figure A10-29: Airbus A380-841 glass cockpit. *Source: Rhys Dudley.*

A traditional cockpit relies on numerous mechanical gauges to display information, whereas a glass cockpit uses several displays driven by flight management systems that can be adjusted to display flight information as needed.

It is felt that the glass cockpit simplifies aircraft operation and navigation and allows pilots to focus only on the most pertinent information. The layout of control panels in modern airliners has become largely unified across the industry.

The majority of the systems-related controls (such as electrical, fuel, hydraulics and pressurisation) are usually located in the ceiling on an overhead panel. Radios are generally placed on a panel between the pilot's seats known as the pedestal. Automatic flight controls such as the autopilot are usually placed just below the windscreen and above the main instrument panel on the glare shield.

CNC LATHE

A computer numerical controlled (CNC) lathe is a machine that performs a number of cutting operations without having to be continuously supervised by an operator. The CNC system will control the lathe to machine a work piece of the desired profile and size, changing tools automatically to carry out a series of cutting operations.

Ergonomics are applied to the use of the keyboard and monitor whereby the operator programmes the machine to undertake the cutting process. The programming part of the operation is very similar to any computer interface.

The monitor, often in a fixed position on the machine control panel, benefits from being height adjustable. If it remains in the fixed position the operator is likely to need to stand for extended times to see the screen while inputting data. This posture will also mean that arms are extended from the shoulders for long periods, which can cause fatigue.

Figure A10-30: CNC lathe. *Source: NE Reihart and Sons.*

Figure A10-31: CNC lathe controls. *Source: NE Reihart and Sons.*

The initiating action on a CNC lathe is loading the material to be machined. This is usually a round stock bar from which multiple components will be machined. This requires the worker stooping and twisting to feed the stock bar into the lathe chuck. Following loading of the material comes the setting of the cutting tools. These are often on a tool cassette (a multiple tool holder) located at the rear of the machine that will also require reaching, twisting and stooping for access. Anthropometrics are critical on machine tools, because access is often restricted or requires the arms to be extended. Replacement tools should be pre-assembled and brought to the machine on a custom designed work trolley to reduce manual handling.

The operator, when in attendance setting or observing the machine, will benefit from standing on a cushion mat floor to reduce foot and leg fatigue. The operators may also be susceptible to fatigue as a consequence of monotony.

The relationship between physical stressors and human reliability

The performance of the human system is only at an optimum within certain environmental limits. Physical conditions such as noise, glare and lighting level, dust and fumes etc. will influence the individual. For example, noise can affect concentration at levels well below the levels dictated by the Control of Noise at Work Regulations (CNWR) 2005.

A noise source at 5dBA above ambient is sufficient to be a 'nuisance' and affect concentration. These, and other physical stressors such as lighting (too much, too little) and high and low temperatures, can affect the accuracy of detection of information and the speed of processing information. It is evident that human beings, unlike machines, rarely breakdown completely. They are more likely to show a slight degradation of performance over a wide range of environmental conditions that can be difficult to detect.

This results in the maintenance of individual's function as they can maintain some sort of functioning long after they have passed the peak of their performance, but it also blurs the point at which they should stop in order to avoid errors. Regulations and good practice on working hours for coach drivers, hospital doctors and others must wrestle with the problem of matching specific limits with the multitude of factors that can affect performance.

Performance is degraded under the following types of situations:

- Working for too lengthy a period which produces fatigue.
- Working at times of day when body mechanisms are not functioning efficiently, i.e. the diurnal rhythm is disturbed.
- Loss of motivation to perform.
- Lack of stimulation resulting in low arousal.
- Working under conditions of conflict, threat, both physical and psychological, or conditions that threaten the body's homeostatic or coping mechanism and cause stress.

The relationship between the length of the working day and performance is not clear. Studies have suggested that there is a highly significant reduction in output and an increased risk of accidents if a working day of about ten hours is exceeded. Obviously this will vary with workload, intensity of work and any outside activities undertaken by the individual. It should be appreciated that this dip in performance applies to mental work as well as physical activities.

The effects of under-stimulation, fatigue and stress on human reliability

Under-stimulation occurs when there's not enough going on in our environments to keep our attention or keep us interested. This can result in 'boredom', a distinct emotional state in which the level of stimulation is perceived as unsatisfactorily low.

Lack of stimulation can lead to increased neural arousal in search of variety. Under stimulation can lead us to engage in challenging behaviours which may be positive (questioning safe systems of work) or disruptive, not wearing PPE and challenging the rules that state it must be worn.

As health is not merely the absence of disease or infirmity but a positive state of complete physical, mental and social well-being (WHO, 1986), a healthy working environment is one in which there is not only an absence of harmful conditions but an abundance of health-promoting ones.

Factors related to under stimulation include meaningless tasks, lack of variety, work load and work pace (too little, to slow), lack of participation and control in decision making. Repetitive or awkward movement or having to stay in the same position for long periods at a time can result in muscular fatigue. Poorly designed tasks and workstations may mean that the operative has to sit awkwardly for long periods or stretch constantly to reach operating buttons or levers. If the task is machine-paced then there is little or no scope for rest and this can result in high levels of stress as well as muscular fatigue.

Medical opinion states that possibly 70% of patients being treated have conditions that have their origins in stress. Physical manifestations of mental stress can be very serious: coronary heart disease, bronchial asthma, etc. Change in the workplace has meant for most people that the pace has speeded up and increasing numbers are unable to cope with it.

Causes of stress can be divided into two groups: environmental and psychological. The environmental stressors are things such as bodily injury, noise, extremes of temperature, etc. The psychological stressors include periods of intense mental activity, being unable to achieve what is required, physical or social isolation, etc.

The body's response to stress is to increase the secretion of the hormone adrenaline. This increases muscle tension and heart rate to prepare the body for fight or flight, a primitive response not really appropriate for modern industrial society.

If this state is maintained for long periods it can result in such conditions as high blood pressure, gastric ulcers, etc. In this state of agitation, the individual is prone to make errors. Ergonomic considerations can go a long way to relieving or even preventing stress. Task design, machine design, operator-controlled pace and appropriate rest periods can all help reduce fatigue and stress, thereby reducing error and accidents.

A10.7 - Behavioural change programmes

Principles of behavioural change programmes

Behaviour change is believed to occur in stages and that movement through these stages is neither unitary nor linear, but rather, cyclical or spiral pattern, involving a process of adoption, maintenance, relapse, and re-adoption over time.

Typical behavioural change programmes at the national level have included the promotion of seat belt use in vehicles, drink awareness campaigns and more recently health and well-being programmes dealing with smoking, eating and exercise.

All such programmes take a long time to be accepted, the use of seat belts being worn as a norm has taken over a generation of promotion and finally a change in legislation to enable non-conformity by individuals to be punished.

Organisational conditions needed for success in behavioural change programmes

Organisations must be committed to a long programme (sometimes many years) and well supported initiative to implement behavioural change, in exactly the same way behaviour change is enabled at the international level.

Prochaska and DiClemente suggested that behavioural change involves both progress and periodic relapse. That is, even with successful behaviour change, people will move back and forth between the five stages of change for some time, experiencing one or more periods of relapse to earlier stages, before moving once again through the stages of contemplation, preparation, action and eventually, maintenance.

In successful behavioural change, while relapses to earlier stages inevitably occur, individuals never remain within the earlier stage for long. The process of behaviour change requires advocacy to secure senior management leadership and commitment to policies and programmes that support health and safety and promote changes.

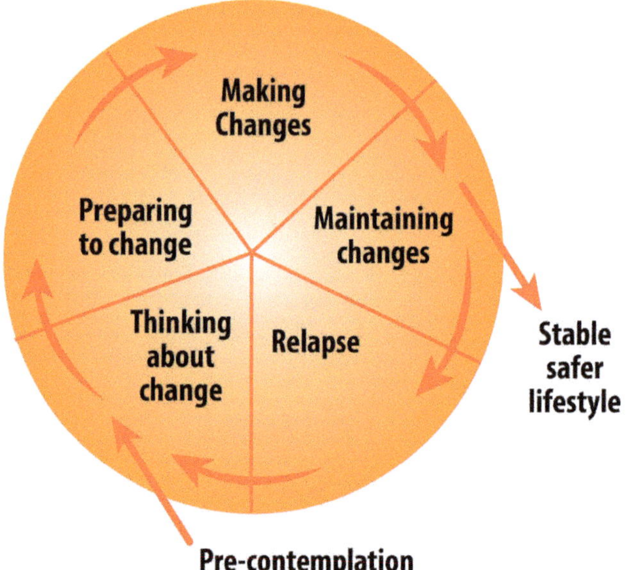

Figure A10-32: Behavioural change model. *Source: RMS.*

Behaviour change requires the promotion of behavioural norms that support collective health and safety objectives and challenge harmful practices. A variety of promotional methods should be used to promote health and safety concerns and stimulate managers and workers to adopt the new behaviour.

The behaviour change programme must be important to the individuals involved in the change. The behavioural change programme is likely to be more effective if rewards and punishment are almost certain for good/bad practices.

Incentive schemes should encourage the desired behaviours and disciplinary schemes should punish undesired behaviour. The reward/punishment must be given as soon after the desired/undesired behaviour as possible, so the individual has an unambiguous link as to the cause of the reward/punishment.

Examples of typical behavioural change programme contents

All workplace behavioural change programmes will use some or all of the stages developed by Prochaska and DiClemente to change manager and worker behaviour. A well-structured behavioural change programme will incorporate:

- Health and safety behaviour observation - this involves observing and assessing workplace practices and identifying unsafe behaviours.
- Enlisting observers - observers should be enlisted on the basis of their commitment to health and safety.
- Devising a checklist of behaviours - observers should be aided by a checklist or other aids for monitoring behaviour. They should, however, be competent enough to observe unsafe or unhealthy behaviours not incorporated in the checklist.
- Create the recording system - the recording system should be able to extrapolate data from the checklists and classify forms of unsafe or unhealthy behaviour.
- Analyse the data - data should be analysed on the basis of classifications that will identify trends and enable specific aspects of unsafe or unhealthy behaviour to be targeted.

- Promote the behavioural change programme - both managers and workers need to understand the benefit of the behaviour change programme. The programme must not be seen as 'just another management initiative'.
- Implementation of the programme - once the programme has been established continuity must be maintained or everyone will revert back to their former behaviour.
- Maintaining the behaviour - maintain the behaviour through continual reinforcement of the behaviour change message, with a sustained effort in preventing behaviour reverting back to its former level.

All behaviour based programmes need to identify behaviours that can be easily observed and assessed.

This page is intentionally blank

Element
A11

The role of the health and safety practitioner

Learning outcomes

On completion of this element, candidates should be able to demonstrate understanding of the content through the application of knowledge to familiar and unfamiliar situations. In particular, they should be able to:

A11.1 Explain the role of the health and safety practitioner.

A11.2 Explain the importance of effective communication and negotiation skills when promoting health and safety.

A11.3 Outline how health and safety practitioners can use financial justification to aid decision making.

Contents

Relevant statutory provisions

N/A

Sources of reference

Reference information provided, in particular web links, was correct at time of publication, but may have changed.

IIRSM Code of Ethics, http://www.iirsm.org/code-ethics

IOSH Code of Conduct, http://www.iosh.co.uk/About-us/Regulating-the-profession.aspx

Leading health and safety at work, INDG417, IOD and HSE publication, HSE Books, http://www.hse.gov.uk/pubns/indg417.pdf

Managing for health and safety, HSG65, HSE Books ISBN: 978-0-717664-56-6, http://www.hse.gov.uk/pubns/priced/hsg65.pdf

Occupational Safety and Health Consultants Register (OSHCR), http://www.oshcr.org/

The above web links along with additional sources of reference, which are additional to the NEBOSH syllabus, are provided on the RMS Publishing website for ease of use - www.rmspublishing.co.uk.

A11.1 - The role of the health and safety practitioner

Role of health and safety practitioners

Health and safety practitioners, also known as health and safety officers, managers, advisers or consultants, strive to minimise the risk of harm or injury at work and to ensure the workplace complies with national laws related to its activities. Health and safety practitioners seek to achieve this by educating colleagues and setting procedures to be followed, aiming to build a culture of safety in the workplace. To achieve a safe working environment, health and safety practitioners need to work closely with managers, employees and often, trade unions. They will also liaise with external contacts, such as customers, contractors, clients and the enforcing agency.

Their role may include:

- Advising managers on how to comply with health and safety legislation.
- Developing strategies for managing safety.
- Developing policies and procedures for the organisation.
- Carrying out site visits, safety inspections and safety audits.
- Identifying potential hazards, risks and control measures.
- Designing and delivering training courses on health and safety issues.
- Investigating accidents and finding ways to prevent them happening again.
- Advising on specialist areas such as fire precautions and control of hazardous substances.
- Keeping records, including accident statistics.

MANAGEMENT OF HEALTH AND SAFETY AT WORK REGULATIONS (MHSWR) - REGULATION 7 (HEALTH AND SAFETY ASSISTANCE)

Regulation 7 of MHSWR 1999 places an absolute duty on the employer to take the following actions:

- Appoint one or more competent persons to assist with compliance with health and safety legislation.
- Make arrangements for the persons to co-operate.
- Ensure the number of people appointed, time available; means at their disposal are adequate. Consider size of organisation, risks and distribution of the risks.
- If persons appointed are not employees ensure they are provided with information.
- Ensure persons appointed are competent.
- Persons appointed are informed of any person working under a fixed-term contract or employed in an employment business.
- Appoint people from within the employer's employment, where they are competent, in preference to others.

The regulations do not require the self-employed, who are themselves competent, to appoint anyone. The regulations do not require individuals in business partnership to appoint a person if one of them is competent.

Dealing with conflicts of interest

A conflict of interest is anything that impedes or might be perceived to impede an individual's ability to act impartially and in the best interest of a client. The 'client' may be an organisation or a person. A conflict of interest can cast doubt on your integrity; it can also have a damaging effect on the profession as a whole.

A conflict of interest may occur where you:

- Are acting for another person with competing interests at the same time (for example, advising both the seller of a machine and the user of the machine simultaneously).
- Have other obligations or pressures that detract from or affect your relationship with the client (for example, acting as an expert witness where you/your firm has an interest in the outcome of the proceedings).
- Are faced with conflicting duties owed to different clients (for example, employer, workers or third parties).

A conflict of interest may occur if, for example, you recognise a dangerous procedure but do not recommend a shutdown in case it ultimately leads to the closure of the company and your redundancy. Conflicts of interest may also arise through the offer or acceptance of inducements. Inducements can include gifts, hospitality, preferential treatment and inappropriate appeals to friendship or loyalty.

It is the duty of the health and safety professional to ensure that they carry out their work with professional integrity at all times. A conflict of interest may arise in a number of ways when a health and safety practitioner is carrying out their professional role.

The way in which the conflict of interest is dealt with may differ in specific detail, but in general should be dealt with in a way that ensures honesty, integrity and in so doing ensures the individual maintains respect.

The health and safety practitioner may find out information that may affect themselves or others they care about. For example, in dealing with health and safety at a senior level, the health and safety practitioner may be provided with privileged information about an option to close part of the organisation instead of making health and safety improvements to it, and this closure could affect members of the health and safety practitioner's family.

This would present a conflict of interest as the possible effects on the health and safety practitioner's family may influence advice or comments made. As the senior managers may not know that the family members could be affected by the decision, the health and safety practitioner would need to disclose this fact and the conflict of interest to the senior managers. This would provide the senior managers with an opportunity to consider and respond to the conflict of interest. They may wish to continue with the health and safety practitioner's involvement, knowing the conflict, or obtain advice from someone else.

Similarly, the health and safety practitioner must not allow their decisions to be influenced to the detriment of health and safety, particularly in a situation where they may gain from it. For example, accepting the pressure from an employer to delay essential expenditure on the maintenance of safety critical equipment at a time that the organisation is going through a difficult financial period, in case it ultimately led to closure of the company and their own redundancy.

When there may be a conflict of interest it is important to bring this to the attention of all parties at the earliest opportunity, for example, when invited to be the health and safety advisor to both parties in a contract, both parties affected must be informed immediately.

The meaning of the term 'competence'

Competence is a combination of knowledge, skills, experience and recognition of the limits of your capabilities. For example, as a health and safety practitioner you will have studied the harmful effects of noise along with prevention and control strategies for noise. You may not, however, have been taught the skills or have had the experience to be competent to undertake a detailed noise survey.

Competence is an attribute of an individual that enables them to perform functions. The term competence is often used in an overall sense to describe an accumulation of competencies. In addition, a competency may be derived from the integration of a number of connected skills, for example, conducting an effective health and safety meeting is a competency that requires a number of skills, for example, good communication, resolving conflict and managing time.

In addition, the competency to conduct an effective health and safety meeting requires an amount of connected knowledge, for example, knowledge of particular hazards and risks being discussed. Therefore, competencies are a combination of knowledge and skill and are observable.

In the field of health and safety, competencies that a person may have could include competency to conduct an accident investigation, inspection or risk assessment. Competencies held may be wide ranging, in that they can be applied to a number of circumstances or very specific. For example, a competency may be related to inspection of a particular type of equipment or limited to only conducting manual handling risk assessments.

Competencies may also be classified related to levels of proficiency, for example, beginner level, learner level, skilled level and expert level.

Training is often an important component of establishing competency; however, competence can be a developed attribute acquired without formal training and established through experience. Training is not usually sufficient on its own, as experience is an important part of developing competency, enabling the consolidation of knowledge and skills through their practical application. This is particularly the case when considering the competency of health and safety practitioners.

Competencies, including those for health and safety practitioners, may be classified as:

- ***Behavioural*** - relating to awareness and interpersonal competencies, for example, awareness of a person's own limitations, initiative, motivation and team work.
- ***Managerial*** - organisational, people-related and role-related competencies, for example, prioritisation and decision making.
- ***Technical*** - competencies relating to functions, for example, risk assessing, inspecting or auditing.

Competency, once acquired, can decline due to lack of maintenance of the skill. This can be because of the absence of applied positive experience or exposure to negative experience that is counter to the competency.

It can also decline due to the knowledge held becoming inappropriate, for example, because important parts have become forgotten or it is out of date.

For these reasons it is important that competence is maintained. For health and safety practitioners, like many professionals, this is achieved through continuing professional development. This promotes an awareness of the need to maintain competency and recognises that it is an on-going process. Many health and safety professional bodies require continuing professional development in order to retain membership.

The Health and Safety Executive (HSE) has commented on their website that competency can be defined as 'the ability to undertake responsibilities and perform activities to a recognised standard on a regular basis. It is a combination of skills, experience and knowledge.

Training is an important component of establishing competency, but is not sufficient on its own, for example, consolidation of knowledge and skills through practical application is a key part of developing competency.' This is particularly the case when considering the competency of health and safety practitioners.

Competence is the ability for every director, manager and worker to recognise the risks in operational activities and then apply the right measures to control and manage those risks.

The MHSWR Regulation 7 states that competence means having sufficient training and experience or knowledge and other qualities. The approved code of practice (ACOP) to Regulation 7 says:

"Competence in the sense it is used in these Regulations does not necessarily depend on the possession of particular skills or qualifications. Simple situations may require only the following:

(a) An understanding of relevant current best practice.

(b) An awareness of the limitations of one's own experience and knowledge.

(c) The willingness and ability to supplement existing experience and knowledge, when necessary by obtaining external help and advice."

Figure A11-1: Approved code of practice (ACOP). *Source: MHSWR Regulation 7.*

It is recognised in the ACOP that where more complicated situations exist a higher level of knowledge and experience would be required, that appropriate health and safety qualifications may be needed and that membership of an appropriate professional body may indicate competence. Judith Hackitt, recently the HSE's Chair, said: *"The essence of competence is relevance to the workplace. What matters is that there is a proper focus on both the risks that occur most often and those with serious consequences."*

Need to evaluate and develop their own practice so as to maintain competence

Health and safety practitioners need to keep abreast of new developments, such as changes in the law and new technical developments, for example, technical developments in guarding. Failure of the health and safety practitioner to keep up to date with changes and failure to develop their own knowledge and expertise could result in the health and safety practitioner giving wrong information or not using the best controls to reduce risk. These failures have moral, legal and economic implications.

The ACOP to Regulation 7 of MHSWR emphasises the need for the health and safety practitioner to have a willingness to supplement existing experience and knowledge. In addition to supplementing knowledge and experience, it is essential to maintain existing knowledge and experience.

In order to evaluate and develop their own practice, the health and safety practitioner could devise a personal development plan stating their aims and objectives and how they plan to achieve them. For example, a health and safety practitioner could decide that they should find out more about the "Work at Height Regulations" and plan to attend a seminar on that subject.

A continuing professional development (CPD) portfolio is a useful way to formalise this, where copies of course programmes and learning outcomes may be held.

Continuous professional development (CPD) is an essential system required by professional bodies like IOSH to ensure the ongoing competence of chartered health and safety practitioners. CPD is a continuous process that concentrates on reviewing and developing an individual's professional knowledge, skills and abilities. IOSH require all members to participate in the CPD process, which is managed on line, in the form of a member CPD log.

Members demonstrate their continuous professional development by undertaking and recording a wide range of personal health and safety development activities. Activities include improvements to their organisation's health and safety, such as, the implementation of a HSMS or review of the health and safety policy, training which they have attended or delivered and contributions they have made to articles in health and safety journals.

Mentoring and supporting the development of health and safety competency in other relevant employees

Mentoring is an all-embracing form of educational support and development mechanism. Mentoring in relation to health and safety has significant potential, the more knowledgeable (competent) practitioner providing information and support to assist the development/evolution and implementation of another person.

The success of the mentoring relationship is founded upon the link forged between two individuals, the mentor and mentee. Mentoring is a developmental process to help and support people to manage their own learning in order to maximise their potential, develop their skills, and improve their performance.

Mentoring includes elements of coaching, facilitating and counselling and is aimed at sharing knowledge and encouraging individual development. There are many different mentoring models.

The following list is adapted from Lord, Atkinson and Mitchell 2008:

- Apprenticeship - the mentor acts as a master teacher conveying rules to be emulated.
- Competence - mentors perform the role of trainer.
- Reflective - mentor adopts the role of 'critical friend' to develop a reflective practitioner.
- Model - mentor inspires and demonstrates best practices.
- Sponsor - mentor 'opens doors' to introduce the mentee to the right people.
- Educator - the mentor provides opportunities for the mentees learning.

Mentoring is designed to develop personal growth and to increase an individual's competence on health and safety.

Distinction between leadership and management

The main distinction between leadership and management is that leaders have people follow them while managers have people who work for them. Leaders set the goals and direction and act as visionaries when motivating and leading in a new direction. Managers, through workers, specialise on conformance to standards (such as MHSWR or PUWER) and managing teams and individuals within the boundaries set by such standards.

Managers also organise and direct activities to achieve objectives, using resources such as people, time, money and equipment to achieve the task. Managerial activities include planning resources and tasks to achieve the objectives often within the constraints of a financial budget.

Managers ensure support functions are aligned for the achievements of tasks and ensure the task is completed within a specified time frame. Managers typically make decisions in instant time and there is an immediacy required from them to be managing in a 'live' environment such as a production process.

Leadership focuses on the long-term vision or goal. Encouraging managers and workers to willingly follow along the desired direction. Leaders inspire others and inject enthusiasm into the common goals. Leaders coach and mentor others, imparting their knowledge and wisdom so that individuals learn and grow.

The successful health and safety practitioner will need to be both a strong leader and manager to get their workers on board to follow them towards their vision of health and safety. Good leadership will help workers to understand and believe in the vision of a safe place to work while managing will get workers making the day-to-day things happen safely. ***See figure ref IA11-2*** which highlights features of managers and leaders.

Figure A11-2: The distinction between leadership and management. *Source: RMS.*

The need to adopt different management styles dependent on any given situation

Many managers assume that leadership style is a function of their personality rather than a strategic decision they can make. The manager should be able to choose the style that has most effect in the situation presented.

There are many different models of management styles. However, Daniel Goleman measured the impact of 6 different styles of management and leadership. The six different styles of leadership following Goleman (2000) are:

- ***Coercive:*** relatively aggressive style that demands compliance, the style works best in crisis situations i.e. immediately following an accident, where decisive management is required.
- ***Authoritative:*** it is considered to be the most effective motivating style of leadership that drives people to improve performance. Works best when a new vision is required or a new safe system of work developed.
- ***Affiliative:*** a style that focuses on people, it values employees and their emotions, rather than their objectives. Works best when the manager is aware the team is operating in stressful situations, such as the aftermath of a fire.
- ***Democratic:*** a democratic leader builds trust and respect by listening to every member's opinions. The style works best when the manager wants everyone in the team to contribute i.e. risk profiling of an organisation.
- ***Pacesetting:*** a leader sets high standards and wants the goal to be reached in an excellent and fast way. Works best with highly motivated, competent teams such as engineers when reviewing a safe system of work in the oil processing industry.

- ***Coaching:*** a coaching leader helps employees to identify their strengths and weaknesses. Works best when the manager is improving the performance of the team i.e. compliance with a standard or convention.

Development, implementation, maintenance and evaluation of health and safety management systems

ROLE (AND FUNCTION) OF HEALTH AND SAFETY PRACTITIONERS

The role of the health and safety practitioner is to support line management in meeting their responsibilities by:

- Providing information and advice.
- Supporting line management with the co-ordination of health and safety effort.
- Monitoring the effectiveness of actions to meet responsibilities.

The function of the health and safety practitioner includes:

- Identify problems (including hazards).
- Assess the need for action.
- Assist with the assessment of risks.
- Design and develop strategies and plans.
- Advise on relevant current best practice for prevention.
- Present themselves and their advice in an independent manner.
- Promote and communicate health, safety and welfare advances and practices.
- Assist with the implementation of these strategies and plans.
- Evaluate their effectiveness.
- Maintain records relating to health and safety performance.

Design of health and safety management systems

The health and safety practitioner's role related to the health and safety management system (HSMS) will vary according to the organisation. They may be involved in the design of the health and safety management systems (HSMS) and therefore help to decide which model to use, for example, HSG65 or BS OHSAS 18001 (soon to become ISO 45001). In doing this, the health and safety practitioner may take advice from others with experience of management systems, for example, quality or environmental practitioners. The health and safety practitioner is well placed to consider all aspects that might influence the HSMS design, including aspects like the culture of the organisation, legal requirements and other driving factors.

It is important that the health and safety practitioner contribute to the detailed design of the system, not just the decision to have a certain overall structure. This will help to ensure the system meets detailed legislative requirements and its continued effectiveness. The health and safety practitioner's role will include co-ordination of consultation on the design of the HSMS and involve discussions with those affected, in particular managers, employee representatives and employees.

Implementation of health and safety management systems

The health and safety practitioner is likely to be fully involved in supporting the implementation of the HSMS. This may include providing the final version of the documented system to those involved in its operation, communication of the HSM at briefings and one-to-one support to managers trying to implement the system. This is an important phase, requiring involvement of the health and safety practitioner's detailed understanding of the intent and structure of the system to aid its smooth implementation.

It will also provide an early opportunity to identify if adjustments or clarification of the system is needed. Whilst the health and safety practitioner may be highly involved in the design and implementation of the HSMS it is important that management and employees embrace the system as theirs and recognise the health and safety practitioner's role as supportive.

Evaluation of health and safety management systems

Once the HSMS has been implemented, competent persons are required to evaluate the HSMS. Though evaluation may include employees and managers, the perspective of the health and safety practitioner could provide a useful and impartial evaluation of the HSMS. This can be provided during formal review, internal audit or as an ongoing observational evaluation of effectiveness.

Maintenance of health and safety management systems

Maintenance may involve corrective action, changes or improvements to the effectiveness of the HSMS. The HSMS may require maintenance due to changes in process, people, and equipment, knowledge of risk or legislation and therefore should be planned for. The health and safety practitioner can be involved in this by providing assessment of the effects of the change and proposing amendments to the HSMS to meet them.

It is important that the proposed changes receive the same careful attention as the original design, therefore the health and safety practitioner may be involved in seeking views on the proposed changes before they are implemented. This may require the health and safety practitioner involving the expertise of others, such as electrical/mechanical/chemical engineers, to consider the proposed amendment in relation to their specialist area.

Sensible risk management including the importance of proportionality when assessing/controlling risk

The concept of proportionality is used as a standard of fairness in statutory interpretation processes as a method intended to assist in discerning the correct balance between the restriction imposed by a corrective measure and the severity of the nature of the prohibited act.

In the case of the health and safety practitioner this is the balance between a safe place of work and a dangerous place of work or what is an acceptable level of risk? Sensible risk management means balancing the level of risk against the measures needed to control the real risk in terms of money, time or trouble. Action is not required when the money, time or trouble would be grossly disproportionate to the level of risk. This is the concept of 'reasonably practicable'.

A risk assessment should be completed by someone with a knowledge of the activity, process or material that is being assessed. Workers and their health and safety representatives are a valuable source of information. If a health and safety practitioner assists with the risk assessment, managers and workers should still be involved.

In some organisations the health and safety risks will be tangible and immediate health and safety issues, for example, machine guarding, where the consequences of no guards can lead to injury.

Other risks may be health related and it could be a long time before the illness becomes apparent such as musculoskeletal conditions related to using display screen equipment. The level of detail in a risk assessment should be proportionate to the risk and appropriate to the nature of the work.

Insignificant risks can usually be ignored, for example, cutting your finger on a piece of paper, as can risks arising from routine activities associated with life in general, for example, crossing the road, unless the work activity compounds or significantly alters those risks.

If you are a road worker, then you are exposed to the hazard of traffic more significantly than you would be as a citizen of a city or town. In this example the risks must be assessed and controls put in place.

The health and safety practitioner will use a risk assessment essentially as a tool to inform decisions by assisting in the understanding of the nature and degree of risk and for extrapolating, from available data, experience of harm. If the probability of harm is low, then control measures would be applied proportionately.

Enabling work activities as part of proportionate and sensible risk management

There are relatively few examples in health and safety legislation of processes or activities that have been banned because the risks they entail are so high and their control inherently so difficult that it is not possible to find any control measure that the health and safety practitioner could feel confident would work in practice.

A fundamental principle underpinning health and safety legislation across the world is that those who create risks from work activity are responsible for protecting workers and the public from the consequences of that activity.

The general approach of health and safety legislation is to set out the objectives to be achieved and to give considerable choice to duty holders as to the measures they should put in place to meet these objectives. For a non-prescriptive regime to work, duty holders must have a clear understanding of what they must do to comply with their legal obligations.

Managing health and safety is an integral part of managing a business. The duty holder or employer need to do a risk assessment to find out about the risks in the workplace, put sensible measures in place to control them, and make sure they stay controlled.

The role of the health and safety practitioner in this exchange is to ensure risk controls adopted are proportionate to the risk. That the risk controls adopted do not create additional risks or a greater level of risk through their adoption and the risk control provides an acceptable balance between time, effort and cost to implement the control and the level of risk reduction achieved.

Organisational risk profiling; its purpose, practicality and organisational context

The risk profile of an organisation informs all aspects of the approach to leading and managing its health and safety risks. Every organisation will have its own risk profile. This is the starting point for determining the greatest health and safety issues for the organisation. In some businesses the risks will be tangible and immediate safety hazards, whereas in other organisations the risks may be health-related and it may be a long time before the illness becomes apparent.

A risk profile examines:

- The nature and level of the threats faced by an organisation.
- The likelihood of adverse effects occurring.
- The level of disruption and costs associated with each type of risk.
- The effectiveness of controls in place to manage those risks.

A risk profile examines the nature and levels of threats faced by an organisation. It examines the likelihood of adverse effects occurring, the level of disruption and costs associated with each type of risk and the effectiveness of the control measures in place.

The purpose of risk profiling is to ensure that managers focus on risks that have been identified and prioritised for action, and that minor risks will not have been given the same level of priority. Risk profiling also informs decisions about what risk control measures are needed. The practicality of risk profiling is to generate an outline of the most important risks faced by the organisation.

The profile will use the many risk assessments generated within the organisation for significant hazards and look at the wider organisational context of the consequences such as loss of reputation, business disruption or the potential for harm to be caused.

A risk presented by an unguarded machine within a factory may impact only the machine operator, whereas not managing hygiene in a food processing plant may impact the many thousands of persons who purchase the product.

The contribution of the health and safety practitioner in achieving the objectives of an organisation

The MHSWR 1999 requires that:

- Regulation 5 (1) Every employer shall make and give effect to such arrangements as are appropriate, having regard to the nature of his activities and the size of his undertaking, for the effective planning, organisation, control, monitoring and review of the preventive and protective measures.
- (2) Where the employer employs five or more employees, he shall record the arrangements referred to in paragraph (1).

Performance measures are usually classified as goals or objectives. Goals are more long range and difficult to measure, while objectives are shorter term and more quantitative. Usually, an objective is a goal broken down into a series of achievable steps.

The health and safety practitioner has a role in contributing to the health and safety goals and objectives of an organisation. The most common types of health and safety goals are based on lost time injury rates or worker absence. Some health and safety goals are vague and difficult to quantify such as 'achieving compliance' or 'provide a safe place of work'.

Other goals can be specific, for example, 'zero accidents in the workplace' or 100% safety compliance, but, if the goal is perceived as unrealistic it will not act as a motivator to the workers involved in its achievement. The health and safety practitioner has the knowledge, skill and experience to establish meaningful broad ranging health and safety goals through their clear understanding of the activities and impact of the organisation.

Goals motivate people to focus their attention and action, to try harder and to persist until a desired future state has been reached. The usefulness of goal setting for increasing task performance has been demonstrated in many countries, across a variety of occupations.

It is critical, however, that a goal matches the performance criteria being monitored, (i.e. attaining the goal of improved health and safety in the workplace is measured by monitoring and assessing several health and safety performance indicators). If a goal is accepted by employees, the health and safety practitioner can measure types of behaviour that that can be improved.

Once the health and safety practitioner has identified an area for improvement local knowledge can be applied to establish appropriate behaviours to achieve the improvement (for example, are plumbers using hot works permits?) The health and safety practitioner can then aggregate activities together to form a goal that can be explained and subdivided into objectives that can be achieved, for example, training for all employees in the use of hot works permits.

Meaning of the term 'ethics'

The term ethics relates to:

'The study of moral standards and how they affect conduct'.

Ethics are therefore values and rules of conduct that society expects of an individual in relation to their position and role.

Ethics are established by society as a norm of expectation and by professional bodies by their consideration of what standards they would expect of their members.

The health and safety practitioner carries additional moral responsibilities to those held by the general public; this is because they are acting in a professional capacity and their ability to make informed decisions that a member of the public would not.

For example, if a health and safety professional were to ignore unsafe work practices, to save the organisation money in difficult times, they would have failed in their moral duty to conduct themselves in a manner that ensured the best interests of health and safety and in so doing may have acted unethically.

Practical application of ethical principles that underpin professional codes of conduct

Ethics is concerned with a system of moral principles, for example, honesty, respect and integrity. They govern the appropriate conduct for a person or group, in particular when they are acting in a professional capacity. It is essential that the health and safety practitioner act honestly, for example, they should not misrepresent the level of risk from a situation in order to gain action to improve health and safety.

They must respect themselves and others, for example the health and safety practitioner must deal with people with an equal level of respect, whatever their role or level in the organisation. The views of workers on health and safety matters are as valuable as any others; a lot has been learned from workers where the health and safety practitioner has treated them with due respect and given time to listen to their views.

In the same way the health and safety practitioner must treat people they deal with respectfully whatever their race, colour, religion, disability, sex, marital status, age, sexual orientation, nationality or ethnic origins. The integrity of the health and safety practitioner is paramount. For example, if a worker discloses information on a confidential basis it would be inappropriate for the health and safety practitioner to pass this on to someone else without the agreement of the worker. Most codes of conduct seek to ensure conduct maintains such ethics as honesty, respect and integrity. IOSH has an established Code of Conduct - ***see figure ref A11-3***.

Code points	***Guidance on code points***
1) Members of the Institution, wherever employed, owe a primary loyalty to the workforce, the community they serve and the environment they affect. Their practice should be performed according to the highest standards and ethical principles, maintaining respect for human dignity. Members shall seek to ensure professional independence in the execution of their functions.	1) The term 'professional independence' relates to your function within the organisation you advise on health and safety. You may have either an advisory or an executive role. You may also have their responsibilities not related to health and safety. As a professional, you should be able to exercise your health and safety function according to your independent professional judgment.

Figure A11-3: Sample of Code of Conduct. *Source: IOSH.*

The Code of Conduct applies to all member categories and to all aspects of the performance of their health and safety functions. It applies equally to those who carry out health and safety duties as employees and those who have particular roles and responsibilities as advisors, consultants, researchers, employers and providers of education and training.

The Code of Conduct establishes rules of ethical practice, contains guidance on a range of questions on ethical matters as they affect members and discusses 'professional conduct'.

A11.2 - The importance of effective communication and negotiation skills when promoting health and safety

Why effective communication is important

Studies of major accidents/incidents have shown that without effective communication, managers and workers are less likely to make the right decisions and errors or violations may occur. It is therefore important to provide effective communication within an organisation in order to help to ensure the correct health and safety behaviour takes place.

Regulation 10 and 11 of MHSWR 1999 require the employer to ensure that procedures are in place to provide employees with adequate information and also to ensure that arrangements are made for regular consultation with employees on health and safety matters,

To achieve success in health and safety management, there needs to be effective communication up, down and across the organisation. Organisations need to communicate information to their workers on the risk to their health and safety identified in their risk assessments, and the preventive and protective measures necessary to control risk.

Communication is not only how a message is conveyed but how the message is received and understood. To be effective the information provided should be communicated appropriately taking into account the workers literacy and language and the size of the organisation. Effective communication is also how you listen to the worker or worker representative to gain the full meaning of what is being said.

Effective communication combines a set of skills including nonverbal communication, for example, body language, voice tone and volume and verbal communication (the actual words used). The topic under discussion can affect communication because workers may form pre judgments (a manager telling a worker how to safely operate a machine they have never actually operated may lack credibility) or there may be a 'communications history' between participants, for example, a lack of trust in what is said.

The role of the manager, with support from the safety practitioner is to lay out clear communications procedures for safety-critical tasks and ensure that written instructions are clear and up to date. The manager must make sure that safety-critical messages have been given attention and are understood so that they are acted upon.

The manager must think about what needs to be communicated and who should be communicated too, especially where this involves the organisations health and safety policy, risk assessment findings and safe systems of work.

The need for health and safety practitioners to consult and negotiate with others

Health and safety at work is about securing the health, safety and welfare of workers by reducing risks, and protecting them and others from harm or illness arising out of work activities by taking the right precautions. As an employer or manager, there is a legal duty to manage health and safety sensibly in the organisation, and involving workers is central to that process.

Those organisations that recognise the importance of establishing a positive health and safety culture have arrangements in place for good management and worker co-operation and consultation. This may be achieved by consulting workers on a one-to-one basis, consulting with worker health and safety representatives or by using health and safety committees.

It is important to establish that there is a difference between informing people and consulting them. Informing is a one-way process involving the provision of relevant information, for example, by management to workers, whereas consulting is a two-way process where account is taken of the views of those being consulted before a decision is taken.

The health and safety practitioner has knowledge, skill and experience on the level and types of risk faced by the organisation and its workers. This needs to be applied to setting objectives that are realistic and measurable. Not everyone in the organisation will have the same level of competence with regards health and safety as the safety practitioner does.

This will mean that some objectives will be challenged by managers as unrealistic or not achievable within the finite resources available to the organisation. Workers, on the other hand, are likely to see the potential for improvement in health and safety by the adoption of the objectives, but not express the same concern when the cost of achieving the objective in terms of time, effort and money is considered.

The safety practitioner will therefore have to achieve a balance between the aspirations of workers and the resources made available by managers. This can only be achieved by consulting representatives of employee's safety so that resource is targeted to areas of highest concern for the worker.

This process is likely to involve negotiating more rather than less resource from managers and maintaining the morale and involvement of the worker while being perceived as providing less resource for safety objectives.

Influencing ownership of health and safety at all levels of an organisation

Effective risk management depends, in part, on the behaviour of individual workers in an organisation. A significant number of accidents can be traced to unsafe behaviours. Poorly designed equipment or operations, poor systems and poor working conditions can all encourage unsafe behaviours, but these behaviours are not inevitable.

An organisation's attitudes and values regarding safe working are important factors that influence its approach to work and ultimately its health and safety performance. Put another way, it's not enough to provide safe equipment, systems and procedures if the culture doesn't encourage healthy and safe working.

While employers have the duty to provide workplaces where risks to health and safety are properly controlled, national and international laws also require employees to help. In addition, employers must consult workers and their representatives as part of the process. Managers do not have the solutions to all health and safety problems. Workers and their representatives have the detailed knowledge and experience of how the job is done and how it affects them. For this reason, workplaces in which workers actively contribute to health and safety often have a lower occupational risk level and accident rates.

Worker participation helps in developing effective ways of protecting workers. By getting involved in an issue at the planning stage, workers are more likely to identify the reasons for taking a particular action, help find practical solutions, and comply with the end result. If workers are given the opportunity to participate in shaping safe work systems, then they can advise, suggest, and request improvements - helping to develop measures to prevent occupational accidents and ill-health in a timely and cost effective manner.

In many situations directors and senior managers (sometimes called top managers) are appointed by employers to ensure employers' health and safety responsibilities are met. The role of directors and senior managers will include ensuring that the aims, objectives and overall approach to health and safety are suitable and sufficient.

Employers give directors and senior managers health and safety responsibilities (an obligation to achieve something) relating to their roles. The directors and senior managers are accountable (personally answerable for what is done or not done) to their employer for meeting these responsibilities.

In practice, the responsibilities of directors and senior managers will include ensuring a suitable management system is established and maintained. This will require the directors and senior managers to establish an appropriate health and safety policy, measurable health and safety objectives and programmes to meet the objectives.

Their responsibilities will also include ensuring risk assessments have been carried out, management arrangements are made and control measures are in place and used. In addition, they will usually have a responsibility to lead by example and participate in health and safety leadership activities like visits to the workplace. It is usual for them to have specific responsibilities to ensure that health and safety needs are resourced, that they receive and consider reports on health and safety and that action is taken to ensure deficiencies are dealt with and objectives are met.

They are also expected to be responsible for arranging competent health and safety assistance and conducting a regular (usually annual) review of health and safety performance. Their responsibilities will also include establishing the roles and responsibilities of middle managers, people with primary health and safety functions, and workers. Directors and senior managers are also responsible for establishing arrangements to hold these people to account for the responsibilities they are given.

In addition to the responsibilities directors and senior managers have in relation to the organisation they will be given individual responsibilities relating to their specific role in the organisation, for example, the management of a region, division or site.

All organisations have management processes or arrangements to deal with payroll, personnel issues, finance and quality control, many of these are also legal responsibilities that are factors in operating a business. The same level of attention that goes into ensuring a profitable organisation or the achievement of quality objectives must go into managing health and safety.

Importance of receiving, and acting on feedback from all stakeholders on health and safety performance

All organisations seek a positive safety culture where working practices and rules are developed that effectively control hazards and employees have a positive attitude towards risk management and compliance with the control processes.

A positive safety culture also requires a capacity to learn from accidents, near misses and safety performance indicators and bring about continual improvement - changes to processes and procedures that demonstrate a weakness because an accident or near miss has occurred.

The organisation is not restricted to accidents and incidents purely from an internal perspective. Where the organisation is selling or providing goods and services for use by others, then a duty of care exists to ensure these goods and services are safe to use.

An organisation can develop standard safe working practices that comply with the national legislation and best practice. However, to work effectively, the organisation needs to learn from what's happening in its workplace and market place. Only by being aware of and analysing accidents and near misses is it possible to develop suitable improvements to safe goods and services and working practices.

Organisations need to react to safety inspections and reports so that managers know where they need to concentrate their efforts. A prerequisite for safe processes, goods and services is good information form workers, consumers and clients.

In order for the information to flow, the workforce needs to be willing to participate and be prepared to report their mistakes, near misses and accidents and the organisation needs to react to market research and customer feedback.

This willingness to give feedback will depend on how the organisation investigates incidents and how it handles blame. A blame culture, one that looks to blame and punish people when things go wrong will encourage very little reporting.

On the other hand, a completely no-blame culture, one that allows all mistakes or errors to go unpunished, including those that are reckless or negligent, is not really feasible either, and probably won't be acceptable to the organisation or to individual workers.

Case Study - The Space Shuttle Challenger disaster

The Space Shuttle Challenger disaster occurred on January 28, 1986, when the NASA Space Shuttle Orbiter Challenger broke apart 73 seconds into its flight, leading to the deaths of its seven crew members. The Rogers Commission found NASA's organisational culture and decision-making processes had been key contributing factors to the accident, with the agency violating its own safety rules.

There was an inadequate response to internal warnings about the faulty design of a seal. A decision was taken to go ahead with the launch in very cold temperatures despite the faulty seal. The decision made was a result of conflicting scheduling/safety goals, the mindset of individuals and the effects of fatigue.

Methods of communication media available to promote the health and safety message

Studies of major accidents/incidents have shown that without effective communication, managers and workers are less likely to make the right decisions and errors or violations may occur. It is therefore important to provide effective communication within an organisation in order to help to ensure the correct health and safety behaviour takes place.

Effective communication involves providing what is required in the right way and at the right time. This will mean using the best communication method for the information being provided and the circumstances. It is therefore necessary to consider the individual receiving the communication to ensure that the information is understood and the correct health and safety behaviour results.

For example, when managers communicate with individuals it is important that managers are aware of the individual's needs, desires, capabilities and expectations in order to communicate with them effectively. In some cases, this will mean ensuring that the communication provides information on why action is to be taken as well as what is to be done.

This is particularly the case with regard to health and safety. For example, a manager may communicate to maintenance workers that they must wear eye protection when doing a task, but without emphasising why they should wear eye protection they may not behave correctly and wear it. It is also important that managers at all levels communicate their commendation of positive health and safety behaviour when they see it, as this will make it more likely that the positive behaviour will continue.

Communication within an organisation should include workers communicating to manager's information on concerns they have regarding health and safety. It is important that there are means for workers to do this in a positive way, so that they are encouraged to report hazards and deficiencies that could be a danger to themselves and others. To encourage workers to continue to report concerns managers should provide workers with feedback on concerns they communicate.

The communication process

Communication may be defined as 'a process by which information is exchanged between individuals through a common system of symbols, signs or behaviour'. Exchanging information may seem a simple task, but communicating in an effective way is often much more difficult than it appears. Poor communication, while common, can result in serious misunderstandings, leading to incorrect health and safety behaviour, accidents/incidents and work-related harm to health.

BARRIERS TO EFFECTIVE COMMUNICATION

It is essential that communications are understood in order for them to be acted on and for individuals to have the correct health and safety behaviour. However, communication is not always effective as there are a number of barriers that may prevent it being understood, including:

- Noise and other similar distractions.
- Sensory impairment (poor hearing or eyesight).
- Complexity of the information.
- Language/dialect of the speaker.
- Illogically presented information.
- Ambiguity of the information.
- Use of technical and local terms or abbreviations.
- The timeliness of the communication.
- Lengthy communication methods.
- Inattention of the person receiving the information.
- Lack of trust or respect.
- Capabilities of the person receiving the information to understand it, due to reduced health, fatigue, stress or limitations on their mental processing ability.

GENERAL PRINCIPLES OF COMMUNICATION

Communication is a skill that people often take for granted. Like any other skill, some individuals are better at it than others. A one-way communication process involves sending information to an individual and not obtaining confirmation of effective communication. This type of process has the limitation that confirmation of receipt and understanding of the information is not achieved.

Therefore, communication is generally a two-way process and is most effective in this form. In a two-way communication process the needs of the person receiving the information are equally as important as the needs of the person sending the information.

The person sending the information should not assume that the person receiving it has understood what was sent. The two-way process requires the person sending the information to confirm that the other person has received and understood it.

To ensure the success of two-way communication the person sending the information should:

- Communicate to the person receiving the information at a time that is suitable for their circumstances.
- Communicate in a form capable of being understood by the person receiving the information.
- Keep the content of the communication concise and relevant to the information being communicated.
- Provide the information in an order that is logical to the person receiving it.
- Use clear and unambiguous terms.
- Ensure that what is communicated matches the way it is communicated. If something is important, the communication of it should not be rushed. Non-verbal communication may communicate something different to what is said. Consider the need for respect and whether anger or humour may distract from what is communicated.
- Plan and encourage feedback to confirm understanding, including time, method and the form of feedback. The person receiving the communication may have to respond in a particular way, for example, repeating the information back to the person who gave the information.
- Use open ended questions to investigate understanding.
- Use closed questions which tend to provide yes/no answers, to confirm understanding.
- Repeat the communication or use a different method of communication if understanding is unclear.

Communication methods

In ensuring effective communication, the correct method of communication should be selected and used for the particular information and circumstances in which it is being provided. This will mean considering the principles of good communication and the merits and limitations of different methods. The different methods used to communicate health and safety typically include verbal, written and graphic methods.

VERBAL COMMUNICATION

Verbal communication is widely used in relation to health and safety and involves direct speech between individuals or through another medium, for example, a telephone, radio, video link or electronic speaker system.

Two-way forms of verbal communication provide the advantage of being able to communicate and obtain confirmation of understanding efficiently and promptly. This allows communication to be done in small amounts where necessary, ensuring understanding after each part of the information is communicated. This is particularly useful where information needs to be communicated accurately and understanding confirmed.

The barriers to communication outlined above are all relevant to verbal communication. In particular, this method of communication can be influenced by the language spoken, dialects, accents and the unintended wrong use of words.

These barriers can lead to misunderstandings or the wrong meaning being communicated. Where clarity of small amounts of important information is to be communicated the convention of spelling the letters of words and using a phonetic spelling alphabet is sometimes adopted. For example, the phonetic spelling alphabet of the International Civil Aviation Organisation (ICAO) assigns code words to letters of the English language, for example, A - Alpha, B - Bravo, C - Charlie, etc. This is particularly useful when communicating by radio or telephone and helps to avoid errors in verbal communication.

Where the information to be communicated is complex it may be preferable to use written communication to ensure clarity or a combination of written and verbal communication.

WRITTEN COMMUNICATION

Written communications are widely used for health and safety purposes, including policy documents, work instructions, procedures, site rules, permits-to-work, contractor contracts, meeting agendas and minutes, and reports. The written method of communication enables the information to be retained by the person receiving it and referred to if they forget what the information was.

It may also enable an individual to receive information prior to needing it and means they can read it at a time that suits them. Obtaining confirmation of understanding of written communications can be as important as for verbal communication. However, written communications are often used as a form of one-way communication, for example, notices or instructions posted on an information board.

One advantage of the written form of communication is that it can provide information to a large number and range of people at the same time. However, this can make it difficult to take account of the range of people reading it and the need to ensure effective communication to each of them.

In particular, it may be hard to obtain or manage feedback from all of the people sufficiently to ensure they understood what was communicated. For example, this difficulty would affect a situation where all workers in an organisation were issued with a new health and safety handbook. It could be difficult to confirm whether they had all read and understood the handbook.

When communicating in the written form it is important to consider the barriers to and principles of communication, this will help to make the communication effective. Particular attention should be paid to the use of appropriate structure, language, writing style and terms used to ensure the person receiving the communication understands it. Clarity and understanding of written communications can be tested by getting a sample of those who will receive the written communication to read it and checking they understood it. For example, procedures may be read by a sample of workers who have to follow them.

GRAPHICAL COMMUNICATION

Graphical communications may be used to replace or support other forms of communication. Verbal presentations and written reports will often include graphical images to illustrate or clarify complex relationships, for example, changes in accident/incident frequency rates over a period of time may be illustrated by a graph.

Graphical communication is used to good effect with health and safety signs, where the colour and pictorial representation of hazards and control measures are used. This avoids problems where workers may have limited language abilities and would struggle to read words. Graphical communication is also used in posters and moving image communications, for example, electronic induction packages and message boards.

Graphical communication in the form of drawings, photographs, still or moving digital images may be used to clarify information on hazards or risk control measures. They can provide an opportunity to make the communication of information more effective by containing examples of hazards and control measures from the actual workplace of those persons with whom they are communicating. The visual effect of graphical communication of this type may enable managers and workers to better understand practical aspects of health and safety. For example, a rule that requires a worker to wear a hard hat correctly could easily be demonstrated and understood by showing an image of a worker wearing a hard hat correctly.

Use and effectiveness of various communication measures

NOTICE BOARDS

A traditional communication technique is to post health and safety information on notice boards. The advantage of this communication measure is that the communication is easily available to everyone in a particular work area where the notice board is located and to those that pass near it. They can be a cheap and effective technique of communication when used to make general statements or to keep workers aware of current information and proposed developments.

Notice board information relies on an individual's ability to read, understand and apply the information correctly. Therefore, particular care should be taken to ensure appropriate language(s) are used for information placed on a notice board and they should not be used as a substitute for more effective two-way communication that can provide confirmation of receipt and understanding of information.

Also, the information posted on the notice board must be kept up to date and maintained in a legible condition if it is to be effective. This usually requires someone being given responsibility for the control of items placed on and removed from notice boards.

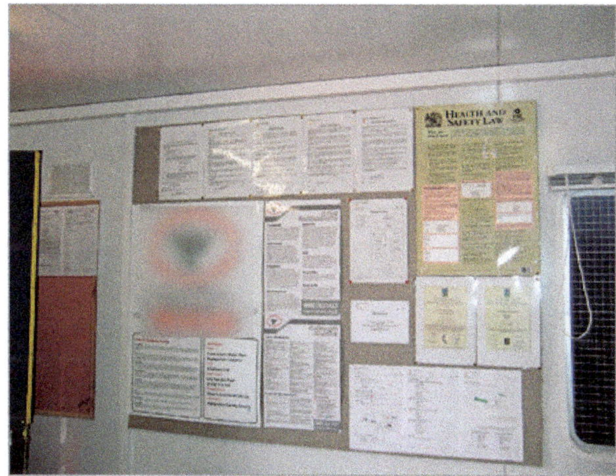

Figure A11-4: Health and safety notice board. *Source: RMS.*

Figure A11-5: Health and safety suggestion scheme. *Source: RMS.*

HEALTH AND SAFETY MEDIA

Moving image media

DVDs of relevant health and safety topics are often used to provide and refresh awareness and understanding of health and safety. They can provide information in a way that improves motivation and leads to better health and safety behaviour. The visual impact in them is also a strong stimulus that maintains attention while learning and enables training to provide practical examples of hazards and measures to minimise risk.

A common use for moving image media is at site induction of new workers and contractors, where it can provide information about the site, its rules and correct health and safety behaviour.

Shocking images of injuries and work-related ill-health are sometimes used to illustrate what might happen if procedures are not followed. It has been found that their effect may not change attitude in the long term, but can gain the attention of those seeing the images and allow them to accept the importance of health and safety measures when this is explained after seeing the images.

However, for some people the shocking images cause them to put up a barrier to receiving further information and can prevent the desired health and safety message being received.

Poster campaigns

Posters displaying health and safety information are sometimes seen as an inexpensive and visible way of showing commitment to health and safety. This approach can be ineffective if management place too much reliance on them and use them to inform workers to take care to avoid hazards that should have been managed by better control measures. To be effective, messages communicated by posters should be positive, aimed at the correct audience and be believable.

Positive - posters warning of the consequences if a particular action is not taken can be ineffective. To be effective, messages should emphasise the positive health and safety benefits of working safely and healthily to those reading the posters.

Aimed at the correct audience - posters quickly lose their impact and blend into the background, therefore, posters should be changed regularly to maintain their impact as a means of communication. There can also be a reduction in impact when the message being communicated is perceived to be irrelevant by those reading the poster.

Therefore, campaigns should be carefully targeted and posters positioned so they are seen by those the campaign is aimed at. For example, a poster reminding workers to wear the correct clothing when working with machinery should be sited close to the relevant machines.

Believable - messages should be convincing and realistic. Care must be taken to avoid offending or distracting the person reading the poster from the message by the use of inappropriate images of people. Similarly, pictures of horrific injuries can lead to a rejection of the intended message as the person reading it may focus on the image rather than the message. Some images may be so distressing that individuals refuse to look at the poster at all.

The advantages of using posters for communicating health and safety include:

- They have a relatively low cost.
- They provide visual impact to health and safety messages, allowing brief messages to be easily understood.
- They are flexible, enabling them to be displayed in the most appropriate positions and moved easily.
- They can be used to reinforce verbal instructions and provide a constant reminder of important health and safety issues.
- They can enable workers to be involved in health and safety locally, by encouraging them to suggest ideas for posters, selecting or designing posters and locating the posters where they will have most effect.

The disadvantages of using posters for communicating health and safety include:

- There is a need to change posters on a regular basis to maintain attention on them.
- They may become soiled, defaced and out of date.
- There is a possibility that they might be seen to trivialise serious matters.
- They might offend people if inappropriate stereotypes are used, for example, by using an image of the 'stupid worker'.
- It is not easy to assess whether the message has been understood.
- They may be used by some employers as an easy, if not particularly effective, way of discharging their responsibility for health and safety by moving the responsibility onto workers.

Toolbox talks

'Toolbox talks' are short communication sessions provided by supervisors and team leaders to explain or remind workers of important health and safety risks and control measures. The term toolbox talk comes from a scheme where short talks on health and safety were provided by supervisors in the workplace, sometimes near where the workers kept their tools.

They were designed to take place in the workplace so that workers were not distracted from their work for too long. The term is now often used to describe any form of communication where workers are gathered together for a short talk in or near to the worker's workplace.

Because toolbox talks are planned and organised in advance, the technique provides fast and reasonably consistent communication of specific subjects. The toolbox talks often cover only one subject to ensure the attention of workers is focused and two-way communication is encouraged.

Subjects covered in talks may provide a reminder of a particular risk related to work activities or discuss recent good or poor health and safety experience. The talks may be used to tell workers about proposed changes, for example, changes in personal protective equipment, work practices or procedures.

Toolbox talks enable health and safety issues to be discussed by supervisors and workers, providing increased awareness/understanding of risks and control measures and reinforcement of rules. They can be regularly used to encourage workers to have appropriate health and safety behaviour.

Memos/e-mails

Memos and e-mails are a useful way to provide health and safety communication on a timely basis. For example, they can give information on proposed changes to procedures before the amendment and distribution of the whole procedure. Memos are often used as a means of one-way communication. One of the difficulties in using memos is that the person sending the memo might assume that everyone to whom they sent the memo received it, however, this might not be correct.

Where the memo is important to health and safety there should be a process that confirms receipt. Similar problems can exist with the use of e-mails, although software is available to check whether a person has received and opened an e-mail. However, confirmation of receipt of a memo or e-mail does not confirm that it has been read, understood or actioned.

Many people complain about the number of emails they receive. Where individuals receive a high volume of e-mails there is a risk that important health and safety e-mails may be missed. This situation is reported to be particularly difficult in situations where individuals are copied in to e-mails in bulk, causing people to receive e-mails that are of very low or no relevance to them. This could reduce the perceived value of e-mails received from these people and lead to important health and safety messages they send being missed.

Worker handbooks

Workers need to know the current rules and procedures so they can follow them and have the correct health and safety behaviour. Issuing a worker handbook will help to provide this information, but if it is not kept up to date it can lead to workers making errors.

Handbooks are often issued to new workers at induction. They are useful in communicating site rules and information such as accident/incident and ill-health reporting procedures. Similarly, they will often contain information relating to on site emergency arrangements such as fire and first-aid. To be effective, the organisation should establish a mechanism to recall and reissue the handbooks when changes occur.

Some organisations operate a loose-leaf folder design to enable amendments to be made. If paper-based handbooks are provided, consideration needs to be given to where each person keeps the handbook so that it is available for use and updating. Some organisations have computer systems available and therefore choose to provide handbooks in an electronic format. This can enable changes to the electronic copy of the handbook to be made quickly.

However, unless workers can be equally quickly informed of the changes they are not likely to follow the new rules and procedures because they will not be aware of them. Toolbox talks or other means of communication may have to be used to inform workers affected by the changes.

If toolbox talks are used, they would provide an opportunity to discuss the changes and ensure people knew why the changes were taking place and that they understood what the changes were. This could provide additional confidence that everyone would follow the new rules and procedures.

Procedures for resolving conflict and introducing change

RESOLVING CONFLICT

Conflict can arise within an organisation as a consequence of personality clashes, poor communication, conflicting interests/priorities, lack of leadership and lack of management control. Conflict can have a harmful or adverse effect on behaviour and human reliability related to health and safety.

The harmful effects of conflict can be resolved by implementing a number of strategies, including:

- ***The clarification of goals and objectives*** - goals and objectives should be clearly defined. Everyone should understand their roles and what they are expected to achieve in order to avoid misunderstanding and conflict.
- ***The distribution of resources*** - a fair distribution of resources should be encouraged through, for example, the ability to make a case for extra materials and finance based on need. Although it is difficult to achieve a totally equitable system, organisations should strive to reduce possible causes of resentment amongst managers and staff.
- ***Personnel policies and procedures*** - a just and equitable system should be in place. Rewards and promotion should be made on merit. Disciplinary/grievance procedures should be fair and even-handed. An arbitration system to resolve employer/employee disputes should be carefully considered.
- ***Non-monetary rewards*** - can be introduced where financial resources are limited, for example, flexible working hours and the use of 'perks' to recognise and reward good performance.
- ***Leadership and management*** - a combative management style is more likely to lead to conflict than one which is supportive of subordinate's efforts. Managers who participate and encourage cooperation are likely to win the respect, trust and commitment of workers.
- ***Group - team building activities*** - encouraging the positive benefits of group behaviour can increase cohesiveness and reduce conflict.
- ***Development of interpersonal/group process skills*** - training on individual and group processes can lead to a better understanding of an individual's own behaviour and that of others. This can lead to improved communication and encourage a constructive approach to resolving conflicts.

Attention to ***organisational processes*** and the ***socio-technical organisation*** can also bring benefits in terms of conflict resolution. In the former the nature and structure of the organisation is moulded to encourage communication, the sharing of information and discourage unnecessary bureaucracy.

The latter encourages the development of psychological and social factors, which are in keeping with the structural and technical requirements, in order to reduce negative sources of dysfunctional conflict.

INTRODUCING CHANGE

The management of change and the need to be flexible are integral to an organisation's ability to adapt to its environment and the needs of its customers.

Therefore, change is both inescapable and inevitable if organisations are to survive. There are two main objectives to planned organisational change that need to be considered by managers.

These are:

1) The improvement of the organisation's ability to cope with change.
2) The need to modify individual behaviour in order to adapt to change.

The human behaviour aspect of introducing change is often underestimated. Senior managers, who may be filled with great expectations related to the change, find it hard to understand the fear and uncertainty of others. Loss and grief are just as normal reactions to change as are excitement and anticipation.

In order to introduce a planned change, a systematic, and three stage process can be used to modify human behaviour:

1) ***Unfreezing*** - encouraging the reduction of those forces that maintain human behaviour and encouraging the recognition of the need for change. Potential problems are identified at this stage.
2) ***Movement*** - this stage involves developing new attitudes or behaviour. This can be achieved by problem diagnosis, action planning, involvement, communication and implementation.
3) ***Refreezing*** - stabilising and reinforcing the new changes. The consequences of change are analysed and any necessary follow-up and stabilisation work carried out through supporting mechanisms such as clarifying policy and structure, providing coaching and using the influence of group norms.

The organisation should recognise that people are not detached from their work, but have a sense of emotional involvement. People are resistant to change for many reasons including security, habit, fear of the unknown, mistrust of management intentions and even because of the inconvenience the change may cause.

Unfreezing involves removing these barriers by identifying them and providing the appropriate stimuli in order to reduce resistance. This can be achieved by the creation of an environment of shared trust and commitment with, for example, new contracts or agreements and the development/use of a consultative processes.

Ensuring roles and responsibilities are clear, understood by all workers and implemented

Large organisations are more likely to have hierarchical structures with many different levels of employees all reporting upwards to various categories of manager. Hierarchy is a reference to an organisation that is structured in layers where each person has clear roles and responsibilities.

Those at the top of the hierarchy have more authority than those lower down. The managers, in a hierarchical structure, are likely to have a relatively narrow span of control with a number of employees reporting to each managerial level in the hierarchy.

Smaller organisations are likely to have much wider spans of controls in a flatter hierarchical structure or a matrix structure that increases the level of understanding required by the manager to undertake their role in the organisation.

Employees need to know the boundaries of authority that they operate in. They need to know what outputs are expected, what machines they are empowered to operate, what processes they are monitoring and what they are expected to produce. The employee effectively needs permission from the employer to do the key elements of their jobs.

Clear job descriptions and personal remits enable workers to focus on their job-specific tasks. It allows employees to prioritise their workload and reduce the chance of work duplication or conflicting instructions. In some industries, for example, offshore oil extraction, where safe operation is a priority, closely defined roles and responsibilities ensure that workers are competent for the tasks they undertake.

Clearly defined roles and responsibilities are of particular importance in large organisations to ensure no part of the workload is overlooked. In smaller organisations job roles may be less structured as employees may be required to take on a variety of tasks and responsibilities.

It is important that everyone in an organisation knows their health and safety role, how they can contribute to achieving health and safety objectives and what the employer expects of them. This will help the organisation to achieve and maintain a good health and safety performance. The senior management of an organisation should identify who needs to do what with regards to the management of health and safety.

In particular, it is essential that senior management clearly and concisely define in writing the roles and responsibilities of managers and others that have a specific role in health and safety. This will include defining roles and responsibilities for the most senior manager (top manager), members of the senior management team and anyone in the senior management team taking a lead role in health and safety.

The roles and responsibilities of all managers should be documented and defined in sufficient detail for them to understand what they are responsible for. When defining responsibilities, simple, summary statements like 'the manager will be responsible for health and safety in their area' should be avoided unless more detail is provided to support what is meant by the summary statement. Responsibilities of managers should at least include requirements for:

- Planning for health and safety - for example, responsibilities for organising meetings and a programme for risk assessments.
- Implementation of health and safety - for example, responsibilities for communication between managers, provision of information to workers, specific actions required to keep the workplace safe and healthy.
- Evaluation of health and safety - for example, responsibilities for carrying out workplace inspections and investigations into accidents/incidents.

It is also important that managers are provided with appropriate authority (power to make decisions and take action independently) to carry out their role and meet their responsibilities.

Section 2 of the HASAW Act 1974 implies this responsibility: 'Except in such cases as may be prescribed, it shall be the duty of every employer to prepare and as often as may be appropriate revise a written statement of his general policy with respect to the health and safety at work of his employees and the ***organisation*** and arrangements ....'

The duty to ***organise*** is amplified in the MHSWR 1999 to ensure that employers appoint competent persons with sufficient resources to manage health and safety effectively.

Where different managers of the organisation work together or share responsibility for something it is important to ensure roles and responsibilities are clear and do not conflict with each other. For example, operations and maintenance managers may share responsibility for the health and safety of work equipment that needs to be taken out of use for maintenance.

The roles and responsibilities of the operations and maintenance managers should clarify who is in control of the equipment at the different stages involved, including when it is being taken out of use, maintained and returned to use.

Other roles that relate directly to the effective management of health and safety that should be defined include:

- Health and safety practitioners.
- Occupational health specialists.
- Those that conduct governance audits, if they include health and safety.
- Human resources managers.
- Training managers.
- Purchasing and contracts managers.
- Workers and their representatives.

A11.3 - The health and safety practitioner's use of financial justification to aid decision making

The significance of budgetary responsibility

Financial management is not simply keeping accounting records, it is an essential part of organisational management and cannot be seen as a separate task. Financial management involves planning, organising, controlling and monitoring financial resources in order to achieve organisational objectives.

A budget is the financial description of an action plan outlining how money will be used in the organisation. Budgets are often based on knowledge and assumptions against which actual performance is measured. In order to set an accurate, realistic budget the budget holder needs to consider the objectives to be reached in spending the money and the activities involved in achieving the objectives.

These activities will require a resource that can be measured in time, people and money. Therefore, the allocation of resource has an associated cost. Good budgetary management can stimulate planning for the support and use of resources so that plans become achievable and realistic. Good budgetary management can also improve the clarity of the manager's vision and be a motivator to staff. The health and safety practitioner will often require resource to investigate accidents and adopt preventative actions. This is certain to involve the expenditure of money as part of the resource allocation. The organisation is likely to have set a financial budget for the proactive and reactive costs of managing safety and the safety practitioner may be accountable for where the budget is allocated.

If the whole budget is spent on reacting to safety failures, it's likely that improvements in safety management will be slow and there will be repeats of the same type of accident. Allocating funding (resource) to proactive measure will reduce the number of accidents.

In allocating funds from a budget on proactive measure, the safety practitioner can consider the 'payback' time period. Payback period budgeting refers to the period of time required to recoup the funds expended in an investment, or to reach the break-even point. For example, a $1000 investment on additional machine guarding which reduces the costs of accidents on that machine by $500 per year would have a two-year payback period.

A 'payback' period measures how long something takes to 'pay for itself'. The shorter the payback period the more likely senior managers will be to agree to fund the project since the financial stability of the business is not adversely affected in the medium or long term.

Each year, significant numbers of workers are injured or made ill by their work. As well as the human cost (in terms of the impact on the individual's quality of life and for fatal injuries, loss of life) financial costs (for example, in terms of lost production and healthcare costs), will affect the organisations profitability. Many of the costs associated with injury and ill health cannot be recovered through insurance schemes.

The cost must be covered by reallocating financial resource from other areas of the business. This reallocation may result in some of the organisations plans not being achieved- a position that often has to be explained to shareholders in the organisation.

Importance of recognising who is the responsible budget holder and how they can be influenced

Some health and safety initiatives will require money in order to achieve their objectives. If you are not the budget holder, the safety practitioner will probably need to make the financial case for why management should invest in the project.

The safety practitioner will have to demonstrate how the time and resources required will offer good value for money. It is worth noting that often, money is a limited resource within the organisation, and the desired safety project may be competing for this scarce resource against other projects that improve productivity or reduce costs - both being seen as having a direct impact on the organisations profit and loss.

The practitioner will need to carry out an investment appraisal to demonstrate how long it will take for the proposal to pay off the costs involved, and any possible profits that it could generate. There are a number of different investment appraisal methods that could be used; often the organisation will use different appraisal strategies for different size projects.

In allocating budgets, the organisation will usually divide the estimated revenues from its activities based on the estimated resources that will be used. This allows for easier identification of expenditure that is specific to a particular programme or organisation function. The safety practitioner will need to identify the manager responsible for the budget where the function in which the proposal to spend money for safety improvements will take place. If the proposal is for segregating vehicles from pedestrians in a warehouse operation, then the manager with budgetary responsibility for the warehouse needs to be approached.

The manager can be influenced by the safety practitioner presenting a case for change that is based on reducing costs through reduced risk. This can be expanded upon by projections on reduced employee turnover, lessened threat of legal actions, an improved standing with suppliers and partners when they visit the well organised warehouse, increased productivity and higher morale.

Cost benefit analysis in relation to risk control cost decisions

PRINCIPLES OF COST BENEFIT ANALYSIS

Cost benefit analysis (CBA) is a formal system that provides a framework that allows the user to identify all the desirable and undesirable consequences of an activity and then quantify them, preferably in objective terms.

Usually the purpose of CBA is to evaluate in common terms the impact of a particular course of action. The technique has three stages:

1) Identify and quantify all positive and negative effects of the proposed action.
2) Apply monetary value to all effects.
3) Compare costs and benefits.

Cost benefit analysis may be conducted at national level by governments and at a local level by organisations. In health and safety, cost benefit analysis may be used when deciding the most cost effective risk control measure. It is important to ensure credibility and acceptance of proposed risk control measures, using cost benefit analysis will assist in ensuring that control measures selected are appropriate to the circumstances and provide maximum benefit for the money expended.

Cost benefit analysis can be used at national government level to evaluate the effects of the introduction of legislation relating to health and safety. On an organisational, rather than governmental level, costs relevant to a CBA will include items such as managerial time, training, and equipment such as guards or personal protective equipment (PPE) and reduced productivity if safer systems of work slowdown production.

Although the cost of preventing occupational accidents/incidents and diseases is borne by employers, the cost of treating the results can rest largely with insurance companies and society as a whole. Therefore, the employer may choose to use CBA having little regard for societal costs.

The employer may use CBA to determine whether the precautions in place meet requirements to ensure health and safety 'so far as is reasonably practicable' (balance of risk against cost), where this is a requirement.

Though cost benefit analysis often relates to financial costs, it is possible for the analysis to take account of other factors than money, for example, the human cost.

The main difficulty that may be encountered when using CBA is therefore putting values on emotive or intangible items such as:

- Public relations.
- Customer perceptions.
- The loss of a human life.
- The reduction in quality of human life.

One way of dealing with the quantification of the effects on human life, for the purposes of cost benefit analysis, is by using the concept of 'Quality Adjusted Life Year' (QALY). This allows for a scale of disability ranging from zero (dead) through to one (perfect health). QALYs are the number of years (i.e. in terms of age) multiplied by the value determined by a disability scale. A monetary value can then be assigned to a QALY.

In the UK the Health and Safety Executive (HSE) uses a measure called the value of a statistical life (VOSL) when analysing cost and benefits. This is an estimate of the monetary benefit that society attaches to reducing the risk of death through injury and ill-health.

For example, if a survey showed that individuals were prepared to pay £10 for a health and safety device that reduced the risk of a fatality by one in 10,000 and if the hazard this related to affected 10,000 people and the device was used, the risk of the number of deaths would be reduced by one.

Therefore, in this example one 'statistical life' would be saved by the use of this health and safety device. This would show that individuals were prepared to pay £100,000 (£10 x 10,000) for this reduction in risk and the VOSL would therefore be £100,000. The type of survey that leads to the establishment of a VOSL is called a stated preference survey.

A stated preference study conducted in the Netherlands in 2002 to establish the value of a statistical life in road safety suggested that the VOSL was 2.2 million Euro. The UK Government have not endorsed any particular VOSL for the appraisal of health and safety at work legislation. However, the HSE uses the VOSL of £850,000, developed for the appraisal of road safety schemes in 1996.

The value of an appropriate VOSL has been subject of a considerable amount of debate in the UK, as it has been in a number of countries, and the government department for trade and industry suggested the valuation was too low and that a range of £1 million to £10 million was more appropriate, but with a narrower range of £2 million to £4 million being applied for routine cases.

The stated preference studies used to decide a VOSL may also reflect a varying willingness of individuals to accept a level of risk to life and this affects their view on what is an acceptable VOSL, making a range of values more appropriate. Values established in the USA suggest an average VOSL of $5 million is appropriate. From the range of VOSLs it can be seen that stated preference surveys of this type are greatly influenced by the specific health and safety device being considered and the perspectives of the individuals involved in the survey.

For non-fatal injuries and incidence of ill-health, other costs are used to indicate the costs to organisations, individuals and society, they can be derived from survey data, calculations and estimates. The following values were calculated in various reports produced in the UK by the HSE and provide a guide to the comparative value of a range of injuries and ill-health.

	(£) Total Cost (Rounded) 2013-2014
Fatal injuries	1,575,000
Non-fatal injuries	7,500
7 or more days absence	27,700
Up to 6 days absence	880
Ill-health	18,700
7 or more days absence	37,400
Up to 6 days absence	850

Figure A11-6: Work injury and ill-health cost values. *Source: HSE Cost to Britain model.*

COST BENEFIT ANALYSIS AT NATIONAL LEVEL

Cost benefit analysis at a national level is useful in deciding if the introduction of new legislation has value, without the effort and consequences of introducing it and seeing what the effects are. This form of analysis may provide positive support for the introduction of the legislation and information that may be used to persuade the government to pass it and the organisations affected by it to accept it.

Cost benefit analysis may be particularly important in situations where proposed legislation is raising the standards that organisations have to work to, for example, by introducing specific requirements to introduce control measures like a formal health and safety management system or to reduce worker exposure levels to a substance.

This approach is well developed in some countries of the world, for example, in the UK where in 1998 the UK Government published new guidance on the form of assessment that must be carried out for all regulatory proposals that lead to new legislation. The UK's Prime Minister's foreword to this guidance set out the key requirements of Regulatory Impact Assessments (RIAs).

> *"The Assessment should include a clear statement of the objectives of the regulatory proposal and its likely effects. It should demonstrate that the proposal is the most effective means of meeting the stated objectives, set out the costs and benefits of the proposal, and identify who will be affected."*

Figure A11-7: UK Prime Minister's forward to guidance. *Source: UK, The better regulation guide and regulatory impact assessment.*

The HSE is required to prepare a cost benefit analysis for all proposals for new health and safety legislation, except those with a zero or minimum cost impact.

The CBA considers the resources to be used and saved from society as a whole. For example, the cost to industry of implementing regulations relating to control of asbestos at work may be offset by the money that industry, the insurance companies and wider society as a whole saves through not having to treat asbestosis and other related diseases.

Example - Regulations for construction health, safety and welfare - UK

The regulatory impact assessment relating to the introduction of regulations to control construction health, safety and welfare in the UK identified a number of negative effects/costs and positive effects/benefits:

Negative effects - costs to employers

The negative effects arise from factors like:

- The requirement to fit an intermediate guard rail to prevent falls from heights.
- The segregation of vehicles from pedestrians on sites.
- The provision of additional welfare facilities.
- Additional inspections of scaffolding and fire and emergency precautions.

Estimated first year cost:	£45 - £54 million.
Subsequent additional costs:	£6 million, occasionally rising to £19 million per year.
Total cost over 25 years:	£113 - £130 million (0.2% of the industry's annual turnover - estimated £50 billion).

Positive effects - benefits

About 60% of accidents/incidents on construction sites were as a result of falls from height or being struck by vehicles on site.

The additional protection measures were expected to achieve a 10% - 15% reduction in these type of accidents/incidents that, in turn, would mean a 6% - 9% overall reduction in construction site accidents/incidents.

Estimated annual benefits:	£18 - £27 million
Total savings over 25 years:	£127 - £210 million

All costs relate to the time of the study that preceded the introduction of the regulations in 1996. The cost benefit analysis showed that although the cost of implementing the regulations outweighed the benefits in the first year, the overall benefits to society over 25 years (£127 million to £210 million) were expected to outweigh the costs over the same period (£113 million to £130 million).

COST BENEFIT ANALYSIS BY ORGANISATIONS

Cost benefit analysis, at the ***organisational*** (board) level, will be a major consideration when deciding on expenditure for the introduction of new technology and methods or making the choice to continue doing work, in the light of legislative changes that require a higher standard of compliance than previously.

For example, reductions in acceptable worker exposure levels associated with workplace exposure limits (WELs) for substances or noise exposure limits.

The analysis may include the costs of defining the ***design*** parameters and specification for new plant and equipment, the ***planning*** work to achieve the change and a prediction of the ***operational*** cost and likely payback period for the investment.

Direct and indirect cost and benefits

The costs of risk control measures can be categorised into direct and indirect costs. Direct costs relate to the cost of purchasing and maintaining risk control measures, such as the provision of first-aid points and a training programme. These are usually reasonably easy to identify. In addition to these capital costs, the cost of running and maintaining systems must be added. These would include the required supervision and maintenance of the first-aid facilities and people.

Indirect costs are often more difficult to quantify as they relate to less tangible issues such as:

- The loss of production incurred by the use of a safety device, such as a guard fitted to a machine.
- The disruption caused when a risk control measure, such as security cameras or a sprinkler system, is installed.
- Reduced efficiency in the movement of goods and materials due to a fire wall being built.

In making the decision to provide a particular control measure, cost benefit analysis may identify some of these indirect costs and provide an opportunity to mitigate them.

For example, disruption from installing a sprinkler system may be mitigated by arranging with the contractor to have the system installed when disruption may be minimal, this may be at the weekend.

The following table summarises some of the ***organisational, design, planning and operational*** costs and benefits associated with risk control:

Direct costs	***Direct benefits***
Capital costs of: ■ Higher standards of construction. ■ Risk control equipment (for example, sprinkler system). ■ Labour and materials. ■ Maintenance of plant, building facilities and risk control equipment. ■ Additional management and supervision. ■ Fire detection and control systems. ■ First-aid provision. ■ Training.	■ Reduction in insurance premiums. ■ Tax benefits related to capital investment and other business taxes. ■ Possible grants available for capital expenditure via incentive schemes such as those intended to stimulate regional development or health and safety improvements.
Indirect costs	***Indirect benefits***
■ Interference to production or provision of services during installation and maintenance of control measures. ■ Loss of production or service during worker training (for example, first-aid courses). ■ Poor worker relations due to the introduction of new control measures (for example, guards slowing down production thereby impacting on payment schemes for workers).	Fewer insured and business interruption losses such as: ■ Worker losses - injury and ill-health. ■ Third party losses (customers, public and neighbours). ■ Business interruption losses. ■ Property losses. Improved relations with: ■ The workforce (including productivity). ■ Customers and suppliers. ■ The general public.

Figure A11-8: Cost and benefits associated with direct and indirect costs of risk control measures. *Source: RMS.*

Break-even analysis

Those risks that are retained within an organisation should be reduced through a programme of risk control measures. Break-even analysis is simply a technique for estimating the minimum financial turnover required to cover the costs of the control measures. These costs can be divided into fixed costs and variable costs. In business finance terms, the break-even point is calculated using the following formula:

Break-even point	=	$\dfrac{F}{1-\dfrac{V}{P}}$

Where:

F	=	fixed costs		These are those costs that must be paid regardless of the amount of production or sales
V	=	variable cost per unit	(or total variable costs)	These costs vary in proportion to sales turnover
P	=	selling price per unit	(or the total sales value - may be substituted by the symbol S)	

For example, consider a situation where fixed costs for a year are £40,000, variable costs per unit are £2 and units sell for £10 each. The break-even point is:

$\dfrac{40{,}000}{1-\dfrac{2}{10}}$	=	£50,000 break-even point

Therefore £50,000 of sales is needed to cover costs (including controls and losses).

A variation on this is the break-even quantity (BEQ).	$BEQ = \dfrac{F}{P - V}$

Because of the effect of break-even values, it is in the interest of organisations to reduce fixed costs to a minimum. An organisation with high fixed costs is less able to withstand a reduction in sales turnover. The break-even point is therefore very dependent on the profit margin, the amount by which the sale price of a product exceeds its overall costs. An organisation has to increase sales in order to cover the costs of the risk control measure or, alternatively, the associated losses.

As can be seen from the table in ***figure ref IA9-15***, the lower the profit margin the higher the amount of sales required to meet the costs of losses or prevention.

Annual costs	***Profit margin***			
£	1%	2%	4%	5%
1,000	100,000	50,000	25,000	20,000
10,000	1,000,000	500,000	250,000	200,000
25,000	2,500,000	1,250,000	625,000	500,000
50,000	2,500,000	2,500,000	1,250,000	1,000,000
100,000	10,000,000	5,000,000	2,500,000	2,000,000
200,000	20,000,000	10,000,000	5,000,000	4,000,000

Figure A11-9: Organisational costs and profit data. *Source: RMS.*

When considering costs, a quotation attributed to a Chief Executive Officer of Du Pont, a company that has a renowned reputation for safety, illustrates the point: *"Had our safety performance been the average for our industry last year we would have had to spend another $90 million to meet the increased costs. To produce another $90 million we would have needed to generate an extra $2 billion of business."*

Internal and external sources of information that should be considered when determining costs

A budget identifies the planned expenditure for a project. It is used as a baseline against which the actual expenditure and predicted eventual cost of the work can be reported. In order to decide if a project is viable the cost of the project has to be evaluated and compared with the potential return from undertaking the project.

Initial cost estimates can be used as a comparator, but these need refining as the feasibility and desirability of the initiative are investigated and a greater understanding of scope, schedule and resources is developed.

Sources of internal information that can be used to develop cost models will usually be found in the financial department of the organisation. The hourly costs of labour can be supplied, together with the costs of overheads such as electricity, gas and other consumable items. It is likely that the financial department will be able to provide cost information for the utilisation of space i.e. the cost per square meter of occupying a location.

Maintenance departments may be able to provide cost information on planned preventative maintenance programmes or the cost associated with making a repair. Maintenance departments are also likely to be able to advice on the replacement frequency for components (if the project being resourced is a new machine) and be able to give energy costs for a given size of gas turbine or electrical motor.

Human Resource departments are likely to be able to advice on the cost of labour to undertake a task. More complex worker roles often attract a higher wage rate than simpler tasks. The organisation is likely to have a wage structure that needs to be taken into account, for example, if the project is replacing an old manually operated machine with a new computer controlled machine the skill level required is higher and the wage rate for the employee may be commensurate with that skill level.

Purchasing departments are likely to be able to provide cost information on a range of consumables that may be required. A new machine may require cutting fluids in its operation, the cost of this consumable item could be obtained from the purchasing department.

External sources of cost information would include manufacturer's catalogues or supplier's price lists. The internet can be used to compare and contrast manufacturers cost data to ensure the resource purchased is competitively priced. When the resource required is a service to be provided, contractors can be asked to quote for the provision of the resource or service. Where the costs are high, it is frequently required that more than one quotation for supply is sought to ensure a competitive price has been achieved.

Necessity of both short and long term budgetary planning

Long term budgetary planning is usually associated with a term usually longer than one year. A long-range budget involves more uncertainty than a short-term budget because global events, for example, earthquakes, typhoons, droughts, wars and political stability are less predictable. Market movements and the business cycle are more easily predictable in the short term.

Long term planning is required for order to ensure sustainable profitability. While planning for the long term is necessary, the plan must be flexible to account for the uncertainty inherent to it. Organisations develop plans to reach their overall goals, and it is common practice to separate planning into phases. This allows tracking immediate improvements while evaluating progress toward goals and targets.

The different time frames of the planning process place the focus on time-sensitive aspects of the company's structure and environment. Planning can be differentiated by the time frames of the inputs and expected outcomes. Short term usually involves processes that show results within a year. Medium-term plans may take longer than a year to achieve their intended outcome. Long-term plans, which are the overall goals the organisation sets, are often four or five years in the future and usually are based on reaching the medium-term targets.

Short-term budgetary planning will involve activities such as current statistics, for example, lost time accidents, and introducing physical control measures to prevent a recurrence, training programmes to improve knowledge or changes that ensure legal compliance.

Typically, these are shorter term measures that have an immediate impact. If the root causes of the accident rates is the attitude of the workers to using safety features of machines or applying information given on training courses then medium term planning needs to take place in the form of a campaign of activities to change behaviours.

If the training courses have had an immediate and lasting effect and solved the problems arising, then a schedule of training programmes can be developed for the medium term. The cost of these courses can be gauged (and budgeted for) as the current cost plus an inflationary figure to account for an increase in fees.

The estimate of how high an inflationary figure to use is somewhat arbitrary, but financial departments are probably already using an organisationally agreed multiplier to estimate future costs. Where a short-term response to equipment failure is to repair the machine, a medium-term solution is to arrange for a service contract. Medium-term planning implements policies and procedures to ensure that short-term problems don't recur.

In the long term, the organisation will seek a permanent solution that will make the achievement of long term goals more certain. Long-term budgetary planning reacts to the competitive situation of the company in its social, economic and political environment and develops strategies for adapting and influencing its position to achieve its long-term goals. Long term budgetary planning will, therefore, typically address major capital expenditures such as purchasing new machinery that is more reliable or easier to maintain.

If the organisation operates a fleet of diesel powered vehicles predictions on the rising/falling cost of diesel fuel will influence the purchasing decision to have more fuel economic vehicles. Long term budgetary planning will implement policies and procedures that shape the company's profile to match management's vision of where the organisation needs to be. When short-term and medium-term planning is successful, long-term planning builds on those achievements to preserve accomplishments and ensure continued progress.

This page is intentionally blank

Relevant statutory provisions

Content

RMS Publishing's technical authors regularly review examiners reports for all NEBOSH awards to ensure that the specific publication content is in keeping with the level of study required for the award.

The review considers the core training materials, assessment criteria and relevant legislation. At each stage care is taken to pitch the level of the content to the examination requirements and, in particular, knowledge of the legislation required at this level.

The syllabus does not require knowledge of all legislation to the same depth, this is reflected in the 'relevant statutory provisions' section of the study book. Relevant statutory provisions abstracts are designed to focus on the specific aspects (often popular with the examiners) of the legislation required to meet the syllabus. The study book provides guidance in the form of 'outline of main points' enabling students to focus on the critical learning points and avoid over studying. In addition, legislation is considered in context in the relevant elements of the study book.

Students are advised to obtain or gain access to statutory documents, approved codes of practice and guidance related to the relevant statutory provisions as part of their personal development programme, for the purpose of the examination and future career development.

NEBOSH do not examine on legislation until it has been in force for 6 months. Students may show knowledge of new legislation in their answers until that point, students referring to the former legislation will not lose marks until the 6 month period has passed.

Corporate Manslaughter and Corporate Homicide Act (CMCHA) 2007

Considered in context in Element A3.

These explanatory notes relate to the Corporate Manslaughter and Corporate Homicide Act which received Royal Assent on 26 July 2007. They have been prepared by the Ministry of Justice in order to assist the reader in understanding the Act. They do not form part of the Act and have not been endorsed by Parliament.

The notes need to be read in conjunction with the Act. They are not, and are not meant to be, a comprehensive description of the Act. So where a section or part of a section does not seem to require any explanation or comment, none is given.

Arrangement of Act

Section 1:	The offence.
Section 2:	Meaning of "relevant duty of care".
Section 3:	Public policy decisions, exclusively public functions and statutory inspections.
Section 4:	Military activities.
Section 5:	Policing and law enforcement.
Section 6:	Emergencies.
Section 7:	Child-protection and probation functions.
Section 8:	Factors for jury.
Section 9:	Remedial Orders.
Section 10:	Power to order conviction etc to be publicised.
Section 11:	Application to Crown bodies.
Section 12:	Application to armed forces.
Section 13:	Application to police forces.
Section 14:	Application to partnerships.
Section 15:	Procedure, evidence and sentencing.
Section 16:	Transfer of functions.
Section 17:	DPP's consent required for proceedings.
Section 18:	No individual liability.
Section 19:	Convictions under this Act and under health and safety legislation.
Section 20:	Abolition of liability of corporations for manslaughter at common law.
Section 21:	Power to extend section 1 to other organisations.
Section 22:	Power to amend Schedule 1.
Section 23:	Power to extend section 2(2).
Section 24:	Orders.
Section 25:	Interpretation.
Section 26:	Minor and consequential amendments.
Section 27:	Commencement and saving.
Section 28:	Extent and territorial application.
Section 29:	Short title.
Schedule 1:	List of Government departments etc.
Schedule 2:	Minor and consequential amendments.

Outline of main points

The Corporate Manslaughter and Corporate Homicide Act introduced a new offence, across the UK, for prosecuting companies and other organisations where there has been a gross failing in the management of health and safety, with fatal consequences.

Section 1 defines the offence and identifies the sorts of organisation to which the Act applies. The offence only applies in circumstances where an organisation owed a duty of care to the victim under the law of negligence. This reflects the position under the common law offence of gross negligence manslaughter and, by defining the necessary relationship between the defendant organisation and victim, sets out the broad scope of the offence. Duties of care commonly owed by corporations include the duty owed by employers to their employees to provide a safe system of work and by an occupier of buildings and land to people in or on, or potentially affected by, the property. Duties of care also arise out of the activities that are conducted by corporations, such as the duty owed by transport companies to their passengers.

Section 5 provides an exemption that applies to the police and other law enforcement bodies in respect of all categories of duty of care referred to in section 2, i.e., including those duties of care owed by an organisation as an employer or the occupier of premises. But this wide exemption is available only in limited circumstances, specifically operations dealing with terrorism, civil unrest or serious disorder in which an authority's officers or employees come under attack or the threat of attack; or where the authority in question is preparing for or supporting such operations; or where it is carrying on training with respect to such operations.

Section 6 clarifies that the offence does not apply to the emergency services when responding to emergencies. This does not exclude the responsibilities these authorities owe to provide a safe system of work for their employees or to secure the safety of their premises. Emergency circumstances are defined in terms of those that are life-threatening or which are causing, or threaten to cause, serious injury or illness or serious harm to the environment or buildings or other property. However, the exemption does not extend to medical treatment itself, or to decisions about this (other than decisions that establish the priority for treating patients).

Section 8 outlines factors for the jury to consider when assessing an organisation's culpability. This sets out the test for assessing whether the breach of duty involved in the management failure was gross. The test asks whether the conduct that constitutes this failure falls far below what could reasonably have been expected. The jury is directed by the Act to consider the extent to which the evidence shows that there were "attitudes, policies, systems or accepted practices within the organisation that were likely to have encouraged any such failure… or to have produced tolerance of it".

Sections 9 and 10 make provision for remedial orders and publicity orders to be made on conviction.

Sections 11 to 13 deal with the application of the offence to the Crown and police forces, where a number of provisions are required to reflect the particular status of Crown bodies and police forces. Section 14 makes provision to accommodate the application of the offence to partnerships. Section 15 makes further supplemental provision to ensure that rules of procedure, evidence and sentencing apply to Crown bodies, police forces and those unincorporated bodies to which the offence applies. Section 16 sets out where liability will fall following machinery of Government changes or other cases where functions are transferred.

Section 19 clarifies that a conviction for corporate manslaughter would not preclude an organisation being convicted for a health and safety offence on the same facts if this were in the interests of justice. It would therefore also be possible to convict an individual on a secondary basis for such an offence under provisions such as section 37 of the Health and Safety at Work etc. Act 1974. This does not impose any new liabilities on individuals but ensures that existing liabilities are not reduced as an unintended consequence of the new offence.

The Schedules to the Act set out the Government departments and other similar bodies to which the offence will apply and make a number of minor and consequential amendments.

Deregulation Act (DA) 2015

Considered in context in Element A3. See also, HASAWA 1974 (Section 3(2)).

The Deregulation Act 2015 received Royal Assent on 26 March 2015. The Act covers a wide range of topics and detailed below are sections that relate to health and safety.

Modification to the general duty of self-employed persons to limit exposure to health and safety risks, by restricting this duty to self-employed persons who 'conduct an undertaking of a prescribed description'

Section 1 Health and safety at work: general duty of self-employed persons amends Section 3(2) of the Health and Safety at Work etc. Act 1974, which places a general duty on employers and the self-employed to conduct their business in a way that ensures, as far as is reasonably practicable, that they and persons (other than their employees) are not exposed to risks to their health or safety.

The amendment limits the scope of this general duty on self-employed persons, by applying the duty to those self-employed persons carrying out certain activities on a prescribed list, rather than all those who are self-employed. This has the effect of exempting self-employed persons who are not covered by the prescribed list, from this duty.

Widening of the exemption for Sikhs from wearing safety helmets, by extending it to all workplaces instead of just construction sites

Section 6 and 7 Requirements to wear safety helmets: exemption for Sikhs. The exemption for turban wearing Sikhs from wearing a safety helmet is extended from construction sites to all workplaces (except in urgent response to hazardous situations such as fire or riots, or if the individual is a member of Her Majesty's Forces and taking part in a military operation). This therefore includes visitors to a workplace as well as workers.

Deregulation Act 2015 (Health and Safety at Work) (General Duties of Self-Employed Persons) (Consequential Amendments) Order (DAHSW) 2015

Considered in context in Element A3.

The Deregulation Act 2015 (Health and Safety at Work) (General Duties of Self-Employed Persons) (Consequential Amendments) Order 2015 makes consequential amendments to various provisions in health and safety legislation to exempt self-employed workers, but only if their work activities post no potential risk of harm to others.

Employers' Health and Safety Policy Statements (Exception) Regulations (EHSPS) 1975

Considered in context in Element A1.

The Employers' Health and Safety Policy Statements (Exception) Regulations (EHSPS) 1975 which exempt 'any employer who carries on an undertaking in which for the time being he employs less than five employees'.

All the employees of an undertaking count for the purposes of this exemption, whether they are employed in one or several sites or establishments. It should also be noted that, for the purposes of these and other health and safety requirements, trainees count as employees of the immediate provider of their workplace training or work experience (unless this is a college or other educational establishment, and the training or work experience is provided on a course run by the college or educational establishment).

Employment Rights Act (ERA) 1996

Considered in context in Element A3. See also, PIDA and ERRA.

All employees are protected by the Employment Rights Act 1996, as amended, against suffering any harm because of any reasonable actions they take on health and safety grounds. This applies regardless of their length of service.

Employees, including health and safety representatives, should not suffer harm, for instance by being denied a promotion or being dismissed unfairly, because they:

- Carry out, or propose to carry out, activities that you have assigned to them in connection with preventing or reducing health and safety risks;
- Perform, or propose to perform, functions they have as union-appointed or employee-elected health and safety representatives, or health and safety committee members;
- Stand as a candidate in an election to be an employee representative or participate in the election by voting;
- Bring to their employers attention, by reasonable means, a concern about situations at work that they reasonably believe are harmful, or potentially harmful, to health and safety;
- Reasonably believe a situation to be of serious and imminent danger and - because they could not reasonably be expected to avert it - they leave or propose to leave the workplace or any dangerous part of it, or if they refuse to return while the danger continues; and
- Reasonably believe a situation to be of serious and imminent danger, and take or propose to take appropriate steps to protect themselves and others. This is to be judged by looking at all knowledge, facilities and advice available at the time.

Employers could be taken to an employment tribunal if they penalise employees in this way.

Enterprise and Regulatory Reform Act (ERRA) 2013

Considered in context in Element A4. See also, HASAWA 1974 (Section 47) and HSWCER.

Section 69 of the Enterprise and Regulatory Reform Act (ERRA) 2013 came into force on 1st October 2013, amending section 47 of the Health and Safety at Work etc. Act 1974. This means that in respect of workplace accidents after that date, an injured party cannot simply base their claim on a breach of statutory duty.

Instead, claimants will need to show that there has been negligence on the part of the employer. Previously some statutory Regulations imposed a 'strict liability' on the employer, who could be liable to pay damages even though there was nothing the employer could have done to prevent the accident.

An exception to this arises from the Health and Safety at Work etc. Act 1974 (Civil Liability) Exceptions) Regulations 2013 which aims to ensure that a pregnant worker continues to have the right to bring a claim for breach of statutory health and safety duty in relation to rights under the Pregnant Workers Directive (EEC 92/85 as amended).

Equality Act (EA) 2010

Considered in context in Element A2.

Arrangement of relevant Parts, Chapters and Sections of Act

Parts of the Act

1) Socio-economic inequalities.
2) Equality: key concepts.
3) Services and public functions.
4) Premises.
5) Work.
6) Education.
7) Association.
8) Prohibited conduct: ancillary.
9) Enforcement.
10) Contracts.
11) Advancement of equality.
12) Disabled person: transport.
13) Disability: miscellaneous.
14) General exemptions.
15) Family property.
16) General and miscellaneous.
17) Schedules 1-28.

Part 2, Chapters

1) Protected characteristics.

2) Prohibited conduct.

Discrimination.

Adjustments for disabled.

Discrimination: supplementary.

Outline of main points

PART 2 EQUALITY: KEY CONCEPTS

Chapter 1 Protected Characteristics

Section 4 the following characteristics are protected characteristics:

- Age.
- Disability.
- Gender reassignment.
- Marriage and civil partnership.
- Pregnancy and maternity.
- Race.
- Religion or belief.
- Sex.
- Sexual orientation.

Section 5 Age

(1) In relation to the protected characteristic of age:

(a) A reference to a person who has a particular protected characteristic is a reference to a person of a particular age group.

(b) A reference to persons who share a protected characteristic is a reference to persons of the same age group.

(2) A reference to an age group is a reference to a group of persons defined by reference to age, whether by reference to a particular age or to a range of ages.

Section 6 Disability

(1) A person (P) has a disability if:

(a) P has a physical or mental impairment.

(b) The impairment has a substantial and long-term adverse effect on P's ability to carry out normal day-to-day activities.

(2) A reference to a disabled person is a reference to a person who has a disability.

(3) In relation to the protected characteristic of disability:

(a) A reference to a person who has a particular protected characteristic is a reference to a person who has a particular disability.

(b) A reference to persons who share a protected characteristic is a reference to persons who have the same disability.

(4) This Act (except Part 12 and section 190) applies in relation to a person who has had a disability as it applies in relation to a person who has the disability; accordingly (except in that Part and that section).

(a) A reference (however expressed) to a person who has a disability includes a reference to a person who has had the disability.

(b) A reference (however expressed) to a person who does not have a disability includes a reference to a person who has not had the disability.

(5) A Minister of the Crown may issue guidance about matters to be taken into account in deciding any question for the purposes of subsection (1).

(6) Schedule 1 (disability: supplementary provision) has effect.

Section 9 Race

(1) Race includes:

(a) Colour.

(b) Nationality.

(c) Ethnic or national origins.

(2) In relation to the protected characteristic of race:

(a) A reference to a person who has a particular protected characteristic is a reference to a person of a particular racial group.

(b) A reference to persons who share a protected characteristic is a reference to persons of the same racial group.

(3) A racial group is a group of persons defined by reference to race; and a reference to a person's racial group is a reference to a racial group into which the person falls.

(4) The fact that a racial group comprises two or more distinct racial groups does not prevent it from constituting a particular racial group.

(5) A Minister of the Crown may by order:

(a) Amend this section so as to provide for caste to be an aspect of race.

(b) Amend this Act so as to provide for an exception to a provision of this Act to apply, or not to apply, to caste or to apply, or not to apply, to caste in specified circumstances.

(6) The power under section 207(4)(b), in its application to subsection (5), includes power to amend this Act.

Section 11 Sex

In relation to the protected characteristic of sex:

(a) A reference to a person who has a particular protected characteristic is a reference to a man or to a woman.

(b) A reference to persons who share a protected characteristic is a reference to persons of the same sex.

Part 2 Prohibited Conduct

Section 13 Direct discrimination

(1) A person (A) discriminates against another (B) if, because of a protected characteristic, A treats B less favourably than A treats or would treat others.

(2) If the protected characteristic is age, A does not discriminate against B if A can show A's treatment of B to be a proportionate means of achieving a legitimate aim.

(3) If the protected characteristic is disability, and B is not a disabled person, A does not discriminate against B only because A treats or would treat disabled persons more favourably than A treats B.

(4) If the protected characteristic is marriage and civil partnership, this section applies to a contravention of Part 5 (work) only if the treatment is because it is B who is married or a civil partner.

(5) If the protected characteristic is race, less favourable treatment includes segregating B from others.

(6) If the protected characteristic is sex:

(a) Less favourable treatment of a woman includes less favourable treatment of her because she is breast-feeding.

(b) In a case where B is a man, no account is to be taken of special treatment afforded to a woman in connection with pregnancy or childbirth.

(7) Subsection (6)(a) does not apply for the purposes of Part 5 (work).

(8) This section is subject to sections 17(6) and 18(7).

Section 15 Discrimination arising from disability

(1) A person (A) discriminates against a disabled person (B) if:

(a) A treats B unfavourably because of something arising in consequence of B's disability.

(b) A cannot show that the treatment is a proportionate means of achieving a legitimate aim.

(2) Subsection (1) does not apply if A shows that A did not know, and could not reasonably have been expected to know, that B had the disability.

Section 19 Indirect discrimination

(1) A person (A) discriminates against another (B) if A applies to B a provision, criterion or practice which is discriminatory in relation to a relevant protected characteristic of B's.

(2) For the purposes of subsection (1), a provision, criterion or practice is discriminatory in relation to a relevant protected characteristic of B's if:

(a) A applies, or would apply, it to persons with whom B does not share the characteristic.

(b) It puts, or would put, persons with whom B shares the characteristic at a particular disadvantage when compared with persons with whom B does not share it.

(c) It puts, or would put, B at that disadvantage.

(d) A cannot show it to be a proportionate means of achieving a legitimate aim.

Section 20 Duty to make adjustments

(1) Where this Act imposes a duty to make reasonable adjustments on a person, this section, sections 21 and 22 and the applicable Schedule apply; and for those purposes, a person on whom the duty is imposed is referred to as A.

(2) The duty comprises the following three requirements.

(3) The first requirement is a requirement, where a provision, criterion or practice of A's puts a disabled person at a substantial disadvantage in relation to a relevant matter in comparison with persons who are not disabled, to take such steps as it is reasonable to have to take to avoid the disadvantage.

(4) The second requirement is a requirement, where a physical feature puts a disabled person at a substantial disadvantage in relation to a relevant matter in comparison with persons who are not disabled, to take such steps as it is reasonable to have to take to avoid the disadvantage.

(5) The third requirement is a requirement, where a disabled person would, but for the provision of an auxiliary aid, be put at a substantial disadvantage in relation to a relevant matter in comparison with persons who are not disabled, to take such steps as it is reasonable to have to take to provide the auxiliary aid.

(6) Where the first or third requirement relates to the provision of information, the steps which it is reasonable for A to have to take include steps for ensuring that in the circumstances concerned the information is provided in an accessible format.

(7) A person (A) who is subject to a duty to make reasonable adjustments is not (subject to express provision to the contrary) entitled to require a disabled person, in relation to whom A is required to comply with the duty, to pay to any extent A's costs of complying with the duty.

(8) A reference in section 21 or 22 or an applicable Schedule to the first, second or third requirement is to be construed in accordance with this section.

(9) In relation to the second requirement, a reference in this section or an applicable Schedule to avoiding a substantial disadvantage includes a reference to:

(a) Removing the physical feature in question.

(b) Altering it.

(c) Providing a reasonable means of avoiding it.

(10) A reference in this section, section 21 or 22 or an applicable Schedule (apart from paragraphs 2 to 4 of Schedule 4) to a physical feature is a reference to:

(a) A feature arising from the design or construction of a building.

(b) A feature of an approach to, exit from or access to a building.

(c) A fixture or fitting, or furniture, furnishings, materials, equipment or other chattels, in or on premises.

(d) Any other physical element or quality.

(11) A reference in this section, section 21 or 22 or an applicable Schedule to an auxiliary aid includes a reference to an auxiliary service.

(12) A reference in this section or an applicable Schedule to chattels is to be read, in relation to Scotland, as a reference to moveable property.

(13) The applicable Schedule is, in relation to the Part of this Act specified in the first column of the Table, the Schedule specified in the second column.

Section 26 Harassment

(1) A person (A) harasses another (B) if:

(a) A engages in unwanted conduct related to a relevant protected characteristic.

(b) The conduct has the purpose or effect of:

(i) Violating B's dignity.

(ii) Creating an intimidating, hostile, degrading, humiliating or offensive environment for B.

(2) A also harasses B if:

(a) A engages in unwanted conduct of a sexual nature.

(b) The conduct has the purpose or effect referred to in subsection (1)(b).

(3) A also harasses B if:

(a) A or another person engages in unwanted conduct of a sexual nature or that is related to gender reassignment or sex.

(b) The conduct has the purpose or effect referred to in subsection (1)(b).

(c) Because of B's rejection of or submission to the conduct, A treats B less favourably than A would treat B if B had not rejected or submitted to the conduct.

(4) In deciding whether conduct has the effect referred to in subsection (1)(b), each of the following must be taken into account:

(a) The perception of B.

(b) The deriving from these hazards and the health and safety policy of their employer and the organisation and arrangements for fulfilling that policy. other circumstances of the case.

(c) Whether it is reasonable for the conduct to have that effect.

Section 27 Victimisation

(1) A person (A) victimises another person (B) if A subjects B to a detriment because:

(a) B does a protected act.

(b) A believes that B has done, or may do, a protected act.

(2) Each of the following is a protected act:

(a) Bringing proceedings under this Act.

(b) Giving evidence or information in connection with proceedings under this Act.

(c) Doing any other thing for the purposes of or in connection with this Act.

(d) Making an allegation (whether or not express) that A or another person has contravened this Act.

(3) Giving false evidence or information, or making a false allegation, is not a protected act if the evidence or information is given, or the allegation is made, in bad faith.

(4) This section applies only where the person subjected to a detriment is an individual.

(5) The reference to contravening this Act includes a reference to committing a breach of an equality clause or rule.

Health and Safety at Work etc. Act (HASAWA) 1974

Considered in context in Elements A3 and A8.

Arrangement of Act

PRELIMINARY

1) Preliminary.

GENERAL DUTIES

2) General duties of employers to the employees.
3) General duties of employers and self-employed to persons other than their employees.
4) General duties of persons concerned with premises to persons other than their employees.
5) [Repealed].

6) General duties of manufacturers etc. as regards articles and substances for use at work.
7) General duties of employees at work.
8) Duty not to interfere with or misuse things provided pursuant to certain provisions.
9) Duty not to charge employees for things done or provided pursuant to certain specific requirements.

THE HEALTH AND SAFETY COMMISSION AND THE HEALTH AND SAFETY EXECUTIVE

10) Establishment of the Commission and the Executive.
11) General functions of the Commission and the Executive.
12) Control of the Commission by the Secretary of State.
13) Other powers of the Commission.
14) Power of the Commission to direct investigations and Inquiries.

HEALTH AND SAFETY REGULATIONS AND APPROVED CODES OF PRACTICE

15) Health and safety regulations.
16) Approval of codes of practice by the Commission.
17) Use of approved codes of practice in criminal proceedings.

ENFORCEMENT

18) Authorities responsible for enforcement of the relevant statutory provisions.
19) Appointment of inspectors.
20) Powers of inspectors.
21) Improvement notices.
22) Prohibition notices.
23) Provisions supplementary to ss. 21 and 22.
24) Appeal against improvement or prohibition notice.
25) Power to deal with cause of imminent danger.
26) Power of enforcing authorities to indemnify their inspectors.

OBTAINING AND DISCLOSURE OF INFORMATION

27) Obtaining of information by the Commission, the Executive, enforcing authorities etc.
28) Restrictions on disclosure of information.

SPECIAL PROVISIONS RELATING TO AGRICULTURE

29-32) [repealed].

PROVISIONS AS TO OFFENCES

33) Offences.
34) Extension of time for bringing summary proceedings.
35) Venue.
36) Offences due to fault of other person.
37) Offences by bodies corporate.
38) Restriction on institution of proceedings in England and Wales.
39) Prosecutions by inspectors.
40) Onus of proving limits of what is practicable etc.
41) Evidence.
42) Power of court to order cause of offence to be remedied or, in certain cases, forfeiture.

FINANCIAL PROVISION

43) Financial provisions.

MISCELLANEOUS AND SUPPLEMENTARY

44) Appeals in connection with licensing provisions in the relevant statutory provisions.
45) Default powers.
46) Service of notices.
47) Civil liability.

Section 69 of the Enterprise and Regulatory Reform Act 2013 (ERRA), which came into force on 1 October 2013, amends the Health and Safety at Work etc Act 1974 (HSWA) Section 47 to the effect that claims for compensation for workplace incidents can only be made when negligence can be proved. Discussed in ***Element A4.2 - Breach of statutory duty, defences and damage***.

48) Application to Crown.
49) Adaptation of enactments to metric units or appropriate metric units.
50) Regulations under the relevant statutory provisions.
51) Exclusion of application to domestic employment.

52) Meaning of work and at work.
53) General interpretation of Part I.
54) Application of Part I to Isles of Scilly.

Outline of main points

OVERALL AIMS OF THE ACT

1) To protect people.
2) To protect the public from risks which may arise from work activities.

THE MAIN PROVISIONS - SECTION 1

a) Securing the health, safety and welfare of people at work.
b) Protecting others against risks arising from workplace activities.
c) Controlling the obtaining, keeping, and use of explosive and highly flammable substances.
d) Controlling emissions into the atmosphere of noxious or offensive substances.

Duties are imposed on:

a) The employer.
b) The self-employed.
c) Employees.
d) Contractors and subcontractors.
e) Designers, manufacturers, suppliers, importers and installers.
f) Specialists - architects, surveyors, engineers, personnel managers, health and safety practitioners, and many more.

EMPLOYER'S DUTIES - [TO EMPLOYEES]

Section 2(1)

To ensure, so far as ***reasonably practicable***, the health, safety and welfare at work of employees.

Section 2(2)

Ensuring health, safety and welfare at work through:

- Safe plant and systems of work e.g. provision of guards on machines.
- Safe use, handling, storage and transport of goods and materials e.g. good manual handling of boxes.
- Provision of information, instruction, training and supervision e.g. provision of induction training.
- Safe place of work including means of access and egress e.g. aisles kept clear.
- Safe and healthy working environment e.g. good lighting.

Further duties are placed on the employer by:

Section 2(3)

Prepare and keep up to date a written safety policy supported by information on the organisation and arrangements for carrying out the policy. The safety policy has to be brought to the notice of employees. If there are fewer than five employees, this section does not apply.

Section 2(4)

Recognised Trade Unions have the right to appoint safety representatives to represent the employees in consultations with the employer about health and safety matters.

Section 2(6)

Employers must consult with any safety representatives appointed by recognised Trade Unions.

Section 2(7)

To establish a safety committee if requested by two or more safety representatives.

EMPLOYER'S DUTIES - [TO PERSONS NOT HIS EMPLOYEES]

Section 3

a) Not to expose them to risk to their heath and safety e.g. contractor work barriered off.
b) To give information about risks which may affect them e.g. location induction for contractors.

SELF EMPLOYED DUTIES

Section 3

a) Not to expose themselves to risks to their health and safety e.g. wear personal protection.
b) Not to expose other persons to risks to their health and safety e.g. keep shared work area tidy.

Some of the practical steps that an organisation might take in order to ensure the safety of visitors to its premises are:

- Identify visitors by signing in, badges etc.
- Provide information regarding the risks present and the site rules and procedures to be followed, particularly in emergencies.
- Provide escorts to supervise visitors throughout the site.
- Restrict access to certain areas.

Figure RSP-1: Risks from roadside work. *Source: RMS.*

Figure RSP-2: Risks from street light repairing or tree felling. *Source: RMS.*

CONTROL OF PREMISES

Section 4

This section places duties on anyone who has control to any extent of non-domestic premises used by people who are not their employees. The duty extends to the provision of safe premises, plant and substances, e.g. maintenance of a boiler in rented out property.

MANUFACTURERS, DESIGNERS, SUPPLIERS, IMPORTERS, INSTALLERS

Section 6

This section places specific duties on those who can ensure that articles and substances are as safe and without risks as is reasonably practicable. The section covers:

- Safe design, installation and testing of equipment (including fairground equipment).
- Safe substances tested for risks.
- Provision of information on safe use and conditions essential to health and safety.
- Research to minimise risks.

EMPLOYEES' DUTIES

Section 7

a) To take reasonable care for themselves and others that may be affected by their acts/omissions e.g. wear eye protection, not obstruct a fire exit.

b) To co-operate with the employer or other to enable them to carry out their duty and/or statutory requirements e.g. report hazards or defects in controls, attend training, provide medical samples.

Additional duties created by the Management of Health and Safety at Work Regulations 1999 employees' duties:

- Every employee shall use any equipment, material or substance provided to them in accordance with any training and instruction.
- Every employee shall inform (via supervisory staff) their employer of any (a) risk situation or (b) shortcoming in the employer's protection arrangements.

OTHER DUTIES

Section 8

No person to interfere with or misuse anything provided to secure health and safety - e.g. wedge fire door open, remove first aid equipment without authority, breach lock off systems.

Section 9

Employees cannot be charged for anything done or provided to comply with a specific legal obligation e.g. personal protective equipment, health surveillance or welfare facilities.

OFFENCES COMMITTED BY OTHER PERSONS

Section 36

- Where the commission by any person of the breach of legislation is due to the act or default of some other person, that other person shall be guilty of the offence and may be charged with and convicted of the offence whether or not proceedings are taken against the first mentioned person.
- Case law indicates that 'other person' refers to persons lower down the corporate tree than mentioned in section 37, e.g. middle managers, safety advisors, training officers; and may extend to people working on contract e.g. architects, consultants or a planning supervisor.

OFFENCES COMMITTED BY THE BODY CORPORATE

Section 37

- Where there has been a breach of legislation on the part of a body corporate (limited company or local authority) and the offence can be proved to have been committed with the consent or connivance of or to be attributable to any neglect on the part of any director, manager, secretary or similar officer of the body corporate, he, as well as the body corporate, can be found guilty and punished accordingly.

ONUS OF PROOF

Section 40

In any proceedings for an offence under any of the relevant statutory involving a failure to comply with a duty or requirement:

- To do something so far as is practicable.
- To do something so far as is reasonably practicable.
- It shall be for the accused to prove that the requirements were met rather than for the prosecution to prove that the requirements were not met.

Health and Safety at Work etc. Act 1974 (Civil Liability) (Exceptions) Regulations (HSWCER) 2013

Considered in context in Element A4. See also, ERRA 2013. HASAWA 1974 and MHSWR 1999 (Regulation 22).

Arrangement of Regulations

1) Citation and commencement.
2) Exception relating to compulsory maternity leave.
3) Exception relating to the Management of Health and Safety at Work Regulations 1999.
4) Revocations.
5) Review.

Outline of main points

These Regulations came into force on 1st October 2013 and are made in accordance with the Health and Safety at Work etc. Act 1974 and create exceptions from the exclusion of civil liability for breach of statutory duty for pregnant workers and workers who have recently given birth or are breastfeeding.

Health and Safety at Work etc. Act 1974 (General Duties of Self-Employed Persons) (Prescribed Undertakings) Regulations (HSWAPU) 2015

Considered in context in Element A3. See also, HASAWA 1974.

The Health and Safety at Work etc Act 1974 (General Duties of Self-Employed Persons) (Prescribed Undertakings) Regulations 2015 specify the list of activities where self-employed workers are required to comply with health and safety duties under s.3(2) of the Health and Safety at Work etc Act 1974, following changes to health and safety laws for self-employed workers.

The list includes activities in agriculture, construction, gas and railways, and working with asbestos. In addition, self-employed workers who carry out an undertaking that involves an activity that poses risks to the health and safety of others are required to comply with the health and safety duties imposed by s.3(2) of the Act.

Health and Safety (Consultation with Employees) Regulations (HSCER) 1996

Considered in context in Element A9.

Arrangement of Regulations

1) Citation, extent and commencement.
2) Interpretation.
3) Duty of employer to consult.
4) Persons to be consulted.
5) Duty of employer to provide information.
6) Functions of representatives of employee safety.
7) Training, time off and facilities for representatives of employee safety and time off for candidates.
8) Amendment of the Employment Rights Act 1996.
9) Exclusion of civil liability.
10) Application of health and safety legislation.
11) Application to the Crown and armed forces.
12) Disapplication to sea-going ships.

13) Amendment of the 1977 Regulations.

Outline of main points

1) The HSCER come into force on 1ST October 1996 and are made under the European Communities Act 1972.

2) "Employees" do not include persons employed in domestic service in private households. Workplaces are defined as "any place where the employee is likely to work, or which he is likely to frequent in the course of his employment or incidentally to it".

3) Where there are employees not represented by the Safety Representatives and Safety Committee Regulations (SRSCR), the employer shall consult those employees in good time on matters relating to their health and safety at work. In particular they must be consulted on:

- The introduction of any new measures which may affect their safety and health.
- Arrangements made by the employer for appointing or nominating competent persons in accordance with regs. 6(1) and 7(1) of the Management of Health and Safety at Work Regs 1999 (MHSWR).
- Any safety information the employer is legally obliged to provide to workers.
- The planning and organisation of any health and safety training required under particular health and safety laws.
- The health and safety consequences for employees of the introduction of new technologies into the workplace.

4) Employers can consult either directly with employees or, in respect of any group of employees, one or more elected representatives of that group. These are referred to as "representatives of employee safety" (RES). If the latter option is chosen, then employers must tell the employees the name of the representative and the group he/she represents. An employer which has been consulting a representative may choose to consult the whole workforce. However, the employer must inform the employees and the representatives of that fact.

5) If the employer consults employees directly then it must make available such information, within the employers' knowledge, as is necessary to enable them to participate fully and effectively in the consultation. If a representative is consulted, then the employer must make available all necessary information to enable them to carry out their functions, and of any record made under the Reporting of Injuries, Diseases and Dangerous Occurrences Regs 2013 which relates to the represented group of employees.

6) Representatives of employee safety have the following functions:

- To make representations to the employer on potential hazards and dangerous occurrences at the workplace which affect, or could affect the represented employees.
- Make representations to the employer on general matters of health and safety.
- To represent the employees in workplace consultations with HSE or local authority inspectors.

7) Representatives of employee safety must be given reasonable training in order to carry out their duties. Employers must meet the costs of the training and any travel and subsistence. They must also permit the representatives to take time off with pay during working hours in order for them to carry out their functions. Time off shall also be given, with pay, where this is required for any person standing as a candidate for election as a representative. Employers must also provide suitable facilities for the representatives to carry out their duties.

8) The Employment Rights Act 1996, which gives protection against unfair dismissal or discrimination on grounds of health and safety, is amended to protect representatives of employee safety and candidates for their election.

9) A breach of the HSCER does not confer any right of action in any civil proceedings.

10) Ensures that certain provisions of health and safety legislation (including enforcement provisions) operate in respect of the HSCER. The Regulations are made under the European Communities Act 1972. Enforcement is by the enforcing authorities appointed under the Health and Safety at Work Act 1974.

11) The HSCER will apply in respect of the armed forces. However, the representatives of employee safety will be appointed by the employer, rather than elected. Furthermore, representatives in the armed forces will not be entitled to time off with pay under regulation 7.

12) The HSCER do not apply to the master or crew of a seagoing ship.

13) The SRSCR are amended so that they now include employees of coal mines.

Health and Safety (Fees) Regulations (HSFR) 2012

Considered in context in Element A3.

These Regulations put a duty on HSE to recover its costs for carrying out its regulatory functions from those found to be in material breach of health and safety law.

Duty holders who are compliant with the law, or where a breach is not material, will not be charged FFI for any work that HSE does with them, including inspection, investigation and taking enforcement action.

Health and Safety Information for Employees Regulations (IER) 1989

Law considered in context in Element A9.

Arrangement of Regulations

1) Citation and commencement.

2) Interpretation and application.

3) Meaning of and revisions to the approved poster and leaflet.

4) Provision of poster or leaflet.

5) Provision of further information.

6) Exemption certificates.
7) Defence.
8) Repeals, revocations and modifications.

The schedule repeals, revocations and modifications.

Part I - Repeals.

Part II - Revocations.

Part III - Modifications.

Outline of main points

The Health and Safety (Information for Employees) Regulations 1989 require that information relating to health and safety at work to be furnished to all employees by means of posters or leaflets in a form approved by the Health and Safety Executive.

The Health and Safety Information for Employees (Modifications and Repeals) Regulations 1995 amended these regulations; this allows the HSE to approve an alternative poster to the basic 'Health and Safety Law' poster.

The Health and Safety Executive (HSE) may approve a particular form of poster or leaflet for use in relation to a particular industry or employment and, where any such form has been approved, the HSE shall publish it.

If a poster is used, the information must be legible and up to date. The poster must be prominently located in an area which all employees have access. If a leaflet is used, revised leaflets must be issued to employees when any similar changes occur.

The HSE has published a new, simplified version of the ***Health and Safety Law Poster; the 2009 poster.*** It tells workers what they and their employers need to do in simple terms, using numbered lists of basic points.

The employer is required by law to either display the HSE-approved poster in a prominent position or to provide each of their workers with the equivalent 'leaflet'.

MODIFICATION TO THE REGULATIONS

It is no longer a requirement for the employer to add the contact details of the enforcing authority and the HSE's Employment Medical Advisory Service.

Details may be added of any employee safety representatives or other health and safety contacts, but this is not compulsory.

The existing ***"Health and Safety Law - what you should know"*** poster can continue to be displayed until 5th April 2014, so long as they are readable and contain up-to-date contact details.

Workers can be given copies of the equivalent 1999 leaflet until 5th April 2014. After this date, the 2009 approved poster must be displayed or workers must be provided with personal copies of the 2009 equivalent.

Legal Aid, Sentencing and Punishment of Offenders Act (LASPO) 2012 and Legal Aid, Sentencing and Punishment of Offenders Act 2012 (Fines on Summary Conviction) Regulations (LASPOFSC) 2015

Considered in context in Element A2.

Legal Aid, Sentencing and Punishment of Offenders Act 2012 (Fines on Summary Conviction) Regulations 2015 have brought into force Section 85 of the 2012 Act.

Section 85 removes the limit on certain fines on summary conviction by magistrates' court. Prior to this fines were restricted to no more than £5,000 (however expressed) or, exceptionally, a specified higher amount (for example £20,000 under sections 2 to 6 of the Health and Safety at Work Act 1974).

The change applies to offences committed on or after 12 March 2015 (but is not retrospective).

Management of Health and Safety at Work Regulations (MHSWR) 1999

Considered in context in Elements A1 and A7.

Arrangement of Regulations

1) Citation, commencement and interpretation.
2) Disapplication of these Regulations.
3) Risk assessment.
4) Principles of prevention to be applied.
5) Health and safety arrangements.
6) Health surveillance.
7) Health and safety assistance.
8) Procedures for serious and imminent danger and for danger areas.
9) Contacts with external services.
10) Information for employees.
11) Co-operation and co-ordination.

12) Persons working in host employers' or self-employed persons' undertakings.
13) Capabilities and training.
14) Employees' duties.
15) Temporary workers.
16) Risk assessment in respect of new or expectant mothers.
17) Certificate from a registered medical practitioner in respect of new or expectant mothers.
18) Notification by new or expectant mothers.
19) Protection of young persons.
20) Exemption certificates.
21) Provisions as to liability.
22) Exclusion of civil liability.
23) Extension outside Great Britain.
24) Amendment of the Health and Safety (First-Aid) Regulations 1981.
25) Amendment of the Offshore Installations and Pipeline Works (First-Aid) Regulations 1989.
26) Amendment of the Mines Miscellaneous Health and Safety Provisions Regulations 1995.
27) Amendment of the Construction (Health, Safety and Welfare) Regulations 1996.
28) Regulations to have effect as health and safety regulations.
29) Revocations and consequential amendments.
30) Transitional provision.

SCHEDULES

Schedule 1 - General principles of prevention.

Schedule 2 - Consequential amendments.

Outline of main points

Management of Health and Safety at Work Regulations (MHSWR) 1999 set out some broad general duties which apply to almost all kinds of work. They are aimed mainly at improving health and safety management. You may already be familiar with broad health and safety law of this kind - as it is the form taken by the Health and Safety at Work Act (HASAWA) 1974. The Regulations work in a similar way, and in fact they can be seen as a way of fleshing out what is already in the HASAWA. The 1999 Regulations replace the Management of Health and Safety at Work Regulations 1992, the Management of Health and Safety at Work (Amendment) Regulations 1994, the Health and Safety (Young Persons) Regulations 1997 and Part III of the Fire Precautions (Workplace) Regulations 1997. ***The Principal Regulations are discussed below.***

RISK ASSESSMENT (REGULATION 3)

The regulations require employers (and the self-employed) to assess the risk to the health and safety of their employees and to anyone else who may be affected by their work activity. This is necessary to ensure that the preventive and protective steps can be identified to control hazards in the workplace.

A ***hazard*** is defined as something with the potential to cause harm and may include machinery, substances or a work practice.

A ***risk*** is defined as the likelihood that a particular hazard will cause harm. Consideration must be given to the population, i.e. the number of persons who might be exposed to harm and the consequence of such exposure.

Where an employer is employing or about to employ young persons (under 18 years of age) he must carry out a risk assessment which takes particular account of:

- The inexperience, lack of awareness of risks and immaturity of young persons.
- The layout of the workplace and workstations.
- Exposure to physical, biological and chemical agents.
- Work equipment and the way in which it is handled.
- The extent of health and safety training to be provided.
- Risks from agents, processes and work listed in the Annex to Council Directive 94/33/EC on the protection of young people at work.

Where 5 or more employees are employed, the significant findings of risk assessments must be recorded in writing (the same threshold that is used in respect of having a written safety policy). This record must include details of any employees being identified as being especially at risk.

PRINCIPLES OF PREVENTION TO BE APPLIED (REGULATION 4)

Regulation 4 requires an employer to implement preventive and protective measures on the basis of general principles of prevention specified in Schedule 1 to the Regulations. These are:

1) Avoiding risks.
2) Evaluating the risks which cannot be avoided.
3) Combating the risks at source.
4) Adapting the work to the individual, especially as regards the design of workplaces, the choice of work equipment and the choice of working and production methods, with a view, in particular, to alleviating monotonous work and work at a predetermined work-rate and to reducing their effect on health.
5) Adapting to technical progress.

6) Replacing the dangerous by the non-dangerous or the less dangerous.
7) Developing a coherent overall prevention policy which covers technology, organisation of work, working conditions, social relationships and the influence of factors relating to the working environment.
8) Giving collective protective measures priority over individual protective measures.
9) Giving appropriate instructions to employees.

HEALTH AND SAFETY ARRANGEMENTS (REGULATION 5)

Appropriate arrangements must be made for the effective planning, organisation, control, monitoring and review of preventative and protective measures (in other words, for the management of health and safety). Again, employers with five or more employees must have their arrangements in writing.

HEALTH SURVEILLANCE (REGULATION 6)

In addition to the requirements of specific regulations such as Control of Substances Hazardous to Health (COSHH) and Asbestos regulations, consideration must be given to carry out health surveillance of employees where there is a disease or adverse health condition identified in risk assessments.

HEALTH AND SAFETY ASSISTANCE (REGULATION 7)

The employer must appoint one or more competent persons to assist him in complying with the legal obligations imposed on the undertaking. The number of persons appointed should reflect the number of employees and the type of hazards in the workplace.

If more than one competent person is appointed, then arrangements must be made for ensuring adequate co-operation between them. The Competent person(s) must be given the necessary time and resources to fulfil their functions. This will depend on the size the undertaking, the risks to which employees are exposed and the distribution of those risks throughout the undertaking.

The employer must ensure that competent person(s) who are not employees are informed of the factors known (or suspected) to affect the health and safety of anyone affected by business activities.

Competent people are defined as those who have sufficient training and experience or knowledge and other qualities to enable them to perform their functions.

Persons may be selected from among existing employees or from outside. Where there is a suitable person in the employer's employment, that person shall be appointed as the 'competent person' in preference to a non-employee.

PROCEDURES FOR SERIOUS AND IMMINENT DANGER AND FOR DANGER AREAS (REGULATION 8)

Employers are required to set up emergency procedures and appoint ***competent persons*** to ensure compliance with identified arrangements, to devise control strategies as appropriate and to limit access to areas of risk to ensure that only those persons with adequate health and safety knowledge and instruction are admitted.

The factors to be considered when preparing a procedure to deal with workplace emergencies such as fire, explosion, bomb scare, chemical leakage or other dangerous occurrence should include:

- The identification and training requirements of persons with specific responsibilities.
- The layout of the premises in relation to escape routes etc.
- The number of persons affected.
- Assessment of special needs (disabled persons, children etc).
- Warning systems.
- Emergency lighting.
- Location of shut-off valves, isolation switches, hydrants etc.
- Equipment required to deal with the emergency.
- Location of assembly points.
- Communication with emergency services.
- Training and/or information to be given to employees, visitors, local residents and anyone else who might be affected.

CONTACTS WITH EXTERNAL SERVICES (REGULATION 9)

Employers must ensure that, where necessary, contacts are made with external services. This particularly applies with regard to first-aid, emergency medical care and rescue work.

INFORMATION FOR EMPLOYEES (REGULATION 10)

Employees must be provided with relevant information about hazards to their health and safety arising from risks identified by the assessments. Clear instruction must be provided concerning any preventative or protective control measures including those relating to serious and imminent danger and fire assessments. Details of any competent persons nominated to discharge specific duties in accordance with the regulations must also be communicated as should risks arising from contact with other employer's activities ***(see Regulation 11).***

Before employing a child (a person who is not over compulsory school age) the employer must provide those with parental responsibility for the child with information on the risks that have been identified and preventative and protective measures to be taken.

CO-OPERATION AND CO-ORDINATION (REGULATION 11)

Employers who work together in a common workplace have a duty to co-operate to discharge their duties under relevant statutory provisions. They must also take all reasonable steps to inform their respective employees of risks to their health or safety which may arise out of their work. Specific arrangements must be made to ensure compliance with fire legislation.

PERSONS WORKING IN HOST EMPLOYERS' OR SELF EMPLOYED PERSONS' UNDERTAKINGS (REGULATION 12)

This regulation extends the requirements of regulation 11 to include employees working as sole occupiers of a workplace under the control of another employer. Such employees would include those working under a service of contract and employees in temporary employment businesses under the control of the first employer.

CAPABILITIES AND TRAINING (REGULATION 13)

Employers need to take into account the capabilities of their employees before entrusting tasks. This is necessary to ensure that they have adequate health and safety training and are capable enough at their jobs to avoid risk. To this end consideration must be given to recruitment including job orientation when transferring between jobs and work departments. Training must also be provided when other factors such as the introduction of new technology and new systems of work or work equipment arise.

Training must:

- Be repeated periodically where appropriate.
- Be adapted to take account of any new or changed risks to the health and safety of the employees concerned.
- Take place during working hours.

EMPLOYEES' DUTIES (REGULATION 14)

Employees are required to follow health and safety instructions by using machinery, substances, transport etc. in accordance with the instructions and training that they have received.

They must also inform their employer (and other employers) of any dangers or shortcoming in the health and safety arrangements, even if there is no risk of imminent danger.

TEMPORARY WORKERS (REGULATION 15)

Consideration is given to the special needs of temporary workers. In particular to the provision of particular health and safety information such as qualifications required to perform the task safely or any special arrangements such as the need to provide health screening.

RISKS ASSESSMENT IN RESPECT OF NEW OR EXPECTANT MOTHERS (REGULATION 16)

Where the work is of a kind which would involve risk to a new or expectant mother or her baby, then the assessment required by regulation 3 should take this into account.

If the risk cannot be avoided, then the employer should take reasonable steps to:

- Adjust the hours worked.
- Offer alternative work.
- Give paid leave for as long as is necessary.

CERTIFICATE FROM A REGISTERED MEDICAL PRACTITIONER IN RESPECT OF NEW OR EXPECTANT MOTHERS (REGULATION 17)

Where the woman is a night shift worker and has a medical certificate identifying night shift work as a risk then the employer must put her on day shift or give paid leave for as long as is necessary.

NOTIFICATION BY NEW OR EXPECTANT MOTHERS (REGULATION 18)

The employer need take no action until he is notified in writing by the woman that she is pregnant, has given birth in the last six months, or is breastfeeding.

PROTECTION OF YOUNG PERSONS (REGULATION 19)

Employers of young persons shall ensure that they are not exposed to risk as a consequence of their lack of experience, lack of awareness or lack of maturity.

No employer shall employ young people for work which:

- Is beyond his physical or psychological capacity.
- Involves exposure to agents which chronically affect human health.
- Involves harmful exposure to radiation.
- Involves a risk to health from extremes of temperature, noise or vibration.
- Involves risks which could not be reasonably foreseen by young persons.

This regulation does not prevent the employment of a young person who is no longer a child for work:

- Where it is necessary for his training.
- Where the young person will be supervised by a competent person.
- Where any risk will be reduced to the lowest level that is reasonably practicable.

(Note: Two HSE publications give guidance on the changes. HSG122 - New and expectant mothers at work: A guide for employers and HSG165 - Young people at work: A guide for employers).

EXEMPTION CERTIFICATES (REGULATION 20)

The Secretary of State for Defence may, in the interests of national security, by a certificate in writing exempt the armed forces, any visiting force or any headquarters from certain obligations imposed by the Regulations.

PROVISIONS AS TO LIABILITY (REGULATION 21)

Employers cannot submit a defence in criminal proceedings that contravention was caused by the act or default either of an employee or the competent person appointed under Regulation 7.

EXCLUSION OF CIVIL LIABILITY (REGULATION 22)

As amended by Health and Safety at Work etc. Act 1974 (Civil Liability) (Exceptions) Regulations 2013):

Regulation 22 specifies:

"(1) Breach of a duty imposed by regulation 16, 16A, 17 or 17A shall, so far as it causes damage, be actionable by the new or expectant mother.

(2) Any term of an agreement which purports to exclude or restrict any liability for such a breach is void."

REVOCATIONS AND AMENDMENTS (REGULATIONS 24-29)

The Regulations:

- Revoked regulation 6 of the Health and Safety (First-Aid) Regulations (FAR) 1981 which confers power on the Health and Safety Executive to grant exemptions from those Regulations.
- Amended the Offshore Installations and Pipeline Works (First-Aid) Regulations 1989.
- Amended the Mines Miscellaneous Health and Safety Provisions Regulations 2013.
- Amended the Construction (Health, Safety and Welfare) Regulations 1996 (replaced by CDM 2007).

The Regulations provided that, with some exceptions, the Fire Precautions (Workplace) Regulations (FPWR) 1997 (now revoked) were to be considered as health and safety regulations within the meaning of the Health and Safety at Work etc Act (HASAWA) 1974. The Regulations also made amendments to the statutory instruments as specified in Schedule 2.

TRANSITIONAL PROVISION (REGULATION 30)

The Regulations contain a transitional provision (regulation 30). The substitution of provisions in the 1999 Regulations for provisions of the Management of Health and Safety at Work Regulations (MHSWR) 1992 shall not affect the continuity of the law; and accordingly anything done under or for the purposes of such provision of the 1992 Regulations shall have effect as if done under or for the purposes of any corresponding provision of these Regulations.

Public Interest Disclosure Act (PIDA) 1998

Considered in context in Element A2.

Protection for 'whistleblowers' was established by the Employment Rights Act 1996 (ERA). It introduced rights of protection for workers not to suffer detriment or dismissal for raising concerns ('blowing the whistle') on their employers' fraudulent, criminal or dangerous activities. ERA was amended by the Public Interest Disclosure Act 1998 (PIDA) and later by the Enterprise and Regulatory Reform Act 2013 (ERRA). The effect has been to clarify what constitute "protected disclosures", what classes of person are protected, and how disclosures can be made.

The PIDA gives legal protection to employees against being dismissed or penalised by their employers as a result of disclosing information which is considered to be in the public interest. The principle of the Act is that where an individual discovers information which he or she believes to show malpractice or wrongdoing within their organisation that this information should be disclosed without fear of reprisal.

This should be facilitated by a process which ensures that the person making the disclosure is afforded proper protection and that the information is acted upon quickly. The Act offers a right to redress in the event of victimisation if individuals raise concerns in the ways specified by the legislation.

Reporting of Injuries, Diseases and Dangerous Occurrences Regulations (RIDDOR) 2013

Law considered in context in Element A5.

Arrangement of Regulations

1) Citation and commencement.
2) Interpretation.
3) Responsible person.
4) Non-fatal injuries to workers.
5) Non-fatal injuries to non-workers.
6) Work-related fatalities.
7) Dangerous occurrences.
8) Occupational diseases.
9) Exposure to carcinogens, mutagens and biological agents.
10) Diseases offshore.
11) Gas-related injuries and hazards.
12) Recording and record-keeping.
13) Mines, quarries and offshore site disturbance.
14) Restrictions on the application of regulations 4 to 10.
15) Restriction on parallel requirements.
16) Defence.
17) Certificates of exemption.
18) Revocations, amendments and savings.
19) Extension outside Great Britain.
20) Review.

SCHEDULES

Schedule 1 - Reporting and Recording Procedures.
Schedule 2 - Dangerous Occurrences.
Schedule 3 - Diseases Reportable Offshore.
Schedule 4 - Revocations and Amendments.

Outline of main points

The Reporting of Injuries, Diseases and Dangerous Occurrences Regulations (RIDDOR) 2013 covers the requirement to report certain categories of injury and disease sustained at work, along with specified dangerous occurrences and gas incidents, to the relevant enforcing authority. These reports are used to compile statistics to show trends and to highlight problem areas, in particular industries or companies.

THE MAIN POINTS OF RIDDOR

Reporting

1) When a person ***dies or suffers any specified injury*** listed in regulation 4 ***(non-fatal injuries to workers)*** as a result of a work-related accident or dies as a result of occupational exposure to a biological agent or an incident occurs of the type listed as a dangerous occurrences in Schedule 2 the responsible person must notify the relevant enforcing authority by the quickest practicable means (usually by telephone) without delay and must send them a report in an approved manner (on-line) within 10 days. This therefore includes accidents connected with work where:
   - An employee or a self-employed person at work is killed or suffers a specified injury (including as a result of physical violence).
   - A member of the public is killed or taken to hospital.

A work-related ***'accident'*** in the context of RIDDOR 2013 includes an act of non-consensual physical violence done to a person at work.

2) In cases of work-related diseases that are listed in regulation 8 and 10 the responsible person must send a report of the diagnosis in an approved manner (on-line) to the relevant enforcing authorities without delay. In cases of diseases related to carcinogens, mutagens and biological agents that are listed in regulation 9 the responsible person must notify the relevant enforcing authority in an approved manner.
3) If personal injury results in ***more than 7 days (excluding the day of the accident) incapacity*** for routine work, but is not one of the specified non-fatal injuries, the responsible person must send a report to the relevant enforcing authority in an approved manner (on-line) as soon as is practicable and in any event within ***15 days*** of the accident. The day of the accident is not counted, but any days which would not have been working days are included.
4) If there is an accident connected with work (including an act of physical violence) and an employee, or a self-employed person at work, suffers an over-three-day injury it must be recorded by the employer.
5) The enforcing authority for most workplaces is either the Health and Safety Executive or the Local Authority, for railway operations it is the Office of Rail Regulation (ORR).

Responsible person

Reportable event	***To***	***Responsible person***
Death, specified non-fatal injury, over 7 day injury, disease.	Employee.	Employer.
	Self-employed person working in someone else's premises.	Person in control of the premises: ■ At the time of the event. ■ In connection with trade, business or undertaking.
Specified injury, over 7 day injury, disease.	Self-employed in own premises.	Self-employed person or someone acting for them.
Death, being taken to hospital or a specified non-fatal injury on hospital premises.	A person not at work.	Person in control of the premises: ■ At the time of the event. ■ In connection with trade, business or undertaking.
Dangerous occurrences - general.		Person in control of the premises where, or in connection with the work going on at which, the dangerous occurrence happened: ■ At the time of the event. ■ In connection with trade, business or undertaking.

Road traffic accidents

Road traffic accidents only have to be reported if:

- Death or injury results from an accident involving a train.
- Death or injury results from exposure to a substance being conveyed by a vehicle.
- Death or injury results from the person being engaged in work connected with loading or unloading of any article/substance or results from another person engaged in these activities.
- Death or injury results from the person being engaged in work on or alongside a road or results from another person engaged in these activities.

Work on or alongside a road means work concerned with the construction, demolition, alteration, repair or maintenance of:

- The road or the markings or equipment on the road.
- The verges, fences, hedges or other boundaries of the road.
- Pipes or cables on, under, over or adjacent to the road.
- Buildings or structures adjacent to or over the road.

Non-employee

The responsible person must not only report non-employee deaths, but also cases that involve major injury or being taken to hospital if caused by an accident out of or in connection with their work.

Employee death

Where an employee dies within one year of the date of an accident, as a result of a reportable injury, as soon as the employer knows they must inform the enforcing authority in writing of the death.

Gas incidents

Specified gas incidents are notified without delay and reported within 14 days to the Health and Safety Executive.

Injury under medical supervision

Reporting and recording requirements do not apply in situations where the injury or death of a person arises out of the conduct of an operation, examination or other medical treatment of that person whilst under the supervision of a registered medical practitioner or dentist.

Self-employed people

If a self-employed person suffers a specified non-fatal injury while working at premises that are owned or occupied by themselves they do not need to notify the enforcing authority immediately. However, they or someone acting for them must report the injury within 10 days.

Where an injury is not a specified non-fatal injury, but causes a self-employed person to be incapacitated from routine work for more than seven consecutive days, the self-employed person or someone acting for them must report it within 15 days of the accident. There is no reporting requirement for situations where a self-employed person suffers a fatal accident or fatal exposure on premises controlled by that self-employed person.

Recording

In the case of an accident at work, the following details must be recorded:

- Date and time.
- Name.
- Occupation.
- Nature of injury.
- Place of accident.
- Brief description of the circumstances in which the accident happened.
- In the case of a person not a work, instead of occupation a record of their status should be made (for example, passenger, customer, visitor or bystander).
- The date on which the accident was first notified or reported to the relevant enforcing authority and the method used.

For non-reportable injuries that incapacitate for more than 3 days a record of notification or reporting is not relevant, but a record of the accident must be maintained.

Similar information should be recorded for dangerous occurrences, except that details of injured persons will not be relevant.

In the case of a diagnosis of a reportable disease, the following details must be recorded:

- The date of diagnosis of the disease.
- The name of the person affected.
- The occupation of the person affected.
- The name or nature of the disease.
- The date on which the disease was first reported to the relevant enforcing authority.
- The method by which the disease was reported.

Records must be kept for at least 3 years and kept at the place where the work it relates to is carried out or at the usual place of business of the responsible person.

Defences

A person must prove that they were not aware of the event requiring reporting and that they had taken all reasonable steps to be made aware, in sufficient time.

SPECIFIED INJURIES (RIDDOR 2013 - REGULATION 4)

The list of specified non-fatal injuries is:

- Any bone fracture diagnosed by a registered medical practitioner, other than to a finger, thumb or toe.
- Amputation of an arm, hand, finger, thumb, leg, foot or toe.
- Any injury diagnosed by a registered medical practitioner as being likely to cause permanent blinding or reduction in sight in one or both eyes.
- Any crush injury to the head or torso causing damage to the brain or internal organs in the chest or abdomen.
- Any burn injury (including scalding) which:
    - Covers more than 10% of the whole body's total surface area.
    - Causes significant damage to the eyes, respiratory system or other vital organs.
- Any degree of scalping requiring hospital treatment.
- Loss of consciousness caused by head injury or asphyxia.

- Any other injury arising from working in an enclosed space which:
  - Leads to hypothermia or heat-induced illness.
  - Requires resuscitation or admittance to hospital for more than 24 hours.

DISEASES (RIDDOR 2013 - REGULATIONS 8, 9 AND 10)

Regulation 8 - Occupational diseases

- Carpal Tunnel Syndrome (CTS), where the person's work involves regular use of percussive or vibrating tools.
- Cramp in the hand or forearm, where the person's work involves prolonged periods of repetitive movement of the fingers, hand or arm.
- Occupational dermatitis, where the person's work involves significant or regular exposure to a known skin sensitiser or irritant.
- Hand Arm Vibration Syndrome (HAVS), where the person's work involves regular use of percussive or vibrating tools, or the holding of materials which are subject to percussive processes, or processes causing vibration.
- Occupational asthma, where the person's work involves significant or regular exposure to a known respiratory sensitizer.
- Tendonitis or tenosynovitis in the hand or forearm, where the person's work is physically demanding and involves frequent, repetitive movements.

Regulation 9 - Exposure to carcinogens, mutagens and biological agents

- Any cancer attributed to an occupational exposure to a known human carcinogen or mutagen (including ionising radiation).
- Any disease attributed to an occupational exposure to a biological agent.

Regulation 10 and Schedule 3 - Diseases offshore

Examples of diseases listed in this Schedule are:

- Chickenpox.
- Cholera.
- Diphtheria.
- Dysentery (amoebic or bacillary).
- Mumps.
- Food poisoning.
- Legionellosis.
- Malaria.
- Measles.
- Meningitis.

DANGEROUS OCCURRENCES (RIDDOR 2013 - SCHEDULE 2)

Dangerous occurrences are events that have the potential to cause death or serious injury and so must be reported whether anyone is injured or not. Examples of dangerous occurrences that must be reported are:

- The failure of any load bearing part of any lifting equipment, other than an accessory for lifting.
- The failure of any pressurised closed vessel or any associated pipework.
- Any unintentional incident in which plant or equipment either:
  - Comes into contact with an uninsulated overhead electric line in which the voltage exceeds 200 volts.
  - Causes an electrical discharge from such an electric line by coming into close proximity to it.
- Electrical short-circuit or overload attended by fire or explosion which results in the stoppage of the plant involved for more than 24 hours.

The schedule also identifies Dangerous Occurrences that are specific to mines, quarries, transport systems, and offshore workplaces.

Safety Representatives and Safety Committees Regulations (SRSC) 1977

Considered in context in Element A9.

Arrangement of Regulations

1) Citation and commencement.
2) Interpretation.
3) Appointment of safety representatives.
4) Functions of safety representatives.
5) Inspections of the workplace.
6) Inspections following notifiable accidents, occurrences and diseases.
7) Inspections of documents and provision of information.
8) Cases where safety representatives need not be employees.
9) Safety committees.
10) Power of Health and Safety Commission to grant exemption.
11) Provision as to industrial tribunals.

Outline of main points

The Safety Representatives and Safety Committees Regulations (SRSC) 1977 are concerned with the appointment by recognised trade unions of safety representatives, the functions of the representatives and the establishment of safety committees. Representatives are appointed when a recognised trade union notifies the employer in writing. Representatives must have been employed throughout the preceding 2 years or, where this is not reasonably practicable, have had at least 2 years' experience in similar employment.

Similarly, employees cease to be representatives when:

- The employer has been notified by writing by the Trade Union.
- The representative ceases to be.
- He/she resigns.

FUNCTIONS OF TRADE UNION - APPOINTED SAFETY REPRESENTATIVES

The SRSCR grant safety representatives the right to carry out certain functions as outlined as follows.

Functions are activities that safety representatives are permitted to carry out by legislation, but do not have a 'duty' to perform and therefore are treated as advisory actions. As a consequence the representatives cannot be held accountable for failing to carry out these activities or for the standard of the advice given, when performing their functions. They are, however, still employees and have the same consequent duties as any other employee (for example their duties under HASAWA Ss 7and 8). Their functions as safety representatives are:

a) To take all reasonably practical steps to keep themselves informed of:
   - The legal requirements relating to the health and safety of persons at work, particularly the group or groups of persons they directly represent.
   - The particular hazards of the workplace and the measures deemed necessary to eliminate or minimise the risk deriving from these hazards and the health and safety policy of their employer and the organisation and arrangements for fulfilling that policy.

b) To encourage co-operation between their employer and his employees in promoting and developing essential measures to ensure the health and safety of employees, and in checking the effectiveness of these measures.

c) To carry out investigations into:
   - Hazards and dangerous occurrences (incl. accidents) at the workplace.
   - Complaints, by any employee he represents, relating to that employee's health, safety or welfare.

d) To carry out inspections of the workplace.

e) To bring to the employer's notice, normally in writing, any unsafe or unhealthy conditions, or unsafe working practices, or unsatisfactory arrangements for welfare at work which comes to their attention whether during an inspection/investigation or day to day observation.

The report does not imply, that all other conditions and working practices are safe and healthy or that the welfare arrangements are satisfactory in all other respects. Making a written report does not preclude the bringing of such matters to the attention of the employer or his representative by a direct oral approach in the first instance, particularly in situations where speedy remedial action is necessary. It will also be appropriate for minor matters to be the subject of direct discussion, without the need for a formal written approach.

f) To represent the employees they were appointed to represent in consultation at the workplace with inspectors of the Health and Safety Executive and of any other enforcing authority within the Act.

g) To receive information from inspectors in accordance with section 28(8) of the 1974 Act.

h) To attend meetings of safety committees during which he/she attends in his capacity as a safety representative in connection with any of the above conditions.

EMPLOYERS DUTIES

The Regulations require employers to make any known information available to safety representatives which is necessary to enable them to fulfil their functions. This should include:

a) Information about the plans and performances of the undertaking and any changes proposed, in so far as they affect the health and safety at work of their employees.

b) Information of a technical nature about hazards to health and safety and precautions deemed necessary to eliminate or minimise them, in respect of machinery, plant, equipment, processes, systems of work and substances in use at work. This should include any relevant information provided by consultants or designers or by the manufacturer, importer or supplier of any article or substance used, or proposed to be used, at work by their employees.

c) Information which the employer keeps relating to the occurrence of any accidents, dangerous occurrences or notifiable industrial disease and any statistical records relating to such accidents, dangerous occurrences or cases of notifiable industrial disease.

d) Any other information specifically related to matters affecting the Health and Safety at work of his employees, including the result of any measurements taken by persons acting on his behalf in the course of checking the effectiveness of his health and safety arrangements.

e) Information on articles or substances which an employer issues to homeworkers.

f) Any other suitable and relevant reasonable facility to enable the representatives to carry out their functions.

TRAINING

The basis of Trades Union Congress (TUC) policy is that the union appointed safety representative will be trained on TUC approved courses. However, there is much to be gained by the employer approaching the trades unions active in his workplace with the objective of holding joint company/industry based courses.

In any event it is prudent for the employer to carry out company/industry orientated training to supplement the wide industry based TUC course. The functions and training of the safety representatives should be carried out during normal working hours. The representative must receive normal earnings, this taking into consideration any bonuses which would have been earned if carrying out their normal work activities.

FUNCTIONS OF HEALTH AND SAFETY COMMITTEES

If two or more appointed safety representatives request in writing the formation of a safety committee, the employer must implement this request within three months. Consultation must take place with the representatives making the request and the appointing trade union. A basic requirement for a successful safety committee is the desire of both employee and management to show honest commitment and a positive approach to a programme of accident prevention and the establishment of a safe and healthy environment and systems of work. For any committee to operate effectively, it is necessary to determine its objectives and functions.

Objectives

a) The promotion of safety, health and welfare at work by providing a forum for discussion and perhaps a pressure group.

b) To promote and support normal employee/employer systems for the reporting and control of workplace problems.

Functions

a) To review accident and occupational health trends.

b) To review recurring problems revealed by safety audits.

c) To consider enforcing authority reports and information releases.

d) To consider reports on matters arising from previous safety committee meetings.

e) To assist in the development of safety rules and systems of work and procedures.

f) To review health and safety aspects of future development and changes in procedure.

g) To review health and safety aspects of purchasing specifications of equipment and materials.

h) To review renewal/maintenance programmes.

i) To monitor safety training programmes and standards achieved.

j) To monitor the effectiveness of safety and health communications within the workplace.

k) To monitor the effectiveness of the Safety Policy.

This may be summarised as review and recommend on the overall direction of the health and safety programme, on specific aspects of the programme, on difficulties encountered in its implementation and to monitor the programme in both a specific and overall manner.

Composition

The membership and structure of the safety committee should be settled in consultation between management and the trade union representatives concerned. This should be aimed at keeping the total size as compact as possible, compatible with the adequate representation of the interests of management and employees. Management representatives will naturally be appointed by the management. Employee representatives will either be appointed by a recognised Trade Union (HASAWA 2(4)) or, in a non-union company, elected by their colleagues. The committee suggested in HASAWA section 2 (7) will probably be the 'Company Safety Committee.' There is nothing to prevent the formation of 'works' or 'office' committees as required in order to maintain the company safety committee at a reasonable size.

Social Action, Responsibility and Heroism (SARAH) Act

Considered in context in Elements A3 and A4.

Volunteers and those helping others in society will now have due consideration of their intent in any negligence claim. In its assessment of meeting a standard of care, courts will consider if the person showed a predominantly responsible approach towards protecting the safety of others or the action was taken in heroic circumstances, offering protection in an emergency.

The SARAH Act applies when a court, considering a claim in negligence or breach of statutory duty, is determining the steps that a person was required to take to discharge a duty of care. The Act sets out the following additional factors that the court "must" now consider, that is, whether:

- The alleged negligence or breach of statutory duty occurred when the defendant was acting for the benefit of society or any of its members.
- The defendant demonstrated a predominately responsible approach towards protecting the safety or interests of others.
- The alleged negligence or breach of statutory duty occurred when the defendant was acting heroically by intervening in an emergency to assist an individual in danger.

The changes have been designed to help volunteers, community groups, businesses and people who step in heroically to help in dangerous circumstances. It follows years of concern that people were being put off from doing simple good deeds for fear of legal action if something went wrong. It also is said by the government to be necessary to combat the so called 'Compensation Culture'.

Index

D

J

K

L

M

This page is intentionally blank